# Encyclopedia of
# Plant Physiology

New Series Volume 15 A

Editors

A. Pirson, Göttingen
M.H. Zimmermann, Harvard

8 General Conclusions . . . . . . . . . . . . . . . . . . . . . . . . 84
References . . . . . . . . . . . . . . . . . . . . . . . . . . . . . 86

## I.3 Modern Solution Culture Techniques

C.J. ASHER and D.G. EDWARDS (With 3 Figures)

1 Major Differences Between Solution Culture and Soil Culture . . . . . . 94
1.1 Mechanical Support . . . . . . . . . . . . . . . . . . . . . . . 94
1.2 Spatial Variation in Root Environment Parameters . . . . . . . . . . 95
1.3 Temporal Variation in Root Environment Parameters . . . . . . . . . 97
1.3.1 Nutrient Depletion . . . . . . . . . . . . . . . . . . . . . 97
1.3.2 pH Shifts . . . . . . . . . . . . . . . . . . . . . . . . . . 98
1.4 Root-Microorganism Interactions . . . . . . . . . . . . . . . . . 98
2 Uses and Limitations of Existing Solution Culture Methods . . . . . . . 99
2.1 Non-Renewed or Intermittently Renewed Water Cultures and Sand Cultures . . . . . . . . . . . . . . . . . . . . . . . . . . . . 99
2.1.1 Use in Teaching, Demonstration, and Diagnosis . . . . . . . . . 99
2.1.2 Production of Roots for Ion Transport Studies . . . . . . . . . 101
2.1.3 Nutrient Essentiality . . . . . . . . . . . . . . . . . . . . . 101
2.1.4 Effects of Root Environment Parameters . . . . . . . . . . . . 101
2.1.5 Establishment of Critical Tissue Concentrations . . . . . . . . . 105
2.1.6 Control of Plant Nutrient Status . . . . . . . . . . . . . . . 105
2.1.7 Study of Symbiotic Associations with Microorganisms . . . . . 106
2.1.8 Commercial Crop Production . . . . . . . . . . . . . . . . . 108
2.2 Mist Culture . . . . . . . . . . . . . . . . . . . . . . . . . . . 108
2.3 Flowing Solution Culture . . . . . . . . . . . . . . . . . . . . . 109
2.3.1 The Flow Rate Problem . . . . . . . . . . . . . . . . . . . . 109
2.3.2 Composition of Flowing Culture Solutions . . . . . . . . . . . 110
2.3.3 Research Applications . . . . . . . . . . . . . . . . . . . . . 111
2.3.4 Likely Future Developments . . . . . . . . . . . . . . . . . . 113
2.3.5 Commercial Crop Production . . . . . . . . . . . . . . . . . 114
3 Summary and Conclusions . . . . . . . . . . . . . . . . . . . . . . 115
References . . . . . . . . . . . . . . . . . . . . . . . . . . . . . 115

## I.4 Diagnosis of Mineral Deficiencies Using Plant Tests

D. BOUMA (With 5 Figures)

1 Introduction . . . . . . . . . . . . . . . . . . . . . . . . . . . 120
2 Plant Analysis . . . . . . . . . . . . . . . . . . . . . . . . . . 121
2.1 Physiological Basis . . . . . . . . . . . . . . . . . . . . . . . 121
2.2 Choice of Tissue . . . . . . . . . . . . . . . . . . . . . . . . 123
2.3 Factors Affecting the Relationship Between Nutrient Concentration and Yield . . . . . . . . . . . . . . . . . . . . . . . . . . . . . 124
2.3.1 Plant Development . . . . . . . . . . . . . . . . . . . . . . 124
2.3.2 Effects of Changes in Age of Tissue . . . . . . . . . . . . . . 125
2.3.3 Plant Age and Critical Levels . . . . . . . . . . . . . . . . . 126
2.3.4 Interactions Between Nutrient Elements . . . . . . . . . . . . 127
2.3.5 Environmental Factors . . . . . . . . . . . . . . . . . . . . 128
2.3.6 Other Factors Affecting Nutrient Composition . . . . . . . . . 130
3 Physiological and Biochemical Approaches to Diagnosis . . . . . . . . 131
3.1 Introductory Remarks . . . . . . . . . . . . . . . . . . . . . . 131
3.2 Physiological Approaches . . . . . . . . . . . . . . . . . . . . 131
3.2.1 Physiological Assessment . . . . . . . . . . . . . . . . . . . 131
3.2.2 Nutrient Stress . . . . . . . . . . . . . . . . . . . . . . . 132
3.2.3 Approaches Based on Photosynthesis . . . . . . . . . . . . . . 132
3.2.4 Other Approaches . . . . . . . . . . . . . . . . . . . . . . 133

3.3 Biochemical Approaches . . . 135
3.3.1 Nitrogen and Molybdenum . . . 135
3.3.2 Phosphorus . . . 136
3.3.3 Potassium and Magnesium . . . 137
3.3.4 Iron and Manganese . . . 138
3.3.5 Copper . . . 139
3.3.6 Zinc . . . 140
4 Prospects for the Future . . . 140
References . . . 141

**I.5 Interactions Between Nutrients in Higher Plants**
A.D. Robson and M.G. Pitman (With 9 Figures)

1 Introduction . . . 147
2 Interactions Between Nutrients in Monoculture . . . 152
2.1 Interactions Between Nutrients Affecting the Absorption of Nutrients 152
2.1.1 Interactions Occurring in the Soil . . . 152
2.1.2 Absorption from Solution at the Root Surface . . . 156
2.2 Interactions Between Nutrients Affecting the Utilization of Nutrients Within the Plant . . . 160
2.2.1 Distribution . . . 161
2.2.2 Function . . . 164
2.3 Complex Interactions Between Nutrients Involving Several Processes 167
2.3.1 Calcium/Aluminium/Phosphate . . . 167
2.3.2 Zinc/Phosphate . . . 169
3 Interactions Between Nutrients in Mixed Communities . . . 170
4 Conclusion . . . 173
References . . . 173

**I.6 Import and Export of Mineral Nutrients in Plant Roots**
U. Lüttge (With 10 Figures)

1 Introduction: The Dual Role of Roots in the Evolution of Higher Land Plants . . . 181
2 Relations Between Structure and Transport Functions Along the Length of Roots . . . 182
2.1 The Phenomenon of Variations in Transport Functions Along the Length of Roots . . . 182
2.2 Structure-Function Relations in Various Root Zones . . . 183
2.2.1 The Root Surface . . . 183
2.2.2 The Cortex . . . 189
2.2.3 The Endodermis . . . 190
2.2.4 The Stele . . . 193
3 Variations of Physiological Activities Along the Length of Roots . . . 199
3.1 Growth, Differentiation and Hormonal Gradients . . . 199
3.2 Bioelectrical Fields Along Roots . . . 200
3.3 Differences in Ion Transport Mechanisms Along Roots . . . 201
4 Root-Shoot Interactions and Circulation in the Whole Plant . . . 202
4.1 Some Examples Illustrating General Aspects of Circulation . . . 202
4.2 Nitrogen, Sulphur and Phosphorus . . . 203
5 Conclusion . . . 204
References . . . 204

**I.7 Cycling of Elements in the Biosphere**
C.C. Delwiche (With 5 Figures)

1 The Sources of Plant Constituents . . . 212
1.1 Soil and Atmospheric Sources . . . 212
1.2 The Weathering Process . . . 212

2 The Nature of Cycles . . . . . 214
2.1 The Hydrologic Cycle . . . . . 214
2.2 The Sedimentary Cycle . . . . . 215
2.3 The Magmatic Cycle . . . . . 217
2.4 The Geobiological Cycles . . . . . 217
3 The Nitrogen Cycle . . . . . 219
3.1 Overall Cycle Features . . . . . 219
3.2 Nitrification . . . . . 221
3.3 Denitrification . . . . . 222
3.4 Nitrogen Fixation . . . . . 223
3.5 Human Influences . . . . . 224
4 The Sulfur Cycle . . . . . 225
4.1 Comparison with the Nitrogen Cycle . . . . . 225
4.2 Microbial Oxidation . . . . . 227
4.3 Sulfate Reduction . . . . . 227
4.4 Patterns of Sulfur Movement . . . . . 228
4.5 Human Influences . . . . . 228
5 The Phosphorus Cycle . . . . . 229
5.1 Oxidation and Reduction . . . . . 229
5.2 Movement and Transport in the Biosphere . . . . . 230
5.3 Human Influences . . . . . 231
6 Other Elements . . . . . 232
6.1 Biological Cycling . . . . . 232
6.2 The Special Significance of Iron and Aluminum . . . . . 232
6.3 Hydrogen Ion . . . . . 233
6.4 Characteristics of Sediments . . . . . 234
6.5 Passive Cycling . . . . . 235
6.6 Possibilities of Deficiency . . . . . 235
7 "Open" Versus "Closed" Agricultural Systems . . . . . 236
References . . . . . 237

## II. Inorganic Nitrogen Nutrition

### II.1 Physiology, Biochemistry and Genetics of Dinitrogen Fixation

H. Bothe, M.G. Yates, and F.C. Cannon (With 3 Figures)

1 The Nitrogen-Fixing Organisms and the Nitrogenase Reactions . . . . . 241
1.1 Introduction . . . . . 241
1.2 Nitrogen Fixation by Free-Living Organisms . . . . . 244
1.3 Symbiotic Nitrogen Fixation . . . . . 245
1.4 Substrates of Nitrogenase . . . . . 247
2 Biochemistry of Nitrogen Fixation . . . . . 248
2.1 Introduction . . . . . 248
2.2 Nomenclature of Nitrogenase Proteins . . . . . 249
2.3 Physicochemical Properties of Nitrogenase Proteins . . . . . 249
2.4 Metal Clusters in Nitrogenase Proteins . . . . . 251
2.5 EPR and Mössbauer Spectroscopy on the MoFe Protein . . . . . 252
2.6 The FeMo Cofactor and the Fe Protein . . . . . 253
2.7 Nitrogenase Proteins in Photosynthetic Organisms . . . . . 254
2.8 The Mechanism of Nitrogenase Activity . . . . . 254
2.8.1 The Roles of the Two Proteins . . . . . 255
2.8.2 Evidence for Interaction of MgATP and MgADP with the MoFe Protein . . . . . 257
2.8.3 The Nature of the Active Site(s) . . . . . 257
2.8.4 Pathways of $N_2$-Reduction . . . . . 258
3 Electron Transport to Nitrogenase . . . . . 259
3.1 Introduction . . . . . 259
3.2 Ferredoxins . . . . . 260

3.3 Flavodoxins . . . 260
3.4 Electron Donors . . . 261
4 Mechanisms to Protect Nitrogenase Against Damage by Oxygen . . . 263
4.1 In Free-Living Organisms . . . 263
4.2 The Heterocysts of Blue-Green Algae . . . 264
4.3 The Role of Leghaemoglobin in Legume Nodules . . . 265
5 Regulation of Nitrogenase Activity and Biosynthesis . . . 265
5.1 Regulation of Nitrogenase Biosynthesis . . . 265
5.2 Regulation of Nitrogenase Activity . . . 267
6 The Hydrogenase-Nitrogenase Relationship . . . 268
7 The Molecular and Genetic Characterization of Nitrogen Fixation Genes 271
7.1 Introduction . . . 271
7.2 The *nif* Genes . . . 272
7.3 *nif* Gene Products . . . 273
7.4 Cloning of *K. pneumoniae nif* Genes . . . 274
7.5 A Physical Map of *nif* Genes . . . 275
7.6 Interspecies Homology of Nitrogenase Genes . . . 276
References . . . 276

**II.2 Dinitrogen-Fixing Symbioses with Legumes, Non-Legume Angiosperms and Associative Symbioses**
A. QUISPEL (With 7 Figures)

1 Introduction . . . 286
2 Description of the Main Symbiotic Dinitrogen-Fixing Systems . . . 287
2.1 Associative Symbioses . . . 287
2.2 Symbioses with Cyanobacteria . . . 287
2.2.1 Distribution . . . 287
2.2.2 Description and Development . . . 287
2.2.3 $N_2$ Fixation ($C_2H_2$ Reduction) . . . 288
2.3 Root Nodules with Actinomycetes: Actinorhizas . . . 288
2.3.1 Distribution . . . 288
2.3.2 Description . . . 289
2.3.3 Infection and Development . . . 291
2.3.4 $N_2$ Fixation ($C_2H_2$ Reduction) . . . 291
2.4 Leguminous Root Nodules with *Rhizobium* . . . 291
2.4.1 Distribution . . . 291
2.4.2 Description . . . 291
2.4.3 Infection and Nodule Development . . . 292
2.4.4 $N_2$ Fixation ($C_2H_2$ Reduction) . . . 295
2.5 Non-Leguminous Root Nodules with *Rhizobium* . . . 295
3 The Dinitrogen-Fixing Micro-Symbionts: Isolates and Cultures . . . 295
3.1 Introduction . . . 295
3.2 Cyanobacteria . . . 296
3.3 Frankia, the Endophyte from the Actinorhizas . . . 297
3.3.1 Isolation and Cultivation . . . 297
3.3.2 Specificity . . . 298
3.3.3 Nutrient Requirements . . . 298
3.3.4 Metabolic Activities . . . 299
3.4 *Rhizobium* . . . 299
3.4.1 Isolation and Description . . . 299
3.4.2 Taxonomy . . . 300
3.4.3 Metabolism . . . 300
3.4.4 $N_2$ Fixation ($C_2H_2$ Reduction) . . . 302
3.4.5 Genetics . . . 303
4 Symbiotic Relations . . . 304
4.1 Chemotaxis and Rhizosphere Accumulation . . . 304
4.2 Binding of *Rhizobium* to Root Hairs . . . 305

4.3 Root Hair Deformation and Infection-Thread Formation . . . . . . 307
4.4 Cell Wall Degrading Enzymes . . . . . . . . . . . . . . . . . . . 307
4.5 The Role of Plant Hormones in Nodule Formation . . . . . . . . . 308
4.6 Miscellaneous Problems . . . . . . . . . . . . . . . . . . . . . 310
5 The $N_2$-Fixing System . . . . . . . . . . . . . . . . . . . . . . 311
5.1 Introduction . . . . . . . . . . . . . . . . . . . . . . . . . . 311
5.2 Bacteroids . . . . . . . . . . . . . . . . . . . . . . . . . . . 312
5.3 The Bacteroid-Containing Plant Cells . . . . . . . . . . . . . . 314
5.4 Nitrogenase . . . . . . . . . . . . . . . . . . . . . . . . . . . 315
5.5 $NH_3$ Assimilation . . . . . . . . . . . . . . . . . . . . . . . 315
5.6 Oxygen Regulation and Leghaemoglobin . . . . . . . . . . . . . . 317
5.7 Hydrogen Production and Hydrogen Uptake . . . . . . . . . . . . 318
6 Root Nodules as Part of the Whole Plant . . . . . . . . . . . . . . 319
7 Concluding Remarks . . . . . . . . . . . . . . . . . . . . . . . . 323
References . . . . . . . . . . . . . . . . . . . . . . . . . . . . . 323

**II.3 Dinitrogen Fixation in Rhizosphere and Phyllosphere Associations**
J. DÖBEREINER (With 2 Figures)

1 Introduction . . . . . . . . . . . . . . . . . . . . . . . . . . . . 330
2 Characterization of Rhizocoenoses . . . . . . . . . . . . . . . . . 330
2.1 Sugar Cane – *Beijerinckia* . . . . . . . . . . . . . . . . . . . 331
2.2 *Paspalum notatum – Azotobacter paspali* . . . . . . . . . . . . 332
2.3 *Azospirillum* Rhizocoenoses . . . . . . . . . . . . . . . . . . 332
2.3.1 Taxonomy of *Azospirillum* spp. . . . . . . . . . . . . . . . 333
2.3.2 Root Infection . . . . . . . . . . . . . . . . . . . . . . . 334
2.3.3 Host Plant Specificity . . . . . . . . . . . . . . . . . . . 336
2.3.4 Physiology of *Azospirillum* . . . . . . . . . . . . . . . . 337
2.4 Associations with Other $N_2$-Fixing Bacteria . . . . . . . . . . 340
3 Agronomic Aspects . . . . . . . . . . . . . . . . . . . . . . . . . 341
3.1 Plant Genotype Effects . . . . . . . . . . . . . . . . . . . . . 341
3.2 Environmental Effects . . . . . . . . . . . . . . . . . . . . . . 342
3.3 Inoculation . . . . . . . . . . . . . . . . . . . . . . . . . . . 342
4 Phyllosphere Associations . . . . . . . . . . . . . . . . . . . . . 343
4.1 Microorganisms in the Phyllosphere . . . . . . . . . . . . . . . 344
4.2 Nitrogen Fixation in the Phyllosphere . . . . . . . . . . . . . . 344
5 General Conclusion . . . . . . . . . . . . . . . . . . . . . . . . . 344
References . . . . . . . . . . . . . . . . . . . . . . . . . . . . . 345

**II.4 Uptake and Reduction of Nitrate: Bacteria and Higher Plants**
L. BEEVERS and R.H. HAGEMAN

1 Introduction . . . . . . . . . . . . . . . . . . . . . . . . . . . . 351
2 Available Nitrogen Sources . . . . . . . . . . . . . . . . . . . . . 352
2.1 Species Differences in Ammonium and Nitrate Utilization . . . . . 352
2.2 Influence of Ammonium or Nitrate on Cation Uptake . . . . . . . 353
2.3 Nitrate Uptake . . . . . . . . . . . . . . . . . . . . . . . . . 354
2.4 Influence of Ammonium on Nitrate Uptake and Utilization . . . . . 355
3 Nitrate Reduction . . . . . . . . . . . . . . . . . . . . . . . . . 356
3.1 Bacteria . . . . . . . . . . . . . . . . . . . . . . . . . . . . 356
3.2 Dissimilatory Nitrate Reductase . . . . . . . . . . . . . . . . 356
3.3 Assimilatory Nitrate Reduction in Bacteria . . . . . . . . . . . 358
3.4 Characterization of Nitrate Reductase from Higher Plants . . . . . 358
4 Molybdenum in Nitrate Reduction . . . . . . . . . . . . . . . . . . 360
5 Nitrite Reduction . . . . . . . . . . . . . . . . . . . . . . . . . 361
5.1 Assimilatory Bacteria . . . . . . . . . . . . . . . . . . . . . . 361
5.2 Dissimilatory Bacteria . . . . . . . . . . . . . . . . . . . . . 361
5.3 Nitrite Reductase in Plants . . . . . . . . . . . . . . . . . . . 361

6 Location of Enzymes of Nitrate Assimilation in Higher Plants . . . . . 363
7 Provision of Reductant for Nitrate Assimilation in Higher Plants . . . . 363
8 Regulation of Nitrate Reductase in Higher Plants . . . . . . . . . . . 364
8.1 Substrate . . . . . . . . . . . . . . . . . . . . . . . . . . 364
8.2 Hormonal . . . . . . . . . . . . . . . . . . . . . . . . . . 365
8.3 Molybdenum . . . . . . . . . . . . . . . . . . . . . . . . 365
8.4 Ammonium . . . . . . . . . . . . . . . . . . . . . . . . . 366
8.5 Light . . . . . . . . . . . . . . . . . . . . . . . . . . . 366
8.6 Genetic . . . . . . . . . . . . . . . . . . . . . . . . . . 366
8.7 In Vivo Controls . . . . . . . . . . . . . . . . . . . . . . 367
9 Concluding Thoughts . . . . . . . . . . . . . . . . . . . . . . 368
References . . . . . . . . . . . . . . . . . . . . . . . . . . . 369

## II.5 Uptake and Reduction of Nitrate: Algae and Fungi

W.R. Ullrich (With 4 Figures)

1 Introduction . . . . . . . . . . . . . . . . . . . . . . . . . . 376
2 Nitrate and Nitrite Reduction in Algae . . . . . . . . . . . . . . . 377
2.1 Nitrate Reductase of Eucaryotic Algae . . . . . . . . . . . . . 377
2.2 Nitrate Reductase in Blue-Green Algae . . . . . . . . . . . . . 380
2.3 Nitrite Reductase in Algae . . . . . . . . . . . . . . . . . . 380
2.4 Location of Nitrate and Nitrite Reduction in Algal Cells . . . . . . 381
2.5 Stoichiometry Between Nitrate Reduction and $O_2$ Exchange . . . . 381
3 Nitrate Uptake in Algae . . . . . . . . . . . . . . . . . . . . . 382
3.1 General Remarks . . . . . . . . . . . . . . . . . . . . . . . 382
3.2 Substrate Affinity . . . . . . . . . . . . . . . . . . . . . . 383
3.3 Light Dependence . . . . . . . . . . . . . . . . . . . . . . 384
3.4 pH-Dependence . . . . . . . . . . . . . . . . . . . . . . . 385
3.5 Dependence on Carbon Sources . . . . . . . . . . . . . . . . 385
3.6 Inhibition by Anions . . . . . . . . . . . . . . . . . . . . . 386
3.7 Inhibition by Ammonia and Amino Compounds . . . . . . . . . . 387
3.8 Effect of Metabolic Inhibitors and Uncouplers . . . . . . . . . . 387
3.9 Stoichiometry Between the Uptake of Nitrate and that of Other Ions 388
3.10 Transport Mechanism . . . . . . . . . . . . . . . . . . . . 388
4 Nitrite Uptake in Algae . . . . . . . . . . . . . . . . . . . . . 389
5 General Remarks on Regulation of Nitrate and Nitrite Uptake . . . . . 390
6 Uptake and Reduction of Nitrate and Nitrite in Fungi . . . . . . . . . 391
References . . . . . . . . . . . . . . . . . . . . . . . . . . . 393

# III. Metabolism of Sulfur and Phosphorus

## III.1 Reduction and Other Metabolic Reactions of Sulfate

J.A. Schiff (With 6 Figures)

1 Introduction . . . . . . . . . . . . . . . . . . . . . . . . . . 401
2 The Place of Sulfate Reduction in the Sulfur Cycle . . . . . . . . . . 402
3 Phylogenetic Distribution of Reactions Involving Sulfate Transfer and Reduction . . . . . . . . . . . . . . . . . . . . . . . . . . 403
4 Sulfate Uptake, Activation and Transfer . . . . . . . . . . . . . . 404
5 Sulfate Reduction . . . . . . . . . . . . . . . . . . . . . . . 406
5.1 Detailed Reactions of the Two Assimilatory Pathways . . . . . . . 408
5.1.1 The APS Pathway . . . . . . . . . . . . . . . . . . . 408
5.1.2 The PAPS Pathway . . . . . . . . . . . . . . . . . . 410
5.2 Location of Sulfate Reduction in Tissues and Organs of Multicellular Plants . . . . . . . . . . . . . . . . . . . . . . . . . 413

6 Speculations on the Origin and Evolution of Pathways of Sulfate Reduction . . . . . . . . . . . . . . . . . . . . . . . . . . . 413
References . . . . . . . . . . . . . . . . . . . . . . . . . . . 416

**III.2 Physiology and Metabolism of Phosphate and Its Compounds**
R.L. BIELESKI and I.B. FERGUSON (With 4 Figures)

1 Introduction . . . . . . . . . . . . . . . . . . . . . . . . . . . 422
2 Uptake and Transport of Phosphate . . . . . . . . . . . . . . . . 424
3 Efflux of Phosphate, and Aspects of Phosphate Deficiency . . . . . . . . 428
4 Phosphorus Compartments and Pools . . . . . . . . . . . . . . . . 431
5 The Form of Phosphorus in the Cell . . . . . . . . . . . . . . . . 433
6 Synthesis and Turnover of Phosphorus Compounds . . . . . . . . . . . 440
7 Dynamics of Phosphate Use in the Plant . . . . . . . . . . . . . . . 443
8 Conclusions . . . . . . . . . . . . . . . . . . . . . . . . . . . 445
References . . . . . . . . . . . . . . . . . . . . . . . . . . . 445

**Author- and Subject Index** (see Part B)

# Contents Part B


**IV. General Function of Inorganic Nutrients in Growth and Metabolism**

IV.1 Genetic Basis of Inorganic Plant Nutrition
G.C. GERLOFF and W.H. GABELMAN (With 2 Figures) . . . . . . . . . . . 453

IV.2 Mineral Nutrition and Growth
J. MOORBY and R.T. BESFORD (With 24 Figures) . . . . . . . . . . . . 481

IV.3 Proteins, Enzymes and Inorganic Ions
R.G. WYN JONES and A. POLLARD (With 5 Figures) . . . . . . . . . . . 528

IV.4 The Enzymological Function of Heavy Metals and Their Role in Electron Transfer Processes of Plants
G. SANDMANN and P. BÖGER (With 4 Figures) . . . . . . . . . . . . . 563

**V. Special Functions of Some Elements**

V.1 Calcium Transport and Function
D. MARMÉ (With 2 Figures) . . . . . . . . . . . . . . . . . . . . . . 599

V.2 Boron in Plant Metabolism
W.M. DUGGER (With 4 Figures) . . . . . . . . . . . . . . . . . . . . 626

V.3 Sodium Versus Potassium: Substitution and Compartmentation
T.J. FLOWERS and A. LÄUCHLI (With 6 Figures) . . . . . . . . . . . . 651

V.4 Silica Metabolism
D. WERNER and R. ROTH (With 3 Figures) . . . . . . . . . . . . . . . 682

V.5 Involvement of Unusual Elements in Plant Growth and Nutrition
E.G. BOLLARD (With 4 Figures) . . . . . . . . . . . . . . . . . . . . 695

**VI. Synthesis and Outlook**
R.L. BIELESKI and A. LÄUCHLI (With 1 Figure) . . . . . . . . . . . . 745

**Author Index** . . . . . . . . . . . . . . . . . . . . . . . . . . . 757

**Subject Index** . . . . . . . . . . . . . . . . . . . . . . . . . . . 829

# List of Contributors Part A and B


C.J. Asher
Dept. of Agriculture
University of Queensland
St. Lucia
Queensland 4067/Australia

L. Beevers
770 Van Vleet Oval
Dept. of Botany and Microbiology
University of Oklahoma
Norman, OK 73019/USA

R.T. Besford
Glasshouse Crops Research Institute
Worthing Road
Littlehampton
West Sussex BN16 3PU/
United Kingdom

R.L. Bieleski
Dept. of Scientific and
Industrial Research
Division of Horticulture and Processing
Private Bag
Auckland/New Zealand

P. Böger
Lehrstuhl für Physiologie und
Biochemie der Pflanzen
Universität Konstanz
Postfach 5560
D-7750 Konstanz/FRG

E.G. Bollard
Dept. of Scientific and
Industrial Research
Division of Horticulture and
Processing
Private Bag
Auckland/New Zealand

H. Bothe
Botanisches Institut
Universität Köln
Gyrhofstr. 15
D-5000 Köln 41/FRG

D. Bouma
CSIRO
Division of Plant Industry
P.O. Box 1600
Canberra, ACT, 2601/Australia

G.D. Bowen
CSIRO Division of Soils
Private Bag 2
Post Office
Glen Osmond
South Australia 5064/Australia

F.C. Cannon
BioTechnica International, Inc.
85 Bolten Street
Cambridge, MA 02140/USA

C.C. Delwiche
University of California
Dept. of Land, Air and Water
Resources
Hoagland Hall
Davis, CA 95616/USA

J. Döbereiner
EMBRAPA PNPBS, Km 47
Seropédica
23460 Rio de Janeiro/Brazil

W.M. Dugger
Dept. of Botany and Plant
Sciences
University of California
Riverside, CA 92521/USA

D.G. Edwards
Dept. of Agriculture
University of Queensland
St. Lucia
Queensland 4067/Australia

E. Epstein
University of California
Dept. of Land, Air and Water
Resources
Hoagland Hall
Davis, CA 95616/USA

I.B. FERGUSON
Division of Horticulture and Processing
DSIR
Private Bag
Auckland/New Zealand

T.J. FLOWERS
School of Biological Sciences
University of Sussex
Falmer
Brighton BN1 9QG/United Kingdom

R.C. FOSTER
CSIRO Division of Soils
Private Bag 2
Post Office
Glen Osmond
South Australia 5064/Australia

W.H. GABELMAN
Department of Horticulture
University of Wisconsin
1575 Linden Dr.
Madison, WI 53706/USA

G.C. GERLOFF
Dept. of Botany
Birge Hall
University of Wisconsin-Madison
Madison, WI 53706/USA

R.H. HAGEMAN
Dept. of Agronomy
University of Illinois
1102 S Goodwin Ave.
Urbana, IL 61801/USA

A. LÄUCHLI
University of California
Dept. of Land, Air and Water Resources
Hoagland Hall
Davis, CA 95616/USA

U. LÜTTGE
Institut für Botanik
Fachbereich Biologie (10)
Technische Hochschule Darmstadt
Schnittspahnstraße 3
D-6100 Darmstadt/FRG

D. MARMÉ
Institut für Biologie III
Schänzlestraße 1
D-7800 Freiburg/FRG

H. MARSCHNER
Institut für Pflanzenernährung
Universität Hohenheim
Postfach 700562
D-7000 Stuttgart 70/FRG

J. MOORBY
Agricultural Research Council
160 Great Portland St.
London WIN 6DT/United Kingdom

M.G. PITMAN
School of Biological Sciences, A12
University of Sydney
NSW 2006/Australia

A. POLLARD
Dept. of Biochemistry and Soil Sciences
University College of North Wales
Bangor, Gwynedd
Wales/United Kingdom

A. QUISPEL
Botanical Laboratory
Dept. of Plant Molecular Biology
University of Leiden
Nonnensteeg 3
2311 VL Leiden/The Netherlands

A.D. ROBSON
Dept. of Soil Science and Plant Nutrition
Institute of Agriculture
University of Western Australia
Nedlands
Western Australia 6009/Australia

R. ROTH
Fachbereich Botanik
der Phillipps-Universität Marburg
Lahnberge
D-3550 Marburg/Lahn/FRG

A.D. ROVIRA
CSIRO Division of Soils
Private Bag 2
Post Office
Glen Osmond
South Australia 5064/Australia

G. SANDMANN
Lehrstuhl für Physiologie und Biochemie der Pflanzen
Universität Konstanz
Postfach 5560
D-7750 Konstanz/FRG

J.A. SCHIFF
Institute for Photobiology
of Cells and Organelles
Brandeis University
Waltham, MA 02254/USA

W.R. ULLRICH
Institut für Botanik
Technische Hochschule Darmstadt
Schnittspahnstraße 3
D-6100 Darmstadt/FRG

D. WERNER
Fachbereich Botanik
der Philipps-Universität Marburg
Lahnberge
D-3550 Marburg/Lahn/FRG

R.G. WYN JONES
Dept. of Biochemistry and Soil
Science
University College of North Wales
Bangor, Gwynedd
Wales/United Kingdom

M.G. YATES
Agricultural Research Council
Unit of Nitrogen Fixation
University of Sussex
Falmer
Brighton BN1 9QG/United Kingdom

# Introduction

A. LÄUCHLI and R.L. BIELESKI

Initial plans for this volume were made in 1974 when manuscripts for Vol. 2A and B on Transport in Cells, Tissues and Organs were in preparation. In discussions involving ULRICH LÜTTGE, MICHAEL PITMAN and A.L. the idea emerged that a volume was needed in the *Encyclopedia of Plant Physiology, New Series* to cover the field of Inorganic Plant Nutrition.

To treat the whole field of plant nutrition in one volume has meant that we have been able to present only a selection of topics. It has not been possible to have a full review of the literature pertaining to each nutrient element. The organization of this volume therefore evolved from the concept that the broad subject should be arranged functionally rather than by elements. Furthermore, this volume complements Vol. 2, for many central functions of inorganic nutrients are intrinsically connected to their transport in the plant. Thus our authors were asked not to attempt comprehensive reviews and exhaustive literature surveys, but rather to integrate the wealth of information on inorganic plant nutrition and incorporate personal viewpoints.

This volume consists of two parts and is organized in a foreword, five main sections, and a final chapter on synthesis and outlook. Section I aims at integrating modern aspects of inorganic plant nutrition and incorporates various features of nutrient transport at the soil–root interface, in the plant, and in the biosphere. Section II is devoted to inorganic nitrogen nutrition and treats symbiotic dinitrogen fixation (including associative symbioses) as well as uptake and reduction of nitrate. The physiology and metabolism of sulfate and phosphate (Sect. III) conclude Part A of the volume. In Part B, Section IV treats the general function of inorganic nutrients in growth and metabolism and includes the genetic basis of inorganic plant nutrition. Finally, Section V highlights novel ideas on the functions of some particular elements – from the nutrient elements calcium and boron and sodium-potassium relationships to silica and unusual elements.

We would like to thank all those who contributed to this volume. Our primary thanks go to Professor PIRSON, one of the Series Editors, whose advice on organization of the volume and continued interest in its progress have been extremely valuable. We also owe thanks to Professor PIRSON for understanding that it was impossible to keep the whole book manuscript to the expected size of a regular volume in this series. Otherwise, the treatment of some topics would have been very sketchy. We also thank the publishers for their support in the production of this volume.

We thank the colleagues whose discussions in the early stages helped us to establish a group of topics that, we hope, reveal something of the ferment and vitality of present-day plant nutrition research. Thus we particularly thank

E. EPSTEIN, A.R. FERGUSON, I.B. FERGUSON, H. GREENWAY, U. LÜTTGE and M.G. PITMAN. Special thanks are due to A.R. FERGUSON for his careful and thorough preparation of the subject index. Above all we thank the authors of this volume for their patience and their understanding that the Editors had to make demands for reorganization and modification of their chapters.

# I. General Chapters of Inorganic Plant Nutrition

# I.1 General Introduction to the Mineral Nutrition of Plants

H. MARSCHNER

## 1 Introduction and Historical Résumé

### 1.1 Essential Mineral Elements – Plant Nutrients

The beneficial effect of adding mineral elements (e.g. plant ash, lime) to soils to improve plant growth has been known in agriculture for more than 2000 years. Nevertheless, even 150 years ago it was still a matter of scientific controversy as to whether mineral elements function as nutrients for plant growth. It was mainly to the credit of JUSTUS VON LIEBIG (1803–1873) that the scattered information concerning the importance of mineral elements for plant growth was collected and summarized, and that mineral nutrition of plants was established as a scientific discipline. This achievement led to a rapid increase in the use of mineral fertilizers. In Europe in particular large amounts of potash, superphosphate and later inorganic nitrogen were used in agriculture and horticulture to improve plant growth.

LIEBIG's conclusion as to the essentiality of the mineral elements, nitrogen, sulfur, phosphorus, potassium, calcium, magnesium, silicon, sodium and iron was based on observations and speculations rather than on precise experimentation. This shaky foundation on which the "mineral element theory" was laid was one of the reasons for the onset of a large number of studies at the end of last century. Both sand and water culture experiments were carried out on higher plants to establish essentiality of the mineral elements for plant growth and their role in plant metabolism. Progress in this research was closely related to developments in analytical chemistry (purification of chemicals, methods for determination). This relationship is well reflected in the time scale of the discovery of the essentiality of the micronutrients (Table 1).

**Table 1.** Discovery of the essentiality of micronutrients for higher plants

| Element | Year | Discovered by |
|---|---|---|
| Iron | 1860 | J. SACHS |
| Manganese | 1922 | J.S. MCHAGUE |
| Boron | 1923 | K. WARINGTON |
| Zinc | 1926 | A.L. SOMMER and C.B. LIPMAN |
| Copper | 1931 | C.B. LIPMAN and MCKINNEY |
| Molybdenum | 1938 | D.I. ARNON and P.R. STOUT |
| Chlorine | 1954 | T.C. BROYER et al. |

As proposed by ARNON (1950), it is generally agreed that the term essential mineral element (=mineral plant nutrient) should be restricted to those elements which fulfil the following criteria: (a) that there is a positive requirement of the element for normal growth or reproduction, (b) that it is not possible to replace the function of this element by other elements, (c) that it has a direct or indirect action in plant metabolism. The term does not include those elements which, for example, compensate the toxic effects of other elements. Thus, a mineral element which only stimulates growth but does not satisfy the above criteria cannot be defined as mineral plant nutrient. It can merely be described as a "beneficial" element (see below).

Six macronutrients are now firmly established for all higher plants. However, the possibility still exists that with improving analytical techniques new micronutrients may be revealed so that the present list of 13 mineral nutrients for higher plants (Table 2) may yet be extended.

It is still difficult to generalize concerning the essentiality of mineral elements for plant growth. This is particularly obvious when higher and lower plants are compared. In some bacteria and fungi, for example, even calcium, boron and chlorine are probably not essential. On the other hand, for higher plants whether also sodium and silicon are essential or just "beneficial elements" appears to be genotypically dependent (see below).

## 1.2 Function of Essential Mineral Elements

For mineral elements that are constituents of organic compounds which act in plants as enzymes, co-enzymes, membrane constituents etc. at least some of their function has been well defined. The reader is referred for detailed information on nitrogen to Chapter II, on sulfur and phosphorus to Chapters III.1 and III.2, on potassium and calcium to Chapter V.1 and on boron to Chapter V.2, all this Volume[1]. A chapter dealing with micronutrients in general is to be found in Chapter IV.4 of the first edition of Encyclopedia of Plant Physiology, Vol. IV. Detailed information on plant nutrients may be also found in recent text books such as EPSTEIN (1972), GAUCH (1972), HEWITT and SMITH (1975), BERGMANN and NEUBERT (1976), MENGEL and KIRKBY (1978), MENGEL (1979).

For this general introduction it is adequate to sum up one or two typical well-established functions of each mineral nutrient (Table 2). Table 2 also refers to relevant chapters of this Volume. The mechanisms of these effects on enzyme reactions are of differing nature: (a) Valency change as an electron transferring agent (e.g. Fe), (b) Attachment of the substrate to the enzyme (e.g. Mg) or (c) Change in molecular configuration of enzyme or substrate or both (e.g. K).

Most results on enzyme activation (Table 2 last column) are based on in vitro experiments using relatively high concentrations of inorganic ions. Some doubt is left therefore as to whether in vivo ion concentrations are sufficiently high in the cell compartments or at the sites of enzyme reaction (e.g. membranes) to act as activators as in the in vitro experiments. This is of particular relevance to the micronutrients. In addition, several morphological, physiological and

1 Further chapters cited in this chapter refer to the present volume, unless specified

**Table 2.** The amounts and examples of the functions of mineral nutrients in higher plants

| Element | µmol per g[a] | Constituent of essential organic compounds (examples) | Activator (+) or inhibitor (−) of enzymes or enzyme reaction |
|---|---|---|---|
| N | 1000 | Proteins, nucleic acid | Nitrate reductase $NO_3^-$ (+), $NH_4^+$ (−) see Chap. II.4 |
| S | 30 | Protein, sulfolipids | (see Chaps. IV.3 and III) |
| P | 60 | Nucleic acids, co-enzymes | ADPG-pyrophosphorylase (−) starch phosphorylase (see Chap. III.2) |
| K | 250 | Unknown | Starch synthase (see Chaps. IV.3 and V.3) |
| Ca | 125 | Pectates | Amylase (+), (see below and Chap. V.1) |
| Mg | 80 | Chlorophyll | Phosphorylation (+) RubP-carboxylase (+) |
| Fe | 2.0 | Ferredoxin, cytochrome | (see Chap. IV.4) |
| Mn | 1.0 | Unknown | "Hill reaction" (+) Amino peptidases (+) |
| Zn | 0.3 | Carbonic anhydrase alkaline phosphatase | Hexokinase (+) Ribonuclease (−) (see Chap. IV.4) |
| Cu | 0.1 | Ascorbic acid oxidase plastocyanine | (see Chap. IV.4) |
| B | 2.0 | Unknown | (see Chap. V.2) |
| Cl | 3.0 | Unknown | (see Chap. IV.4) |
| Mo | 0.001 | Nitrate reductase | (see Chap. II.4) |

[a] Average concentration in dry matter sufficient for adequate growth (see for example BERGMANN and NEUBERT 1976)

biochemical changes in higher plants caused by deficiencies of certain mineral nutrients, micronutrients in particular, either cannot or can only indirectly be causally related to the known functions of the essential mineral elements. Within the last 20 years research activities have thus centered more on integrated metabolic processes influenced by essential mineral elements rather than on purely biochemical in vitro studies (see below).

## 1.3 Beneficial Mineral Elements

Mineral elements which either stimulate growth without being essential (see definition above) or which are essential only for certain plant species or families are usually defined as beneficial elements. This applies particularly to sodium, silicon and cobalt. The difficulties in this definition of a beneficial mineral element are readily demonstrated in the following examples.

*Sodium.* The use of the term "natrophilic" and "natrophobic" plant species indicates the wide range of response to sodium in the plant kingdom (MARSCHNER 1971, JENNINGS 1976). Sodium is essential for certain halophytes (BROWNELL and JACKMAN 1966), and is considered also as a micronutrient at least for $C_4$ plant species (BROWNELL 1979). In addition it has a beneficial effect on growth of the so-called halophytic species such as *Beta vulgaris,* particularly under conditions of low potassium supply (EL-SHEIKH et al. 1967, MARSCHNER and POSSINGHAM 1975), but is either without effect or is even harmful on growth of natrophobic species such as *Phaseolus vulgaris* (HAWKER et al. 1974). This differentiation between families and species in response to sodium, which is considered in more detail in Chapters IV.3 and V.3, opens interesting possibilities into basic research on the compartmentation and function of mineral elements which are not constituents of organic compounds (e.g. potassium). It is also of interest in applied research, since sodium is abundant in most natural ecosystems and in particular in the irrigation water of arid and semi-arid climates. This must be considered in relation to the fairly close positive correlation between the beneficial effect of sodium on growth and the "salt tolerance" of certain plant species. It is urgently necessary to obtain more information on the role of sodium in plants in general and on its ability to replace potassium in particular. This would allow a more straightforward strategy in selection and breeding programmes for crop plants with high "salt tolerance" in order to adapt these plants to saline soils, as has been sucessfully demonstrated by EPSTEIN et al. (1980).

*Silicon.* Silicon is an essential mineral element for certain lower plants (see Chap. V 4). It is also well established as a beneficial mineral element for higher plants, the monocotyledons in particular. In higher plants uptake of silicon is closely related to the transpiration rate of the plants (JONES and HANDRECK 1965) as well as to the transpiration rate of individual organs (HUTTON and NORRISH 1974). It is deposited mainly in the cell walls of the epidermal layers, and causes an increase in mechanical resistance. This can increase resistance to certain fungal infections (VOLK 1958, FIDANOVSKI 1969). This deposition of silicon also increases the rigidity of leaf blades and thus alters the leaf angle, thereby increasing the efficiency of light utilization for photosynthesis, particularly in dense plant populations (YOSHIDA et al. 1969, YUAN and CHANG 1978).

Silicon is particularly effective in counteracting manganese toxicity (VLAMIS and WILLIAMS 1967). Growth stimulation by silicon is therefore often limited to conditions of excessive manganese supply (HORST and MARSCHNER 1978a, VORM and DIEST 1979). This effect of silicon seems to be either primarily the result of restricted manganese uptake as in rice plants (YUAN and CHANG 1978, VORM and DIEST 1979) or, as in other species such as barley (WILLIAMS and VLAMIS 1957) and bean (HORST AND MARSCHNER 1978a), caused by more uniform distribution of manganese within the leaf tissue. These effects of silicon fully justify its classification as a beneficial mineral element. At least in *Equisetum arvense,* however, silicon is essential for growth (CHEN and LEWIN 1969). There is also good evidence for the essentiality of silicon for rice (YUAN and CHANG 1978). The surprising results of MIYAKE and TAKAHASHI (1978) showing the essentiality of silicon also for tomato plants (*Lycopersicon esculentum* Mill) needs to be critically re-examined. Unfortunately, these authors did not mention the variety used for these studies, information which is essential for all studies on mineral metabolism of higher plants in general (see Vol. 2/B.9, this Series) and on mineral elements such as silicon in particular. The situation with silicon seems to be similar to that of sodium, where classification as either an essential or beneficial mineral element depends upon the genotypes. For further details on silicon see Chapter V.4.

*Other Mineral Elements.* The essentiality of other mineral elements such as cobalt is well documented for certain higher and lower plants. For cobalt this essentiality is coupled with the mode of nitrogen supply, as this metal is a constituent of the vitamin $B_{12}$ and thus essential for all higher and lower plants which rely on $N_2$-fixation for their nitrogen supply. For the same plant species supplied with bound nitrogen either in inorganic or organic form, cobalt does not have the characteristics of a beneficial mineral element. These special cases of essentiality of cobalt and other mineral elements are discussed in detail in Chapter V.5.

## 1.4 Recent Developments

Despite the early recognition of the essentiality of certain mineral elements for plant metabolism and growth, the interest of plant physiologists and plant biochemists in detailed studies on mineral nutrition was relatively small until about 30 years ago. There were, of course, notable exceptions, including HOAGLAND and his school in California, but in general much greater attention was paid to photosynthesis, respiration, or synthetic pathways of carbohydrates, lipids and other organic components. After 1950, however, rapid development started in research on mineral nutrition. This was mainly by the use of radioisotopes in short-term studies with excised roots on the ion uptake mechanism and the approach of applying enzyme kinetics and the carrier concept to membrane transport processes of inorganic ions in plant cells and tissues (EPSTEIN and HAGEN 1952). This development was associated with a deeper understanding of membrane transport of inorganic ions in plant cells and tissues (see Vol. 2 A/B, this Series). During this period it was also realized that mineral elements are not only excellent tools for studies of the mechanism or energetics of membrane transport processes, but that essential mineral elements are also causally involved in regulating biochemical and metabolic processes of fundamental importance in the various compartments of living cells. Some of the more recent important developments in studies relating to the functions of some essential mineral elements in integrated systems such as plant cells or plant tissues are presented below.

**Table 3.** Effect of calcium chloride ($10^{-2}$ M) and benzyladenine ($10^{-7}$ M BA) on corn leaf segments kept for 4 days in darkness. (After POOVAIAH and LEOPOLD 1973a, POOVAIAH 1979)

| Treatment | Chlorophyll OD 665 | Protein % initial | Hydraulic permeability half time in min for equilibrium |
|---|---|---|---|
| Control | 0.21 | 58 | 0.8 |
| Ca | 0.32 | 76 | 2.5 |
| BA | 0.38 | 78 | 3.1 |
| Ca+BA | 0.66 | 90 | 8.0 |

### 1.4.1 Calcium

Although not a constituent of cell membranes, calcium ions are of fundamental importance for membrane permeability and transport and the maintenance of cell integrity (see Chap. V.1). The regulating effect of calcium at the plasmalemma on influx and efflux of other ions and selectivity in ion uptake has been one of the major research topics in ion uptake studies for more than 20 years (see Chaps. I.6; I.7 and V.1). It is now accepted that ion uptake studies, for example, with plant roots in the absence of externally supplied $Ca^{2+}$, are of rather limited value.

The role of $Ca^{2+}$ for membrane stability, and thus compartmentation of a single cell or a plant tissue, has consequences for several processes other than ion transport. It affects the efflux rate of sugars from plant roots (SHAY and HALE 1973) or leaf segments (HAWKER et al. 1974) into the external solution, as well as the flux rate of endogenous respiratory substrates from the vacuole to the respiratory enzymes in the cytoplasm (BANGERTH et al. 1972). An increase in the calcium content of a tissue therefore decreases the respiration rate (FAUST and KLEIN 1974). Typical features of senescence include an increase in the rate of respiration together with enhanced permeability and thus a decline in compartmentation. Calcium is thus directly involved in the regulation of senescence in addition to the role played by phytohormones such as ethylene or cytokinins (Table 3).

These results show that senescence of the leaf segments was retarded by calcium. A similar influence was found with cytokinin and the effect of both substances was additive. Furthermore, in *Phaseolus vulgaris* the abscission of leaf blades, caused by senescence of the pulvinar tissue, can be dramatically delayed or even prevented by high $Ca^{2+}$ concentrations (POOVAIAH and LEOPOLD 1973b).

The widespread occurrence of calcium-related physiological disorders, particularly in fruits (see below), has also initiated intensive research on the role of $Ca^{2+}$ in fruit development and fruit ripening (for a comprehensive review see BANGERTH 1979). Fruit ripening as a special case of senescence is correlated with increased ethylene production. This is stimulated in calcium-deficient tissue (FAUST and SHEAR 1969) and can be depressed by calcium treatment of the fruit (LOUGHEED et al. 1979). According to MATTOO and LIEBERMAN (1977), the causal connection between calcium-regulated membrane permeability and

**Table 4.** Calcium content (μmol $g^{-1}$ dry weight) in non ripening (*rin*) and normal Rutgers tomato pericarp tissues of different stages of fruit development. (After POOVAIAH 1979)

| Days after anthesis | Soluble Ca | | Bound Ca | |
|---|---|---|---|---|
| | rin | Rutgers | rin | Rutgers |
| 40 | 7.5 | 8.7 | 13.3 | 14.1 |
| 50 | 10.3 | 15.1 | 16.7 | 6.2 |
| 60 | 12.3 | 15.6 | 33.9 | 7.3 |

ethylene production and thus senescence of a tissue can be seen in the compartmentation of the enzyme system for ethylene synthesis in a cell wall–membrane complex, i.e. outside the cytoplasm. A fall in calcium content of a fruit due to "dilution" by growth (see below) or by a decrease in the concentration of physiologically active calcium (for example by precipitation as oxalate, chelation or compartmentation in the vacuole) can be considered as an essential step in the regulation of phytohormone-dependent fruit ripening. Fruit ripening certainly requires removal of part of the calcium from the middle lamella (bound calcium). This causal involvement of calcium is well reflected in comparative studies with a ripening and a non-ripening mutant (rin) of tomato (Table 4).

In the rin mutant the calcium content and particularly the bound fraction continuously increased after anthesis whereas in Rutgers variety the total calcium content remained constant and the bound calcium declined. This decline was correlated with an increase in polygalacturonase activity, the enzyme which is responsible for decomposing pectates of the middle lamella and finally "softening" this tissue.

Calcium in the middle lamella is not only important for the structure of the pectins itself, but also for the resistance of the tissue to fungal infections. There is a strict negative correlation between the calcium content of the tissue and the infection rate of lettuce leaves with *Botrytis cinera* Pers. (KRAUSS 1971) or apple fruits with *Gloeosporium* spp. (SHARPLES and JOHNSON 1977). The causality of this negative correlation is the result of either the inability of polygalacturonase to act on calcium pectates (BATEMAN and LUMSDEN 1965) or the direct inhibition of this enzyme by calcium ions (CORDEN 1965). Plant cell walls containing a high proportion of calcium pectates should therefore be expected to show increased resistance to fungal infections. These various functions of calcium, particularly in integrated systems, considered together with the increase in occurrence of calcium-related physiological disorders in crop plants are reflected in the pronounced interest of crop physiologists in calcium nutrition of plants (Sect. 3).

### 1.4.2 Potassium

With a few exceptions (halophytes), potassium is the most important inorganic solute in plants. This explains at least in part the particular role of potassium in osmoregulation and in water relations in plants as well as in enzyme activation

**Table 5.** Effects of potassium supply on carbon dioxide assimilation in *Medicago sativa*. (PEOPLES and KOCH 1979)

| Substrate | $CO_2$ diffusion Resistance (s $cm^{-1}$) | | Efficiency of photosystem and $CO_2$ fixation (µmol acceptor red $mg^{-1}$ chl.) | | Ribulosecarbox. µmol $CO_2$ $mg^{-1}$ protein $h^{-1}$ |
|---|---|---|---|---|---|
| mM $K^+$ | Stomata | Mesophyll | System I | System II | |
| 0.0 | 9.3 | 8.4 | 238 | 282 | 1.79 |
| 0.6 | 6.8 | 7.0 | 262 | 294 | 4.51 |
| 4.8 | 5.8 | 3.0 | 241 | 285 | 6.06 |

(Chaps. IV.3; V.3; see also Vol. II A/B, this Series, and a review by LÄUCHLI and PFLÜGER 1978). The high mobility of potassium in transport through membranes enables it also to act as counterion for anions both in short- and long-distance transport (Chaps. I.6 and I.7) and in ATPase-stimulated proton efflux, for example at the plasmalemma of root cells (Chap. I.7). This proton-coupled potassium transport has consequences for both the pH and osmotic pressure in the cell and its compartments. The role of potassium in the water balance of higher plants can now be directly related to its function in regulating osmotic pressure of the guard cells and thus stomatal movement (HUMBLE and RASCHKE 1971; HSIAO 1976, RASCHKE 1979).

Potassium-mediated changes in turgor and thus induction of movements are also responsible for the phytochrome-controlled movements, as for example in the hypocotyl hook of mung bean (BROWNLEE and KENDRICK 1979) or the nyctinasty in *Albizzia* (SATTER and GALSTON 1971) and the seismonastic reactions in *Mimosa* (CAMPBELL et al. 1979).

The varied functions of potassium enable this element to influence complex processes such as photosynthesis of higher plants at various stages. With potassium deficiency even the $CO_2$ diffusion through the stomata can be limited, and this is reflected in an increase in stomatal resistance (COOPER et al. 1967). The increase in mesophyll resistance when potassium supply is limited appears however, to be primarily related to inhibited protein synthesis and in particular to RubP-carboxylase synthesis (Table 5).

Incubating potassium-deficient leaves in potassium salt solution rapidly increased RubP-carboxylase activity, an increase which could be prevented by actinomycin (PEOPLES and KOCH 1979). This close correlation between potassium supply and synthesis of an enzyme in the chloroplasts has also been demonstrated with nitrate reductase in spinach chloroplasts (PFLÜGER and WIEDEMANN 1977).

The stimulation of bound starch synthase activity from chloroplasts by potassium (HAWKER et al. 1974) on the one hand and the enhanced export of sucrose from leaf cells as "source" (DOMAN and GEIGER 1979) on the other are further examples of the action of potassium in an integrated system. This also applies to the effect of potassium on the $N_2$-fixation rate of legumes (MENGEL et al. 1974) and growth rate of fruits and storage organs, i.e. yield formation (see Sect. 4).

The mineral nutritional status of plants can strongly affect the "quality" of crop plants and its organs not only for human and animal nutrition, but also as substrate for parasitic attack. When potassium supply is inadequate, for example, the synthesis of highly polymerized compounds (starch, proteins) is inhibited and the resulting plant composition is of lowered "quality" from the agricultural and horticultural view point. However, the corresponding increase in low molecular weight compounds (sugars, amino acids) in these plants increases its "quality" for various parasites, fungi, bacteria or insects. As an example, the close negative correlation between infection of soybean seeds with *Diaporthe sojae* and the potassium level (CRITTENDEN and SVEC 1974) may be cited. A comprehensive and more general review on this subject has been published elsewhere (PERRENOUD 1977) with impressive examples showing the close relationships between mineral nutrition of plants and plant diseases and plant protection.

### 1.4.3 Phosphorus

Phosphorus is essential not only as a constituent of various organic compounds (see Chap. III.2) but also inorganic phosphorus ($P_i$) exerts important regulatory functions in energy transfer. Another instructive example of its regulatory function has been described recently in chloroplasts where a concentration of approximately 10 mM in the stroma almost completely inhibits starch synthesis (HELDT et al. 1977). The control seems to be exerted mainly via the allosteric regulation of ADPG-pyrophosphorylase by $P_i$. From these results obtained with chloroplasts one might speculate that in starch-storing seeds such as cereal grains, phytin (Chap. II.2) might not only have the function as storage for phosphate and inorganic cations (LOTT and BUTTROSE 1978) but also take part in the regulation of starch synthesis. This aspect along with other, more indirect effects of phosphorus, are discussed later (Sect. 3).

### 1.4.4 Nitrogen

Recent developments in research on inorganic nitrogen nutrition are characterized by particular attention to the various aspects of dinitrogen fixation (Chap. II. 1, 2, 3), nitrate reduction and its regulation (Chap. II.4) and the consequences of intensity and sites (plant organs) of nitrate reduction for various other aspects of mineral nutrition such as cation/anion balance (Chap. I.6) or the interaction between plant roots and their substrate (see next section). Ammonium as compared to nitrate also seems to exert some distinct regulatory functions, not only in induction of flower formation (see Sect. 4) but also, for example, in chloroplasts by enhancing the direction of photosynthetic carbon flow towards the Krebs cycle (MOHAMED and GUANAM 1979).

The main effect of both ammonium and nitrate nitrogen on plant growth and development (e.g. flowering and senescence), however, seems to be causally related to direct interactions between nitrogen nutrition and phytohormones. The close correlation between nitrogen supply (nitrate) and the formation of cytokinins in the roots and their export to the shoot has been clearly demon-

**Table 6.** Effect of nitrate supply to potato plants on exudation rate and cytokinin export from the roots over a 24-h period. (SATTELMACHER and MARSCHNER 1978)

| Nitrogen supply | Days | Exudation rate (ml 24 $h^{-1}$) | Exudate | |
|---|---|---|---|---|
| | | | Concentration of cytokinins (ng $ml^{-1}$) | Amount of cytokinins (ng 24 $h^{-1}$) |
| +N | 0 | 33 | 6 | 196 |
| | 3 | 30 | 14 | 420 |
| | 6 | 33 | 17 | 561 |
| −N | 3 | 13 | 2 | 26 |
| | 6 | 17 | 1 | 17 |
| +N[a] | 9 | 22 | 6 | 132 |

[a] Renewed nitrate supply after 6 days −N

**Table 7.** Effect of nitrate supply on ABA content in different parts of potato plants. (KRAUSS 1978)

| | ng ABA $g^{-1}$ fresh weight and $ml^{-1}$ respectively | | |
|---|---|---|---|
| | Shoot | Root | Exudate |
| Beginning of experiment | 12.6 | 0.3 | 0.1 |
| After 6 days +N | 12.2 | 0.5 | 0.1 |
| After 6 days −N | 23.6 | 4.1 | 4.4 |

strated in sunflower plants by WAGNER and MICHAEL (1971) and SALAMA and WAREING (1979). This nitrogen-induced change in cytokinin export from roots is quite impressive in both the speed and extent of the response to nitrogen supply as shown in Table 6.

Nitrogen supply also affects other phytohormones, and abscisic acid (ABA) in particular. However, the effect is the reverse of that of the cytokinins. Nitrogen deficiency increases ABA in leaves (GOLDBACH et al. 1975), roots and exudate (Table 7).

The importance of these phytohormones in cell metabolism (e.g. proteinsynthesis, stomatal regulation) and developmental processes (e.g. flower initiation, senescence) indicates the necessity of considering the nutritional status of plants in general and of nitrogen in particular in the relation to hormonal activity. This applies not only to research in the physiology of yield formation (see Chap. IV.2), but also in developmental physiology or stress physiology (GÖRING and THIEN 1979).

### 1.4.5 Copper

Typical symptoms of copper deficiency in higher plants such as development of "pendula" forms in trees (OLDENKAMP and SMILDE 1966) or wilting of young

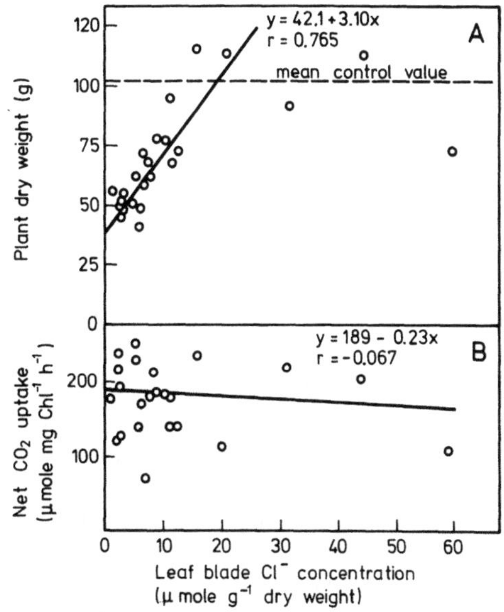


**Fig. 1A, B.** Effects of chlorine deficiency on growth and photosynthesis of sugar beet (*Beta vulgaris* L.). (TERRY 1977)

leaves (RAHIMI and BUSSLER 1973a, GRAHAM 1976) are related to the role of copper in polyphenol-oxidase activity and thus the synthesis of lignins. The extent to which lignification is inhibited in copper-deficient tissue is correlated with inadequate development of xylem vessels (RAHIMI and BUSSLER 1973b). A test for lignins can thus be used as a rapid method for diagnosis of the copper nutritional status of plants (RAHIMI and BUSSLER 1973c, see also Chap. 1.5). This role of copper also explains the positive correlation in cereals, for example, between copper nutritional status, haulm stability and resistance to lodging, and the interaction between copper and nitrogen fertilizer application (VETTER and TEICHMANN 1968). Another typical symptom of copper deficiency, non-viability of pollen grains and the consequences for yield formation, is discussed in Section 4.

### 1.4.6 Chlorine

In the absence of chlorine, deficiency symptoms and severe growth depressions have been demonstrated in higher (BROYER et al. 1954, JOHNSON et al. 1957) as well as in lower plants (MARTIN 1965). Whereas in studies with isolated chloroplasts a role of chlorine in photosystem II electron transport has been confirmed (KELLEY and IZAWA 1978), results of TERRY (1977) with intact sugar beet plants contradict its direct involvement in photosynthesis in vivo, as only growth but not uptake of $CO_2$ was affected by chlorine (Fig. 1).

The relatively high chlorine demand for optimal leaf growth in sugar beet ($>20$ µmol $g^{-1}$ dry matter) might indicate genotypic differences in demand for chlorine, with perhaps positive correlation with the halophytic nature of a species. The high demand for chlorine for sugar beet certainly opens possibilities for more detailed studies concerning the functions of this mineral element in higher plants.

# 2 Uptake and Long-Distance Transport of Mineral Elements

## 2.1 Ion Concentration at the Root Surface, Role of the "Rhizosphere"

In the past, studies on ion uptake by roots have mainly concentrated on the role of the free space, carrier mechanisms, ion selectivity, and occurrence of single- or multiple-binding sites at the plasmalemma or the "dual mechanism" of ion uptake. However, in the last decade increasing attention has been paid to electrophysiological aspects (Chaps. I.6 and I.7), as well as to physiological and anatomical gradients along the roots and their consequences for uptake and transport of ions (Chap. I.7).

Another field of new interest in ion uptake and mineral nutrition in general is the consequences which arise at the root surface as the result for example of either removal of ions (development of a "depletion zone" at the root surface) or difference in cation/anion uptake and the corresponding induced change in pH around the root (Chap. I.6). These factors are of less importance in a stirred nutrient solution, but can be dominant in ion uptake and in the mineral nutrition of plants growing in solid substrates such as soils. An example for the role of the source of nitrogen on pH in the rhizosphere and its consequences for phosphorus uptake by the plants is shown in Table 8.

Changes in the pH at the root surface are the result of both different cation/anion uptake ($NH_4^+ > SO_4^{2-}$, $NO_3^- > Ca^{2+}$) and corresponding changes in net efflux of $H^+$ and $OH^-/HCO_3^-$. Nitrate reduction in the roots also leads to additional $OH^-$ or $HCO_3^-$ efflux according to the equation: $NO_3^- + 8\,H^+ + 8\,e^- = NH_3 + OH^- + 2\,H_2O$ (KIRKBY and MENGEL 1967).

Additionally certain plant species and genotypes of particular species suffering from iron deficiency are able to decrease the pH of the substrate by increasing proton efflux (BROWN 1978). In sunflower, for example, this iron deficiency-induced decrease in the substrate pH is mainly restricted to the apical zone and is correlated with anatomical changes in the roots (see Chap. I.7) and a marked increase in iron uptake (Table 9).

Supplying a sparingly soluble inorganic iron-III-compound (e.g. iron-III-hydroxide) this iron-stress-induced pH decrease leads to the mobilization of iron in the substrate with correspondingly periodical changes in both the iron nutritional status of the plants and pH fluctuations in the substrate (VENKAT-RAJU and MARSCHNER 1972). These effects can be depressed in amplitude but not prevented by an increase in phosphate concentration (Fig. 2).

The ability of a plant to increase the uptake rate of iron in response to iron stress is causally related to an increase in the reducing capacity of its roots (BROWN 1978, RÖMHELD and MARSCHNER 1979). Both response phenomena are typical for so-called iron-efficient plant species (see also Chap. IV.1). This stress response offers an interesting insight into a regulatory mechanism of ion uptake controlled by the nutritional status ("internal concentration") (more details in Chap. I.6 and I.7).

In addition to the pH in the rhizosphere of the root–solid substrate interface, ion concentration at the root surface can also be quite different from the average concentration in the substrate. For mineral nutrients such as phosphorus and

**Table 8.** Effect of soil pH and supply of either ammonium or nitrate salts on pH of the bulk soil and rhizosphere soil and on phosphorus uptake by soybean plants. (After RILEY and BARBER 1971)

| Soil without salts without plants | pH in the soil | | | | P uptake (mmol pot$^{-1}$) | |
|---|---|---|---|---|---|---|
| | Bulk soil (outside of the rhizosphere) | | Rhizosphere soil | | | |
| | $NH_4^+$ | $NO_3^-$ | $NH_4^+$ | $NO_3^-$ | $NH_4^+$ | $NO_3^-$ |
| 5.2 | 4.98 | 5.43 | 4.71 | 6.60 | 0.27 | 0.16 |
| 6.3 | 5.90 | 7.00 | 5.60 | 7.05 | 0.16 | 0.09 |
| 6.7 | 6.64 | 7.01 | 6.25 | 7.19 | 0.14 | 0.07 |

[a] $(NH_4)_2SO_4$ and $Ca(NO_3)_2$ respectively

**Table 9.** Effect of iron nutritional status of sunflower plants (preculture ± Fe) on iron accumulation in various root zones. (RÖMHELD and MARSCHNER 1979)

| Preculture | cm from the root tip | | |
|---|---|---|---|
| | 0–1.5 (root tip) | 1.5–3.5 | 3.5–5.5 |
| | $^{59}Fe$ content (μmol g$^{-1}$ dry wt. 10 h$^{-1}$) | | |
| +Fe | 0.3 | 0.6 | 0.6 |
| −Fe | 21.8 | 2.5 | 1.7 |

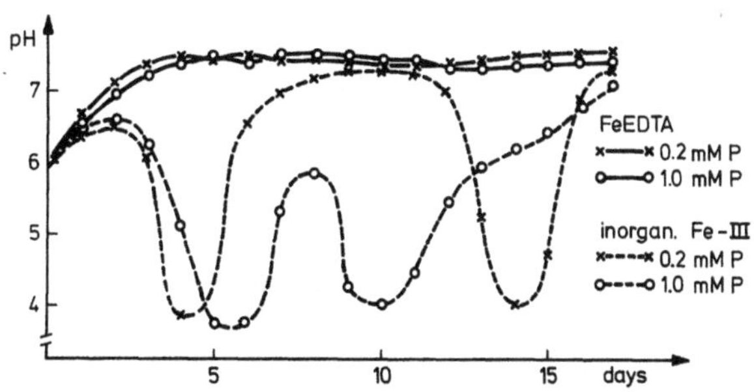


**Fig. 2.** Interaction between source of iron and phosphate concentration in iron stress reactions (fluctuations in the pH of the nutrient solution) of sunflower plants. (MARSCHNER et al. 1978)

potassium, the diffusion rate to the root surface usually limits uptake rate, and a depletion zone can occur around the root (BREWSTER and TINKER 1970). This may restrict the uptake of these ions to the apical region of the roots regardless of a potentially high uptake rate in the basal regions (Chap. I.7). Both root growth in general and root hair formation in particular (BHAT and NYE 1974) are therefore of overwhelming importance for mineral nutrition in solid substrates. In contrast they are of much less significance in solution culture where exhaustion zones are of limited importance.

Ion concentration at the root surface can also be much higher than that in the average substrate. This occurs when the uptake rate of an ion (e.g. calcium) is relatively low as compared to the mass flow of solution to the root surface induced by transpiration (MENGEL et al. 1969). As result of this process the availability of both mineral nutrients (e.g. by precipitation of Ca phosphates) and water (e.g. saline soils) can be depressed. For plants growing in solid substrates and transpiring at a high rate, the concentration of soluble salts at the root surface can be several times higher than in the bulk substrate (SINHA and SINGH 1976).

Results from studies in solution culture investigating the effect of high salt concentrations, e.g. NaCl, on ion uptake and water status of plants cannot therefore be extrapolated to solid substrate conditions even when average iso-osmotic concentrations in both substrates are considered.

Ion concentration and ion availability at the root surface can also be affected indirectly by microorganisms in the rhizosphere. This subject is of increasing interest in mineral nutrition in general and especially in relation to the actual and potential role of certain associations for $N_2$ fixation (Chaps. II.2 and II.3) and symbiosis for phosphate uptake (e.g. mycorrhiza by crop plants (Chap. I.2).

## 2.2 Long-Distance Transport in the Xylem

### 2.2.1 From the Roots to the Shoot

Our understanding of long-distance transport from roots to shoots is based mainly on analysis of xylem exudate from decapitated plants. The composition of this exudate, however, does not necessarily reflect conditions occurring at the sites of excretion of the ions into the root xylem, as resorption by the surrounding xylem can induce considerable changes in ion concentration and relation of ions in the xylem during long-distance transport (Chap. I.7). Furthermore the contribution made by phloem in vivo to long-distance transport of mineral nutrients from roots to shoots cannot be determined by this method. Only indirect evidence is available for this in vivo transport, as shown in studies comparing forms of nitrogen supply (MARTIN 1971). The possibility must also be considered that diurnal fluctuations (day–night) occur in the relative importance of mineral element transport in the phloem from roots to shoots, as has been demonstrated for phloem transport into fruits (HOCKING et al. 1978).

For some mineral nutrients such as phosphorus (DE JAGER 1979) and potassium (BARBER 1979) the uptake rate by the roots seems to be controlled by the shoot rather than by the roots (Table 10).

The mechanism of this feed-back effect from the shoots is not fully understood as yet (Chap. I.6). For phloem-mobile mineral elements such as phosphorus and potassium, the transmission of information concerning the nutritional status of the shoot may be achieved by the intensity of retranslocation of these mineral elements in the phloem to the roots. In plants where circulation of potassium from the shoots into the roots as salts of organic acids is important (Chaps. I.6 and I.7), a regulation of potassium uptake by the potassium content

**Table 10.** Effect of K pretreatment of corn plants (split root experiment) on K content of shoots and roots and $V_{max}$ for subsequent K influx. (After BARBER 1979)

| Percent of root exposed to K | K content ($mmol\ g^{-1}$ dry wt.) | | $V_{max}$ $pmol\ cm^{-1}\ s^{-1}$ |
|---|---|---|---|
| | Shoots | Roots | |
| 100 | 2.05 | 1.48 | 15.8 |
| 50 | 1.64 | 1.41 | 28.0 |
| 25 | 1.10 | 1.28 | 33.8 |
| 15 | 1.05 | 1.41 | 36.8 |

of the shoots and thus intensity of potassium retranslocation would be a consequence of this model. In the case of phloem-immobile mineral elements such as calcium the removal from exchange sites of the xylem might offer another type of control mechanism (see Sect. 3).

### 2.2.2 Into Fruits, Seeds and Storage Organs

The transpiration rate of a plant organ (e.g. water loss $g^{-1}$ dry wt.) is one of the determining factors in the ratio of mineral element import via xylem versus phloem. Compared to fully expanded leaves, the transpiration rate and thus mineral element import via the xylem can be expected to be much lower in developing leaves and particularly fruits, seeds and storage organs (e.g. potato tubers).

During recent years PATE and co-workers (PATE et al. 1974a, b, PATE and HOCKING 1978) have developed a method to calculate the relative importance of xylem and phloem in the transport of mineral elements into the fruits of various legumes. The question of whether uncontaminated phloem sap is obtained by this method is discussed later (Sect. 3). In general (PATE and HOCKING 1978), mineral element import (% of total import) into legume fruits occurs predominantly via the phloem: 80% C, N, S; 70%–80% P, K, Mg, Zn; 62%–66% Fe, Mn, Cu and approximately 30% Ca. However, for certain elements (e.g. Cu and Fe) considerable diurnal fluctuations occur in the ratio of xylem/phloem import (HOCKING et al. 1978), and the ratio may also be dependent on the availability of particular mineral elements in the root medium. This has been demonstrated for manganese (Table 11).

It is obvious from Table 11 that when manganese availability is low, xylem import of manganese into fruits is small relative to phloem import, but becomes important at high manganese availability in the rooting medium. Manganese concentrations in the phloem sap below 0.5 ppm are associated with a marked increase in occurrence of a "split seed" disorder, involving discoloration, splitting, and deformity of seeds (HOCKING et al. 1977).

In this legume fruit system no data are presented for the two micronutrients boron and molybdenum. For boron at least there is evidence that this element is quite mobile in the phloem of stems and leaves (OERTLI and RICHARDSON

**Table 11.** Effect of manganese level in the rooting medium on plant growth, manganese level in seeds (dry matter) and in transport fluid serving fruits of *Lupinus angustifolius*. (After HOCKING et al. 1977)

| Concentration in rooting medium (ppm Mn) | Dry weight (g plant$^{-1}$) | Mn content in seeds (ppm) | Manganese concentration | |
|---|---|---|---|---|
| | | | Xylem (ppm) | Phloem (ppm) |
| 0.1 | 7.1 | 0.56 | 0.18 | 0.56 |
| 0.15 | 12.6 | 0.66 | 0.19 | 0.92 |
| 0.60 | 13.8 | 1.08 | 0.36 | 1.28 |
| 20.0 | 14.6 | 11.65 | 5.88 | 6.94 |

1970) and that fruits and storage organs may be supplied mainly via the phloem. The same is true for growing potato tubers where the influx of xylem fluid is small or unimportant. Only calcium, and no other mineral element, including boron, needs to be supplied directly to the tuber surface to maintain a high growth rate (KRAUSS and MARSCHNER 1975). Similarly, during fruit development of *Arachis hypogaea* and *Trifolium subterraneum* net influx of xylem fluid is very low, but the boron demand for fruit and seed growth is completely met by internal supply, most probably by phloem import (CAMPELL et al. 1975).

### 2.2.3 Retranslocation of Mineral Elements from Leaves

This process is particularly important (a) in perennials in autumn, (b) in both perennials and annuals when uptake is inadequate (c) during reproductive stages or storage organ growth. The extent of retranslocation also depends upon the particular mineral element. Occurrence of deficiency symptoms on older leaves (N, K, Mg, P) reflects a high retranslocation rate, whereas the occurrence of deficiency symptoms on only the young leaves (Ca, Fe, Mn, Cu, Zn, Mo, B) demonstrates insufficient retranslocation rates of these mineral elements. This retranslocation from the older leaves has to take place in the phloem. It cannot be concluded, however, that the occurrence of the deficiency symptoms on younger leaves reflects inadequate phloem mobility of these mineral elements. Retranslocation from the older leaves involves several steps: (1) "Mobilization" within the leaf cells, (2) transport to the sieve elements, (3) phloem loading, (4) mobility within the sieve tubes.

Phloem mobility is thus only one factor which determines the retranslocation rate of a mineral element from the leaves.

Retranslocation from leaves is particularly important during seed development, as in this growth stage the carbohydrate supply to the roots and thus root activity and ion uptake generally decline rapidly. The extent of retranslocation of certain mineral elements during fruit and storage growth depends upon various factors, such as uptake rate by the roots and "sink-size", particularly in relation to "source-size" (see Sect. 4). An example for the range of redistribution is given in Table 12.

**Table 12.** Retranslocation during the reproductive stage (percentage of the proportional net loss from the time when maximum content was recorded) of mineral elements from leaflets of the main axis of *Lupinus albus*, corresponding loss in dry weight 39%. (After HOCKING and PATE 1978)

| Element | Loss in % | Element | Loss in % | Element | Loss in % |
|---|---|---|---|---|---|
| Nitrogen | 73 | Calcium | 18 | Manganese | 50 |
| Potassium | 73 | Iron | 25 | Copper | 43 |
| Phosphorus | 75 | Zinc | 45 | Sodium | 4 |
| Magnesium | 57 | | | | |

The high retranslocation rate of N, P, K, and Mg from leaves is in accordance with the occurrence of deficiency symptoms on (older) leaves insufficiently supplied by the root system. Although somewhat lower than for the macronutrients, the retranslocation rate of the micronutrients Fe, Zn, Mn and Cu (Table 12) is still surprisingly high, considering the fact that deficiency symptoms of these elements occur only on young, growing parts (leaves), but not on fully expanded leaves.

For the micronutrients only a certain fraction can be mobilized from fully expanded leaves (LONERAGAN et al. 1976, LONERAGAN 1978). It appears that when the shoot apex or fruits are inadequately supplied, the demand cannot be transformed into a "signal" strong enough to bring about extensive mobilization and retranslocation from fully expanded leaves. These leaves therefore do not show micronutrient deficiency symptoms. In contrast, in the case of the macronutrients nitrogen, potassium and magnesium, this "signal" seems to be strong enough to induce exhaustion of the older leaves in these particular elements.

A close correlation between concentration and retranslocation rate of mineral elements from older leaves has been clearly demonstrated by HILL et al. (1978, 1979). In wheat, for example, during grain development, leaves with a high copper content lost more than 70% of their copper, whereas for leaves of copper-deficient plants the value was less than 20%. This proportion of copper retranslocation can be increased, however, particularly from copper-deficient leaves, for example, by shading (HILL et al. 1979). During senescence, induced either by shading or by nitrogen deficiency (HILL et al. 1978), a further fraction of copper is mobilized in these older leaves and is retranslocated. The relatively high retranslocation rate of micronutrients during fruit development (Table 12) is perhaps the result of leaf senescence. The extent of retranslocation may also be causally involved in the development of sulfur deficiency symptoms preferentially either on young or old leaves depending on the level of nitrogen supply (LONERAGAN et al. 1976). This variability in extent of retranslocation of a mineral element and its consequences for plant diagnosis are discussed in more detail in Chap. I.5.

The extent of remobilization of macro- and micronutrients is attracting increasing interest also in connection with selection and breeding of genotypes of high "nutrient efficiency". Genotypes which are able to grow well on substrates (soils) with relatively low nutrient availability may differ not only in the uptake and translocation rate of a particular mineral nutrient, but also in a higher mineral nutrient effectivity at cellular level (compartmentation, solubility etc.) and higher retranslocation rates from older to younger leaves and to seeds and storage organs. (For further details see Chaps. IV.1 and IV.2 and a comprehensive review article by PATE 1976).

## 3 Calcium Nutrition of Higher Plants

### 3.1 Introduction

Calcium nutrition of higher plants is attracting increasing interest due to the widespread occurrence of calcium deficiency or the so-called calcium-related disorders which are particularly associated with intensive horticulture and agriculture. Typical examples are tipburn in lettuce, blackheart in celery, blossom end rot in tomato and water melon or bitter pit in apple. For detailed information the reader is referred to comprehensive review articles (SHEAR 1975, BANGERTH 1979) and an International Symposium Report (SHEAR 1979).

Besides its economic importance, this subject is also of interest for plant physiologists, as these Ca-related disorders often occur on soils or substrates high in available calcium, and thus are only in part caused by insufficient calcium uptake. It is well established now that these disorders are primarily caused by the limited capability of plants to regulate calcium distribution internally in relation to the demand of the various parts and organs and particularly of fast-growing leafy organs, tubers and fruits.

### 3.2 Calcium Demand of Higher Plants

There are considerable genotypical differences in the demand for calcium at a cellular level, causally connected with differences in cation exchange capacity (CEC) of the cell wall system (middle lamella) i.e. amount of free carboxylic groups of pectins (polygalacturonic acid). Thus dicotyledons, as compared to monocotyledons, are characterized by both higher CEC and also much higher calcium demand. A typical example for these differences is presented in Table 13. In agreement with this, Ca-related disorders are typical for dicotyledons but not for monocotyledons.

As the calcium supply falls, the proportion of calcium bound in the cell wall fraction (pectates) increases to more than 50% (MOSTAFA and ULRICH 1976, ARMSTRONG and KIRKBY 1979). In plant species with intensive oxalate formation, the availability of calcium for the other functions, membrane stabilization in particular (Chap. V.1) is further decreased (ARMSTRONG and KIRKBY 1979). Genotypical differences to calcium deficiency within a certain species are therefore at least in part related to differences in oxalate formation, particu-

**Table 13.** Effect of calcium concentration in nutrient solution (μM) on relative growth rate of plants and calcium content (μmol $g^{-1}$ dry wt.) in the shoots. (After LONERAGAN et al. 1968 and LONERAGAN and SNOWBALL 1969)

| Species | Ca supply (μM) | | | | |
|---|---|---|---|---|---|
| | 0.8 | 2.5 | 10 | 100 | 1000 |
| | Relative growth rate (%) | | | | |
| *Lolium perenne* | 42 | 100 | 94 | 94 | 93 |
| *Lycopersicon esculentum* | 3 | 19 | 52 | 100 | 80 |
| | Calcium content (μmol $g^{-1}$ dry wt.) | | | | |
| *Lolium perenne* | 15.0 | 17.5 | 37.4 | 92.3 | 269.5 |
| *Lycopersicon esculentum* | 49.9 | 32.4 | 74.9 | 321.9 | 621.3 |

larly in certain plant parts such as the stem (BRUMAGEN and HIATT 1966) and perhaps also the pedicels (LIEGEL 1970).

Besides this special case of oxalate formation, genotypical differences to calcium deficiency within a certain species may also be related to differences in uptake, long- and short-distance transport and/or compartmentation within the cells (see below and Chap. IV.1).

### 3.3 Calcium Uptake by the Roots

The mobility of calcium is much greater in the apoplast than in the symplast. In roots the uptake and radial transport of calcium into the stele is therefore more intensive in the root tip regions (Chap. I.7). Calcium transport into the shoot therefore seems to be more closely related to root growth rate (VISSER et al. 1971, SCAIFE and CLARKSON 1978) than to the shoot growth rate (RAPPER et al. 1977). Certainly the classical approach of a general close relationship between uptake of calcium and transpiration rate has to be reexamined (ARMSTRONG and KIRKBY 1979).

### 3.4 Long-Distance Transport of Calcium

#### 3.4.1 Xylem Transport

Although the direction of water flux within the xylem determines the direction of long-distance transport of calcium in this conducting system, the transport rate of calcium is only poorly correlated to the flux rate of water in the xylem. The transport rate of calcium in the xylem is regulated more by exchange adsorption than by mass flow (FERGUSON and BOLLARD 1976, ISERMANN 1978). Thus at high flux rates of water in the xylem the discrepancy to the transport rate of calcium can be expected to be particularly wide (see Figs. 3 and 4). It is also important to realize that this exchange adsorption is not restricted to the

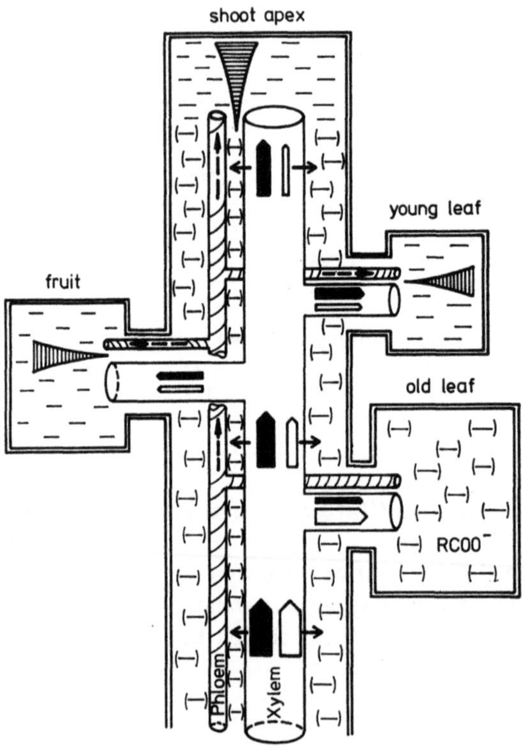


**Fig. 3.** Model for calcium translocation (→ decreasing rate) within the shoot. Mass flow of water in the xylem (⇨, decreasing rate). Fixed negative charges in cells walls (−) and formation of new exchange sites −, sites of IAA synthesis and transport

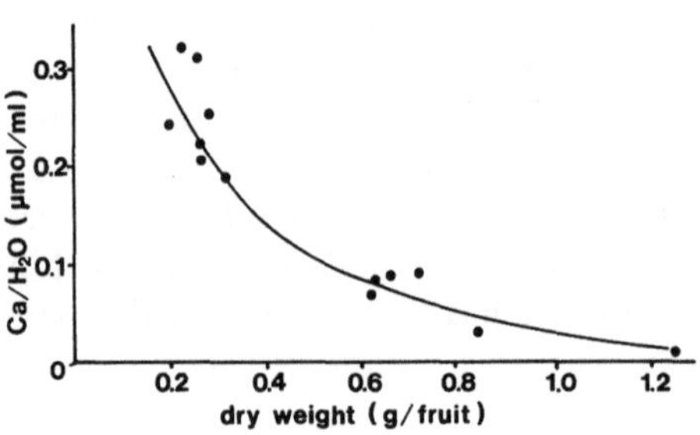


**Fig. 4.** Ratio between calcium accumulation (µmol) and water loss (ml) in paprica fruits (*Capsicum annuum*) during fruit growth. (MIX and MARSCHNER 1976b)

xylem vessels, but can take place in the apoplast (Free Space) of the whole stem, for example (VAN DE GEIJN et al. 1979). It may be supposed that the mobilization of calcium from the stem in apple trees during the early stages of fruit growth (LÜDDERS et al. 1976), for example, takes place at least in part at the expense of this exchangeable fraction. Thus the transport of calcium within the xylem can be controlled to a considerable extent by charge density (dicotyledons > monocotyledons), concentration of cations other than calcium, and the removal of calcium from the exchange sites by adjacent cells along the conducting vessels and particularly in dividing and expanding tissues (e.g. fruits, shoot apex, see Fig. 3).

**Table 14.** Effect of environmental factors and growth rate on Ca and K content of paprika fruits (MIX and MARSCHNER 1976a, b, MARSCHNER unpublished)

| | Treatment | Ca | K |
|---|---|---|---|
| | | (μmol, $g^{-1}$ dry wt. fruit) | |
| Ca supply to the roots | 0.5 mM | 26.9 | 1315 |
| | 5.0 mM | 33.2 | 1228 |
| Relative humidity around the fruits | 40% | 55.4 | 1892 |
| | 90% | 32.7 | 1918 |
| Growth rate of the fruits ($mg^{-1}$ dry wt. increase per day) | 20.1 | 28.2 | 1772 |
| | 29.9 | 20.7 | 1846 |
| | 38.5 | 17.2 | 1813 |

### *3.4.1.1 Correlations with the Transpiration Rate*

Inhibition of the transpiration rate decreases the calcium translocation to the shoot apex, but not necessarily to the shoot in total (ARMSTRONG and KIRKBY 1979). Inhibition of transpiration rate of individual organs like fruits is also correlated with a decrease in calcium translocation into these organs (WIERSUM 1966, Table 14).

Individual organs or plant parts with relatively large surface areas (e.g. leaves in contrast to fleshy fruits) or those directly exposed to the free atmosphere (e.g. outer leaves in contrast to inner leaves of lettuce) are preferentially supplied with water from the xylem and thus also with calcium. As water availability in the shoot is depressed either by a high osmotic pressure of the soil solution (ENDE et al. 1975) or by conditions which favor a high rate of transpiration of the shoot as a whole (KRUG et al. 1972), the net flux of water and calcium within the xylem into fruits or inner leaves can either be temporarily (light period) or permanently (saline substrates) restricted. Such restrictions are associated with an increase in calcium-related disorders (KRUG et al. 1972, ENDE et al. 1975). This transpiration-directed flux in the xylem of both water and calcium can in part be compensated or superimposed by a high root pressure and lead to influx of water and calcium via the xylem into these low-transpiring organs and plant parts (PALZKILL et al. 1976, BRADFIELD and GUTTRIDGE 1979). Diurnal fluctuations in the water status of the shoot and its individual organs (KRUG et al. 1972, WIEBE et al. 1977) are correlated with corresponding fluctuations in the direction of the calcium transport within the xylem (KRUG et al. 1972, KRAUSS and MARSCHNER 1974, WIEBE et al. 1977).

### *3.4.1.2 Role of Binding Sites in the Apoplast*

Regardless of the transpiration rate of the whole plant (WIENEKE and FÜHR 1973a) or of individual leaves (KOONTZ and FOOTE 1966) or of fruits (Table 14) the translocation rate of calcium declines as the plants or organs mature (Fig. 4). Two factors appear to be responsible for this (see Fig. 3): (a) In fruits an increase in water influx via the phloem ("Druckstrom" according to MÜNCH) depresses influx via the xylem, (b) in the case of fully developed leaves due to a decline in removal of $Ca^{2+}$ from the binding sites of the apoplast. This removal of calcium depends mainly on storage as calcium salts in the vacuoles. The storage capacity of vacuoles for soluble calcium salt (nitrate, malate etc.) is, however, restricted for osmotic reasons. In the case of the formation of calcium oxalate as a result of nitrate reduction (EGMOND and HOUBA 1970), xylem transport of calcium into fully grown leaves might be maintained for a longer period depending on nitrate reductase capacity. The effect of the form of nitrogen supply (nitrate or ammonium) on calcium distribution within apple shoots (SHEAR and FAUST 1970) might at least in part be the mechanism described above. Removal of calcium from exchange sites in the apoplast might also be a dominating factor for calcium transport into growing tissue such as shoot apices, young leaves or fruits (Fig. 3). Cell division, and thus the formation of new binding sites for calcium, should decrease calcium ion activity, thus maintaining calcium transport within the apoplast towards this tissue. This mechanism, of course, at least temporarily requires a net flux of solutes within the xylem in the direction of this tissue. Such conditions can be demonstrated, for example, at high root pressure (PALZKILL et al. 1976, BRADFIELD and GUTTRIDGE 1979). The importance of these binding sites in the apoplast for calcium transport can be demonstrated by several examples. A decrease in CEC of a tissue, for example as result of manganese toxicity (HORST and MARSCHNER 1978b) or application of TIBA (MARSCHNER and OSSENBERG-NEUHAUS 1977) is correlated with a decrease in calcium transport into the tissue and the development of calcium-deficiency symptoms. Even the calcium distribution within an organ is correlated with the CEC. In flowers of *Antirrhinum*, for example, there is a simultaneous increase in CEC, calcium content and Ca/K ratio from stigma → style → ovary (KNIGHT and CROOKE 1973).

### 3.4.2 Phloem Transport

Compared to the xylem sap, the calcium concentration in the phloem sap is very low, particularly in relation to potassium. In phloem sap of *Ricinus* stems the following values have been obtained: 0.1–0.7 mM calcium as compared to 50–100 mM potassium (VAN GOOR and WIERSMA 1974, MENGEL and HAEDER 1977). Considering the high pH (7–8) and the high phosphate concentration of the phloem sap, the solubility of ionic calcium can be calculated to be lower than 0.5 mM (VAN GOOR and WIERSMA 1974). According to RAVEN (1977), however, concentrations of up to 3 mM calcium should theoretically be possible. The calcium concentration might also depend upon complexing agents in the phloem sap (BRADFIELD 1976).

Low calcium concentrations in the phloem sap (see also PATE Vol. 1, this Series) and relatively high concentrations in the area of phloem strands (WIENEKE 1979) respectively are presumably the result of either precipitation of calcium (LÄUCHLI 1967, WIENECKE and FÜHR 1973b) or of a low membrane permeability of the sieve tube cells to calcium. Also the existence of efflux pumps for calcium at the sieve tube membranes is discussed (Chap. V.1, this Vol.).

**Table 15.** Estimated deliveries of mineral elements by phloem and xylem and the proportions of actual intake in fruits of *Lupinus angustifolius* L. attributable to vascular transport within a 36-h period. (After HOCKING et al. 1978)

| Element | Delivery in μmol | | % of total (actual) intake |
|---|---|---|---|
| | Phloem | Xylem | |
| K | 8.21 | 2.14 | 103.5 |
| P | 2.30 | 0.12 | 97.7 |
| Ca | 0.12 | 0.85 | 34.0 |
| Fe | 0.04 | 0.01 | 103.6 |

Calculations for phloem transport of calcium are usually based on analysis of phloem sap collected by incision of the stem or removal of the distal end of legume fruits (PATE et al. 1978, PATE and HOCKING 1978). With this method, however, "contamination" by calcium from the apoplast cannot be excluded. In legume fruits, for example, calcium is highly mobile in the apoplast of the pod tissue (MIX and MARSCHNER 1976c). The difficulties in calculating the calcium transport into legume fruits based on xylem and phloem exudate analysis are reflected in Table 15.

Another approach used in the determination of calcium transport in the phloem is by ringing plant stems. Such experiments have led to the conclusion in apple shoots (PRIESTLEY 1976) and cotton stems (WIENECKE 1979) that calcium is transported preferentially in the phloem. It has to be critically re-examined whether such a severe disturbance of the integrity of the whole plant system would not induce side effects as, for example, on calcium movement in the apoplast.

At least in intact roots (MARSCHNER and RICHTER 1974) or stolons of growing potato tubers (KRAUSS and MARSCHNER 1973) it is not possible to observe long-distance transport of calcium within the phloem in the direction of the apical meristems. Retranslocation of calcium after leaf application of $^{45}Ca$ can be demonstrated only when high concentrations of calcium are applied (RINGOET et al. 1968) or when calcium is supplied combined with a complexing agent directly to the leaf veins (MILLIKAN and HANGER 1969). On the other hand, in studies with intact legume plants, between flowering and ripening a substantial net loss of calcium occurred from the upper leaves, associated with an increase in calcium content of the fruits (HOCKING and PATE 1978). It is certainly an interesting aspect for further studies on phloem transport of calcium as to whether retranslocation of calcium is in some way related to tissue senescence, as has been demonstrated with copper (HILL et al. 1978).

### 3.4.3 Xylem Versus Phloem Transport

Despite these contradictory results and differing views about the importance of calcium transport in the phloem, there is general agreement that the calcium concentration in the xylem is several times or even several orders of magnitude

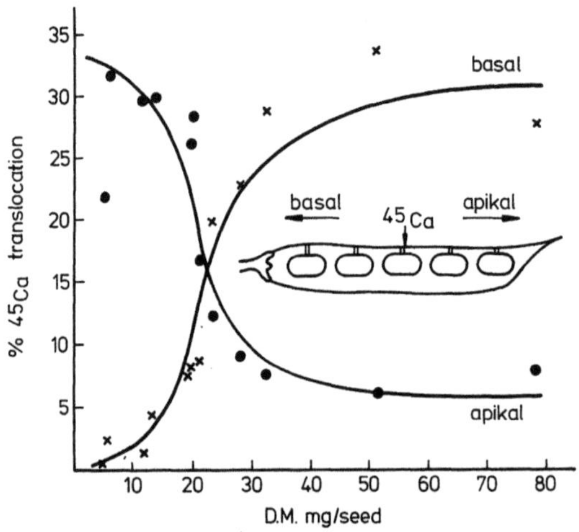


**Fig. 5.** Correlation between seed weight and direction of $^{45}Ca$ transport within bean fruits (*Phaseolus vulgaris*) 3 days after application of $^{45}Ca$ (↓) on the dorsal vascular strand in the middle of the fruit. (MIX and MARSCHNER 1976a)

higher than in the phloem. Consequently, plant organs or parts of plants mainly supplied by the phloem stream are generally high in potassium but low in calcium. In leafy tissue (e.g. lettuce), fleshy fruits (e.g. tomato), storage roots (e.g. celery) or tubers (e.g. potato) for the average of the whole period of growth the relative importance of the phloem/xylem influx ratio remains more or less constant. Thus either direct uptake of calcium from the external medium, as shown for peanut fruits (HALLOCK and GARREN 1968) or potato tubers (KRAUSS and MARSCHNER 1975), or short-term periodical changes in the water influx via the xylem (see Sect. 1.1) are essential to meet the calcium demand of these growing tissues.

In fleshy fruits such externally regulated fluctuations of the xylem fluid influx rate are superimposed by an internal regulation. During fruit growth, the rate of calcium influx distinctly declines (WIENEKE 1974; Fig. 5). Several factors are responsible for this decline: (a) change in the volume/surface area ratio (b) a decline in cell division rate and thus formation of new binding sites for calcium and (c) an increase in the relative importance of solute influx via the phloem.

The ratio of solute influx of xylem/phloem also depends on the growth rate of the fruit. Fast-growing fruits are therefore usually lower in calcium and have a wider K/Ca ratio than slow-growing ones. This has been demonstrated for tomato (WIERSUM 1966), apple (BANGERTH and MOSTAFAWI 1969) and paprika (Table 14). In fleshy fruits at the later stages of growth, even a net efflux of calcium from the fruits seems to occur (WILKINSON 1968, TROMP and OELE 1972, MIX and MARSCHNER 1976a).

In fruits with distinct phases of growth such as legume fruits, the onset of rapid seed growth is correlated with a reversal from apical to basal direction of net solute transport in the xylem of the dorsal vein of the pod (Fig. 5). This change leads to a distinct decline in calcium transport into these fruits. At these later stages of fruit growth, net influx of calcium either ceases or is restricted to a low influx rate via the phloem, as supposed by PATE and HOCKING (1978) or TROMP (1979).

In *Phaseolus vulgaris* the developing seeds are supplied directly with calcium formerly accumulated in the pod tissue (MIX and MARSCHNER 1976c), a way of internal redistribution which presumably also occurs in other legume fruits (HOCKING and PATE 1977). This uptake from the surrounding apoplast into the growing seed is severely restricted (regulated) by the seed coat (MIX and MARSCHNER 1976c), where most of this calcium is immobilized and is unavailable even for subsequent seed germination (HELMS and DAVID 1973).

### 3.5 Role of Phytohormones and Growth Regulators

Phytohormones and growth regulators also influence the long-distance transport of calcium. WIENEKE et al. (1971) assume a direct depressing effect of gibberellins on calcium transport to the shoot apex or fruits of apple. The most marked effects on calcium transport are exerted by auxins. Evidence for this is the distinct inhibitory effect of TIBA (2,3,5-triiodobenzoic acid) on calcium transport into fruits (BANGERTH 1976) or to the upper parts of shoots (WIENEKE et al. 1971) and the apical meristem in particular (MARSCHNER and OSSENBERG-NEUHAUS 1977). As TIBA is known to be an inhibitor of basipetal auxin transport (DE LA FUENTE and LEOPOLD 1972), inhibition of this transport from the sites of synthesis, the apical meristem and developing seeds (NITSCH 1950), can be expected to be correlated with a corresponding decrease in calcium transport into these tissues and organs (Fig. 3). Causal interactions between auxins and calcium mobility have been shown in other tissue (JANISTYN 1973).

The importance of seeds as sites of auxin synthesis for calcium transport into developing fruits has been demonstrated by BANGERTH (1976). The effect of boron sprays in increasing calcium transport into the fruits (DIXON et al. 1973) may relate to conditions where boron deficiency restricts seed development and thus auxin production. Furthermore, competition for calcium between different plant parts (e.g. developing fruits and shoot apices) might be related to a capability for auxin synthesis. Removal of shoot apices (QUINLAN and PRESTON 1973) or inhibition of shoot growth by growth regulators (NAUMANN 1971) can therefore indirectly affect partitioning of calcium within the shoot in favour of calcium transport into fruits.

The linkage between basipetal auxin transport and acropetal calcium transport is still obscure. Auxins might increase mobility of calcium both in the apoplast and in the phloem (BANGERTH 1976), or enhance formation of new binding sites in the apoplast (MARSCHNER and OSSENBERG-NEUHAUS 1977). On the other hand, supraoptimal concentrations of auxins and a correspondingly particular high demand for calcium for elongation growth are assumed to be the explanation for the negative correlations obtained between auxin levels and calcium-related disorders in lettuce cultivars (COLLIER et al. 1979).

### 3.6 Conclusion and Outlook

Calcium transport into rapid-growing leafy tissues, tubers or fleshy fruits is obviously characterized by mechanisms which restrict this transport. Considering some of the effects of calcium – decreasing membrane permeability or inhibiting auxin-induced cell elongation (POOVAIAH and LEOPOLD 1976) – it is necessary that in rapidly expanding tissue and storage tissue in particular, the concentration of calcium is kept low to maintain high membrane permeability for solute flux. In agreement with this, an artificial increase in calcium concentration of fruits rapidly decreases their growth (MIX and MARSCHNER 1976a). Furthermore, "ripening" of fruits (onset of senescence) requires both an increase in membrane permeability and a dissolving of the middle lamella, both processes which are delayed or even prevented by high concentrations of calcium (see Sect. 1).

Impressive crop yield increases have been achieved in recent years by improving agricultural and horticultural techniques together with progress made in plant breeding. Many of those plant organs or plant parts which are considered as "yield", however, are also often characterized genetically by restricted calcium influx. Increasing calcium-deficiency symptoms and economic loss due to Ca-related disorders are the consequence. Although our present knowledge of calcium transport and its regulation is still rather limited, the causal connection directly with the water economy of a plant and its organs, and indirectly with the phloem to xylem flux ratio into an organ, enables some control of calcium distribution. The involvement of phytohormones and auxins in particular, enables new approaches to be made. In the long run, to overcome Ca-related disorders, it will certainly also be necessary to make more use of genotypical differences within species in uptake, transport and distribution of calcium at tissue and cellular level. These latter aspects are discussed in detail in Chap. IV.1.

## 4 Mineral Nutrition and Physiology of Yield Formation – Sink–Source Relationship

### 4.1 Introduction

In determining plant yields, as defined by the amount of either fresh or dry matter or certain constituents (e.g. sugar, starch, protein) produced per unit area, photosynthesis is the most important prerequisite. This holds true in particular for forage plants which show a close relationship between an increase in photosynthetic active surface area and an increase in fresh and dry weight. Here mineral nutrients determine yield, depending on their function as constituents of organic compounds, enzyme activators, osmoregulators etc. (see Chaps. II–V).

In most crop species, however, only certain plant parts or organs (fruits, roots and tubers for example) are considered as "yield". As the photosynthetic activity of these plant parts or organs is generally low or even absent, they act as a "sink" for assimilates produced in other plant parts ("source"). Increasing evidence has been presented during the past 20 years that under these circumstances the sink itself exerts an important regulatory function and can be rate-limiting for yield formation. This source–sink relationship is determined by phytohormones in a more or less specific manner, and this has been demonstrated in a number of cases. For further information the reader is referred to a comprehensive review article by MICHAEL and BERINGER (1980).

It is the aim of this section to attract attention to the fact that mineral nutrients can also exert a distinct regulatory function in this source–sink relationship. This may occur directly or indirectly by changing the phytohormone balance.

### 4.2 Effect of Mineral Nutrition on Phytohormone Level and Sink Formation

Environmental factors such as day length and temperature affect flower formation (i.e. the initiation of seeds and fruits as sink). This occurs primarily by changes in the phytohormone balance, as for example the ratio of gibberellic acid (GA) to cytokinins (CYT). CYT play a particular role in flower initiation (BRUINSMA 1977). As the main sites of CYT synthesis in higher plants are the root tips (see review article of VAN STADEN and DAVEY 1979), all environmental factors – including supply of mineral elements – which affect root growth are closely related to changes in the export of CYT towards the shoot. This can be demonstrated for nitrogen (see below) and phosphorus (MENARY and VAN STADEN 1976, SALAMA and WAREING 1979). The close correlation between phosphorus supply, CYT production and flower formation in tomato is shown in Table 16. From the results of Table 16, it may also be supposed that the increase in spikelet initiation of wheat with increasing phosphorus supply (RAHMAN and WILSON 1977) is causally connected with the CYT level. The same applies to the close positive correlation which has been observed in apple trees between phosphorus nutritional status at the time of flower differentiation and the number of flowers and fruit yield per tree in the subsequent year (BOULD and PARFITT 1973).

The nutritional factor which exerts the most distinct effect on phytohormone level and flower formation seems to be nitrogen. Inorganic nitrogen supply to the roots stimulates the production and export of CYT to the shoots (WAGNER and MICHAEL 1971, SATTELMACHER and MARSCHNER 1978, see also Sect. 1). In apple, ammonium-N as compared to nitrate-N is somewhat more effective in this respect (BUBAN et al. 1978). It is questionable, however, whether the

**Table 16.** Effect of short-term phosphorus deficiency (−P) on flowering and cytokinin in stem exudate of tomato. (MENARY and VAN STADEN 1976)

| | −P | +P | +P +Kinetin[a] |
|---|---|---|---|
| No. flowers (first truss) | 5 | 7 | (16) |
| g callus $flask^{-1}$ [b] | 0.78 | 1.12 | |

[a] Direct application to the shoot
[b] CYT activity in stem exudate determined by tobacco callus test

**Table 17.** Effect of ammonium versus nitrate supply on flowering of Jonathan apple trees in the flowering growing season. (After Grasmanis and Edwards 1974)

| Treatment: duration and time | Mean number of buds per tree | | Percentage of flower buds |
|---|---|---|---|
| | Emerged | Flower | |
| Continuous nitrate | 35.3 | 12.8 | 38.7 |
| 8 weeks ammonium (Dec–Jan) | 33.3 | 21.2 | 63.7 |
| 24 h ammonium (Feb) | 38.8 | 30.3 | 77.5 |

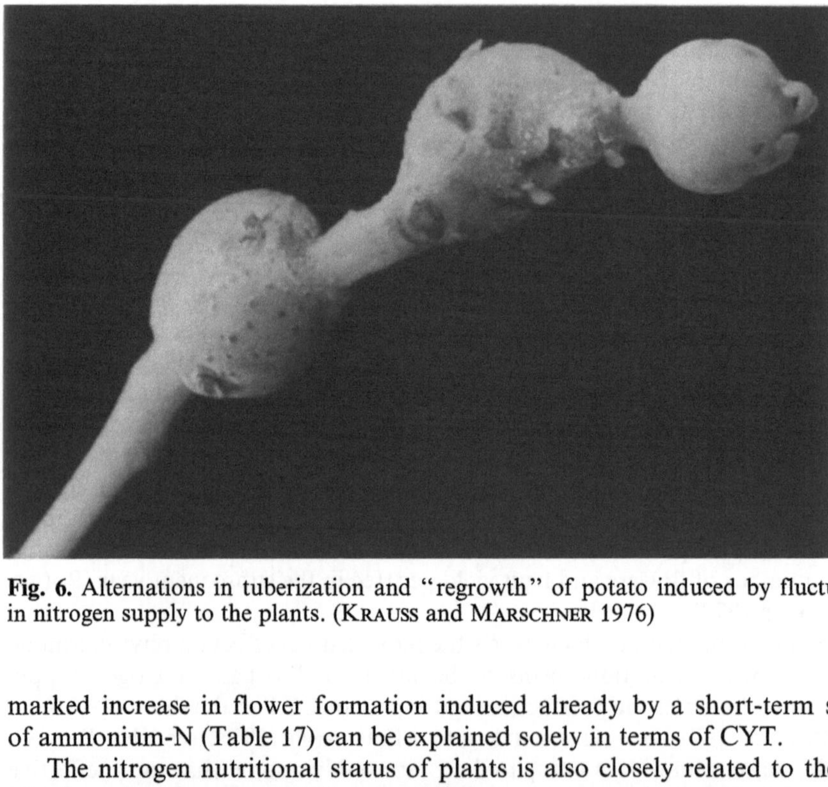

**Fig. 6.** Alternations in tuberization and "regrowth" of potato induced by fluctuations in nitrogen supply to the plants. (Krauss and Marschner 1976)

marked increase in flower formation induced already by a short-term supply of ammonium-N (Table 17) can be explained solely in terms of CYT.

The nitrogen nutritional status of plants is also closely related to the level of other phytohormones. Nitrogen deficiency or even short-term interruption in nitrogen supply to the roots increases the export of abscisic acid (ABA) from roots (Krauss 1978) and the ABA content of the shoots (Goldbach et al. 1975). This is correlated with decrease in GA content of the shoots (Rajagopal and Rao 1974). The interaction between nitrogen nutrition, CYT level and flower initiation as described above might therefore be only one example in a more complex system of regulation of developmental processes by nitrogen. Regardless of this uncertainty in causal interpretations, application of nitrogen at the time of flower initiation can be a quite effective tool in increasing the number of flowers (Trobisch and Schilling 1970) and in decreasing the yearly fluctuations in yield of perennials (Lüdders and Bünemann 1970).

Changes in phytohormone balance are also responsible for the regulatory role of nitrogen in tuber initiation in potato. In this plant species, fluctuations in nitrogen supply either as nitrate or as ammonium are closely correlated not only with corresponding changes in CYT and ABA export from the roots but also with alterations in tuberization and "regrowth" of the tubers (Fig. 6). The same effects resulting from nitrogen supply can be obtained by direct application of phytohormones (ABA, GA) to the stolons and tubers respectively (KRAUSS and MARSCHNER 1976). This result further supports the idea of the regulatory role of nitrogen on induction of both flowers and tubers via a change in the phytohormone balance. Changes in the C/N ratio are unlikely to be the causative factor, as only nitrogen supplied to the roots and not that supplied to the leaves is effective in regulation of tuberization (SATTELMACHER and MARSCHNER 1979).

### 4.3 Effect of Mineral Nutrients on Fertilization

The number of seeds and/or fruits per plant or both can also be directly influenced by mineral nutrients. This is most clearly seen with copper and boron. In cereals in particular copper deficiency affects the reproductive phase more than the vegetative phase. The primary cause of failure in grain setting is the non-viability of pollen of copper-deficient plants (Table 18). The synchronous meiotic division of a large number of pollen mother cells and the corresponding high localized copper demand are likely to be the primary reason for this symptom of copper deficiency (GRAHAM 1975). This might be a useful tool for plant breeding purposes to induce male sterility by a nutritional factor.

The necessity of boron for pollen tube growth (see Chap. V.2) is reflected for example in a decrease in grain number per head in rice (GARG et al. 1979) or even a total lack of fertilization in barley (SIMOJOKI 1972). As demonstrated in cross-pollinating experiments (VAUGHAN 1977) the failure of seed formation in maize suffering from boron deficiency is not due to lack of pollen viability but caused by non-receptiveness of the silks. Dry matter production and its distribution in maize as affected by boron supply (Fig. 7) provided an excellent example of sink limitation. Increasing the boron supply led to an increase in sink size (seed formation) correlated to a corresponding increase in photosynthetic capacity of the leaves.

**Table 18.** Grain set by cross-pollination between Cu-deficient and Cu-sufficient wheat plants. (GRAHAM 1975)

| Cross | | | No. of grain set | Total No. | |
|---|---|---|---|---|---|
| female | × | male | | floats | heads |
| −Cu | × | −Cu | 0 | 73 | 3 |
| +Cu | × | −Cu | 2 | 76 | 3 |
| −Cu | × | +Cu | 47 | 157 | 7 |

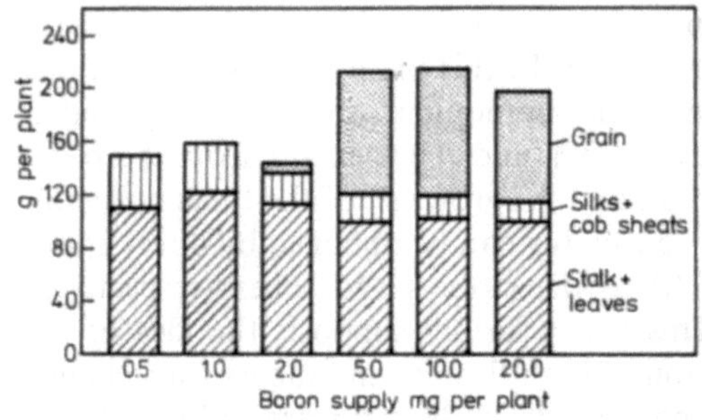


**Fig. 7.** Effect of boron supply to maize plants (*Zea mays* L.) on the production and distribution of dry matter. (Data from VAUGHAN 1977)

## 4.4 Source–Sink Interactions in Relation to Mineral Nutrition

In higher plants after fruit setting or tuber initiation, these newly formed storage organs represent the dominating sink. However, other plant parts can also act as competing sinks of different strengths. In annual plant species mainly the roots are effective, whereas in perennials both roots and apical parts of vegetative tissues play a role as competing sinks.

Usually in crop species the dominance of the fruits in sink competition is quite obvious and certainly a result of successful selection and breeding for higher crop yield. Examples showing extreme competition have been presented by LENZ (1970) for egg plants (*Solanum melongena* L.) and are demonstrated for citrus (LENZ and DÖRING 1975). Increase in fruit number from 0 to 100 per citrus tree strongly depressed further growth of leaves, stems and roots. The decrease in leaf area was correlated with a threefold increase in the transpiration rate per unit leaf area, a decrease in stomatal diffusion resistance (see also LENZ and WILLIAMS 1973) and a corresponding stimulation in net photosynthesis (see also Chap. IV.2).

CYT produced in the roots is distributed within the shoots, depending mainly on the transpiration rate of the various organs of the shoot (MICHAEL et al. 1969a, b, MONSELISE et al. 1978). As CYT is involved in both delaying senescence and in increasing sink activity and capacity (e.g. number of cells per organ; SEILER-KELBITSCH et al. 1974) the transpiration stream-directed distribution of root-borne CYT may be considered as an important regulatory step in the source–sink relationship (see also CARMI and KOLLER 1979). The higher content of nitrogen and calcium in the leaves of plants with heavy fruit load and lower leaf area (LENZ and BÜNEMANN 1969) might therefore not only be a necessary result of the increased transpiration rate, but might also be considered of importance in delaying senescence (see Sect. 1) and maintaining the source. On the other hand, partial defoliation favors CYT import into fruits (MONSELISE et al. 1978) and strongly stimulates the rate of fruit growth (sink activity) in seedless tomato fruits which mainly rely on root-borne CYT (VARGA and BRUINSMA 1974). In cereals such as barley the beneficial effect of awns as transpiring organs on the single grain weight (MICHAEL et al. 1969) is another example for this type of regulatory mechanism.

The importance of root growth for CYT supply to the source (leaves) so that it can compete more effectively with the sink becomes understandable

**Table 19.** Effect of leaf application of growth substances on dry matter distribution in *Daucus carota*. (LINSER et al. 1974)

| Treatment | g dry wt. per plant | | |
|---|---|---|---|
| | Shoot | Root | Total |
| $H_2O$ | 3.2 | 10.9 | 14.1 |
| CYT | 7.3 | 8.8 | 16.1 |
| GA | 9.9 | 5.7 | 15.6 |
| CCC | 2.8 | 10.8 | 13.6 |

in the context of the whole plant. The well-known effect of late nitrogen application to cereals causing delay in leaf senescence and increase in length of ripening (MICHAEL et al. 1974, MICHAEL and BERINGER 1980) has to be considered in relation to increased CYT export from the roots into the leaves. A similar interaction might be assumed in potatoes, where a sudden increase in nitrogen supply leads to a temporary "switching-off" of the tubers as a sink (KRAUSS and MARSCHNER 1971).

The mode of nitrogen supply also presumably affects source–sink interactions via the action of other phytohormones. Interruption of the nitrogen supply to the roots, for example, also causes a marked increase in ABA content of the root exudate as well as the leaves (GOLDBACH et al. 1975, KRAUSS 1978). As ABA is very effective in inducing stomatal closure (ITAI and MEIDNER 1978) and in decreasing the sink activity (WAGNER 1974), shortage of nitrogen can effectively regulate the source–sink interactions primarily at the expense of the source. ABA, however, is also quite phloem-mobile and can be transported from the leaves into grains, i.e. the sink (GOLDBACH et al. 1977), where it seems to be causally involved in termination of the storage process, presumably by inhibition of RNA synthesis in the embryo (HSU 1979).

Of the other phytohormones causally involved in the sink activity and affected by mineral nutrition, data are available only for GA. A close positive correlation occurs between GA activity in shoots and nitrogen supply to roots (RAJAGOPAL and RAO 1974). The stimulation of shoot growth at the expense of root growth, which is a typical response to nitrogen supply (MOTHES 1932, GÖRING and MARDANOV 1976), is presumably mainly the result of a higher sink activity of various shoot parts with high GA concentrations (GARROD 1974). In crop species where certain vegetative parts are considered as "yield", such as storage roots of *Daucus carota*, this phytohormone-regulated sink competition is quite obvious (Table 19). It is a well-known phenomenon that a high nitrogen supply in the later stages of growth induces vigorous shoot growth at the expense of storage root growth (e.g. in carrots or sugar beet). This effect is presumably caused by a shift in the phytohormone balance as discussed above, with corresponding effects on the source–sink-relationship and sink competitions (MICHAEL et al. 1974).

In the source–sink relationship the important steps, regulated at least in the sink by phytohormones (SETH and WAREING 1967, WAGNER 1974), are

**Table 20.** Role of potassium and magnesium in sucrose accumulation in vacuoles isolated from red beet tissue. (DOLL et al. 1979)

| $K^+$ | $Mg^{2+}$ | nmol sucrose $h^{-1}$ |
|---|---|---|
| – | – | 4.9 |
| – | + | 42.3 |
| + | + | 55.5 |

**Table 21.** Time course of the potassium content (mmol $g^{-1}$ dry wt.) in petioles of two tomato varieties. (After LINGLE and LORENZ 1969)

| Variety | Growing stage of the plants | | | | |
|---|---|---|---|---|---|
| | 6 true leaves | 3[rd] cluster full bloom | 1[st] cluster mature green | 1[st] cluster fruit pink | 50% fruits ripe |
| Pearson | 1.47 | 1.64 | 1.85 | 0.78 | 0.39 |
| VF-13 L | 1.47 | 1.41 | 1.58 | 0.54 | 0.10[a] |

[a] Severe potassium deficiency symptoms

phloem loading at the source, long-distance transport in the sieve tubes and unloading at the sink. The involvement of potassium in the sucrose-proton cotransport at the loading site is still an open question (HUTCHINGS 1978, DOMAN and GEIGER 1979). It seems more likely, however, that the favorable effect of potassium on the export of sucrose from the leaves (MENGEL and HAEDER 1977) is primarily the result of a higher rate of sucrose release from the leaf cells into the apoplast and thus an indirect support to phloem loading (DOMAN and GEIGER 1979). A direct role of potassium on sucrose transport within the sieve tubes has still to be elucidated (WILLENBRINK and SCHUSTER 1978). On the other hand, a direct role for potassium has been demonstrated in the storage cells itself (Table 20). This is presumably a reflection of a $Mg^{2+} + K^+$ stimulated ATPase at the inner side of the tonoplast.

From the regulation of starch synthesis in chloroplasts by inorganic phosphate ($P_i$) as demonstrated by HELDT et al. (1977) (see Sect. 1), it may be supposed that a comparable effect also occurs in the amyloplasts of starch-storing cells. The marked increase in formation of phytin in cereal grains in later stages of grain growth may thus perhaps be regarded as necessary in order to maintain a fairly constant concentration of $P_i$ under conditions of decreasing grain water content (MICHAEL et al. 1980), and thus avoid any increase in inhibition of starch synthesis by $P_i$.

Sink activity of fruits or storage organs and thus final yield can also be limited in quite another way, namely by the supply of mineral elements from the leaves as source. An excellent example for this limitation has been shown by LINGLE and LORENZ (1969) comparing tomato genotypes differing in ripening pattern (Table 21). In the genotype which ripened more rapidly and uniformly, severe potassium deficiency symptoms occurred in the leaves. This deficiency could not be corrected by supplying a high concentration of potassium to the roots, presumably because of the dominance of the fruits which were also acting as sink for carbohydrates and thus decreasing root growth.

A similar situation of source limitation caused by shortage of another plant nutrient has been demonstrated by TROBISCH and SCHILLING (1969, 1970). In *Sinapis alba,* the developing seeds compete for nitrogen supplied from the leaves, and seed growth and final yield are primarily determined by the pool size of nitrogen in the vegetative parts. Consequently, additional nitrogen application at flowering markedly increases seed number and yield. Source limitation by shortage of nitrogen but not carbohydrates was also shown when the effect of darkening the leaves was compared with that of removal of the leaves. Darkening reduced seed weight by 20%, but removal of the leaves, which also removed the source for nitrogen, depressed seed weight by nearly 50% (TROBISCH and SCHILLING 1969).

A special type of source limitation can occur in legumes particularly with the onset of seed growth, where a sink competition for carbohydrates arises between seeds and root nodules. Seeds seem to be the dominating sink and are responsible for the rapid fall in $N_2$ fixation in nodules which occurs with the onset of seed growth (HERRIDGE and PATE 1977). Thus also in legumes seed growth can be limited by the amount of nitrogen available in the leaves, particularly when senescence is enhanced by the export of bound nitrogen from the leaves (SINCLAIR and DEWIT 1976). Supplying inorganic nitrogen after flowering to the roots can therefore also be quite a useful means of increasing the protein content of legume seed (SCHILLING and SCHALLDACH 1966); leaf spray during seed filling with nitrogen in combination with other nutrients is particularly effective in increasing yield in soybean (RAMON GARCIA and HANWAY 1976).

These examples showing the interactions between mineral nutrition and the source–sink relationship clearly indicate that this particular area needs much more attention in future research. This holds true especially for direct or indirect effects of plant nutrients other than nitrogen on phytohormone levels. More attention needs to be given to the consequences arising from restricted mineral nutrient uptake by the roots in the period when the fruits or storage organs are established as dominating sinks. Selection and breeding for genotypes with a "high harvest index" and short periods for fruit growth or ripening ("filling period" in cereals for example) might become more and more difficult, not because of a limited source capacity for carbohydrates but rather by limitation of mineral nutrients such as potassium, magnesium, nitrogen or phosphorus. This limitation could be at the source and due to a lack of capacity (amounts) or the capability for retranslocation, or it could be caused by the low activity of roots for uptake during the period of fruit and storage organ formation. For further details the reader is referred to Chap. IV.2.

# 5 Environmental Aspects of Mineral Nutrition

## 5.1 Introduction

Mineral nutrient cycles in terrestrial natural ecosystems are never entirely closed, but only reach a quasi-equilibrium between loss and supply. Contributing factors of this system include the atmosphere, ground and surface water, weathering of rocks and minerals, and biological dinitrogen fixation. Man has influenced this natural cycle of plant nutrients by removing mineral nutrients in the process of crop production. In "classical agriculture" – which still occurs in many developing countries where the population is mainly rural – the "loss" of mineral nutrients has been relatively small. With increasing urbanization, however, this "loss" has markedly increased and has had to be compensated for by the introduction of legumes in the crop rotation and an increase in the application of mineral fertilizers (for more details see Chap. I.8). These changes simultaneously have led to the "production" of large amounts of "waste material" in urban areas and increases in mineral nutrient concentration in ground and surface waters.

This development, considered together with the limited resources (e.g. phosphate deposits) and increasing costs for the production of mineral fertilizers (nitrogen in particular), strongly indicates a need for a more effective utilization of plant nutrients in crop production. This holds true in particular for increasing the contribution of biological dinitrogen fixation (Chap. II.1–3), an improvement in the utilization of mineral nutrients by selection and breeding of crop plants for higher efficiency in general (Chap. IV.1) and for phosphate in particular (Chap. I.2) and an increase in the "re-cycling" of nutrients especially in densely populated urban areas (see below).

### 5.1.1 Nitrogen

The environmental aspect of mineral nutrition, which at present attracts particular interest and is also the subject of most controversial discussions, is the application of nitrogenous fertilizers either in organic (e.g. manure) or inorganic forms (e.g. nitrate- or ammonium-N). The potential losses of nitrogen from soils by leaching of nitrate, denitrification or volatilization of $NH_3$ and hence a corresponding increase in pollution of groundwater and atmosphere are certainly enhanced when nitrogen fertilizers are added to the soil. In agroecosystems both total loss and type of loss of bound nitrogen are, however, more dependent upon climatic and soil conditions and factors such as the supply/demand ratio, timing of fertilizer application, crop rotation etc. than whether the nitrogen has been supplied in inorganic or organic form (for further details see a review by FRISSEL 1977). In this context it is necessary to stress that these potential losses of nitrogen from agroecosystems are not necessarily decreased, for example when the contribution of biological dinitrogen fixation is increased at the expense of the use of nitrate fertilizers. It must be remembered that the application of nitrate as a fertilizer can be better adjusted to the actual demand of plants, whereas mineralization of organically bound nitrogen is regulated primarily by environmental factors such as temperature and soil moisture.

As nitrate is an ideal storage form of nitrogen for higher plants, high concentrations of nitrate in the soils not only increase potential losses by denitrification and leaching but can also lead to high nitrate levels in the plants. This occurs regardless of whether nitrate fertilizers (KNAUER and SIMON 1968) or organically bound nitrogen (LUND et al. 1975) was the original form supplied to the soil. In plants used for human consumption (vegetables such as spinach for example), this nitrate accumulation is undesirable for plant quality. Unskilled handling or processing of plant material high in nitrate can cause nitrite formation, which is potentially toxic to infants (SCHUPHAN 1976), and in addition can increase the possibility of nitrosamine formation, a potentially carcinogenic substance (HILDEBRANDT 1977). Thus control of nitrate formation in soils and uptake by plants are important in both relationships of mineral nutrition and environment as well as mineral nutrition and plant quality. Therefore in soils it is important to control both the amount of nitrate (WEHRMANN and SCHARPF 1979) and to restrict nitrification as, for example, by the use of naturally occurring substances (SAHRAWAT et al. 1977) or synthetic compounds such as dicyandiamide (AMBERGER and GUTSER 1978).

### 5.1.2 Heavy Metals

Heavy metals are integrated components of the biosphere and thus occur naturally in soils and plants. Five heavy metals (Fe, Mn, Zn, Cu, Mo) are essential for all higher plants (see Sect. 1) and in addition others like V and Co are essential for animal and man. The demand of these five heavy metals for plants is rather low (micronutrients). However, the use of chemically pure fertilizers, together with the production of higher yields, increases the necessity of supplying these micronutrients more systematically in fertilizer practice in intensive agriculture and horticulture. Except under soil conditions of low redox potential where iron toxicity may occur in lowland rice (HOWELER 1974) and manganese toxicity may appear in sensitive genotypes (see Chap. IV.1), deficiency is much more widespread as a result of limited availability due to high soil pH (Fe, Zn, Mn) or, as in the case of Cu, high organic matter content or extreme soil leaching.

Increasing the supply of at least some or even all of these five heavy metals in fertilizers is not only necessary to increase plant yield and to prevent deficiencies in plants, but is also important from the view point of plant quality. This holds true for Zn in particular. There is increasing evidence that hidden or acute Zn deficiency in both monogastric animals (WETZEL et al. 1979) as well as man (HAMBIDGE and WALRAVENS 1976) is quite widespread, especially when cereals are the main diet and where the availability of Zn for plants is low due to the specific ecological conditions (e.g. high soil pH in arid/semi-arid regions). Zinc deficiency in animals and man is furthermore accentuated by inadequate availability from cereals. The presence of phytin, i.e. the main storage product for phosphorus in cereal grain and other seeds, seems to be causally responsible for this low Zn availability (REINHOLD et al. 1973). This fact has led human nutritionists to the questionable classification of phytin into the group of "phytotoxicants" (OVERLEAS 1973).

## 5.2 Heavy Metal Toxicity

With the exception of the occurrence of Fe and Mn toxicity in crop plants as mentioned above, heavy metal toxicity was not generally regarded as an agricultural or horticultural problem. It was predominantly only a research object for geobotanists and plant ecologists. The distinct "heavy metal flora" in mining areas, adapted for example especially to high soil levels of Zn and Cu, and the mechanisms of tolerance – either by exclusion, or inactivation within roots or leaves by compartmentation, complex formation etc. – was always a fascinating research subject on environmental aspects of mineral nutrition (KRAUSE 1958, ERNST 1974). The reports of tolerance of crop plants grown in these mining areas (or on serpentine soils) with high levels of Zn, Cu and also with Ni or Cr were quite rare (for further information see ANTONOVICS et al. 1971, ERNST 1974) and presented without considering the possible implications and risks which may arise when these plants are used for animal and human nutrition. An exception to this was the study of heavy metal toxicity for animals and man as a result of direct exposure particularly for miners (e.g. lead, mercury). However, it was only a decade ago that the role of elevated heavy metal contents of plants as part of the food chain for heavy metal toxicity of animals and man suddenly became of overwhelming importance. The "classical" example for this was the occurrence of the human sickness Itai-Itai in Japan, caused by cadmium (Cd) released from industrial areas and passed at least in part from the soil into the food chain via plants (YAMAGATA and SHIGEMATSU 1970). This "accident" stimulated intensive world-wide research on environmental aspects of heavy metals in the mineral metabolism of plants.

### 5.3 Heavy Metals in the Food Chain

In the first edition of this Encyclopedia in Volume IV (Mineral Nutrition) published in 1956, the heavy metal Cd was not even mentioned, and others like Pb, Ni or Hg only briefly discussed. Since then the position has changed dramatically and it is now extremely difficult to keep up with the hundreds of publications which appear each year on Cd and other heavy metals in the soil/plant/animal/man system. It can therefore only be the aim of this section to outline this development and the consequences arising from it, both for future research on mineral nutrition and strategies to control entry of heavy metals in the food chain via the soil and their subsequent transfer into plants, animals and finally to man. An example illustrating the accumulation of Cd in the food chain is shown in Table 22.

Considering that only part of the Cd in the top soil is immediately available to plants the progressive increase in Cd accumulation in the food chain is obvious and can reach particularly high levels (more than 300 ppm) in the kidneys of carnivores in contaminated areas (MARTIN and COUGHTREY 1975). This positive correlation between the Cd content in soils/plants and animals generally occurs (CHANEY et al. 1978a), but not always (Table 23).

The increase in Cd content of the plants and thus the diet from the MSS-amended soils did not increase the Cd content of the kidneys, presumably as a result of restricted Cd resorption in the presence of the simultaneously elevated Zn levels in the diet (CHANEY et al. 1978b).

These two examples stress the necessity for enhanced interdisciplinary research on heavy metals in the food chain. The aim of this section is to present

**Table 22.** Levels of Cd (ppm in dry matter) in various component species of woodland ecosystems. (After MARTIN and COUGHTREY 1975)

| | Collecting areas | |
|---|---|---|
| | Relatively uncontaminated | Contaminated (3 km from a Pb-Zn smelter) |
| Top soil | 2.0 | 42 |
| Moss | 1.3 | 25 |
| *Hedera helix* | 2.1 | 18 |
| Earthworms | 25.2 | 123 |

**Table 23.** Tissue content (ppm in dry matter) of Cd, Ni and Zn in Guinea pig fed with plants grown on municipal sewage sludge (MSS)-amended soil. (After CHANEY et al. 1978b)

| | Cd | Ni | Zn |
|---|---|---|---|
| *Soil* (extractable) | | | |
| Control | 0.02 | 0.4 | 1.6 |
| MSS | 0.30 | 1.1 | 65.0 |
| *Plants* (Swiss chard) | | | |
| Control | 0.5 | 1.4 | 70 |
| MSS | 2.7 | 3.5 | 580 |
| *Kidneys* | | | |
| Control | 14.9 | 0.15 | (No significant |
| MSS | 14.9 | 0.05 | differences) |

some examples from the soil/plant system with special references to Cd, which is taken up quite readily by many crop species and which is one of the most toxic heavy metals for animals and man. For more detailed information the reader is referred to PAGE and BINGHAM (1973), VETTER et al. (1974), DOYLE (1977) and CHANEY and HORNICK (1978).

## 5.4 Heavy Metals in the Soil/Plant System

### 5.4.1 Content of Soils

The average concentration of Cd in the Earth's crust is 0.15 ppm. Depending upon the parent material, the Cd content of soils, however, can vary considerably: 0.1–0.3 ppm in soils deriving from igneous rocks and 0.1–10 ppm and 0.3–11 ppm respectively in soils deriving from metamorphic and sedimentary rocks (WAKITA and SCHMITT 1970). As Cd is usually associated with Zn, Pb-Zn, and Pb-Zn-Cu ores, the Cd content of soils in these mining areas can be several times higher than in average soils (PAGE and BINGHAM 1973, ERNST 1974).

**Table 24.** Heavy metal content of soils and pasture plants at different distances from a lead-zinc smelter. (VETTER et al. 1974)

| Element | Distances in km | | | | |
|---|---|---|---|---|---|
| | 0.75 | 1.5 | 3 | 6 | 12 |
| | Amount kg ha$^{-1}$ in 0–80 cm soil depth | | | | |
| Zn | 7500 | 1850 | 620 | 350 | 240 |
| Pb | 2700 | 730 | 360 | 160 | 130 |
| Cd | 71 | 14 | 4.0 | 1.9 | 0.8 |
| | ppm in pasture plants (dry wt.) | | | | |
| Zn | 523 | 118 | 116 | 74 | 78 |
| Pb | 98 | 55 | 11 | 12 | 7 |
| Cd | 2.1 | 1.1 | 0.6 | 0.4 | 0.2 |

Of much more importance, however, are the man-made elevated environmental levels of Cd, and in particular from emissions from smelting and refining industries (Zn, Pb, Cu). Depending upon the prevailing direction of winds, elevated levels of heavy metals in soils can be detected even up to distances of 40–65 km from a smelter (CARTWRIGHT et al. 1976). In the immediate neighborhood of large smelters soils and plants (uptake from the soil and adhering dust on the leaves) can reach heavy metal levels which make agricultural production impossible, not primarily because of toxicity to the plants but rather of the hazards which might arise from using these plants for animal and human nutrition. This situation is accentuated because in these areas contamination by several heavy metals usually occurs simultaneously (Table 24).

Another important site of contamination of soils and plants with heavy metals, and particularly Pb and Cd, occurs along roadsides. The source of Pb is the lead-tetraethyl in petrol, whereas the Cd content is mainly due to attrition of tires (LAGERWERFF and SPECHT 1970). The Pb content of soils and plants along highways can reach 100 ppm (KLOKE et al. 1966). However, most of this Pb content of roadside plants is not the result of uptake from the soil but rather of surface contamination (HÖLL and HAMPP 1975) as the availability of Pb in soils for uptake by plants is usually rather low.

Cd is also present in phosphate fertilizers. Depending upon origin, the Cd content of rock phosphates can vary between 9 and 130 ppm (REUSS et al. 1978). Amounts of Cd supplied to soils with fertilizer phosphates thus depend largely on the type of rock phosphate, but also on the processing procedure (CHANEY and HORNICK 1978, JUNG et al. 1979). Amounts added to the soil vary from less than 1 g and up to 100 g Cd ha$^{-1}$, based on a fertilizer rate of 50 kg P ha$^{-1}$. Although only a small proportion of this Cd is immediately available for uptake (JUNG et al. 1979), the positive correlation between added phosphate fertilizer (and thus also Cd) and the Cd content of plants can clearly be demonstrated (REUSS et al. 1978). As a consequence of this, long-term calculations on world reserve of rock phosphates and comparisons between different types of phosphate fertilizer processing have to take Cd into account and its implication for the environment.

The main concern of increasing the heavy metal content of soils and thus also the food chain is related to the use of municipal liquid and solid sewage sludge. Here, clearly contradictory environmental aspects and goals exist, as it is also necessary to restrict enlargement of urban trash disposal, prevent further contamination of surface and ground waters and finally to bring about a more efficient recycling of organic matter and plant nutrients also in agroecosystems with predominantly urban populations. However, in industrialized areas in particular, the contamination of municipal liquid and sewage sludge with heavy metals has reached levels which prohibit even short-term usage in agriculture and horticulture. Again, here Cd is of major concern and a rather recent problem, as is indicated by the production and usage of this element (electroplating, pigments and chemicals) which increased in world-wide usage more than 25-fold from 1920 to 1960 (PAGE and BINGHAM 1973). The increased usage of municipal liquid and solid sewage sludge, together with contamination by other heavy metal sources (see above), has resulted in large areas having elevated heavy metal contents in soils and plants. This has stimulated intensive research on the fate of these heavy metals in soils and the uptake and distribution in plants.

### 5.4.2 Soil Factors Affecting Heavy Metal Accumulation in Plants

Similarly to other heavy metal cations such as Mn and Zn, the availability of Pb (ZIMDAHL and FOSTER 1976) and Cd (HERMS and BRÜMMER 1979) decreases with increasing pH. The availability to plants also decreases with increasing cation exchange capacity of the soil. In contrast to Zn or Mn, the binding of Cd and Pb is closely related to the organic matter content of the soil (ANDERSSON 1977b). This can lead to a decrease in Pb uptake by plants despite an increase in Pb supply to the soil by Pb-contaminated compost (ANDERSSON 1977a).

This effect of organic matter on plant availability of Pb has been also demonstrated for Cd in a model experiment (Fig. 8).

Interactions between heavy metal availability and soil organic matter are complicated by changes in redox potential and by the formation of metal-organic complexes of different electrical charges (LAGERWERFF and MILBERG 1978). In general a decrease in redox potential increases the possibility of $H_2S$ formation and thus precipitation and immobilization of heavy metals such as Zn or Cd (HERMS and BRÜMMER 1979). In the absence of drastic decreases in redox potential, the effect of organic matter in decreasing availability by formation of complexes seems to decrease in the order Cu > Pb > Zn > Cd (LAGERWERFF et al. 1977).

These few examples demonstrate that predictions concerning the contamination of the food chain by heavy metals require careful consideration, not only of the soil factors, but also of whether a heavy metal such as Cd is supplied as an inorganic cation (SYMEONIDES and MCRAE 1977) or as a contaminant of sewage sludge along with high contents of organic matter.

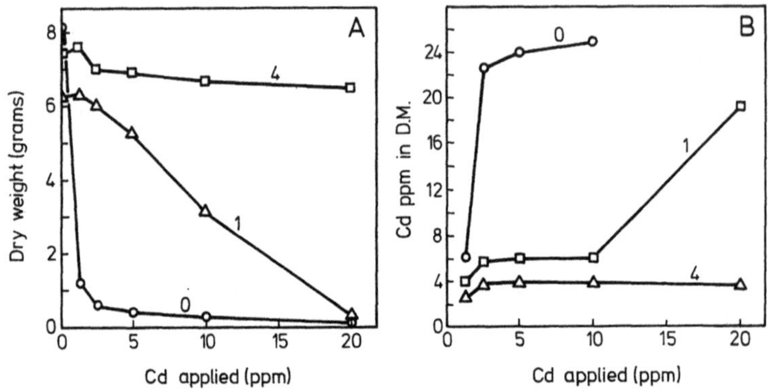


**Fig. 8.** Effect of organic matter (0%–4% as peat) addition to the substrate (quartz sand) on growth of soybean plants (**A**) and cadmium content of the leaves (**B**) at increasing cadmium concentrations in the substrate. (STRICKLAND et al. 1979)

### 5.4.3 Genotypic Differences in Heavy Metal Uptake

In crops, appreciable genotypic differences exist in uptake of heavy metals such as Cd, Ni, Hg or Cr, although these are less dramatic than in the case of the "heavy metal flora" for example from serpentine soils or mining areas (KRAUSE 1958, ANTONOVICS et al. 1971, ERNST 1974). With a few exceptions (see below), Cd is the one of these four heavy metals which in general is most readily taken up and translocated into the shoots (DIEZ and ROSOPULO 1976). Nevertheless, pronounced genotypical differences also exist for Cd in accumulation between crop species. These differences have been summarized by CHANEY and HORNICK (1978) and are shown in Fig. 9.

Compared to these extended and systematic studies of Cd uptake by different crop species much fewer comparative data are available for other elements such as Ni, Cr or Hg. The uptake of Ni seems to be astonishingly high in the vegetative parts of a number of species (DIEZ and ROSOPULO 1976) from 9 ppm in oats up to 21 ppm in wheat. In vegetables, particularly high Ni accumulation has been found in lettuce (FRITZ et al. 1977), cabbage (HARA and SONODA 1979) and spinach (FOROUGHI et al. 1979). Of the vegetable crops spinach (KICK and BRAUN 1977) or bean seem to have a particular ability to accumulate Cr in the shoots as compared to lettuce (HUFFMAN and ALLAWAY 1973).

In higher plants, accumulation of Hg is usually quite low even in heavily contaminated soils (DIEZ and ROSOPULO 1976). On the other hand amongst the lower plants certain fungi (Basidiomycetes) are distinct Hg accumulators (DOMSCH et al. 1976).

Certain of these fungi also have a high affinity for Cd. Collected from an uncontaminated area in Yugoslavia, the range of Cd (ppm in dry matter) varied from 0.62 in *Cantharellus uberius* up to 5.72 in *Macrolepiota* and 10.30 in *Lactarius piperatus* (BYRNE and RAVNIK 1976). Between species of the genus *Agaricus* (champignon) a particular high variability occurs in Cd accumulation. As compared to the Zn content in the dry matter of the fungi (300–500 ppm), the Cd content varied between 0.1 and 170 ppm with corresponding Cd accumu-

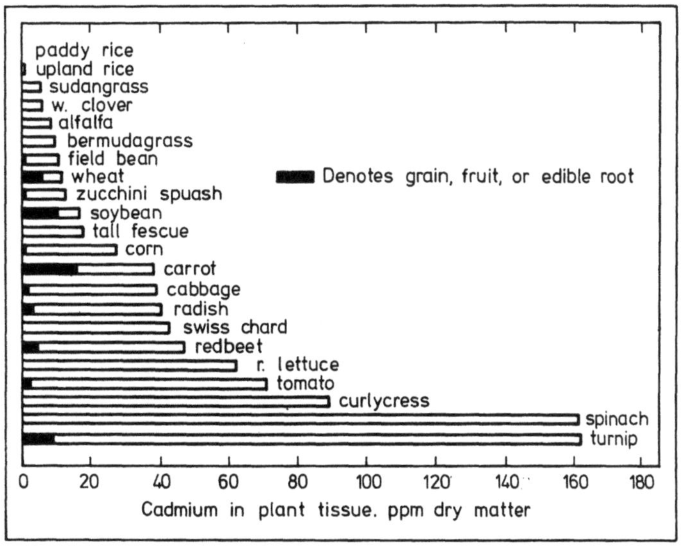


**Fig. 9.** Cadmium uptake by different crop species grown on a soil enriched with 10 ppm cadmium by sewage sludge. (After BINGHAM et al. 1975, 1976, from CHANEY and HORNICK 1978)

lation ratios (fungi/soil) between 3.1 in *A. abruptibulbus* and 292.5 in *A. macrosporus* (MEISCH et al. 1977). Because of the generally high level of Cd and Hg in edible fungi, the excessive consumption should be avoided even of fungi grown on soil not contaminated by man (LORENZ et al. 1978).

Despite the benefical effects of vanadium (V) on plant growth (see Chap. V. 5) and its functions in human and animal metabolism, it is still an open question whether, for example, the excessive accumulation of V in certain species of *Allium* of up to more than 100 ppm in the dry matter (CANNON 1963) compared to less than 1 ppm in the dry matter in most other plant species (CANNON 1963) is harmful in human and animal nutrition. Most of the V is supplied to soils by phosphate fertilizers which can contain up to 5000 ppm V (for more details see LASKE 1979).

The classification of plant species into heavy metal accumulators and non-accumulators as has been shown for Cd in Fig. 9 has to be reexamined critically in view of the varietal differences within certain species. An impressive example for this is shown with lettuce (Fig. 10). In plants grown on the same substrate the Cd content of the tops varied by the factor 20.

Cadmium contents in the shoots of soybean varieties grown on the same soil even varied by a factor of 40 (BOGGESS et al. 1978). Also in cereals such as barley and wheat the varietal differences in Cd accumulation of the seeds are considerable and consistent even when the varieties are grown on different soils where both the amounts and binding states of Cd can be expected to be different (PETTERSSON 1977).

These varietal differences in Cd uptake fit well in the general picture of genotypic differences in mineral nutrition of higher plants (see Chap. IV.1 and

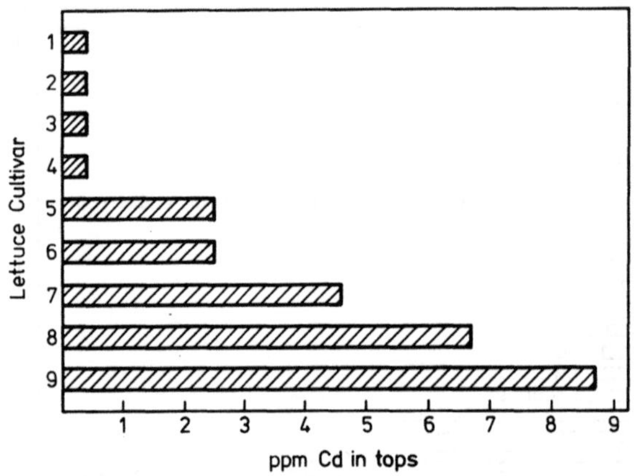


**Fig. 10.** Cadmium content in the tops of nine lettuce cultivars grown in solution culture with 0.1 ppm cadmium. (JOHN and VAN LAERHOVEN 1976)

**Table 25.** Average values for heavy metal content (ppm in dry matter) in different organs of vegetables grown on various soils. (FRITZ et al. 1977)

| Organs | Cu | Zn | Pb | Cr | Cd | Ni |
|---|---|---|---|---|---|---|
| Leaves | 11.3 | 118 | 13.3 | 3.9 | 1.0 | 5.9 |
| Storage roots, tubers | 11.0 | 43 | 12.1 | 1.3 | 0.1 | 3.9 |
| Fruits | 9.1 | 40 | 9.7 | 0.1 | 0.1 | 2.7 |

Vol. 2/B, Chap. III.9, this Vol.) and open up new aspects for selection and breeding programmes for low uptake of certain heavy metals like Cd.

### 5.4.4 Distribution Within the Plants and Their Organs

For heavy metal contamination of the food chain not the total uptake is of primary importance, but rather the content in those parts which are used directly for animal and human consumption. In many crop species, for example, this includes only the fruits, seeds or storage roots, whereas in others it may consist of the whole shoot (e.g. lettuce). As a general rule the heavy metal content per unit dry matter decreases in the order leaves, storage roots, tubers, fleshy fruits and seeds. An example from a range of vegetable crops is shown in Table 25.

The generally high Pb content in all organs is presumably mainly caused by surface contamination. In the case of Ni, however, high contents are found even in the fruits in comparison to the leaves and appear to reflect a particularly high mobility of this heavy metal in phloem transport. Nickel is readily translocated also into seeds (Table 26). This has also been demonstrated in wheat, where with increasing Ni content of the soil a similar increase in Ni content of leaves and seeds has been reported (MITCHELL et al. 1978).

Fortunately this distribution between leaves and fruits or leaves and seeds is much more in favour of the leaves for Cd than for Ni. On the other hand in most plant species uptake and transport of Cd into the shoots is usually

**Table 26.** Distribution of nickel with time following 24 h uptake from 1.0 µM $NiCl_2$ by 94-day-old soybean plants. (After CATALDO et al. 1978)

| Tissue | Distribution (%) following $^{63}Ni^{2+}$ pulse | |
|---|---|---|
| | 24 h | 21 days |
| Roots | 85 | 51 |
| Stems | 5 | 4 |
| Leaves | 6 | 4 |
| Pods | 2 | 2 |
| Leeds | 2 | 39 |

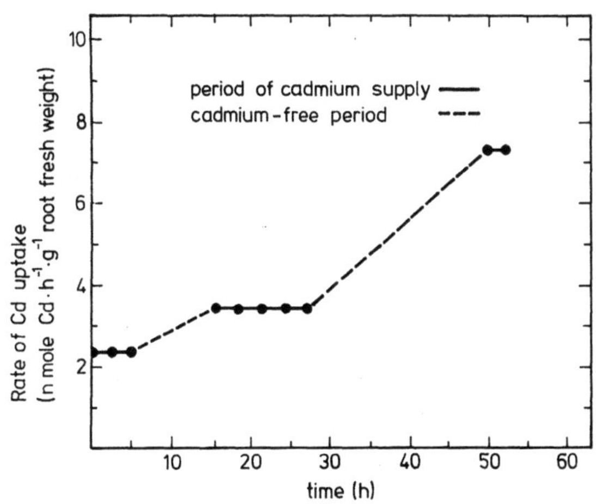


**Fig. 11.** Time course of initial cadmium uptake by tomato plants (*Lycopersicon esculentum*) based on successive runs of partial cadmium depletion from nutrient solutions containing 0.1 µM cadmium. Each period of cadmium supply (—) is separated by a period with a cadmium-free solution (- - -). (PETIT et al. 1978)

directly proportionally to the external concentration (PETTERSSON 1976). In a detailed study on Cd uptake by tomato plants grown under controlled conditions, PETIT et al. (1978) found evidence (Fig. 11) that supplying Cd to the roots induces the formation of new binding sites for Cd within the roots and thus further stimulates uptake. This mechanism needs further attention and investigation, and comparisons should be made with other plant species and genotypes differing in their ability to take up Cd.

Although the translocation of Cd from leaves into seeds as compared to fleshy fruits is usually low and the Cd content of cereal grains only exceeds 1 ppm on highly contaminated soils (PIETZ et al. 1978), in the average diet of U.S. citizens for example more than 50% of the total Cd intake is derived from cereals (CHANEY and HORNICK 1978). As the distribution of Cd at least in wheat grains is rather uniform, the possibility of decreasing the Cd content by processing procedures is limited (OCKER 1977/1978).

In addition to lowland rice (Fig. 9) maize appears to be the other main cereal crop in which is a distinct restriction of Cd translocation into the seeds (PIETZ et al. 1978). However, as already shown for Cd in lettuce (Fig. 10), experiments with only one variety do not permit generalizations on the behavior of Cd in a species. For example, when 20 inbred lines of maize were grown on a sludge-amended soil, the Cd content of the seeds (mg Cd $kg^{-1}$ dry matter) varied between 0.08 in one line to 3.87 in another (HINSLEY et al. 1978).

### 5.4.5 Heavy Metal Tolerance

The increased use of municipal sewage sludge on soils and the subsequent increase in heavy metal content of crops and soil has initiated extensive comparative studies between the heavy metal toxicity and possible yield reduction. As may be expected, results vary widely depending on experimental conditions. The choice of plant species and varieties, soil type, soil pH, form of heavy metal (e.g. inorganic $Cd^{2+}$ as compared to Cd in sewage sludge), composition of the nutrient solution, duration of the experiment etc. all influence the results obtained. For details of results of some comparative studies in nutrient solution using increasing concentrations of a range of heavy metals, the reader is referred for example to FOROUGHI et al. (1979) and HARA and SONODA (1979). Most papers, however, are exclusively concerned with uptake and phytotoxicity of Cd as this heavy metal, as compared to Pb, Ni, Hg or Cr, is particularly toxic for most crop species. Of the vegetables, spinach is much more Cd-sensitive than, for example, tomato (BINGHAM et al. 1975). Cadmium toxicity in plants is often correlated with chlorosis and wilting, symptoms which might be related to inhibited Fe uptake (FOROUGHI and VENTER 1978) or the inhibition of xylem differentiation and thus a disturbance of the water balance of the plants (LAMOREAUX and CHANEY 1977, KIRKHAM 1978).

For causal interpretations of genotypic differences in heavy metal tolerance, it is necessary to obtain more information concerning the mechanisms involved, such as exclusion in uptake, restricted transport into the shoot or high tissue tolerance (e.g. physiological "inactivation" by compartmentation). Research on tolerance mechanisms for Cd, Ni, Cr or Pb of crop plants is certainly necessary and should utilize the extensive knowledge of heavy metal tolerance in the natural vegetation as for Zn and Cu (MATHYS 1975, WU et al. 1975, WAINWRIGHT and WOOLHOUSE 1977) or Ni (LEE et al. 1978). The high mobility of Cr or Ni in plants might be causally related to the formation of complexes with organic acids (LEE et al. 1978) and explain, for example, the particularly high rates of Ni transport into fruits and seeds of crop plants (see Tables 25 and 26).

Without doubt a better understanding of the tolerance mechanism for heavy metals such as Cd in crop plants is also of importance for basic research in plant physiology and biochemistry. It should be borne in mind, however, that phytotoxicity of Cd, for example, is irrelevant from the environmental viewpoint in relation to heavy metal accumulation in the food chain. This discrepancy in crop species between Cd toxicity as measured by yield reduction and Cd accumulation in edible plant parts is shown in Table 27.

In future the strategy of selection and breeding programmes of crop species has to concentrate not on increasing Cd tolerance but in restricting Cd uptake and transfer into edible plant parts. The same strategy is necessary for forage species where Cd tolerance seems to be closely related to high Cd accumulation within the shoot (BINGHAM et al. 1976).

**Table 27.** Soil and plant tissue Cd levels associated with a 25% yield decrement. (After BINGHAM et al. 1975)

| Species | Soil Cd levels producing a 25% yield decrement | Tissue Cd levels at 25% yield decrement | |
|---|---|---|---|
| | | Edible parts | (leaves) |
| | ppm Cd | ppm Cd | |
| *Spinacea oleracea* | 4 | 75 | – |
| *Glycine max.* | 5 | 7 | (7) |
| *Lepidium sativum* | 8 | 80 | – |
| *Lactuca sativa* | 13 | 70 | – |
| *Zea mays* | 18 | 2 | (35) |
| *Daucus carota* | 20 | 19 | (32) |
| *Brassica rapa* | 28 | 15 | (121) |
| *Phaseolus vulgaris* | 40 | 2 | (15) |
| *Triticum aestivum* | 50 | 11 | (33) |
| *Raphanus sativus* | 96 | 21 | (75) |
| *Lycopersicon esculentum* | 160 | 7 | (125) |
| *Brassica oleracea* | 170 | 11 | – |
| *Oryza sativa* | 640 | 2 | (3) |

## 5.5 Concluding Remarks

Concern about the accumulation of certain heavy metals in the food chain, with its consequent potential danger for human and animal nutrition is fully justifiable. Interdisciplinary research programmes are necessary, as well as governmental restrictive measures for the better control of Cd contamination at the source site, including the use of municipal sewage sludge. It is necessary, however, on the other hand to bear in mind that heavy metals are natural components of all ecosystems and cannot be simply considered as environmental pollutants. Increasing the amounts and availability of essential heavy metals such as Zn, Fe, Mo or Mn in soils and plants can be important for both increased plant production and plant quality for animal and human nutrition (see Sect. 1). Research and discussion on heavy metals have to differentiate between these various aspects and take this into consideration in comparative evaluations.

*Acknowledgement.* The author thanks Mr. E.A. KIRKBY for valuable comments on the manuscript.

# References

Allen RD (1969) Mechanism of the seismonastic reaction in *Mimosa pudica*. Plant Physiol 44:1101–1107

Amberger A, Gutser R (1978) Umsatz und Wirkung von Harnstoff-Dicyandiamid sowie Ammonsulfat-Dicyandiamid-Produkten zu Weidelgras und Reis. Z Pflanzenernaehr Bodenkd 141:553–566

Andersson A (1977a) The distribution of heavy metals in soils and soil material as influenced by the ionic radius. Swed J Agric Res 7:79–93
Andersson A (1977b) Some aspects on the significance of heavy metals in sewage sludge and related products used as fertilizers. Swed J Agric Res 7:1–5
Antonivics J, Bradshaw AD, Turner RG (1971) Heavy metal tolerance in plants. Adv Ecol Res 7:1–85
Armstrong MJ, Kirkby EA (1979) The influence of humidity on the mineral composition of tomato plants with special reference to calcium distribution. Plant Soil 52:427–435
Arnon DI (1950) Criteria of essentiality of inorganic micronutrients for plants. In: McElroy WD, Glass B (eds) Trace elements in plant physiology. Chronica Botanica, Waltham, Mass
Bangerth F (1976) A role of auxin and auxin transport inhibitors on the Ca content of artificially induced parthenocarpic fruits. Physiol Plant 37:191–194
Bangerth F (1979) Calcium-related physiological disorders of plants. Annu Rev Phytopathol 1:97–122
Bangerth F, Mostafawi M (1969) Einfluß der Wasserversorgung und des Fruchtgewichtes auf den Mineralstoffgehalt und die Stippigkeit von Apfelfrüchten. Erwerbsobstbau 11:101–104
Bangerth F, Dilley DR, Dewey DH (1972) Effect of postharvest calcium treatment on internal breakdown and respiration of apple fruit. J Am Soc Hortic Sci 97:679–682
Barber SA (1979) Growth experiments for nutrients in relation to demand at the root surface. In: Harley JL, Scott Russell R (eds) The soil–root interface. Academic Press, London New York
Bateman DF, Lumsden RD (1965) Relation between calcium content and nature of the pectic substances in bean hypocotyls of different ages to susceptibility to an isolate of *Rhizoctonia solani*. Phytopathology 55:734–738
Bergmann W, Neubert P (1976) Pflanzendiagnose und Pflanzenanalyse. Fischer, Jena
Bhat KKS, Nye PH (1974) Diffusion of phosphorus to plant roots in soil. III. Depletion around onion roots without root hairs. Plant Soil 41:383–394
Bingham FT, Page AL, Mahler RJ, Ganje TJ (1975) Growth and cadmium accumulation of plants grown on a soil treated with a cadmium-enriched sewage sludge. J Environ Qual 4:207–211
Bingham FT, Page AL, Mahler RJ, Ganje TJ (1976) Yield and cadmium accumulation of forage species in relation to cadmium content of sludge-amended soil. J Environ Qual 5:57–60
Boggess SF, Willavize S, Koeppe DW (1978) Differential response of soybean varieties to soil cadmium. Agron J 70:756–760
Bould D, Parfitt RI (1973) Leaf analysis as a guide to the nutrition of fruit crops. X. Magnesium and phosphorus sand culture experiments with apple. J Sci Food Agric 24:175–185
Bradfield EG (1976) Calcium complexes in the xylem sap of apple shoots. Plant Soil 44:495–499
Bradfield EG, Guttridge CG (1979) Dependence of calcium transport and leaf tipburn in strawberry on relative humidity and nutrient solution concentration. Ann Bot 43:363–372
Brewster JL, Tinker PB (1970) Nutrient cation flows in soil around plant roots. Soil Sci Soc Am Proc 34:421–426
Brown JC (1978) Mechanism of iron uptake by plants. Plant Cell Environ 1:249–257
Brownell PF (1979) Sodium as an essential micronutrient element for plants and its possible role in metabolism. Adv Bot Res 7:117–224
Brownell PF, Jackman ME (1966) Changes during recovery from sodium deficiency in *Atriplex*. Plant Physiol 41:617–622
Brownlee C, Kendrick RE (1979) Ion fluxes and phytochrome protons in mung bean hypocotyl segments. I. Fluxes of potassium. Plant Physiol 64:206–210
Broyer TC, Carlton AB, Johnson CM, Stout PR (1954) Chlorine – a micronutrient element for higher plants. Plant Physiol 29:526–532
Bruinsma J (1977) Rolle der Cytokinine bei Blüten- und Fruchtentwicklung. Z Pflanzenernähr Bodenkd 140:15–23

Brumagen DM, Hiatt AH (1966) The relationship of oxalic acid to the translocation and utilization of calcium in *Nicotiana tabacum*. Plant Soil 24:239–249
Buban T, Vaega A, Traub J, Kuegt E, Bruinsma J (1978) Effects of ammonium and nitrate nutrition on the level of xylem sap of apple rootstocks. Z Pflanzenphysiol 89:289–295
Byrne AE, Ravnik V (1976) Trace element concentration in higher fungi. Sci Total Environ 6:65–78
Campell LG, Miller MH, Loneragan JF (1975) Translocation of boron to plant fruits. Aust J Plant Physiol 2:481–487
Campbell NA, Stika KM, Morrison GH (1979) Calcium and potassium in the motor organ of sensitive plants: Localization by ion microscopy. Science 204:185–187
Cannon HL (1963) The biochemistry of vanadium. Soil Sci 96:196–204
Carmi A, Koller D (1979) Regulation of photosynthetic activity in the primary leaves of bean (*Phaseolus vulgaris* L.) by materials moving in the water-conducting system. Plant Physiol 64:285–288
Cartwright B, Merry RH, Tiller KG (1976) Heavy metal contamination of soils around a lead smelter at Port Pirie, South Australia. Aust J Soil Res 15:69–81
Cataldo DA, Garland TR, Wildung RE, Drucker H (1978) Nickel in plants. II. Distribution and chemical form in soybean plants. Plant Physiol 62:566–570
Chaney RL, Hornick SB (1978) Accumulation and effects of cadmium on crops. In: Proc 1st Int Cadmium Conf Met Bull London
Chaney RL, Stoewsand GS, Bache CA, Lisk DJ (1978a) Cadmium deposition and hepatic microsomal induction in mice fed lettuce grown on municipal sludge-amended soil. J Agric Food Chem 26:992–994
Chaney RL, Stoewsand GS, Furv AK, Backe CA, Lisk DJ (1978b) Elemental content of tissue of guinea pig fed swiss chard grown on municipal sewage sludge-amended soil. J Agric Food Chem 26:994–997
Chen CH, Lewin J (1969) Silicon as a nutrient element for *Equisetum arvense*. Can J Bot 47:125–131
Collier GF, Huntington VD, Cox EF (1979) A possible role of chlorogeneic acid in calcium-related disorders of vegetable crops. Commun Soil Sci Plant Anal 10:481–490
Cooper RB, Blaser RE, Brown RH (1967) Potassium nutrition effects on net photosynthesis and morphology of alfalfa. Soil Sci Soc Am Proc 31:231–235
Corden ME (1965) Influence of calcium nutrition of fusarium wilt of tomato and polygalacturonase activity. Phytopathology 55:222–224
Crittenden HW, Svec CV (1974) Effect of potassium on the incidence of *Diaporthe sojae* in soybean. Agron J 66:696–698
Diez Th, Rosopulo A (1976) Schwermetallgehalte in Böden und Pflanzen nach extrem hohen Klärschlammgaben. Landwirtsch Forsch Sonderheft 22/1:236–248
Dixon B, Sagar GR, Shorrocks VM (1973) Effect of calcium and boron on the incidence of tree and storage pit in apples of the cultivar Egremont Russel. Hort Sci 48:403–411
Doll S, Rodier F, Willenbrink J (1979) Accumulation of sucrose in vacuoles isolated from red beet tissue. Planta 144:407–411
Doman DC, Geiger DR (1979) Effect of exogenously supplied foliar potassium on phloem-loading in *Beta vulgaris* L. Plant Physiol 64:528–533
Domsch KH, Grabbe K, Fleckenstein J (1976) Schwermetallgehalte im Kultursubstrat und Erntegut des Champignons, *Agaricus bisporus* (Lange) Singer, beim Einsatz von Müllklärschlammkompost. Z Pflanzenernaehr Bodenkd 139:487–501
Doyle JJ (1977) Effects of low levels of dietary cadmium in animals – a review. J Environ Qual 6:111–116
Egmond van F, Houba VJG (1970) Production of carboxylates (C-A) by young sugar-beet plants grown in nutrient solution. Neth J Agric Sci 18:182–187
El-Sheikh AM, Ulrich A, Broyer TC (1967) Sodium and rubidium as possible nutrients for sugar beet plants. Plant Physiol 42:1202–1208
Ende van der J, Koornneef P, Sonneveld C (1975) Osmotic pressure of the soil solution. Determination and effects on some glass-house crops. Neth J Agric Sci 23:181–190
Epstein E (1972) Mineral nutrition of plants: Principles and perspectives. Wiley and Sons, New York

Epstein E, Hagen CE (1952) A kinetic study of the absorption of alkali cations by barley roots. Plant Physiol 27:457–474
Epstein E, Norlyn JD, Rush EW, Kingsburg RW, Kelley DB, Cunningham GA, Wrona AF (1980) Saline culture of crops: A genetic approach. Science 210:399–404
Ernst W (1974) Schwermetallvegetation der Erde. Fischer, Stuttgart
Faust M, Klein JD (1974) Levels and sites of metabolically active calcium in apple fruit. J Am Soc Hort Sci 99:93–94
Faust M, Shear CB (1969) Biochemical changes during the development of cork spot of apples. Qual Plant Mater Veg 19:225–265
Ferguson IB, Bollard EG (1976) The movement of calcium in woody stems. Ann Bot (London) 40:1057–1065
Fidanovski F (1869) Der Einfluß von Silicium auf Pflanzen. Diss, TU Berlin, D 83 Nr 256
Foroughi M, Venter F (1978) Die Wirkung unterschiedlicher Cadmiumgaben auf die Eisenversorgung einiger Gemüsearten. Landwirtsch Forsch 31:300–308
Foroughi M, Teicher K, Venter F (1979) Die Wirkung steigender Gaben von Blei, Cadmium, Nickel oder Zink auf Spinat in Nährlösung. Landwirtsch Forsch Sonderheft 35:599–606
Frissel MJ (1977) Cycling of mineral nutrients in agricultural ecosystems. Agro-Ecosyst 4:1–354
Fritz PD, Foroughi M, Venter F (1977) Schwermetallgehalte in einigen Gemüsearten. Landwirtsch Forsch Sonderheft 33/II:335–343
Fuente de la RK, Leopold AC (1972) Two components of auxin transport. Plant Physiol 50:491–495
Garg OK, Sharma AN, Kona GRSS (1979) Effect of boron on the pollen vitality and yield of rice plant (*Oryza sativa* L. var. Jaya) Plant Soil 52:591–594
Garrod JF (1974) The role of gibberellins in early growth and development of sugar beet. J Exp Bot 25:945–954
Gauch HG (1972) Inorganic plant nutrition. Dowden Hutchinson & Ross, Strauchburg, Pa
Geijn van de SC, Petit CM, Roelfsen H (1979) Measurement of the cation exchange capacity of the transport system in intact stems. Methodology and preliminary results. Commun Soil Sci Plant Anal 10:225–236
Göring H, Mardanov AA (1976) Einfluß von Stickstoffmangel auf das K/Ca-Verhältnis und den Zytokiningehalt junger Kürbispflanzen. Biochem Physiol Pflanz 170:261–264
Göring H, Thien BH (1979) Influence of nutrient deficiency on proline accumulation in the cytoplasm of *Zea mays* L. seedlings. Biochem Physiol Pflanz 174:9–16
Goldbach E, Goldbach H, Wagner H, Michael G (1975) Influence of N-deficiency on the abscisic acid content of sunflower plant. Physiol Plant 34:138–140
Goldbach H, Goldbach E, Michael G (1977) Transport of abscisic acid from leaves to grains in wheat and barley plant. Naturwissenschaften 64:488
Goor van BJ, Wiersma D (1974) Redistribution of potassium, calcium, magnesium and manganese in the plant. Physiol Plant 31:163–168
Graham RD (1975) Male sterility in wheat plants deficient in copper. Nature (London) 254:514–515
Graham RD (1976) Anomalous water relations in copper-deficient wheat plants. Aust J Plant Physiol 3:229–236
Grasmanis VO, Edwards GE (1974) Promotion of flower initiation in apple trees by short exposure to the ammonium ion. Aust J Plant Physiol 1:99–105
Hallock DL, Garren KH (1968) Pod breakdown, yield and grade of Virginia-type peanuts as affected by Ca, Mg and K sulfates. Agron J 60:253–257
Hambidge KM, Wallravens PA (1976) Zinc deficiency in infants and preadolescent children. In: Prasad AA, Overleas D (eds) Trace elements in human health and disease, vol I. Zinc and copper. Academic Press, London New York
Hara T, Sonoda Y (1979) Comparison of the toxicity of heavy metals to cabbage growth. Plant Soil 51:127–133

Hawker JS, Marschner H, Downton WJS (1974) Effects of sodium and potassium on starch synthesis in leaves. Aust J Plant Physiol 1:491–501
Heldt HW, Chon CH, Maronde D, Herold A, Stankovic ZS, Walker DA, Kraminer A, Kirk MR, Heber U (1977) Role of orthophosphate and other factors in the regulation of starch formation in leaves and isolated chloroplasts. Plant Physiol 59:1146–1155
Helms K, David DJ (1973) Far red and white light-promoted utilization of calcium by seedlings of *Phaseolus vulgaris* L. Plant Physiol 51:37–42
Herms U, Brümmer G (1979) Einfluß der Redoxbedingungen auf die Löslichkeit von Schwermetallen in Böden und Sedimenten. Mitt Dtsch Bodenkdl Ges 29:533–544
Herridge DF, Pate JS (1977) Utilization of net photosynthesis for nitrogen fixation and protein production in an annual legume. Plant Physiol 60:759–764
Hewitt EJ, Smith TA (1975) Plant mineral nutrition. Engl Univ Press, London
Hildebrandt EA (1977) Das Vorkommen von Nitrosaminen in Nahrungsmitteln und die Folgerungen für die Pflanzenernährung. Z Pflanzenernaehr Bodenkd 140:397–408
Hill J, Robson AD, Loneragan JF (1978) The effect of copper and nitrogen supply on the retranslocation of copper in four cultivars of wheat. Austr J Agric Res 29:925–939
Hill J, Robson AD, Loneragan JF (1979) The effects of Cu supply and shading on Cu retranslocation from old wheat leaves. Ann Bot 43:449–457
Hinsley TD, Alexander DE, Ziegler EC, Barrett GL (1978) Zinc and cadmium accumulation by corn inbreds grown on sludge-amended soil. Agron J 70:425–428
Hocking PJ, Pate JS (1977) Mobilization of minerals to developing seeds of legumes. Ann Bot 41:1259–1278
Hocking PJ, Pate JS (1978) Accumulation and distribution of mineral elements in annual lupins *Lupinus albus* and *Lupinus angustifolius* L. Aust J Agric Res 29:267–280
Hocking PJ, Pate JS, Wee GC, McComb AJ (1977) Manganese nutrition of *Lupinus* ssp. especially in relation to developing seeds. Ann Bot 41:677–688
Hocking PJ, Pate JS, Atkins CA, Sharkey PJ (1978) Diurnal patterns of transport and accumulation of minerals in fruiting plants of *Lupinus angustifolius* L. Ann Bot 42:1277–1290
Höll W, Hampp R (1975) Lead and plants. Residue Rev 54:79–111
Horst WJ, Marschner H (1978a) Effect of silicon on manganese tolerance of bean plants (*Phaseolus vulgaris* L.). Plant Soil 50:287–303
Horst WJ, Marschner H (1978b) Effect of excessive manganese supply on uptake and translocation of calcium in bean plants (*Phaseolus vulgaris* L.). Z Pflanzenphysiol 87:137–148
Howeler RH (1974) Iron-induced oranging disease of rice in relation to physical and chemical changes in a flooded oxisol. Soil. Sci Soc Am Proc 37:898–903
Hsiao TC (1976) Stomatal ion transport. In: Lüttge U, Pitman MG (eds) Transport in plants II, Part B. Encyclopedia of plant physiology new ser, vol II. Springer, Berlin Heidelberg New York
Hsu FC (1979) Abscisic acid accumulation in developing seeds of *Phaseolus vulgaris* L. Plant Physiol 63:552–556
Huffman EWD, Allaway WH (1973) Growth of plants in solution culture containing low levels of chromium. Plant Physiol 52:72–75
Humble GD, Raschke K (1971) Stomatal opening quantitatively related to potassium transport. Evidence from electron probe analysis. Plant Physiol 48:447–453
Hutchings VM (1978) Sucrose and proton cotransport in Ricinus cotyledons. II. $H^+$ efflux and associated $K^+$ uptake. Planta 138:237–241
Hutton JT, Norrish K (1974) Silicon content of wheat husks in relation to water transpired. Aust J Agric Res 25:203–212
Isermann I (1978) Einfluß von Chelatoren auf die Calcium-Verlagerung im Sproß höherer Pflanzen. Z Pflanzenernaehr Bodenkd 141:285–298
Itai C, Meidner H (1978) Functional epidermal cells are necessary for abscisic acid effects on guard cells. J Exp Bot 29:765–770

Jager de A (1979) Localized stimulation of root growth and phosphate uptake in *Zea mays* L. resulting from restricted phosphate supply. In: Harley JL, Scott Russell R (eds) The soil-root interface. Academic Press, London New York
Janistyn B (1973) Indol-3-Essigsäure induzierte Calciumionenabgabe bei Maiskoleoptilzylindern. Z Naturforsch 28c:777–778
Jennings DH (1976) The effects of sodium chloride on higher plants. Biol Rev 51:453–486
John MK, Laerhoven van CJ (1976) Differential effects of cadmium on lettuce varieties. Environ Pollut 10:163–173
Johnson CM, Stout PR, Broyer TC, Carlton AB (1957) Comparative chlorine requirements of different plants species. Plant Soil 8:337–353
Jones LHP, Handreck KA (1965) Studies of silica in the oat plant. III. Uptake of silica from soils by the plant. Plant Soil 23:79–96
Jung J, Isermann K, Henges G (1979) Einfluß von cadmiumhaltigen Düngerphosphaten auf die Cadmiumanreicherung von Kulturböden und Nutzpflanzen. Landwirtsch Forsch 32:262–274
Kelley PM, Izawa S (1978) The role of chloride ion in photosystem-II. I. Effects of chloride ion on photosystem-II electron transport and on hydroxylamine inhibition. Biochim Biophys Acta 502:198–210
Kick H, Braun B (1977) Wirkung von chromhaltigen Gerbereischlämmen auf Wachstum und Chromaufnahme bei verschiedenen Nutzpflanzen. Landw Forsch 30:160–173
Kirkby EA, Mengel K (1967) Ionic balance in different tissues of the tomato plant in relation to nitrate, urea or ammonium nutrition. Plant Physiol 42:6–14
Kirkham MB (1978) Water relations of cadmium-treated plants. J Environ Qual 7:334–336
Kloke A, Riebartsch K, Leh HO (1966) Verunreinigungen von Kulturpflanzen mit Blei aus Kraftfahrzeugabgasen. Landwirtsch Forsch Sonderheft 20:119–123
Knauer N, Simon C (1968) Über den Einfluß der Stickstoffdüngung auf den Ertrag sowie Nitrat-Mineralstoff- und Oxalsäuregehalt von Spinat. Z Acker Pflanzenbau 128:197–220
Knight AH, Crooke WM (1973) Cation exchange capacity and chemical composition of the floral parts of *Antirrhinum* and *Lilium*. Ann Bot 37:155–157
Koontz HV, Foote RE (1966) Transpiration and calcium deposition by unifoliate leaves of *Phaseolus vulgaris* differing in maturity. Physiol Plant 14:313–321
Krause W (1958) Andere Bodenspezialisten. In: Michael G (ed) Handbuch der Pflanzenphysiologie, vol IV Springer, Berlin Göttingen Heidelberg
Krauss A (1971) Einfluß der Ernährung des Salates mit Massennährstoffen auf den Befall mit *Botrytis cinera* Pers. Z Pflanzenernaehr Bodenkd 128:12–23
Krauss A (1978) Tuberization and abscisic acid content in *Solanum tuberosum* as affected by nitrogen nutrition. Potato Res 21:183–193
Krauss A, Marschner H (1971) Einfluß der Stickstoffernährung der Kartoffeln auf Induktion und Wachstumsrate der Knollen. Z Pflanzenernaehr Bodenkd 128:153–168
Krauss A, Marschner H (1973) Langstreckentransport von Calcium in Stolonen von Kartoffelpflanzen. Z Pflanzenernaehr Bodenkd 136:229–240
Krauss A, Marschner H (1974) Einfluß der Tag/Nacht-Periodik auf Knollengewicht und Ca-Verlagerung in Stolonen von Kartoffelpflanzen. Z Pflanzenernaehr Bodenkd 137:116–123
Krauss A, Marschner H (1975) Einfluß des Calciumangebots auf Wachstumsrate und Calciumgehalt von Kartoffelknollen. Z Pflanzenernaehr Bodenkd 139:317–326
Krauss A, Marschner H (1976) Einfluß von Stickstoffernährung und Wuchsstoffapplikation auf die Knolleninduktion bei Kartoffelpflanzen. Z Pflanzenernaehr Bodenkd 139:143–155
Krug H, Wiebe H-J, Jungk A (1972) Calciummangel an Blumenkohl unter konstanten Klimabedingungen. Z Pflanzenernaehr Bodenkd 133:213–226
Lagerwerff JV, Milberg RP (1978) Sign of charge of species of Cu, Cd and Zn extracted from sewage sludge and effect of plants. Plant and Soil 49:117–125
Lagerwerff JV, Specht AW (1970) Contamination of roadside soil and vegetation with cadmium, nickel, lead and zinc. Environ Sci Technol 4:583–586

Lagerwerff JV, Biersdorf GT, Milberg RP, Brower DL (1977) Effects of incubation and liming on yield and heavy metal uptake by rye from sewage-sludged soil. J Environ Qual 6:427–431
Lamoreaux RJ, Chaney WC (1977) Growth and water movement in silver maple seedlings affected by cadmium. J Environ Qual 6:201–205
Laske P (1979) Gehalt gärtnerischer Böden und Erden an löslichem Vanadium. Kali-Briefe 14:747–752
Läuchli A (1967) Nachweis von Calcium- und Strontium-Ablagerungen im Fruchtstiel von *Pisum sativum* mit der Röntgen-Mikrosonde. Planta 73:221–227
Läuchli A, Pflüger R (1978) Potassium transport through plant cell membranes and metabolic role of potassium in plants. In: Proc 11th Congr Int Potash Inst, Bern
Lee J, Reeves RD, Brooks RR, Jafre T (1978) The relation between nickel and citric acid in some nickel-accumulating plants. Phytochemistry 17:1033–1035
Lenz F (1970) Einfluß der Früchte auf das Wachstum, den Wasserverbrauch und die Nährstoffaufnahme von Auberginen. Gartenbauwissenschaft 35:281–292
Lenz F, Bünemann G (1969) Einfluß von Ernährung, Fruchtbehang und Ausläufern auf die Trockenstubstanzbildung und die Wasser- und Nährstoffaufnahme bei Erdbeeren. Erwerbsobstbau 11:185–188
Lenz F, Döring HW (1975) Fruit effects on growth and water consumption in citrus. Gartenbauwissenschaft 40:257–260
Lenz F, Williams CN (1973) Effect of fruit removal on net assimilation and gaseous diffusive resistance of soybean leaves. Angew Bot 47:57–63
Liegel W (1970) Calciumoxalat-Abscheidung in Fruchtstielen einiger Apfelvaritäten. Angew Bot 44:223–232
Lingle JC, Lorenz OA (1969) Potassium nutrition of tomatoes. J Am Soc Hortic Sci 94:679–683
Linser H, Raafat A, Zeid FA (1974) Reinprotein und Chlorophyll bei *Daucus carota* im Verlauf der Vegetationsperiode des ersten Jahres unter dem Einfluß von Wachstumsregulatoren (CCC, GA, BA). Z Pflanzenernaehr Bodenkd 137:36–48
Loneragan JF (1978) Anomalies in the relationship of nutrient concentrations to plant yield. 8th Int Colloq Plant Anal Fert Probl Auckland, NZ
Loneragan JF, Snowball K (1969) Calcium requirements of plants. Aust J Agric Res 20:465–478
Loneragan JF, Snowball K, Simmons WJ (1968) Response of plants to calcium concentration in solution culture. Aust J Agric Res 19:845–857
Loneragan JF, Snowball K, Robson AD (1976) Remobilization of nutrients and its significance in plant nutrition. In: Transport and transfer processes in plants, Chap 39. Academic Press, London New York
Lorenz H, Kossen MTh, Käferstein FK (1978) Blei-, Cadmium- und Quecksilbergehalte in Speisepilzen. Bundesgesundheitsblatt 21:202–204
Lott JNA, Buttrose MS (1978) Globoids in protein bodies of legume seed cotyledons. Aust J Plant Physiol 5:98–111
Lougheed EC, Murr DP, Miller SR (1979) Effects of calcium and daminozide on ethylene production and softening of apple fruits. Experimenta 35:43–45
Lüdders P, Bünemann G (1970) Die Wirkung des Zeitpunktes von Harnstoffspritzungen auf Apfelbäume. Z Pflanzenernaehr Bodenkd 125:144–155
Lüdders P, Bünemann G, Ortmann U (1976) Die Wirkung jahreszeitlich unterschiedlicher Kaliumverfügbarkeit auf Apfelbäume. IX. Einfluß auf den Mineralstoffgehalt ausdauernder Baumteile. Gartenbauwissenschaft 41:260–269
Lund ZF, Doss BD, Lowry FE (1975) Dairy cattle manure – Its effect on yield and quality of coastal bermuda grass. J Environ Qual 4:358–362
Marschner H (1971) Why can sodium replace potassium in plants? Proc 8th Colloq Int Potash Inst Bern
Marschner H, Ossenberg-Neuhaus H (1977) Wirkung von 2,3,5-Trijodbenzoesäure (TIBA) auf den Calciumtransport und die Kationenaustauschkapazität in Sonnenblumen. Z Pflanzenphysiol 85:29–44
Marschner H, Possingham JV (1975) Effect of $K^+$ and $Na^+$ on growth of leaf discs of sugar beet and spinach. Z Pflanzenphysiol 85:6–16

Marschner H, Richter Ch (1974) Calciumtransport in Wurzeln von Mais- und Bohnenkeimpflanzen. Plant Soil 40:193–210
Marschner H, Römheld V, Azarabadi S (1978) Iron stress response of efficient and inefficient plant species. In: Proc 8th Colloq Plant Anal Fert Probl. Auckland, NZ
Martin G (1965) Nécessité du chlore dans nutrition de *Spirodela polyrhiza* cultivé en conditions hétérotrophes. CR Acad Sci 260:5928–5930
Martin MH, Coughtrey PJ (1978) Preliminary observations on the level of cadmium in a contaminated environment. Chemosphere 3:155–160
Martin P (1971) Wanderwege des Stickstoffs in Buschbohnenpflanzen bei Aufwärtstransport nach der Aufnahme durch die Wurzeln. Z Pflanzenphysiol 64:206–222
Mathys W (1975) Enzymes of heavy metal resistant and non-resistant populations of *Silene cucubalus* and their interaction with some heavy metals in vitro and in vivo. Physiol Plant 33:161–165
Mattoo AK, Liebermann M (1977) Localization of the ethylene-synthesizing system in apple tissue. Plant Physiol 60:794–799
Meisch HK, Schmitt JA, Reinle W (1977) Schwermetalle in höheren Pilzen. Cadmium, Zink und Kupfer. Z Naturforsch 32c:172–181
Menary RC, Staden van J (1976) Effect of phosphorus nutrition and cytokinins on flowering in the tomato, *Lycopersicon esculentum* L. Aust J Plant Physiol 3:201–205
Mengel K (1979) Ernährung und Stoffwechsel der Pflanzen, 5th edn. Fischer, Jena
Mengel K, Haeder H-E (1977) Effect of potassium supply on the rate of phloem sap exudation and the composition of phloem sap of *Ricinus communis*. Plant Physiol 59:282–284
Mengel K, Kirby EA (1978) Principles of plant nutrition. Int Potash Inst, Bern
Mengel K, Grimme H, Nemeth K (1969) Potentielle und effektive Verfügbarkeit von Pflanzennährstoffen im Boden. Landwirtsch Forsch 23:(16. Sonderheft) 79–91
Mengel K, Haghparast MR, Koch K (1974) The effect of potassium on the fixation of molecular nitrogen by root nodules of *Vicia faba*. Plant Physiol 54:535–538
Michael G, Beringer H (1980) The role of hormones in yield formation. In: Proc 15th Int Potash Colloq, Wageningen
Michael G, Koushiahi-Tork K, Wilberg E (1969a) Einfluß unterschiedlicher Luftfeuchtigkeit auf Chlorophyll- und Eiweißabbau in Blättern von Tabakpflanzen. Flora A 160:186–195
Michael G, Mounla MAKh, Goldbach H (1974) Nitrogen fertilizer application, hormone activity and crop production. Plant Anal Fert Probl Proc 7th Int Colloq, Hannover
Michael G, Schultz R, Wagner H (1969b) Modellversuch zur Bedeutung der Granne für die Versorgung des Gerstenkornes mit Cytokininen. Flora A 160:306–316
Micheal B, Zink F, Lantzsch HJ (1980) Effect of phosphate application on phytin-P and other phosphate fractions in developing wheat grains. Z Pflanzenernaehr Bodenkd 369–376
Millikan CR, Hanger BC (1969) Movement of a foliar-applied $^{45}$Ca in Brussels sprouts. Austr J Biol Sci 22:545–558
Mitchell GA, Bingham ET, Page AL (1978) Yield and metal composition of lettuce and wheat grown on soils amended with sewage sludge enriched with cadmium, copper, nickel and zinc. J Environ Qual 7:165–171
Mix GP, Marschner H (1976a) Calciumgehalte in Früchten von Paprika, Bohne, Quitte und Hagebutte im Verlauf des Fruchtwachstums. Z Pflanzenernaehr Bodenkd 139:537–549
Mix GP, Marschner H (1976b) Einfluß exogener und endogener Faktoren auf den Calciumgehalt von Paprika- und Bohnenfrüchten. Z Pflanzenernaehr Bodenkd 139:551–563
Mix GP, Marschner H (1976c) Calcium-Umlagerung in Bohnenfrüchten während des Samenwachstums. Z Pflanzenphysiol 80:354–366
Miyake Y, Takahashi E (1978) Silicon deficiency of tomato plants. Soil Sci Plant Nutr 24:175–189
Mohamed AH, Gnanam A (1979) A possible mechanism of ammonium ion regulation of photosynthetic carbon flow in higher plants. Plant Physiol 64:263–268
Monselise SP, Varga A, Knegt E, Bruinsma J (1978) Course of the zeatin content in

tomato fruits and seeds developing on intact or partially defoliated plants. Z Pflanzenphysiol 90:451–460
Mostafa MAE, Ulrich A (1976) Absorption, distribution and form of Ca in relation to Ca deficiency (tip burn) of sugar beet. Crop Sci 16:27–30
Mothes K (1932) Ernährung, Struktur und Transpiration. Ein Beitrag zur kausalen Analyse der Xeromorphosen. Biol Zentralbl 52:193–223
Naumann WD (1971) Calciumverteilung in Äpfeln und Entwicklung von Stippigkeit unter dem Einfluß von Wachstumsregulatoren. Gartenbauwissenschaft 36:63–68
Nitsch JP (1950) Growth and morphogenesis of strawberries as related to auxin. Annu J Bot 37:211–215
Ocker H-D (1977/78) Gehalte und Verteilung von Schwermetallen in Getreide und Getreideprodukten. Ber Landwirtsch 55:796–808
Oertli JJ, Richardson WF (1970) The mechanism of boron immobility in plants. Physiol Plant 23:108–116
Oldenkamp L, Smilde KW (1966) Copper deficiency in douglas fir (*Pseudotsuga menziesii* Mirb. Franco). Plant Soil 25:150–152
Overleas D (1973) Phytates. In: Toxicants occurring naturally in foods. Comm Food Protec Nutr Board, Natl Res Counc. Natl Acad Sci, Washington
Page AL, Bingham FT (1973) Cadmium residues in the environment. Residue Rev 48:1–44
Palzkill DA, Tibbitts TW, Williams PH (1976) Enhancement of calcium transport to inner leaves of cabbage for prevention of tipburn. J Am Soc Hortic Sci 101:645–648
Pate JS (1976) Exchange of solutes between phloem and xylem and circulation in the whole plant. In: Zimmermann MH, Milburn JA (eds) Phloem transport. Encyclopedia of plant physiology, new ser vol I. Springer, Berlin Heidelberg New York
Pate JS, Hocking PJ (1978) Phloem and xylem transport in the supply of minerals to a developing legume (*Lupinus albus* L.) fruit. Ann Bot 42:911–921
Pate JS, Sharkey PJ, Lewis OAM (1976a) Phloem bleeding from legume fruits – a technique for study of fruit nutrition. Planta 120:229–243
Pate JS, Sharkey PJ, Lewis OAM (1974b) Xylem to phloem transfer of solutes in fruiting shoots of legumes, studies by a phloem-bleeding technique. Planta 122:11–26
Pate JS, Kuo J, Hocking PJ (1978) Functioning of conducting elements of phloem and xylem in the stalk of the developing fruit of *Lupinus albus* L. Aust J Plant Physiol 5:321–326
Peoples TR, Koch DW (1979) Role of potassium in carbon dioxide assimilation in *Medicago sativa* L. Plant Physiol 63:878–881
Perrenoud S (1977) Potassium and plant-health. Potash Inst Bern
Petit CM, Ringoet A, Myttenaere C (1978) Stimulation of cadmium uptake in relation to the cadmium content of plants. Plant Physiol 62:554–557
Pettersson A (1976) Heavy metal ion uptake by plants from solutions with metal ion, plant species and growth period variations. Plant Soil 45:445–459
Pettersson A (1977) Differences in cadmium uptake between plant species and cultivars. Swed J Agric Res 7:21–24
Pflüger R, Wiedemann R (1977) Der Einfluß monovalenter Kationen auf die Nitratreduktion von *Spinacia oleracea* L. Z Pflanzenphysiol 85:125–133
Pietz RI, Vetter RJ, Masarik D, McFee WW (1978) Zinc and cadmium contents of agricultural soils and corn in Northwestern Indiana. J Environ Qual 7:381–385
Poovaiah BW (1979) Role of calcium in ripening and senescence. Commun Soil Sci Plant Anal 10:83–88
Poovaiah BW, Leopold AC (1973a) Deferral of leaf senescence with calcium. Plant Physiol 52:236–239
Poovaiah BW, Leopold AC (1973b) Inhibition of abscission by calcium. Plant Physiol 51:848–851
Poovaiah BW, Leopold AC (1976) Effect of inorganic solutes on the bindings of auxin. Plant Physiol 58:783–785
Priestley CA (1976) Some effects of ringing and deblossoming branches of young apple trees in leaf composition. J Exp Bot 27:1325–1332

Quinlan JD, Preston AP (1973) Effect of shoot tipping on fruiting and apple tree growth. East Malling Res Stn Rep
Rahimi A, Bussler W (1973a) Die Diagnose des Kupfermangels mittels sichtbarer Symptome an höheren Pflanzen. Z Pflanzenernaehr Bodenkd 135:267–283
Rahimi A, Bussler W (1973b) Physiologische Voraussetzungen für die Bildung der Kupfermangelsymptome. Z Pflanzenernaehr Bodenkd 136:25–32
Rahimi A, Bussler W (1973c) Die Wirkung von Kupfermangel auf die Gewebestruktur höherer Pflanzen. Z Pflanzenernaehr Bodenkd 135:183–195
Rahman MS, Wilson JH (1977) Effect of phosphorus applied as superphosphate on rate of development and spikelet number. Austr J Agric Res 28:183–186
Rajagopal V, Rao IM (1974) Changes in the endogenous level of auxins and gibberellin-like substances in the shoot apices of nitrogen-deficient tomato plants (*Lycopersicon esculentum* Mill). Aust J Bot 22:429–435
Ramon Garcia L, Hanway JJ (1976) Foliar fertilization of soybean during the seed-filling period. Agron J 68:653–657
Rapper CD, Patterson DT, Parsons LR, Kramer PJ (1977) Relative growth and nutrient accumulation rates for tobacco. Plant Soil 46:473–486
Raschke K (1979) Movements of stomata. In: Haupt W, Feinleib ME (eds) Physiology of movements. Encyclopedia of plant physiology new ser vol VII. Springer, Berlin Heidelberg New York
Raven JA (1977) $H^+$ and $Ca^{2+}$ in phloem and symplast. Relation of relative immobility of the ions to the cytoplasmic nature of the transport path. New Phytol 79:465–480
Reinhold JG, Nasr K, Lahimgarzadeh A, Hedayati H (1973) Effect of purified phytate and phytate-rich bread upon metabolism of zinc, calcium, phosphorus and nitrogen in man. Lancet 1:283–291
Reuss JO, Dooley HL, Grittis W (1978) Uptake of cadmium from phosphate fertilizers by pear, radishes and lettuce. J Environ Qual 7:128–133
Riley D, Barber SA (1971) Effect of ammonium and nitrate fertilization on phosphorus uptake as related to root-induced pH changes at the root–soil interface. Soil Sci Soc Am Proc 35:301–306
Ringoet A, Sauer G, Gielink AJ (1968) Phloem transport of calcium in oat leaves. Planta 80:15–20
Römheld V, Marschner H (1979) Fine regulation of iron uptake by the Fe-efficient plant species *Helianthus annuus*. In: Harley JL, Scott Russell R (eds) The soil-root interface. Academic Press, London New York
Sahrawat KL, Mukarjee SK, Gulat KC (1977) Nitrification inhibitors. II. Studies with furano compounds. Plant Soil 47:687–691
Salama El-D AMS, Wareing PF (1979) Effects of mineral nutrition on endogenous cytokinins in plants of sunflowers (*Helianthus annuus* L.). J Exp Bot 30:971–981
Sattelmacher B, Marschner H (1978) Nitrogen nutrition and cytokinin activity in *Solanum tuberosum*. Physiol Plant 42:185–198
Sattelmacher B, Marschner H (1979) Tuberization in potato plants as affected by application of nitrogen to the roots and leaves. Potato Res 22:39–47
Satter RL, Galston AW (1971) Phytochrome-controlled nyctinasty in *Albizzia librium*. III. Interactions between endogenous rhythm and phytochrome in control of potassium flux and leaflet movement. Plant Physiol 48:740–746
Scaife MA, Clarkson DT (1978) Calcium-related disorders in plants – a possible explanation for the effect of weather. Plant Soil 50:723–725
Schilling G, Schalldach I (1966) Untersuchungen über Transport, Einbau und Verwertung spät gedüngten Mineralstickstoffs bei *Pisum sativum* L. Thaer Arch 10:895–907
Schuphan W (1976) Mensch und Nahrungspflanze. Jung, Den Haag
Seiler-Kelbitsch H, Michael G, Wilberg E (1974) Beziehungen zwischen Korngewicht und Cytokinin-Aktivität bei Sommergerste, untersucht am Beispiel des Abschneidens von Bestockungstrieben. Angew Bot 48:299–307
Seth AK, Wareing PF (1967) Hormone-directed transport of metabolites and its possible role in plant senescence. J Exp Bot 18:65–77

Sharples RO, Johnson DS (1977) The influence of calcium on senescence changes in apple. Ann Appl Biol 85:450–453
Shay FJ, Hale MG (1973) Effect of low levels of calcium on exudation of sugar and sugar derivates from intact peanut roots under axenic conditions. Plant Physiol 51:1061–1063
Shear CB (1975) Calcium-related disorders of fruits and vegetables. Hort Sci 10:361–365
Shear CB (1979) International symposium on calcium nutrition of economic crops. Commun Soil Sci Plant Anal 10:1–501
Shear CB, Faust M (1970) Calcium transport in apple trees. Plant Physiol 45:670–674
Simojoki P (1972) Results of boron fertilizer experiments in barley. Ann Agric Fenniae 11:333–341
Sinclair TR, Dewit CT (1976) Analysis of the carbon and nitrogen limitations to soybean yield. Agron J 68:319–324
Sinha BK, Shingh NT (1976) Chloride accumulation near corn roots under different transpiration, soil moisture and soil salinity regimes. Agron J 88:346–348
Staden van J, Davey JE (1979) The synthesis, transport and metabolism of endogenous cytokinins. Plant Cell Environ 2:93–106
Strickland RC, Chaney WR, Lamoreaux RJ (1979) Organic matter influences phytotoxicity of cadmium to soybeans. Plant Soil 52:393–402
Symeonides C, McRae SG (1977) The assessment of plant available cadmium in soils. J Environ Qual 6:120–123
Terry N (1977) Photosynthesis, growth and the role of chloride. Plant Physiol 60:69–75
Trobisch S, Schilling G (1969) Untersuchungen über Zusammenhänge zwischen Massenentwicklung und N-Umsatz während der generativen Phase bei *Sinapis alba* L. Thaer Arch 13:867–878
Trobisch S, Schilling G (1970) Beitrag zur Klärung der physiologischen Grundlage der Samenbildung bei einjährigen Pflanzen und zur Wirkung später zusätzlicher N-Gaben auf diesen Prozeß am Beispiel von *Sinapis alba* L. Thaer Arch 14:253–265
Tromp J (1979) The intake curve for calcium into apple fruits under various environmental conditions. Commun Soil Sci Plant Anal 10:325–335
Tromp J, Oele J (1972) Shoot growth and mineral composition of leaves and fruits of apple as affected by relative air humidity. Physiol Plant 27:253–258
Varga A, Bruinsma J (1974) The growth and ripening of tomato fruits at different levels of endogenous cytokinins. J Hort Sci 49:135–142
Vaughan AKF (1977) The relation between the concentration of boron in the reproductive and vegetative organs of maize plants and their development. Rhod J Agric Res 15:163–170
Venkat-Raju K, Marschner H (1972) Regulation of iron uptake from relatively insoluble iron compounds by sunflower plants. Z Pflanzenernaehr Bodenkd 133:227–241
Vetter H, Teichmann W (1968) Feldversuche mit gestaffelten Kupfer- und Stickstoff-Düngergaben in Weser-Ems. Z Pflanzenernaehr Bodenkd 121:97–111
Vetter H, Mählop R, Fürchtenicht K (1974) Immissionsstoffbelastung in der Nachbarschaft einer Blei- und Zinkhütte. Ber Landwirtsch 52:327–350
Visser J, Locher JTh, Brouwer R (1971) Effects of aeration and mineral supply on growth and mineral content of shoots and roots of apple trees (var. Golden Delicious on M IX). Neth J Agric Sci 19:125–137
Vlamis J, Williams D (1967) Manganese and silicon interaction in the Gramineae. Plant Soil 27:131–140
Volk RJ (1958) Silicon content of the rice plant as a factor influencing its resistance to infections by the blast fungus, *Piricularia oryzae*. Phytopathology 48:179–184
Vorm van der PDJ, Diest van A (1979) Aspects of the Fe and Mn nutrition of rice plants. II. Iron and manganese uptake by rice plants, grown on aerobic water cultures. Plant Soil 52:19–29
Wagner HC (1974) Wuchsstoffgesteuerte Assimilateverlagerung bei Gerste. Angew Bot 48:331–338

Wagner H, Michael G (1971) Der Einfluß unterschiedlicher Stickstoffversorgung auf die Cytokininbildung in Wurzeln von Sonnenblumenpflanzen. Biochem Physiol Pflanzen 162:147–158
Wainwright SJ, Woolhouse HW (1977) Some physiological aspects of copper and zinc tolerance in *Agrostis tenuis* Sibth: Cell elongation and membrane damage. J Exp Bot 28:1029–1036
Wakita H, Schmitt RA (1970) Cadmium. In: Wedepohl KH (ed) Handbook of geochemistry Vol II-4/48. Springer, Berlin Heidelberg New York
Wehrmann J, Scharpf HC (1979) Der Mineralstickstoffgehalt des Bodens als Maßstab für den Stickstoffdüngerbedarf ($N_{min}$-Methode). Plant Soil 52:109–126
Wetzel R, Menke KH, Oelschläger W (1979) Gehalte an Spuren- und Mengenelementen in verschiedenen Mais- und Hafersorten sowie in einigen Heuherkünften. Landwirtsch Forsch 32:101–109
Wiebe HJ, Schätzler HP, Kühn W (1977) On the movement and distribution of calcium in white cabbage in dependence on the water status. Plant Soil 48:409–416
Wieneke J (1974) Untersuchungen zur Translokation von $^{45}$Ca im Apfelbaum. Gartenbauwissenschaft 39:57–67
Wieneke J (1979) Calcium transport and its microautoradiographic localization in the tissue. Commun Soil Sci Plant Anal 10:237–250
Wieneke J, Führ F (1973a) Untersuchungen zur Translokation von $^{45}$Ca im Apfelbaum. I. Transport und Verteilung in Abhängigkeit vom Aufnahmezeitpunkt. Gartenbauwissenschaft 38:91–108
Wieneke J, Führ F (1973b) Mikroautoradiographischer Nachweis von Ca$^{45}$ Kristallablagerungen im Apfelstiel- und Fruchtgewebe. Angew Bot 47:107–112
Wieneke J, Biddulph O, Woodbridge CG (1971) Influence of growth-regulating substances on absorption and translocation of calcium in pea and bean. J Am Soc Hortic Sci 96:721–724
Wiersum LK (1966) Calcium content of fruits and storage tissue in relation to the mode of water supply. Acta Bot Neerl 15:406–418
Wilkinson BG (1968) Mineral composition of apples. IX. Uptake of calcium by the fruit. J Sci Food Agric 19:646–647
Willenbrink J, Schuster WB (1978) Localized inhibition of translocation of $^{14}$C-assimilates in the phloem by valinomycin and other metabolic inhibitors. Planta 319:261–265
Williams DE, Vlamis J (1957) The effect of silicon on yield and manganese-54 uptake and distribution in the leaves of barley plants grown in culture solutions. Plant Physiol 32:404–409
Wu L, Thurman DA, Bradshaw AD (1975) The uptake of copper and its effect upon respiratory processes of roots of copper-tolerant and non-tolerant clones of *Agrostis stolonifera*. New Phytol 75:225–229
Yamagata N, Shigematsu J (1970) Cadmium pollution in perspective. Bull Inst Publ Health 19:1–27
Yoshida S, Nasavero SA, Ramirez EA (1969) Effect of silica and nitrogen supply on some leaf characters of the rice plant. Plant Soil 31:48–56
Yuan HF, Chang YS (1978) Effect of available silicon in paddy soil on the growth of rice plants. Bot Bull Acad Sinica 19:125–138
Zimdahl RL, Foster JM (1976) The influence of applied phosphorus, manure or lime on uptake of lead from soil. J Environ Qual 5:31–34

# I.2 The Significance of Rhizosphere Microflora and Mycorrhizas in Plant Nutrition

A.D. ROVIRA, G.D. BOWEN and R.C. FOSTER

## 1 Introduction

Plant roots are always associated with micro-organisms, ranging from organisms external to the root to root-infecting micro-organisms. The nature and significance of these associations have interested soil microbiologists and, to a much lesser extent, plant physiologists, for decades. The term "rhizosphere" was proposed by HILTNER (1904) to define that soil influenced by living roots. The rhizosphere is variable both in extent and composition, and from the root surface to the bulk soil there is a gradient of many chemical, physical and biological properties. To accommodate this gradient, microbiologists have proposed terms such as "outer rhizosphere", "inner rhizosphere" and "rhizoplane" (the root surface itself). For this paper we treat the rhizosphere as that zone of soil extending from the root–soil interface to the point in the soil where the microflora is unaffected by the root. The non-infective rhizosphere micro-organisms can have a large effect on plant nutrition, as do the infective associations such as mycorrhizal fungi (beneficial) and root diseases (detrimental).

## 2 Energy Supplies in the Rhizosphere

The populations of micro-organisms which build up in and around roots arise because of the greater supply of energy-rich nutrients in the rhizosphere coming from the roots than in the bulk soil; see HALE et al. (1979) for the most recent review on root exudates.

ROVIRA et al. (1979) defined the origin and nature of these organic materials derived from the root as in Fig. 1.

### 2.1 Exudates

Compounds of low molecular weight which leak from all cells into either the intercellular spaces and then to the soil via the cell junctions or directly through the epidermal cell walls into the soil. The release of these compounds is not metabolically mediated.

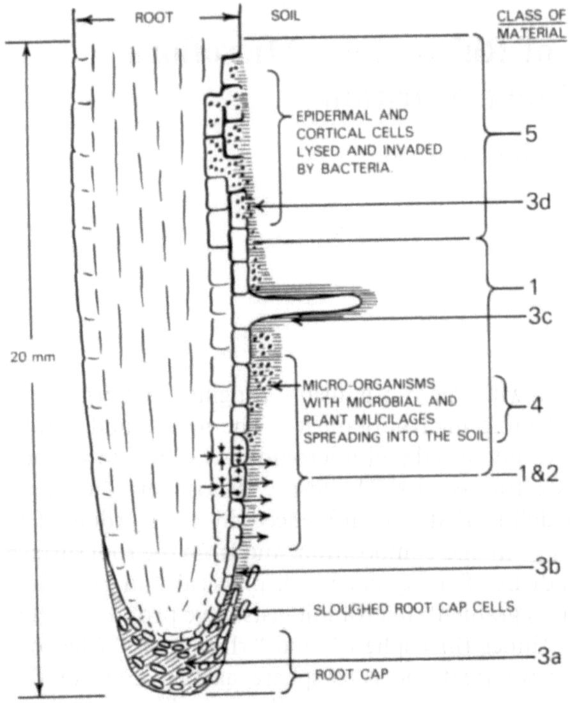


**Fig. 1.** Diagram of a root showing the origins of organic materials in the rhizosphere (see text). (By courtesy of Academic Press)

## 2.2 Secretions

Compounds of low or high molecular weight which are released as a result of metabolic processes.

## 2.3 Plant Mucilages

Four sources of plant mucilages which contribute to the organic materials in the rhizosphere are:

a) Mucilage originating in the root cap and secreted by Golgi vesicles.

b) Hydrolysates of the polysaccharide of the primary cell wall between the epidermal cells and sloughed root cap cells.

c) Mucilage secreted by the epidermal cells which still only have primary walls. This includes mucilages secreted by root hairs. The development of secondary walls in these cells further from the tip means that no further secretions or mucilages are released, only exudates.
At this part of the root the primary plant mucilage is confined to the intercellular grooves where bacteria are unable to enter.

d) Mucilage produced by bacterial degradation of the outer multilamellate primary cell walls of old, dead epidermal cells.

## 2.4 Mucigel

Following JENNY and GROSSENBACHER (1963) this term means the gelatinous material at the surface of roots grown in normal non-sterile soils. It includes natural and modified plant mucilages, bacterial cells and their metabolic products (such as capsules and slimes) as well as colloidal mineral and organic matter from the soil. Whereas root mucilages entirely of plant origin and microbially produced mucilages can only be studied and characterized in axenic cultures, mucigel is a product of the entire root-soil-microbial complex with its own distinctive morphological and biochemical properties. The mucigel could be important in maintaining contact between the root and the soil.

## 2.5 Lysates

These are the compounds released from autolysis of older epidermal cells when the plasmalemma fails. With further time the walls of these epidermal cells are digested by micro-organisms and the cells become heavily colonized releasing the products of microbial activity into the rhizosphere.

In studies on soluble exudates released by roots growing in nutrient solution, ROVIRA (1969) estimated from 0.1% to 0.5% of the carbon fixed by photosynthesis was released into the solution. However, results from solution culture can be misleading, and in simplified or "model" systems some attempt should be made to simulate soil characteristics. Thus, BARBER and GUNN (1974) found that the release of soluble carbon from roots growing in Ballotini (glass beads) increased to 9% of root weight as the resistance to root penetration was increased. In soil, the amounts can be higher still for BARBER and MARTIN (1976) found that roots of wheat seedlings released 5% to 10% of the carbon assimilated by the plant when grown without microorganisms and 12% to 18% in the presence of a microflora over three weeks. These results are consistent with those of MARTIN (1975, 1977), VANCURA (1964) – 7% to 10% of the biomass under water stress; SHAMOOT et al. (1960) – 9% to 42% of root; BREISCH (1974) – 10% to 20% of photosynthate; WAREMBOURG (1975), WAREMBOURG and BILLES (1979) – 12% to 16% of photosynthate; BILLES and CORTEZ (1975) – 4% to 8% of biomass; BALANDREAU and HAMAD-FARES (1975) – 10% to 20% of biomass. The release of such a large amount of carbon into the rhizosphere will affect not only the microflora but also soil processes and the availability of nutrients.

Interactions between exudation and the physiology of the plant as affected by environmental conditions and nutrition have been demonstrated. Temporary wilting of plants increased the exudation of amino acids from roots into sand or solution (KATZNELSON et al. 1954, 1955). Exudation of amino acids and sugars increased with increasing light intensity and temperature (ROVIRA 1959). The link between the exudation and composition of cell contents and cell permeability is demonstrated in the results of BOWEN (1969) in which the greater amounts of amino acids and amides were released from plants on low phosphate nutrition, this greater exudation being due to the higher content of free amino acid and amide. RATNAYAKE et al. (1978) reported that more amino acids and reducing sugars were released from roots grown under low P nutrition due to a change in cell permeability related to changes in phospholipid content resulting from the P deficiency.

## 3 Microbiology of the Rhizosphere

### 3.1 Populations of Micro-Organisms

Bacteria, fungi and actinomycetes are more abundant in the rhizosphere than in soil away from roots – this information is based upon counts of colonies of the organisms growing in nutrient agar from serial dilutions of soil and roots (KATZNELSON et al. 1948, ROVIRA 1965a).

Estimates of populations in the rhizospheres of different plants range from $5 \times 10^8$ to $3 \times 10^9$ $g^{-1}$ soil (ROVIRA and DAVEY 1974). Such figures give the impression that the root surface is covered by a mass of micro-organisms. However, direct microscope observations and counts of microbial populations of roots reveal otherwise. BOWEN and THEODOROU (1973) showed usually less than 10% of the root surface of *Pinus radiata* up to 21 days old was occupied by micro-organisms. In a later study with eight plant species from grassland ROVIRA et al. (1974) found from 4% to 10% cover of roots by bacteria and a density of fungal hyphae of 12 to 14 mm $mm^{-2}$ of root. SCHIPPERS and VAN VUURDE (1978) used fluorochromes and fluorescence microscopy to measure the cover of micro-organisms on wheat roots and found the cover to increase from 1% in that portion of root less than 4 days old and 2% to 4% in the 4- to 8-day-old segments.

### 3.2 Colonization of Roots by Micro-Organisms

The distribution of micro-organisms on the surfaces of roots is non-random; NEWMAN and H.J. BOWEN (1974) showed that bacteria are aggregated in micro-colonies – a result confirmed with subterranean clover by transmission electron microscopy of ultra-thin sections of the root–soil interface (FOSTER and ROVIRA 1978). This non-random distribution occurs for two reasons, firstly, the non-random distribution of micro-organisms in soil due to their association with fragments of organic matter and, secondly the initial colonization of roots at the junctions of epidermal cells due to these being better sites for microbial growth (BOWEN and ROVIRA 1976). This is probably due to greater exudation from the junctions of cells which act as "drainage outlets" for the materials which accumulate in the cell-free spaces of the epidermal and cortical cells of the root (see Fig. 1).

Studies based upon the isolation and characterization of rhizosphere and soil bacteria have shown a selective stimulation of Gram-negative non-sporing rods with characteristic nutritional requirements (LOCHHEAD 1940, LOCHHEAD and CHASE 1943).

The dynamics of this selective stimulation of Gram-negative bacteria was indicated by CHAN and KATZNELSON (1961) when they found that while *Pseudomonas* sp. and *Arthrobacter* sp. colonized roots and culture solution equally well in pure culture, the *Arthrobacter* level was reduced to one hundredth of this level in the presence of *Pseudomonas*. BOWEN and ROVIRA (1976) found that *Pseudomonas* sp. and *Bacillus* sp. had generation times of 5.2 and 39 h respectively on roots (77 and >100 h respectively in soil). Such a difference explains the reported dominance of the former in natural rhizospheres. Even within different *Pseudomonas* species there were up to 500-fold differences in populations on roots (BOWEN 1979).

The interactions between bacteria and mycorrhizal fungi on roots of *Pinus radiata* were demonstrated by BOWEN and THEODOROU (1973); they found that the length of root colonized by *Rhizopogon luteolus* was not affected by *Bacillus* sp. but reduced to one third by *Pseudomonas* sp. Another *Bacillus* sp. stimulated the growth of mycorrhizal fungi significantly.

# 4 Mathematical Modelling of the Rhizosphere

The experimental approach which yielded the above results contrasts with the mathematical modelling of NEWMAN and WATSON (1977), who assumed that the growth and distribution of micro-organisms around roots was controlled by the concentration of soluble carbon compounds which diffused into soil and depended on such factors as soil moisture. Using the computer model, they predicted the effects of changes in soil moisture, initial microbial population and root exudation upon the concentrations of bacteria at different distances from roots. According to the model, a change in soil moisture from $-0.1$ to $-1.0$ bar matric potential would lead to a 31-fold increase in the level of bacteria at the surface of the root; at 0.3 mm and 1.8 mm from the root the concentration of micro-organisms would fall by 75% and 50% respectively at the lower moisture content. Newman and Watson predicted that a 10-fold increase in carbon exuded would lead to a 120-fold increase in population at the root surface and a 3-fold increase in the population of bacteria in the rhizosphere. Such a model could give a better identification of factors controlling microbial growth at and near the root, and in the context of plant nutrition, could indicate the extent of competition between roots and micro-organisms for nutrients diffusing towards roots. A simulation computer model needs validation by experimentation and more hard data is needed to confirm the Newman-Watson model.

# 5 Microscopy of the Rhizosphere

## 5.1 Light Microscopy

Direct microscopy of the rhizosphere was used by STARKEY (1939) when he examined glass slides which had been buried in soil across the paths of roots; these studies revealed the increased population and the diversity of micro-organisms associated with roots. Using the aniline blue stain of JONES and MOLLISON (1948) for bacteria enabled ROVIRA (1956, 1965a) to observe that for roots growing in nutrient solution and glass beads the initial colonization occurred along junctions of epidermal cells followed by the development of micro-colonies. These observations were quantified by MALAJCZUK and BOWEN (unpublished) to show that during the early colonization of roots of *Eucalyptus calophylla* there were 50 times more bacteria along cell junctions than between the junctions (BOWEN and ROVIRA 1976).

A further tool in light microscopy is the use of fluorescence microscopy; TROLLDENIER (1965) used fluorescence with acridine orange to distinguish living from dead bacteria and revealed a heavy coating of bacteria on the roots and root hairs of clover grown in sand or agar. VAN VUURDE and ELENBAAS (1978) found a good contrast between the fluorescence of micro-organisms and roots and soil with successive staining with coriphosphine, congo red and acridine orange.

The use of fluorescent antisera to detect specific microorganisms on roots was used by TRINICK (1970) to follow competition between strains of *Rhizobium* and by SCHANK et al. (1979) to identify *Azospirillum brasilense* associated with the roots of grass.

Although light microscopy of roots greatly helped to develop our concepts of ecology of micro-organisms on roots, the technique is limited to the surfaces of washed roots – a disadvantage overcome by the application of electron microscopy to the rhizosphere, a technique which demonstrates the complexity of this environment through which all nutrients and water must pass to reach the root.

## 5.2 Scanning Electron Microscopy (S.E.M.)

The application of S.E.M. to the root–soil interface (GRAY 1967, DART 1971, CAMPBELL and ROVIRA 1973, ROVIRA and CAMPBELL 1974, ELLIOTT et al. 1979) has greatly improved our concept of the nature of the rhizosphere. S.E.M. has the advantage over Transmission Electron Microscopy that relatively large areas can be viewed and that the technique for preparing specimens is simpler. However, one disadvantage of S.E.M. is that the dehydrated mucilages on roots are opaque (Fig. 2) and hence the actual root surface can be viewed only when this mucilage lifts off (Fig. 3).

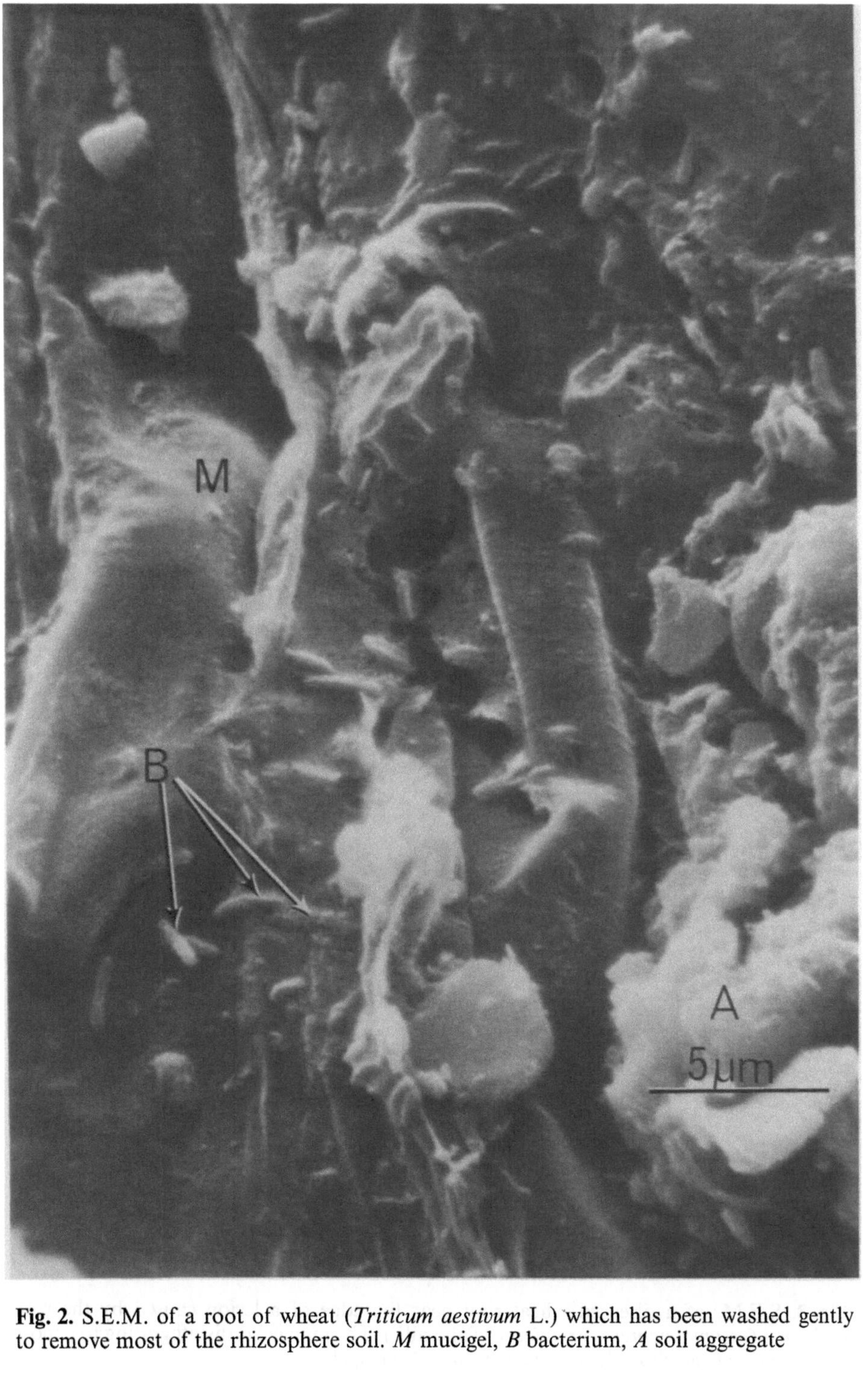


**Fig. 2.** S.E.M. of a root of wheat (*Triticum aestivum* L.) which has been washed gently to remove most of the rhizosphere soil. *M* mucigel, *B* bacterium, *A* soil aggregate

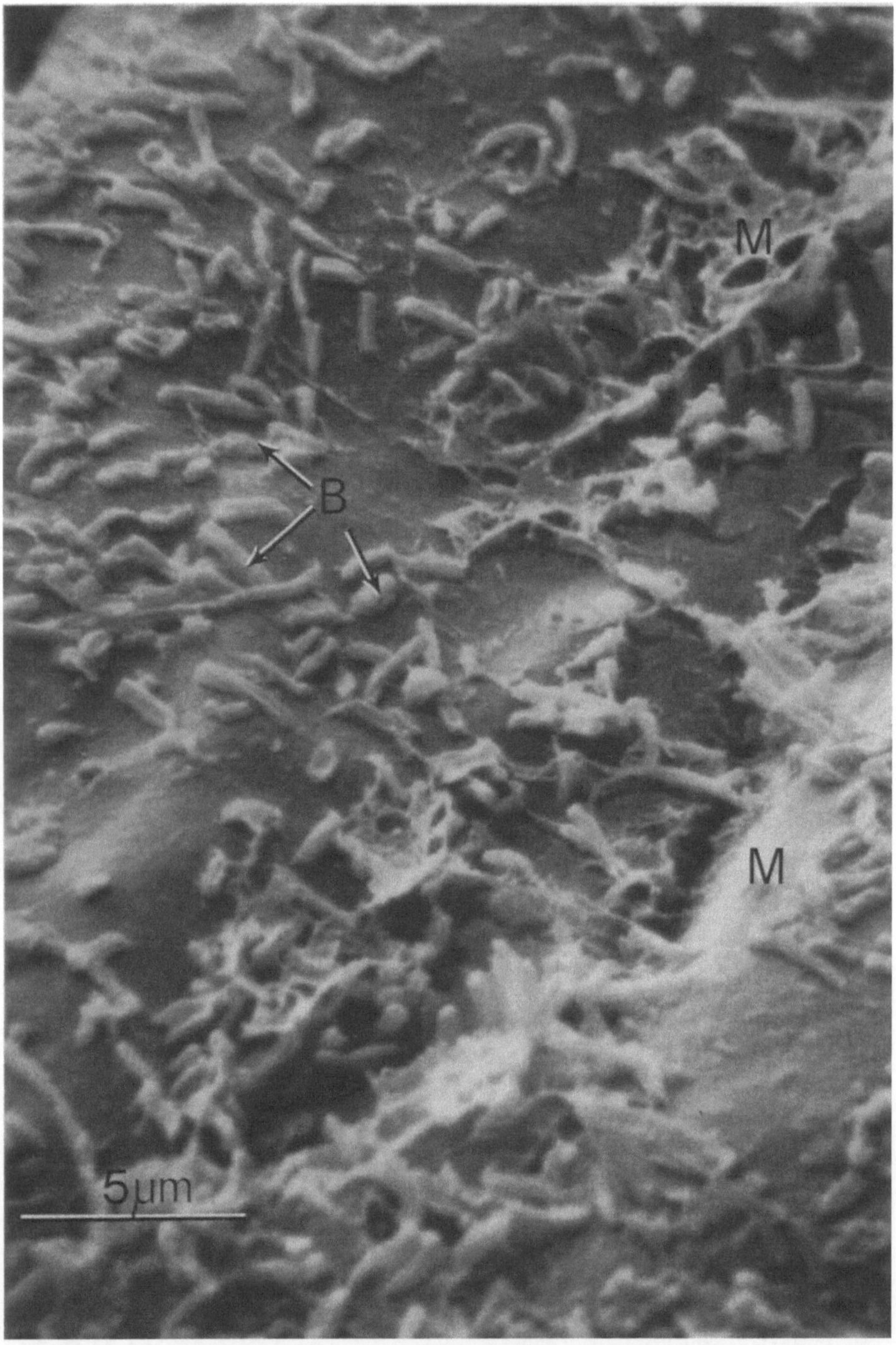


**Fig. 3.** S.E.M. of a diseased wheat root near the edge of a lesion caused by the root pathogen *Gaeumannomyces graminis* ("Take-all"). The mucigel is disintegrating to expose the bacteria at the rhizoplane. *B* bacterium, *M* mucigel

## 5.3 Transmission Electron Microscopy (T.E.M.)

### 5.3.1 General Description

Although SEM can be used to examine large areas of root, it can only reveal ultrastructural details of materials near surfaces and cannot distinguish different rhizosphere microorganisms unless they have distinctive shapes, sizes or characteristic surface markings. In TEM of ultra-thin sections of root surfaces, however, different kinds of organism may be distinguished (FOSTER and ROVIRA 1976, 1978). By examining serial sections the distribution of micro-organisms in the soil fabric from outside the rhizosphere into the root tissue can be determined. SEM and TEM therefore provide complementary views of rhizosphere organization and both are needed (OLD and NICHOLSON 1975, BALENDRAU and KNOWLES 1978). It is difficult to cut ultra-thin sections of roots and adhering clay and sand particles; there are further problems of fixing and embedding such heterogeneous material.

The ultrastructure of the root surface has been widely investigated by TEM (SCOTT et al. 1958, JENNY and GROSSENBACHER 1963, DART and MERCER 1964, LEPPARD 1974, FOSTER and ROVIRA 1976, 1978), but few of these studies have been made on roots of field-grown plants. Usually plants have been grown in reconstituted soils in pots and often the soils have been sterilized and known organisms have been introduced (GREAVES and DARBYSHIRE 1972, DART and MERCER 1964). In view of the complexity of natural rhizospheres, these artificial root–soil interfaces may oversimplify rhizosphere ultrastructure; in this paper we refer only to rhizospheres of roots from field soils.

It has been known for a long time that the epidermal cells of roots have a complex multilamellate structure (BELFORD and PRESTON 1961, SETTERFIELD and BAYLEY 1957, FOSTER 1962, PRESTON 1974). The outer layer of the cell wall is largely mucilaginous but contains a few cellulose microfibrils disposed in open networks arranged in widely spaced lamellae. The inner layer is compact and consists of dense lamellae of cellulose microfibrils. The precise nature of the root–soil interface as observed by SEM or TEM depends on which of these layers is exposed at the root surface either by the lytic action of soil bacteria or by the methods used to extract the roots from soil and to prepare them for electron microscopy. This has meant that the exact nature of the root–soil interface has been a matter of some controversy. In the classical study of JENNY and GROSSENBACHER (1963) the outer wall layer was thought to become increasingly tenuous and it was necessary to use electron-dense markers to demonstrate the surface. LEPPARD (1974) and LEPPARD and RAMAMOORTHY (1975) demonstrated that under some conditions the surface of the root is decorated with tufts of fibres at right angles to the root axis. However, SCOTT et al. (1958), GREAVES and DARBYSHIRE (1972), DART and MERCER (1964), showed that mucilage is bounded by a thin but well-defined electron-dense membrane, which SCOTT et al. (1958) termed a "cuticle".

Experiments with roots of several species grown in natural soils have shown that in the absence of mechanical abrasion from the surrounding soil, the mucilage is bounded externally by a delicate trilamellate electron-dense "cuticle"

(Fig. 4). This layer may be quite thick in the Fabaceae; even in the presence of bacteria, the root at this stage has a smooth surface and the outline of the individual root cells is quite distinct as seen in S.E.M.

Under natural conditions, however, where the root has to force its way into the soil fabric, and where soil bacteria lyse its surface, the fine "cuticle" is ruptured and the mucilages of the outer cell wall layer are released into the soil fabric (Fig. 5). At this stage the root surface appears granular and indistinct in both S.E.M. and T.E.M. and cell outlines are often obscure.

It has long been known that roots are covered with a mucilaginous material (SCHWARTZ 1883, DAWES and BOWLER 1959). Roots of plants from some families (Ericaceae, Fabaceae) secrete copious amounts of mucilage, but other species produce little or none (*Rosa, Dianthus, Chrysanthemum*) (LEISER 1968). Within the same soil, different species in the same family (e.g. Poaceae) may secrete quite different amounts of mucilage (FOSTER 1978).

Most plant-derived mucilage is associated with the root cap and epidermal cells in the root hair zone.

### 5.3.2 Origin and Fine Structure of Root Mucilage

Plant-derived mucilage is secreted by the dictyosomes (MOLLENHAUER and MORRE 1966, LEECH et al. 1963), and pulse-labelling techniques have shown that mucilage is transported through the cytoplasm in enlarged Golgi (dictyosome) vesicles to the plasmalemma where the contents of the vesicles are discharged into the cell wall and thence onto the root surface (NORTHCOTE and PICKETT-HEAPS 1966).

The precise ultrastructure of mucilage as seen by T.E.M. depends on the conditions under which roots were grown and on methods used in the preparation for electron microscopy. Conventional glutaraldehyde/$OsO_4$ double fixation reveals a granular material containing two populations of particles differing in size and electron density. When plants are water-stressed a particularly sticky mucilage may form at root surfaces (NAMBIAR 1976, MARTIN 1977) and when field-grown roots are treated with electron-dense colloids, changes in mucilage structure may be detected as localized patches or layers of electron dense material. The chemical composition of mucilage may vary according to the environmental conditions in the soil, and this may in turn affect the ion exchange properties of the root surface (see below).

**Fig. 4.** T.E.M. of an ultrathin section of a clover (*Trifolium subterraneum* L.) root showing the granular, mucilaginous, outer primary layer of the epidermal cell wall enclosed in an electron dense cuticle. Conventional fixation. *Cl* clay, *R* Rhizobium, *P* primary cell wall, *C* cuticle, *S* secondary cell wall

**Fig. 5.** T.E.M. of *Paspalum* root stained with lanthanum hydroxide. The mucilage of the primary wall surrounds and embeds soil minerals and becomes infected by fungi and bacteria. *B* Bacterium, *W* cell wall material, *M* mucigel

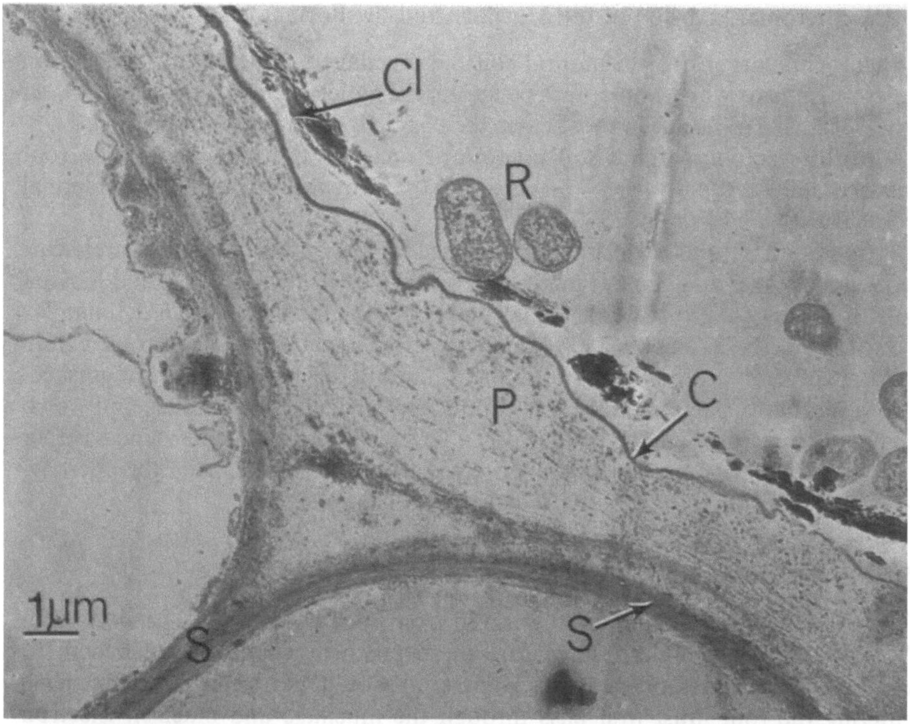


4

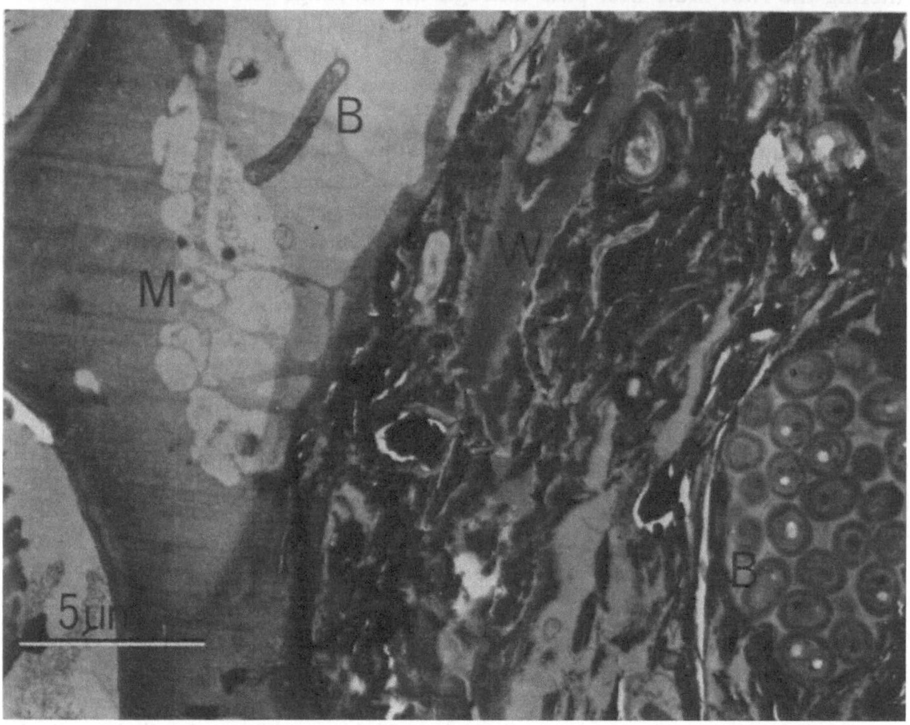


5

### 5.3.3 Microbial Invasion of the Mucilage and the Formation of Mucigel

When plants are grown in natural soil the mucilage is quickly invaded by soil microorganisms which sometimes penetrate the cuticle more or less intact (DART and MERCER 1964), but in most cases the cuticle is soon removed (Fig. 5).

Further activity of the soil microflora removes the mucilage at the root surface and eventually it is restricted to the grooves between the epidermal cells (BOWEN and FOSTER 1978).

Conditions in the rhizosphere encourage microorganisms which secrete carbohydrate slimes and capsule materials (WEBLEY et al. 1965, FOSTER and ROVIRA 1978). Some of these materials stain with Luft's ruthenium red/$OsO_4$ complex (CAGLE et al. 1972, FOSTER 1978) and many have a characteristic fibrous texture so that carbohydrates from roots and microorganisms can be distinguished. However, many bacteria and fungi produce gels which are granular, and which may not react with Ru/Os, so that it is no longer possible to determine whether the gel at the root surface is mainly derived from the root or from the microorganisms; this complex carbohydrate mix is called mucigel.

### 5.3.4 Functions of Root Mucilage and Mucigel

The amount of organic material released from roots is high (SAMTSEVICH 1965, MARTIN 1975) and unless the release of so much mucilage was of benefit to the host such a loss of energy-rich material would be rejected. All nutrients entering the root must first pass through the mucilage and mucigel. Electron probe and heavy metal precipitation techniques have shown that mucigel adsorbs many ions including $Cl^-$, $PO_4^{3-}$, $Ca^{2+}$ and $Fe^{2+/3+}$, and can hold them against mild leaching. Mucigel may serve to trap the transient flux of nutrients released from dead microorganisms: this may be a function of the $COO^-$ groups in acidic polysaccharides which JENNY and GROSSENBACHER (1963) and BALANDREAU and KNOWLES (1978) suggest are important in plant nutrition. LEPPARD (1974) and LEPPARD and RAMAMOORTHY (1975) suggest that fibrils of acidic polysaccharides may be involved in the extraction of ions from minerals and their movement into the root. The ion exchange properties of roots may be in the root surface carbohydrates derived from roots and microbes.

Mucilage may support nitrogen-fixing bacteria (DÖBEREINER 1974, KNOWLES 1977) as well as microorganisms which produce substances which are antagonistic to root pathogens. Transfer of water in the vapour phase is very slow, and one role of root mucigel may be to maintain contact between the minimal voids zone near the root and the root surface and thus maintain a continuous phase for water movement to the root as they shrink and swell with changes in diurnal water stress.

### 5.3.5 The Outer Rhizosphere

The outflow of energy-rich mucilages, cell exudates and (later) cell lysates results in a proliferation of microorganisms at the root–soil interface. Microbial numbers may rise in favourable microsites to $200 \times 10^9$ $cm^{-3}$ (FOSTER and ROVIRA 1978). There is also a great diversity of forms which can be distinguished on the basis of cytoplasmic structure. Many forms in the grooves between the epidermal cells may be restricted to this specialized habitat (FOSTER and ROVIRA 1978, OLD and NICHOLSON 1975).

Some stratification of bacteria occurs within the rhizosphere and many of the unusual forms which are not obtained by traditional methods of plate counting occur in the first 5 microns from the root surface. Further out from the root the rhizosphere may be dominated by one or two morphological types. The density of the bacterial population falls off within 20 μm from the surface, a finding which supports the NEWMAN-WATSON (1977) model.

The main distinction between bacteria in the rhizosphere and those in the bulk soil is that the rhizosphere forms are filled with storage materials such as polyhydroxybutyrate (PHB), polysaccharide granules and polyphosphate granules.

### 5.3.6 Invasion of the Root by Microorganisms

Bacteria attack the empty cells of the epidermis and cortex behind the root hair zone (FOSTER and ROVIRA 1973, 1976, 1978, OLD and NICHOLSON 1975) and the cortex of a normal healthy root may be inhabited by a wide range of microorganisms without any sign of disease in the root. The cells fill with bacteria, many of which are rich in PHB and polyphosphate. Some bacteria attack the cell wall and eventually only the inner cell wall layers may remain, so that the cellulose fibrils are then exposed to the surface. Microorganisms within the root show a narrow range of morphology and some small bacteria seem to be found exclusively within the root. This specialized environment has been called the endorhizosphere (BALANDREAU and KNOWLES 1978) and has also been proposed as a site of non-symbiotic $N_2$ fixation (DÖBEREINER and DAY 1975).

The mixed microbial population of the endorhizosphere quickly causes the collapse and decay of the outer cortical cells of the root; and in some annual crops, such as wheat, the endodermis may be the functional surface of older parts of the root system at anthesis.

## 6 The Role of Rhizosphere Microorganisms in Plant Nutrition

In view of the colonization of roots and rhizosphere by microorganisms the question asked is "Does this microbial population influence the nutrition of the plant?" Unless special precautions for axenic culture (viz. plants grown under sterile conditions) are observed, it is not possible to study nutrient uptake by plants without the microflora. The axenic culture of plants involves the surface sterilization of seed followed by germination under sterile conditions using agar on which contaminated seeds are detected. The germ-free seedlings are transferred to sterile plant nutrient solution, sand or soil (BARBER 1967, 1978, ROVIRA and BOWEN 1969, REUSZER 1949, 1962).

Rhizosphere microorganisms can affect plant nutrition through their influence on (a) availability of nutrients, (b) growth and morphology of roots, (c) nutrient uptake processes and (d) physiology and development of plants.

## 6.1 Availability of Nutrients

### 6.1.1 Nutrient Release and Immobilization

KATZNELSON and ROUATT (1957) found that the respiration of rhizosphere soil was two to four times that of non-rhizosphere soil; this differential was similar for both unamended soil and soil with added amino acids. Whether this greater metabolic activity in rhizosphere soils releases or immobilizes plant nutrients will depend upon the nutrient status of the soil. An example of this is the finding by BARBER (1966) that under low phosphate conditions the microflora competes with the root for phosphate and reduces the uptake of phosphate into the plant.

### 6.1.2 Nitrification and Denitrification

There is now considerable evidence that roots of some plants, especially grasses and trees, inhibit nitrification. This inhibition will influence the whole nitrogen balance of the system (THERON 1966, WOLDENDORP 1963, RICE 1974).

### 6.1.3 Nitrogen Fixation

The fixation of nitrogen under temperate swards was found by PARKER (1957) to average 60 kg $ha^{-1}$ p.a. over a 3-year period of undisturbed sward. Considerable interest in the nitrogen fixation in the rhizospheres of tropical grasses has been aroused by the findings of Döbereiner and associates (DÖBEREINER et al. 1972, DART and DAY 1975). The fixation of nitrogen in undisturbed grass swards in tropical soils is significant in maintaining production of pastures free from legumes and is discussed in Chap. II, 3, this Volume.

### 6.1.4 Phosphate Availability

The ability of rhizosphere bacteria to dissolve water-insoluble forms of phosphate has been demonstrated by GERRETSON (1948), SPERBER (1958) and DUFF et al. (1963), but it is unlikely this mechanism (which is dependent upon the production of acid) will operate to any great extent in the rhizosphere. The production of acid is dependent upon an abundant supply of carbohydrate and population densities found in nutrient media which are much higher than those which occur on roots. In his review of the role of microorganisms in the phosphorus cycle in soil, HAYMAN (1975) points out that those symbiotic organisms involved in phosphate uptake, e.g. mycorrhizas (discussed later in this chapter) have an enormous advantage over free-living organisms and hence will have a much greater impact on phosphate uptake.

### 6.1.5 Minor Nutrients

#### *6.1.5.1 Manganese*

TIMONIN (1946) found that the susceptibility of different oat varieties to manganese deficiency was related to the abundance of manganese-oxidizing bacteria supported in the rhizosphere. This influence of the microflora could be eliminated by partial soil sterilization.

#### *6.1.5.2 Molybdenum*

In a study of the causes for different levels of dental caries in residents in two districts in New Zealand, it was found that the vegetables from these districts differed in their content of molybdenum. These differences were shown by LOUTIT (1968) and LOUTIT et al. (1972) to be due to differences in the rhizosphere microfloras: those organisms from the soil producing low molybdenum plants accumulated more molybdenum than organisms from the soil producing vegetables with higher molybdenum levels.

## 6.2 Growth and Morphology of Roots

### 6.2.1 Root Length and Root Hairs

Rhizosphere microorganisms can reduce root length (BOWEN and ROVIRA 1961 a) and root hair length and density (BOWEN and ROVIRA 1961 b, WELTE and TROLLDENIER 1965, DARBYSHIRE and GREAVES 1970). Such effects will influence the uptake of slowly diffusible nutrients such as phosphate from soil by roots. The reductions in root length and root hair growth caused by microorganisms could lead to a 50% to 60% reduction in the uptake of ions of low mobility in soil.

### 6.2.2 Proteoid Roots

PURNELL (1960) reported localized intense lateral root production on many Proteaceae occurring in low nutrient soils and referred to them as "proteoid" roots. Clumps of proteoid roots form dense mats in the surface soil and, although they are particularly well developed in pockets of humus-rich soil, they are not restricted to these parts of the profile. Similar roots are now known to occur in other families, e.g. on the legume *Viminaria juncea* (LAMONT 1972), species of lupin (TRINICK 1977) and on *Casuarina equisetifolia* (H.G. DIEM personal communication); future studies may show them to be much more widespread. MALAJCZUK and BOWEN (1974) and LAMONT and MCCOMB (1974) showed proteoid roots to be caused by rhizosphere microorganisms.

In phosphate-poor soils, the production of proteoid roots has resulted in a doubling of growth of 4-month-old seedlings of *Banksia grandis* probably due to the intense exploration of soil for poorly mobile ions such as phosphate and a higher phosphate influx to proteoid roots (MALAJCZUK and BOWEN 1974). The physical nature of proteoid roots and their position near the soil surface make them ideally suited for uptake of nutrients leached from leaves and litter and they probably assist in the tightly coupled nutrient cycles so important in ecosystems of low fertility soils. It is highly probable that further study of plants in low nutrient soils, will produce other instances of microbial modification of roots which lead to increased nutrient uptake.

### 6.3 Nutrient Uptake Processes

The microorganisms associated with plant roots have been shown to affect the actual processes of nutrient uptake and translocation as described in reviews by BARBER 1978, BOWEN and ROVIRA 1969. The influence of the rhizosphere population on phosphate uptake and translocation is affected by the plant age; with seedlings the microflora increases uptake, translocation and incorporation (BOWEN and ROVIRA 1966, ROVIRA and BOWEN 1969) while with older plants the microorganisms compete with the roots for phosphate and reduce its uptake into the plant (BARBER 1966, BARBER et al. 1976). With 8-day-old barley seedlings the presence of rhizoplane microflora increased the translocation of phosphate to tops from 15 to 20 p mol $mg^{-1}$ in high phosphate plants and 8 to 22 p mol $mg^{-1}$ in low phosphate plants during a 30-min uptake period. By contrast in 3-week-old barley plants, plants with microorganisms on their roots took up 50% more phosphate over 24 h than did sterile plants, but the proportion of total phosphate translocated to tops was 48% and 34% for sterile and non-sterile plants respectively.

### 6.4 Physiology and Development

The rate of plant development, including flowering time, is affected by certain components of the rhizosphere microflora and this effect has been demonstrated by the application of cultures of known bacteria onto seed (MISHUSTIN and NAUMOVA 1962, ROVIRA 1965b, BROWN et al. 1968, BROWN 1974, 1976). These effects are now attributed to the production of plant growth-regulating substances such as indolyl acetic acid and gibberellic acid (BROWN 1972). The inoculation of crops was practised extensively in the U.S.S.R. but the unpredictability of crop responses to such seed inoculation has thrown doubt upon the practical value of this technique for broad-scale farming (MISHUSTIN and NAUMOVA 1962, RIDGE and ROVIRA 1968).

## 7 Mycorrhizas

The roots of most higher plants are usually mycorrhizal, the three main types of mycorrhizal association involved with inorganic nutrition being: (1) ectotrophic mycorrhizas, restricted mainly to a few tree families (e.g. Pinaceae, Myrtaceae, Fagaceae); (2) ericaceous mycorrhizas of Ericaceae, Epacridaceae, and Empetraceae and (3) vesicular-arbuscular (v.a.) mycorrhizas infecting most other higher plants except for the majority of species in a few families, primarily in the order Centrospermae and in the families Brassicaceae, Fumariaceae, Cyperaceae, Urticaceae, Commelinaceae (GERDEMANN 1975).

So far only phycomycetes of the family Endogonaceae have been shown to cause v.a. mycorrhizas; the ascomycete *Pezizella ericae* has been identified as one of the endophytes of *Vaccinium macrocarpon* (READ 1974) forming erica-

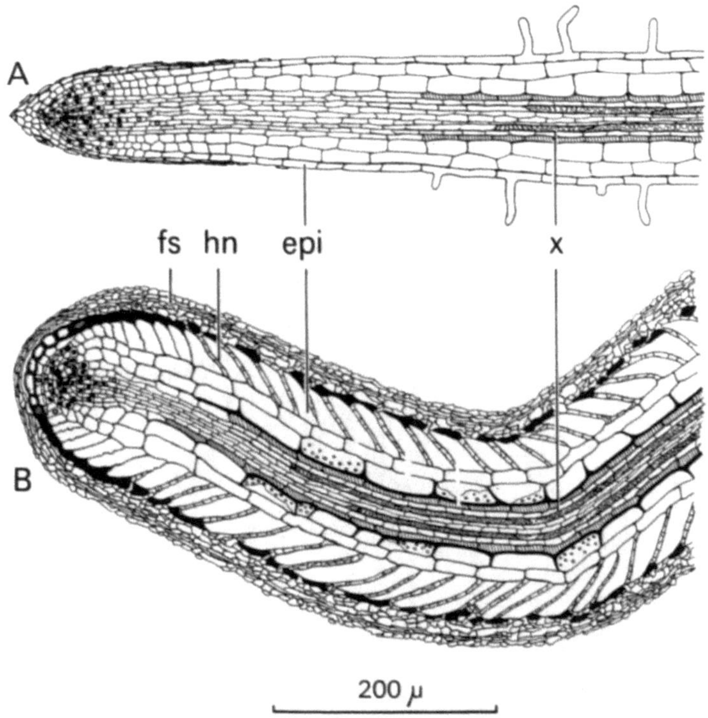


**Fig. 6.** Comparison of **A** uninfected root and **B** ectotrophic mycorrhiza. *fs* fungal sheath; *hn* Hartig net; *epi* epidermis; *x* xylem. (Adapted from CHILVERS and PRYOR 1965. By courtesy of the Australian Journal of Botany)

ceous mycorrhizas. However, a wide range of basidiomycetes, ascomycetes and at least one zygomycete (*Endogone*) are the cause of ectotrophic mycorrhizas.

Ectotrophic mycorrhizal fungi usually form a sheath of hyphae (Fig. 6) which encloses short, branching lateral roots. From the sheath, root-like bundles of hyphae (mycelial strands) pass *outwards* into the soil fabric often permeating it and organic matter (see BOWEN 1973). Hyphae also grow *inwards* into the host tissue to form the Hartig-net in the root cortex, thus giving an intimate connection between soil distant from the root and the cortical cells of the root. The fine structure of ectomycorrhizas is discussed by FOSTER and MARKS (1973).

In the endotrophic associations, hyphae actually enter the cells of the outer cortex of the roots. Here they become much divided to form tree-like "arbuscules" each branch of which is surrounded by the plasmalemma of the host. In v.a. mycorrhizas, the arbuscule increases the surface area of contact between the fungus and root cell by two to three times and some 23-fold increase in cytoplasm occurs in the plant cell (COX and TINKER 1976), indicating an effect of the micro-organisms on the plant's metabolism. Vesicular-arbuscular mycorrhizas also often have large ellipsoid vesicles between or within the cell. No well-defined fungus sheath occurs with v.a. mycorrhizas, but all can occur with ericaceous mycorrhizas. In endomycorrhizas, as with ectomycorrhizas, extensive fungal growth into soil and litter can occur (Fig. 7).

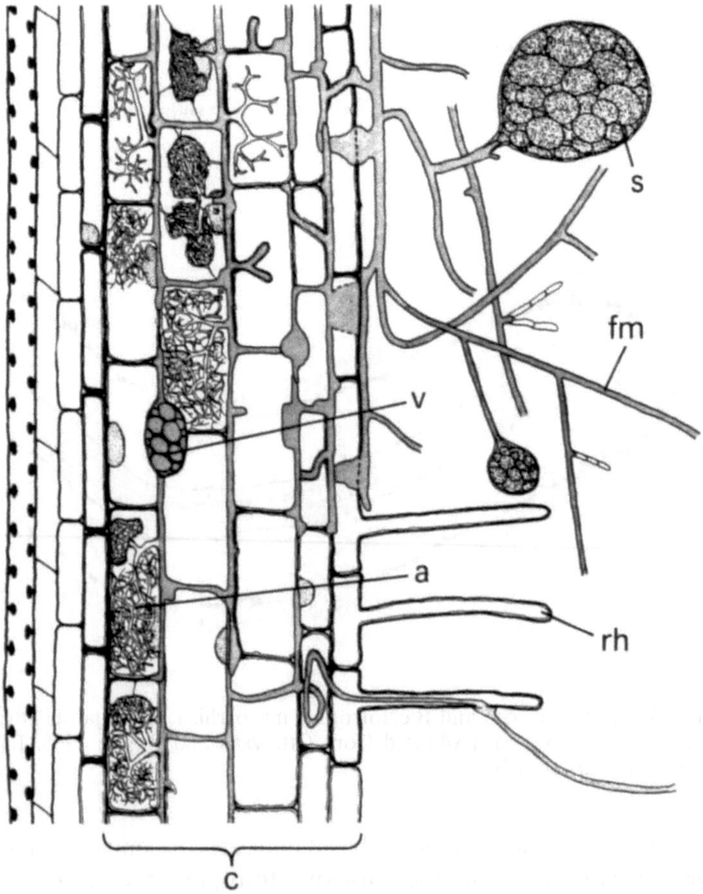


**Fig. 7.** Structure of v.a. mycorrhiza. *c* cortex; *r.h.* root hair; *f.m.* fungal mycelium; *s* spore; *v* vesicle; *a* arbuscule. (SANDERS et al. 1975. By courtesy of Academic Press)

### 7.1 Plant Responses to Infection

Table 1 illustrates the increases in plant growth and differences in nutrient status caused by mycorrhizas of the three main types; there is now a voluminous literature reporting such results. Increases in absorption of a number of ions (Table 1) can result from the relief of a primary deficiency. Thus, HATCH (1937) found ectomycorrhizal *Pinus strobus* to have twice the nitrogen uptake of non-mycorrhizal plants. A recalculation of his data shows nitrogen absorption $g^{-1}$ mycorrhizal roots was 1.6 times that of non-mycorrhizal roots. However, the phosphate absorption of mycorrhizal roots $g^{-1}$ was 2.9 times that of non-mycorrhizal root systems and he may have been dealing primarily with a phosphate deficiency.

The concentration of non-limiting elements, but not the total uptake, may decrease in mycorrhizal plants (Table 1, lines 4, 5) by dilution caused by the increased growth of mycorrhizal plants. Mycorrhizal increases of uptake of

**Table 1.** Plant response to mycorrhizal infection

| Type of mycorrhiza and host | Dry wt. | N | P | K | Ca |
|---|---|---|---|---|---|
| 1. Ectomycorrhizas (*Pinus elliottii* var. *elliotti*[a]) | | | | | |
| Uninoculated | 0.26[d] | 2.4[e] | 0.22[e] | 2.27[e] | |
| *Rhizopogon* sp. | 1.96 | 25.2 | 2.2 | 23.8 | |
| Isolate R8122 | 2.81 | 38.9 | 3.9 | 35.6 | |
| 2. v.a. Mycorrhizas (Rough lemon[b]) | | | | | |
| Uninoculated | 0.67[f] | | 0.10% | 1.51% | 3.15% |
| Inoculated | 11.29 | | 0.17% | 0.67% | 3.05% |
| 3. Ericaceous mycorrhiza (*Vaccinium macrocarpon*[c]) | | | | | |
| Uninoculated | 184[e] | 0.81[e] | | | |
| Inoculated | 235 | 1.82 | | | |

[a] LAMB and RICHARDS (1971)
[b] KLEINSCHMIDT and GERDEMANN (1972)
[c] READ and STRIBLEY (1973)
[d] g plant$^{-1}$
[e] mg plant$^{-1}$
[f] g pot$^{-1}$

a nutrient in limited supply have been shown with phosphate (ectomycorrhizas and v.a. mycorrhizas; see BOWEN 1973, SANDERS et al. 1975, and ABBOTT and ROBSON 1977), with nitrogen (ericaceous mycorrhizas; STRIBLEY and READ 1976, see also BOWEN and SMITH 1981), and zinc (v.a. mycorrhizas; GILMORE 1971); and increases in absorption of potassium, sulphate and $^{90}$Sr (a calcium tracer) have also been reported (BOWEN 1980). Enhanced ion absorption from soil occurs more commonly with poorly mobile ions (see below).

Increases in growth due to mycorrhizas are often greatest at sub-optimal levels of the limiting nutrient. At very low levels of nutrients there can be altogether too little for growth of the particular plant species and at high levels the mycorrhiza may be superfluous to the plant's needs; e.g. ABBOTT and ROBSON (1977) and MENGE et al. (1978) (v.a. mycorrhizas and phosphate nutrition of clover and citrus respectively); STRIBLEY and READ (1976) (ericaceous mycorrhizas and nitrogen nutrition of *Vaccinium macrocarpon*). Indeed, at high nutrient levels mycorrhiza development can be reduced or inhibited (as in MOSSE 1973). In at least one instance depression of mycorrhizal plants at supra-optimal levels of phosphate application has been ascribed to toxicity caused by enhanced absorption by the mycorrhizas (MOSSE 1973).

The large increases in plant growth from mycorrhizal infection suggest estimates of production potentials of soils may be seriously underrated, for almost no agronomic-nutrition studies of plants have ensured adequate mycorrhizal infection with highly efficient fungus. The search for plants and genotypes capable of tolerating nutritionally poor conditions may be assisted by the use of appropriate mycorrhizal fungi.

## 7.2 Mechanisms of the Response

Inflows of phosphate to v.a. mycorrhizal plants have been measured at up to 14 times that into non-mycorrhizal plants in soil (SANDERS and TINKER 1973,

SMITH et al. 1979). The specific absorption rates of ammonium are greater with mycorrhizal than with non-mycorrhizal *V. macrocarpon* (ericaceous mycorrhizas) (STRIBLEY et al. 1975). However, these data themselves do not indicate the mechanisms involved. The three general areas which could affect uptake are: the availability of the nutrients in soil, the effects of the fungi on nutrient absorption by the root itself, and the roles of the fungi in nutrient absorption by the symbiosis.

### 7.2.1 Nutrient Availability

Several studies indicate that v.a. mycorrhizas use the same sources of inorganic phosphate as uninfected plants (e.g. HAYMAN and MOSSE 1972, BARROW et al. 1977), and this is probably also the case for ectomycorrhizas (BOWEN 1973). Both the plants, mycorrhizas and uninfected roots have surface phosphatases (BOWEN 1973, BARTLETT and LEWIS 1973) and, given the opportunity, can absorb simple organic phosphates.

The position with nitrogen nutrition (reviewed by BOWEN and SMITH 1981) is less clear. Although some mycorrhizal fungi possess nitrate reductase and can use nitrate, most prefer ammonium sources; little, if any, nitrate absorption by mycorrhizas has been demonstrated so far. The fungi and higher plants can each use a wide range of relatively simple organic nitrogen compounds, provided they can compete effectively with soil microflora for them; but complex organic nitrogen sources are generally not used. Claims for atmospheric nitrogen fixation by mycorrhizal fungi can be interpreted more satisfactorily as fixation by free-living prokaryotic organisms on the surface of non-sterile mycorrhizal roots.

### 7.2.2 Absorption Characteristics of the Root

Most studies of ion absorption from solution do not distinguish between absorption by the fungus component and effects of the fungus on the absorption characters of the higher plant. The probable production of cytokinins or other growth-regulating compounds in the mycorrhizal association are manifested in enlargement of the cells or the nucleus, increase of host cytoplasm (COX and TINKER 1976) and, at least in ectomycorrhizas, the plant cells have increased longevity. However, the existing data do indicate increases of phosphate absorption by ectomycorrhizas and by v.a. mycorrhizas from solutions are due primarily to absorption by the fungus component (HARLEY 1969, BOWEN et al. 1975).

### 7.2.3 Absorption by the Fungus Component

As indicated above, enhancement of ion uptake (in monocultures) by mycorrhizas occurs more commonly with poorly mobile ions in soil. There is a close relation between rooting intensity and absorption of such ions (BARLEY 1970, NYE and TINKER 1977). Uninfected roots quickly deplete a narrow zone of soil around them for ions such as phosphate, a zone more or less delineated by the root hairs (see NYE and TINKER 1977) after which replenishment is slow. The growth of fungal hyphae into soil and litter can dramatically increase the "rooting density" and the volume of soil being effectively used by the plant. Uptake of nutrients by the mycorrhizal hyphae and their translocation to the root has been experimentally demonstrated for phosphate (MELIN and NILSSON 1950, SKINNER and BOWEN 1974a, HATTINGH et al. 1973 and PARSON and REID 1973), for ammonium (MELIN and NILSSON 1952, STRIBLEY and READ 1974), calcium (MELIN and NILSSON 1955) and for zinc and sulphate (COOPER and TINKER 1978). There could also be an enhanced absorption of potentially toxic heavy metals such as lead and arsenic (see BOWEN 1978) but this requires further study. SANDERS and TINKER (1973) calculated the phosphate fluxes in hyphae of v.a. mycorrhizas to account for the increased inflow to mycorrhizal roots. SANDERS et al. (1977) found that inflow of phosphate to onions growing in soil was closely related to the development of external hyphae with four v.a. mycorrhizal fungi; but COOPER and TINKER (1978) recorded little close relation between influx of phosphate, zinc and sulphate and length of mycorrhizal hyphae, and considered uptake by the fungus to be controlled by the host. Obviously a great deal more study is required of such possible factors, but it is necessary also to know whether uptake occurs along all of the hyphae. Individual hyphae of v.a. mycorrhizal fungi have been observed for up to 8 cm from roots (RHODES and GERDEMANN 1975) and mycelial strands of ectomycorrhizal fungi can intensively grow through soil and litter for up to 12 cm from the root (SKINNER and BOWEN 1974b). SANDERS and TINKER (1973) estimated there were up to 80 cm hyphae $cm^{-1}$ of v.a. mycorrhizal root. However, there is a serious lack of experimental data on soil and plant factors affecting fungal growth through soil, and on the epidemiology of the infection: both these parameters would be expected to affect response greatly, with soil, fungus strain, and plant species being important variables (SKINNER and BOWEN 1974b, BEVEGE and BOWEN 1975, SANDERS et al. 1977).

Influx of ions at the surface of the mycorrhiza may be no higher than that at the extending zone of an uninfected root, but the influx is much more sustained (BOWEN 1973). An enhanced absorption by mycorrhizas has been recorded from solutions (where transfer from the medium to the root surface is not limiting) for phosphate (HARLEY 1969, BOWEN 1973, CRESS et al. 1979), sulphate (MORRISON 1963) and zinc (BOWEN et al. 1974). Absorption of these ions is metabolically mediated and for phosphate at least, kinetic analyses have suggested at least two carrier systems (BOWEN and THEODOROU 1967, for both ectomycorrhizas; and CRESS et al. 1979, for v.a. mycorrhizas). The ectomycorrhiza studies showed mycorrhizas caused by different fungi differed both in the number of carrier sites and their affinity for phosphate. The studies on v.a. mycorrhizas indicated that at low solution concentrations of phosphate (1–20 µm), the carrier systems of mycorrhizas (plus roots) had a greater affinity for phosphate than those of roots alone, but that the mycorrhizas had the same number of carrier sites. Unfortunately, the Hofstee plot interpretation approach is not a unique interpretation of uptake analyses and furthermore the kinetics may have been affected by the uninoculated roots having a higher phosphate concentration than the mycorrhizal roots did. Suggestions that mycorrhizas may be able to absorb phosphate at lower concentrations than the uninfected root can (CRESS et al. 1979) need further evaluation but may be of only limited relevance in phosphate uptake from soil.

In all types of mycorrhizas, phosphate is converted to inorganic polyphosphates (COX et al. 1975, LING-LEE et al. 1975, CALLOW et al. 1978), in which form it is translocated in v.a. mycorrhizas (COX et al. 1975). There is considerable storage of phosphate in ectomycorrhizas, probably as polyphosphate, during periods of plentiful phosphate supply (HARLEY 1969). However, v.a. mycorrhizas do not appear to have as high a storage capacity (BOWEN et al. 1975) and after a short lag period (COOPER and TINKER 1978) phosphate is transferred rapidly to the higher plant via the arbuscules. Electron-probe analysis indicates the arbuscules to be high in phosphorus (SCHOKNECHT and HATTINGH 1976). It has been suggested that phosphate release depends on degeneration of the arbuscule (KINDEN and BROWN 1975) but it is likely that the major part of the phosphate is transferred from the living arbuscule to the plant as orthophosphate before arbuscule degeneration (COX and TINKER 1976, BOWEN et al. 1975, COOPER and TINKER 1978). GIANINAZZI et al. (1979) have speculated on the roles of acid and alkaline phosphatases within arbuscules of v.a. mycorrhizal fungi, and suggested vacuolar alkaline phosphatase may be involved in the transport of phosphate within v.a. fungi. They have also described an alkaline phosphatase in v.a. fungi specific to mycorrhizal infection (GIANINAZZI-PEARSON and GIANINAZZI 1978).

More is needed to be known about ammonium-assimilating enzymes of roots and their mycorrhizal fungi before authoritative statements can be made about the relative rates of uptake in the symbionts (BOWEN and SMITH 1981). Incorporation of ammonium into amino acids results in the production of excess protons (one $H^+$ per ammonium, RAVEN and SMITH 1976). All plants and microorganisms, including mycorrhizas, regulate their intracellular pH by $H^+$ extrusion to the bathing medium or soil, thus reducing the pH in culture medium or rhizosphere (NORKRANS 1950, SMILEY 1974). In ectotrophic beech mycorrhizas, and possibly other mycorrhizas, $NH_4^+$ is assimilated to glutamine in the fungal sheath, and is then transferred to the root. Proton extrusion must take place from the fungal cells, which may assume both assimilatory and pH regulatory roles from the root (discussed by RAVEN et al. 1978). S.E. SMITH (personal communication) has suggested that if nitrate uptake and assimilation by mycorrhizal fungi were to occur and pH regulation were by $OH^-$ extrusion rather than by cation/carboxylate storage, mycorrhizal development might result in considerable savings of photosynthate used for carboxylate production to regulate pH. It is significant in this regard that potassium content of many v.a. mycorrhizal plants is often considerably lower than that of non-mycorrhizal plants (S.E. SMITH personal communication).

### 7.3 Energy Requirements of Mycorrhizas

The energy sources for the fungus are derived mainly from the higher plant. In ectomycorrhizas, sucrose in the higher plant is hydrolyzed to glucose and fructose, and the concentration gradient is maintained by their conversion into trehalose, mannitol and glycogen (LEWIS and HARLEY 1965, BEVEGE et al. 1975). Such fungal sugars do not occur with v.a. mycorrhizas (BEVEGE et al. 1975, HAYMAN 1974), and concentration of sugars may be kept low by their use in growth, conversion to organic acids or storage as glycogen. The assimilate cost of v.a. mycorrhizas has received little study. Sometimes, but not always, ectomycorrhizas can be an assimilate sink; it has been suggested that the growing root and ectomycorrhizas both act as assimilate sinks, the mycorrhizal sink becoming major when root elongation slows and its sink strength declines (BOWEN 1978). Usually the benefits of enhanced nutrition by mycorrhizas far outweighs the assimilate costs but there are suggestions that in some highly fertile soils mycorrhizas may be superfluous but still use up assimilate. This was proposed as the reason for occasional growth reductions by v.a. mycorrhizas in highly fertile soils (CRUSH 1976).

### 7.4 Overview of Mycorrhizas

Most studies of mycorrhizas have focussed on phosphate uptake from nutrient-poor soils. Although their key role is in nutrient (and possibly water) uptake from soils there is little doubt that some aspects of the physiology of the mycorrhizal plant are different from that of the uninfected plant, e.g. the infection causes a marked stimulation of polyphenols, and terpenes in roots (FOSTER and MARKS 1967, KRUPA and FRIES 1971). However, lack of appropriate controls has precluded a good comparison of the physiology of mycorrhizal and non-mycorrhizal plants in most studies. There is some evidence that differences occur, such as a decreased nematode reproduction or a decreased length of mycorrhiza root in some studies, (SCHENCK et al. 1975); SCHÖNBECK (1979) reported an increased susceptibility of endomycorrhizal plants to foliage diseases but decreased susceptibility to root diseases.

Even in nutrient absorption the overview of mycorrhizas may have been too restricted. BOWEN (1980) has suggested they are more probably seen as an alternate root strategy for ion absorption in a great many situations. In many situations other than nutrient-poor soils, it is now well accepted that mycorrhizal responses and plant dependence on mycorrhizas are maximal with species developing low root densities, often with poor root hair formation (BAYLIS 1975, CRUSH 1974). However, the growth of fungi into soil and litter also enables mycorrhizal plants to compete effectively with soil micro-organisms for available nutrients (BOWEN 1968, 1980, GADGIL and GADGIL 1975). Furthermore, the dual uptake system, each member with its own properties, allows more flexibility in handling potentially deleterious situations. As long as the mycorrhizal fungus is not as susceptible to a deleterious soil factor as the root, the mycorrhizal mode of nutrition can be used by the plant to combat the deleterious factor. Such factors can include root disease, high soil salinity, disadvantageous soil pH, and high soil temperatures (BOWEN 1980), e.g. this may explain differences in mycorrhizal and non mycorrhizal plant growth with regard to soil pH and soil temperature demonstrated by MOAWAD (1979). The reasons for differing sensitivities to environmental factors between different mycorrhizal fungi and between the fungi and their plants is an area awaiting more study by physiologists.

Recently, SUTTON and SHEPPARD (1976) showed, by inoculating pots of sterilized dune sand with *Glomus,* that endomycorrhizal *Phaseolus vulgaris* had some five times (by weight) more sand aggregates $kg^{-1}$ than did non-mycorrhizal plants of similar size, the sand grains being held together by the fungal hyphae. TISDALL and OADES (1979) ascribed the greater efficiency of ryegrass compared with clover roots in stabilizing soil aggregates in a loam, to the ryegrass supporting a larger population of v.a. mycorrhizal hyphae in soil. The hyphae were covered with a layer of amorphous material, probably polysaccharide, to which clay particles were firmly attached.

The increase in ion uptake from endomycorrhizas can have large consequences for nitrogen fixation by legumes and by non-legumes such as *Casuarina* spp.. v.a. mycorrhizas have been demonstrated to increase uptake of phosphate, sulfate and zinc, and probably would increase uptake of cobalt, molybdenum, copper, and iron, all of which are involved in nitrogen fixation. MOSSE et al. (1976) found clover, *Centrosema* and *Stylosanthes* did not nodulate in grossly phosphate-deficient soil if they were not also mycorrhizal. SMITH et al. (1979) demonstrated an almost quadrupling of nitrogenase activity in mycorrhizal clover compared with non-mycorrhizal clover of the same nutrient content.

The biological variability between mycorrhizal fungi, both in their infection of the plant under various soil conditions and growth into soils (BEVEGE and BOWEN 1975, SANDERS et al. 1977) with different soil–plant–fungus combinations, leads to very great differences between isolates of mycorrhizal fungi in their stimulation of the plant (Table 1, lines 2, 3). Modification of the one soil similarly leads to different responses to different fungi, e.g. MOSSE (1972) found that stimulation of plant growth with one v.a. mycorrhizal fungus was three times greater than with another in a soil modified to pH 5.8 although both behaved similarly at pH 4.8. Similarly, management practice will change the dominant type of fungus present. MOSSE (1977) recorded that the addition of phosphate to some phosphate-deficient soils advantaged the introduced v.a. mycorrhizal fungi in competition with the indigenous mycorrhizal fungi.

In brief, the soil-mycorrhizal-plant system leads to a much more dynamic picture of nutrition of plants and their physiology, and increased scope for management of nutrition systems, than the simple two-membered concept of plant and soil.

## 8 General Conclusions

"Traditional" plant physiology has usually considered only the higher plant. However, the wide range of microorganisms associating with roots and interacting intimately with them adds a new dimension to our concept of "normal" plants. These microorganisms, especially those that function symbiotically with the higher plant, e.g. mycorrhizal fungi, can have a large impact on uptake of plant nutrients from soils: they can be regarded as an extension of the root, and another agent of the plant plasticity, complementary to genetic plasticity, selecting species and genotypes coping with low fertility sites.

The last 10 years has seen considerable conceptual and experimental developments in rhizosphere biology. These have been helped considerably by the application of electron microscopy of the root–soil interface (including the application of histochemical methods), by the application of traditional population dynamics and simulation modelling and, to a limited extent, by physiological approaches. This microbial dimension to studies of plant biology can be a rewarding field for plant physiologists. For example, rhizosphere micro-organisms can double the losses of assimilates from "healthy" roots and to a level where it is immediately apparent that they must be considered in any attempt to construct an assimilate budget for the plant. Questions can be raised on the mechanisms of assimilate loss from roots and the relation between loss and composition of the root cells, which is in turn affected by the rest of the plant's physiology. With mycorrhizal systems it is particularly appropriate to consider the energy costs of the symbiosis, the mechanisms of transfer of ions and metabolites between fungus and host, not to mention the physiology of infection and the fine balance between symbiosis and parasitism.

Interesting questions are whether micro-organisms associated with roots change the plants' physiology and the biochemical bases of such effects. What are the nature of the compounds produced by microorganisms which enhance assimilate loss, increase phosphate translocation, induce earlier flowering and cause extensive root modification important in nutrient uptake?

In view of these effects of rhizosphere micro-organisms on plants is it possible to manipulate the microflora in the field? The extensive Russian work on inoculation of seed with *Azotobacter* spp. and *Bacillus* spp. with claims of large yield increases has been questioned; MISHUSTIN and NAUMOVA (1962) found that inoculation increased crop yields significantly in one third of field trials. Similar results by RIDGE and ROVIRA (1968) and BROWN (1976) demonstrate that some manipulation of the microflora is possible. However, studies on the distribution of the introduced bacteria show little spread from the seed down the roots in unsterilized soil, so the effects result from a localized concentration of the bacteria around the seed. Most colonization of roots results from contact with micro-organisms in the soil; these organisms will certainly be the primary colonizers, making it difficult for introduced bacteria to spread and find sites which provide energy materials for multiplication. An alternative strategy for establishing favourable rhizosphere microflora may be to grow crops which build up the desired organisms which will then be placed strategically through the soil to colonize roots of subsequent crops.

The successful introduction of efficient species of ectomycorrhizal fungi by inoculation of pine seed has been demonstrated in both field and glasshouse trials with corresponding growth responses in the trees (THEODOROU and BOWEN 1970). The complex interactions between introduced micro-organisms in the rhizosphere are demonstrated in the experiments by BOWEN and THEODOROU (1979), who found that the growth of ectotrophic mycorrhizas on pine roots was increased, decreased or not affected by different species of bacteria. More studies of this type are needed to understand the basic principles of population dynamics in the rhizosphere before manipulation of the rhizosphere becomes a reality in the field.

Finally, we wish to emphasize that most studies on plant physiology and the uptake of nutrients are conducted on plants which support microorganisms in and around roots. It will be necessary, in future, in critical studies on the uptake of nutrients by roots to consider, and if necessary, allow for the involvement of microorganisms in the uptake of nutrients.

*Acknowledgements.* We thank Dr. R. CAMPBELL for the preparation of the SEM specimens, Miss Y. MCEWAN and Mrs. P. UDOMPONGSANON for preparation of the TEM specimens and Mr. T.W. COCK and Mr. J.A. COPPI for preparing the plates for publication.

## References

Abbott LK, Robson AD (1977) Growth stimulation of subterranean clover with vesicular arbuscular mycorrhizas. Aust J Agric Res 28:639–649

Balandreau J, Hamad-Fares I (1975) Importance de la fixation d'azote dans la rhizosphère du riz. Soc Bot Fr, Colloq Rhizosphere:109–119

Balandreau J, Knowles R (1978) The rhizosphere. In: Dommergues YR, Krupa SV (eds) Interactions between non-pathogenic soil microorganisms and plants. Elsevier, Amsterdam New York

Barber DA (1966) Effect of microorganisms on nutrient absorption by plants. Nature 212:638–640

Barber DA (1967) The effects of microorganisms on the absorption of inorganic nutrients by intact plants. I. Apparatus and culture techniques. J Exp Bot 18:163–169

Barber DA (1978) Nutrient uptake. In: Dommergues YR, Krupa SV (eds) Interactions between non-pathogenic soil microorganisms and plants. Elsevier, Amsterdam New York

Barber DA, Gunn KB (1974) The effect of mechanical forces on the exudation of organic substances by the roots of cereal plants grown under sterile conditions. New Phytol 73:39–45

Barber DA, Martin JK (1976) The release of organic substances by cereal roots into soil. New Phytol 76:69–80

Barber DA, Bowen GD, Rovira AD (1976) Effects of microorganisms on the absorption and distribution of phosphate in barley. Aust J Plant Physiol 3:801–808

Barley KP (1970) The configuration of the root system in relation to nutrient uptake. Adv Agron 22:159–201

Barrow NJ, Malajczuk N, Shaw TC (1977) A direct test of the ability of vesicular arbuscular mycorrhiza to help plants take up fixed soil phosphate. New Phytol 78:269–276

Bartlett EM, Lewis DH (1973) Surface phosphatase activity of mycorrhizal roots of beech. Soil Biol Biochem 5:249–257

Baylis GTS (1975) The magnoloid root and mycotrophy in root systems derived from it. In: Sanders FE, Mosse B, Tinker PB (eds) Endomycorrhizas. Academic Press, London New York

Belford DS, Preston RD (1961) The structure and growth of root hairs. J Exp Bot 12:157–168

Bevege DI, Bowen GD (1975) *Endogone* strain and host plant differences in development of vesicular-arbuscular mycorrhizas. In: Sanders FE, Mosse B, Tinker PB (eds) Endomycorrhizas. Academic Press, London New York

Bevege DI, Bowen GD, Skinner MF (1975) Comparative carbohydrate physiology of ecto- and endomycorrhizas. In: Sanders FE, Mosse B, Tinker PB (eds) Endomycorrhizas. Academic Press, London New York

Billes G, Cortez J (1975) Étude de la production de polysaccharides bactériens au niveau de la rhizosphère de *Festuca arundinacea.* Soc Bot Fr Colloq Rhizosphere:41–45

Bowen GD (1968) Phosphate uptake by mycorrhizas and uninfected roots of *Pinus radiata* in relation to root distribution. Proc 9th Int Cong Soil Sci 2:219–228

Bowen GD (1969) Nutrient status effects on loss of amides and amino acids from pine roots. Plant Soil 30:139–142

Bowen GD (1973) Mineral nutrition in Ectomycorrhizae. In: Marks GC, Kozlowski TT (eds) Ectomycorrhizae. Academic Press, London New York

Bowen GD (1978) Dysfunction and shortfalls in symbiotic response. In: Horsfall JD, Cowling EB (eds) Plant disease – an advanced treatise, vol III. Academic Press, London New York

Bowen GD (1979) Integrated and experimental approaches to study the growth of organisms around root and seeds. In: Schippers B, Gams W (eds) Soil-borne pathogens. Academic Press, London New York

Bowen GD (1980) Mycorrhizal roles in tropical plants and ecosystems. In: Mikola P (ed) Tropical mycorrhiza research. Oxford Univ Press, Oxford

Bowen GD, Foster RC (1978) Dynamics of microbial colonisation of plant roots. In: Broughton WJ, John CK (eds) Soil microbiology and plant nutrition. Univ Malaya Press, Malaya

Bowen GD, Rovira AD (1961a) Plant growth in irradiated soil. Nature (London) 211:665–666

Bowen GD, Rovira AD (1961b) Effects of microorganisms on plant growth. I. Development of roots and root hairs in sand and agar. Plant Soil 15:166–188

Bowen GD, Rovira AD (1966) Microbial factor in short-term uptake studies with plant roots. Nature 211:655–666

Bowen GD, Rovira AD (1969) The influence of microorganisms on growth and metabolism of plant roots. In: Whittington WJ (ed) Root growth. Butterworths, London

Bowen GD, Rovira AD (1976) Microbial colonization of plant roots. Annu Rev Phytopathol 14:121–144

Bowen GD, Smith SE (1981) The effects of mycorrhizas on nitrogen uptake by plants. In: Clark FE, Rosswall T (eds) Nitrogen cycling in terrestrial ecosystems. Ecol Bull (Stockholm)

Bowen GD, Theodorou C (1973) Studies on phosphate uptake by mycorrhizas. Proc 14th IUFRO Congr, Munich

Bowen GD, Theodorou C (1973) Growth of ectomycorrhizal fungi around seeds and roots. In: Marks GC, Kozlowski TT (eds) Ectomycorrhizae: their ecology and physiology. Academic Press, London New York

Bowen GD, Theodorou C (1979) Interactions between bacteria and ectomycorrhizal fungi. Soil Biol Biochem 11:119–126

Bowen GD, Skinner MF, Bevege DC (1974) Zinc uptake by ecto- and endo-mycorrhizas and uninfected roots of conifers. Soil Biol Biochem 6:141–144

Bowen GD, Bevege DI, Mosse B (1975) Phosphate physiology of vesicular-arbuscular mycorrhizas. In: Sanders FE, Mosse B, Tinker PB (eds) Endomycorrhizas. Academic Press, London New York

Breisch H (1974) Contribution à l'étude du rôle des exudats racinaires dans les processes d'aggrégation des sols. Thèse Doct Spéc Univ Nancy

Brown ME (1972) Plant growth substances produced by microorganisms of soil and rhizosphere. J Appl Bact 35:443–451

Brown ME (1974) Seed and root bacterization. Annu Rev Phytopathol 12:181–197

Brown ME (1976) Microbial manipulation and plant performance. In: Skinner FA, Carr JG (eds) Microbiology in agriculture fisheries and food. Soc Appl Bacteriol Symp 4. Academic Press, London New York

Brown ME, Jackson RM, Burlington AK (1968) Effects on tomato plants, *Lycopersicon esculentum*, by seed or root treatment with gibberellic acid and indolyl-3-acetic acid. J Exp Bot 19:544–552

Cagle GD, Pfister RM, Vela GR (1972) Improved staining of extracellular polymers for electron microscopy. Appl Microbiol 24:477–487

Callow JA, Capaccio LCM, Parish G, Tinker PB (1978) Detection and estimation of polyphosphate in vesicular-arbuscular mycorrhizas. New Phytol 80:125–134

Campbell R, Rovira AD (1973) A study of the rhizosphere by scanning electron microscopy. Soil Biol Biochem 5:747–752

Chan ECS, Katznelson H (1961) Growth interactions of *Arthrobacter globiformis* and *Pseudomonas* sp. in relation to the rhizosphere effect. Can J Microbiol 7:759–767
Chilvers GA, Pryor LD (1965) The structure of eucalypt mycorrhizas. Aust J Bot 13:245–259
Cooper KM, Tinker PB (1978) Transfer and translocation of nutrients in vesicular-arbuscular mycorrhizas II. Uptake and translocation of phosphorus, zinc and sulphur. New Phytol 81:43–52
Cox G, Tinker PB (1976) Translocation and transfer of nutrients in vesicular arbuscular mycorrhizas. I. The arbuscule and phosphorus transfer: a quantitative ultrastructure study. New Phytol 77:371–378
Cox G, Sanders FE, Tinker PB, Wild JA (1975) Ultrastructural evidence relating to host-endophyte transfer in a vesicular-arbuscular mycorrhiza. In: Sanders FE, Mosse B, Tinker PB (eds) Endomycorrhizas. Academic Press, London New York
Cress WA, Throneberry AO, Londsey DL (1979) Kinetics of phosphorus absorption by mycorrhizal and nonmycorrhizal tomato roots. Plant Physiol 64:484–487
Crush JR (1974) Plant growth responses to vesicular-arbuscular mycorrhizas. VII. Growth and nodulation of some herbage legumes. New Phytol 73:743–749
Crush JR (1976) Endomycorrhizas and legume growth in some soils of the Mackenzie Basin, Canterbury, New Zealand. NZ J Agric Res 19:473–476
Darbyshire JF, Greaves MP (1970) An improved method for the study of the interrelationships of soil microorganisms and plant roots. Soil Biol Biochem 2:63–71
Dart PJ (1971) Scanning electron microscopy of plant roots. J Exp Bot 22:163–168
Dart PJ, Day JM (1975) Non-symbiotic nitrogen fixation in soil. In: Walker N (ed) Soil microbiology. Butterworth, London
Dart PJ, Mercer FV (1964) The legume rhizosphere. Arch Mikrobiol 47:344–378
Dawes CJ, Bowler E (1959) Light and electron microscope studies of the cell wall structure of the root hairs of *Raphanus sativus*. Am J Bot 46:561–565
Döbereiner J (1974) Nitrogen-fixing bacteria in the rhizosphere. In: Quispel A (ed) The biology of nitrogen fixation. Elsevier/North-Holland, Biomedical Press, Amsterdam New York
Döbereiner J, Day JM (1975) Associative symbiosis in tropical grasses: characterization of microorganisms and dinitrogen fixation sites. In: Proc Int Symp $N_2$ fixation. Interdisciplinary discussions. Pullman, Washington State Univ
Döbereiner J, Day JM, Dart PJ (1972) Nitrogenase activity and oxygen sensitivity of the *Paspalum notatum – Azotobacter paspali* association. J Gen Microbiol 71:103–116
Duff RB, Webley DM, Scott RO (1963) Solubilization of minerals and related materials by 2-Betagluconic acid-producing bacteria. Soil Sci 95:105–114
Elliott LF, Gilmour CM, Cochran VL, Coby C, Bennett D (1979) Influence of tillage and residues on wheat root microflora and root colonization by nitrogen-fixing bacteria. In: Harley JL, Scott-Russell R (eds) The soil–root interface. Academic Press, London New York
Foster RC (1962) Cell wall structure and growth. PhD thesis, Univ Leeds, England
Foster RC (1978) Ultramicromorphology of some South Australian Soils. In: Emerson WW, Bond RD, Dexter AR (eds). Modifications of soil structure. Wiley and Sons, New York
Foster RC, Marks GC (1967) Observations on the mycorrhizae of forest trees. II. The rhizosphere of *Pinus radiata* D. Don. Aust J Biol Sci 20:915–926
Foster RC, Marks GC (1973) Structure, morphogenesis and ultrastructure of ectomycorrhizae. In: Marks GC, Kozlowski TT (eds) Ectomycorrhizae. Academic Press, London New York
Foster RC, Rovira AD (1973) The rhizosphere of wheat roots studied by electron microscopy. Bull Ecol Res Commun (Stockholm) 17:93–95
Foster RC, Rovira AD (1976) Ultrastructure of wheat rhizosphere. New Phytol 76:343–352
Foster RC, Rovira AD (1978) The ultrastructure of the rhizosphere of *Trifolium subterraneum* L. In: Loutit MW, Miles JAR (eds) Microbial ecology. Springer, Berlin Heidelberg New York

Gadgil RL, Gadgil PD (1975) Suppression of litter decomposition by mycorrhizal roots of *Pinus radiata*. NZ J For Sci 5:33–41

Gerdemann JW (1975) Vesicular-arbuscular mycorrhizae. In: Torrey JG, Clarkson DT (eds) The development and function of roots. Academic Press, London New York

Gerretson FC (1948) The influence of microorganisms on the phosphate intake by the plant. Plant Soil 1:207–230

Gianinazzi S, Gianinazzi-Pearson Y, Dexheimer J (1979) Enzymatic studies on the metabolism of vesicular-arbuscular mycorrhiza. III. Ultrastructural localization of acid and alkaline phosphatase in onion roots infected by *Glomus mosseae* (Nicol. & Gerd). New Phytol 82:127–132

Gianinazzi-Pearson A, Gianinazzi S (1978) Enzymatic studies on the metabolism of vesicular-arbuscular mycorrhiza. II. Soluble alkaline phosphatase specific to mycorrhizal infection in onion roots. Physiol Plant Pathol 12:45–53

Gilmore AE (1971) The influence of endotrophic mycorrhizae on the growth of peach seedlings. J Am Soc Hort Sci 96:35–38

Gray TRG (1967) Stereoscan electron microscopy of soil microorganisms. Science 155:1668–1670

Greaves MP, Darbyshire JF (1972) The ultrastructure of the mucilaginous layer of plant roots. Soil Biol Biochem 4:443–449

Hale MG, Moore LD, Griffin GJ (1979) Root exudates and exudation. In: Dommergues YR, Krupa SV (eds) Interactions between non-pathogenic soil micro-organisms and plants. Elsevier, North Holland Biomedical Press, Amsterdam New York

Harley JL (1969) The biology of mycorrhiza. Hill, London

Hatch AB (1937) The physical basis of mycotrophy in the genus *Pinus*. Black Rock For Bull 6:1–168

Hattingh MJ, Gray LE, Gerdemann JW (1973) Uptake and translocation of $^{32}P$-labelled phosphate to onion roots by mycorrhizal fungi. Soil Sci 116:383–387

Hayman DS (1974) Plant growth responses to vesicular-arbuscular mycorrhiza. VI Effect of light and temperature. New Phytol 73:71–80

Hayman DS (1975) Phosphorus cycling by soil microorganisms and plant roots. In: Walker N (ed) Soil microbiology. Butterworth, London

Hayman DS, Mosse B (1972) Plant growth responses to vesicular-arbuscular mycorrhizas. IV. Increased intake of labile P from soil. New Phytol 71:41–47

Hiltner L (1904) Über neuere Erfahrungen und Probleme auf dem Gebiet der Bodenmikrobiologie und unter besonderer Berücksichtigung der Gründüngung und Brache. Arb Dtsch Landwirtsch Ber 98:59–78

Jenny H, Grossenbacher KA (1963) Root soil boundary zone as seen in the electron microscope. Proc Soil Soc Am 27:273–277

Jones PCT, Mollison JE (1948) A technique for the quantitative estimation of soil microorganisms. J Gen Microbiol 2:54–69

Katznelson H, Rouatt JW (1957) Manometric studies with rhizosphere and non-rhizosphere soil. Can J Microbiol 3:673–678

Katznelson H, Lochhead AG, Timonin MJ (1948) Soil microorganisms and the rhizosphere. Bot Rev 14:543–587

Katznelson H, Rouatt JW, Payne TMB (1954) Liberation of amino acids by plant roots in relation to desiccation. Nature 174:1110–1111

Katznelson H, Rouatt JW, Payne TMB (1955) The liberation of amino acids and reducing compounds by plant roots. Plant Soil 7:35–48

Kinden DA, Brown MF (1975) Electron microscopy of vesicular-arbuscular mycorrhizas of yellow poplar. II. Intercellular hyphae and vesicles. Can J Microbiol 21:1768–1780

Kleinschmidt GD, Gerdemann JW (1972) Stunting of citrus seedlings in fumigated nursery soils related to the absence of ectomycorrhizas. Phytopathology 62:1447–1453

Knowles R (1977) The significance of asymbiotic dinitrogen fixation in bacteria. In: Hardy RWF, Gibson AH (eds) A treatise on dinitrogen fixation, sect IV: Agronomy and ecology. Wiley and Sons, New York

Krupa S, Fries N (1971) Studies on the ectomycorrhizae of pine. I. Production of volatile organic compounds. Can J Bot 49:1425–1431

Lamb RJ, Richards BN (1971) Effect of mycorrhizal fungi on the growth and nutrient status of slash and radiata pine seedlings. Aust For 35:1–7
Lamont BB (1972) "Proteoid" roots in the legume *Viminaria juncea*. Search 3:91–92
Lamont BB, McComb AJ (1974) Soil microorganisms and the formation of proteoid roots. Aust J Bot 22:681–688
Leech JH, Mollenhauer HH, Whaley WG (1963) Ultrastructural changes in the root apex. Symp Soc Exp Biol 17:74–84
Leiser AT (1968) A mucilaginous root sheath in Ericaceae. Am J Bot 55:392–398
Leppard GG (1974) Rhizoplane fibrils in wheat: demonstration and derivation. Science 185:1066–1067
Leppard GG, Ramamoorthy S (1975) The aggregation of wheat rhizoplane fibrils and the accumulation of soil-bound cations. Can J Bot 53:1729–1735
Lewis DA, Harley JL (1965) Carbohydrate physiology of mycorrhizal roots of beech. III. Movement of sugars between host and fungus. New Phytol 64:256–259
Ling-Lee M, Chilvers GA, Ashford AE (1975) Polyphosphate in three different kinds of tree mycorrhiza. New Phytol 75:551–554
Lochhead AG (1940) Qualitative studies of soil microorganisms. III. Influence of plant growth on the character of the bacterial flora. Can J Res Sec C 18:42–58
Lochhead AG, Chase FE (1943) Qualitative studies of soil microorganisms. V. Nutritional requirements of the predominant bacterial flora. Soil Sci 55:185–195
Loutit MW (1968) Soil microorganisms and molybdenum concentration in plants. Trans 9th Int Congr Soil Sci 3:491–498
Loutit MW, Hillas J, Spears GFS (1972) Studies on rhizosphere microorganisms and molybdenum concentration in plants. II Comparison of isolates from the rhizosphere of plants grown in two soils under the same conditions. Soil Biol Biochem 4:267–270
Malajczuk N, Bowen GD (1974) Proteoid roots are microbially induced. Nature 251:316–317
Martin JK (1975) $^{14}$C-labelled material leached from the rhizosphere of plants supplied continuously with $^{14}CO_2$. Soil Biol Biochem 7:395–399
Martin JK (1977) Factors influencing the loss of organic carbon from wheat roots. Soil Biol Biochem 9:1–9
Melin E, Nilsson H (1950) Transfer of radioactive phosphorus to pine seedlings by means of mycorrhizal hyphae. Physiol Plant 3:88–92
Melin E, Nilsson H (1952) Transport of labelled nitrogen from an ammonium source to pine seedlings through mycorrhizal mycelium. Svensk Bot Tidskr 46:281–285
Melin E, Nilsson H (1955) $Ca^{45}$ used as an indicator of transport of cations to pine seedlings by means of mycorrhizal mycelia. Svensk Bot Tidskr 49:119–122
Menge JA, Labanauskas CK, Johnson ELV, Platt RG (1978) Partial substitution of mycorrhizal fungi for phosphorus fertilization in the greenhouse culture of citrus. Soil Sci Soc Am 42:926–930
Mishustin EN, Naumova AN (1962) Bacterial fertilizers: their effectiveness and mode of action. Mikrobiologiya 31:543–555
Moawad M (1979) Ecophysiology of v.a. mycorrhiza in the tropics. In: Harley JL, Russell RS (eds) The soil–root interface. Academic Press, London New York
Mollenhauer HH, Morre DJ (1966) Golgi apparatus and plant secretion. Annu Rev Plant Physiol 17:27–46
Morrison TM (1963) Uptake of sulphur by excised beech mycorrhizas. New Phytol 62:44–49
Mosse B (1972) The influence of soil type and *Endogone* strain on the growth of mycorrhizal plants in phosphate deficient soils. Rev Ecol Biol Sol 9:529–537
Mosse B (1973) Plant growth responses to vesicular-arbuscular mycorrhiza. IV. In soil given additional phosphate. New Phytol 72:127–136
Mosse B (1977) Plant growth responses to vesicular-arbuscular mycorrhiza. X. Responses of stylosanthes and maize to inoculation in unsterile soils. New Phytol 78:277–288
Mosse B, Powell CW, Hayman DW (1976) Plant growth responses to vesicular-arbuscular mycorrhiza. IX. Interactions between v.a. mycorrhiza, rock phosphate and symbiotic nitrogen fixation. New Phytol 76:331–342

Nambiar EKS (1976) The uptake of zinc-65 by oats in relation to soil water content and root growth. Aust J Soil Res 14:67–74
Newman EI, Bowen HJ (1974) Patterns of distribution of bacteria on root surfaces. Soil Biol Biochem 6:205–209
Newman EI, Watson A (1977) Microbial abundance in the rhizosphere. A computer model. Plant Soil 48:17–56
Norkrans B (1950) Studies in growth and cellulolytic enzymes of *Tricholoma* with special reference to mycorrhiza formation. Symb Bot Upsal 11(1):1–126
Northcote DH, Pickett-Heaps JD (1966) A function of the Golgi apparatus in polysaccharide synthesis and transport in the root cap cells of wheat. Biochem J 98:159–167
Nye PH, Tinker PB (1977) Solute movement in the soil-root system. Blackwell, Oxford
Old KM, Nicholson TH (1975) Electron microscopical studies of the microflora of roots of sand dune grasses. New Phytol 74:51–58
Parker CA (1957) Non-symbiotic nitrogen-fixing bacteria in soil. III. Total nitrogen changes in a field soil. J Soil Sci 8:48–59
Pearson V, Read DJ (1973) The biology of mycorrhiza in the Ericaceae. II. The transport of carbon and phosphorus by the endophyte and the mycorrhiza. New Phytol 72:1325–1331
Preston RD (1974) Physical biology of plant cell walls. Chapman and Hall, London
Purnell HM (1960) Studies of the family Proteaceae. I. Anatomy and morphology of the roots of some Victorian species. Aust J Bot 8:38–50
Ratnayake M, Leonard RT, Menge JA (1978) Root exudation in relation to the supply of phosphorus and its possible relevance to mycorrhizal formation. New Phytol 81:543–552
Raven JA, Smith FA (1976) Nitrogen assimilation and transport in vascular land plants in relation to intracellular pH regulation. New Phytol 76:415–431
Raven JA, Smith SE, Smith FA (1978) Ammonium assimilation and the role of mycorrhizas in climax communities in Scotland. Trans Bot Soc Edinburgh 43:27–35
Read DJ (1974) *Pezizella ericae* sp. nov. the perfect state of a typical mycorrhizal endophyte of Ericaceae. Trans Br Mycol Soc 63:381–383
Read DJ, Stribley DP (1973) Effect of mycorrhizal infection on nitrogen and phosphorus nutrition of Ericaceous plants. Nature 244:81–82
Reuszer HW (1949) A method for determining the carbon dioxide production of sterile and non-sterile roots. Soil Sci Soc Am Proc 14:175–179
Reuszer HW (1962) Axenic culture in the determination of root functions and the interrelationships of microorganisms. Soil Sci 93:56–61
Rhodes LH, Gerdemann JW (1975) Phosphate uptake zones of mycorrhizal and non-mycorrhizal onions. New Phytol 75:555–561
Rice EL (1974) Allelopathy. Academic Press, London New York
Ridge EH, Rovira AD (1968) Microbial inoculation of wheat. Trans 9th Int Congr Soil Sci 3:473–481
Rovira AD (1956) A study of the developments of the root surface microflora during the initial stages of plant growth. J Appl Bacteriol 19:72–79
Rovira AD (1959) Root excretions in relation to the rhizosphere effect. IV. The influence of plant species, age of plant, light, temperature and calcium nutrition on exudation. Plant Soil 11:53–64
Rovira AD (1965a) Plant root exudates and their influence upon soil microorganisms. In: Baker KE, Snyder WC (eds) Ecology of soil-borne plant pathogens. Berkeley Univ
Rovira AD (1965b) Effects of *Azotobacter, Bacillus* and *Clostridium* on the growth of wheat. In: Macura J, Vancura V (eds) Plant microbe relationships. Czech Acad Sci, Prague
Rovira AD (1969) Plant root exudates. Bot Rev 35:35–57
Rovira AD, Bowen GD (1969) Phosphate incorporation by sterile and non-sterile roots. Aust J Biol Sci 19:1167–1169
Rovira AD, Campbell R (1974) Scanning electron microscopy of microorganisms on the roots of wheat. Microb Ecol 1:15–23

Rovira AD, Davey CB (1974) Biology of the rhizosphere. In: Carson EW (ed) The plant root and its environment. Univ Press, Charlottesville, Virg
Rovira AD, Newman EI, Bowen HJ, Campbell R (1974) Quantitative assessment of the rhizoplane microflora by direct microscopy. Soil Biol Biochem 6:211–216
Rovira AD, Foster RC, Martin JK (1979) Note on terminology: Origin, nature and nomenclature of the organic materials in the rhizosphere. In: Harley JL, Scott-Russell R (eds) The soil-root interface. Academic Press, London New York
Samtsevich SA (1965) Active excretions of plant roots and their significance. Sov Plant Physiol 12:731–740
Sanders FE, Tinker PB (1973) Phosphate flow into mycorrhizal roots. Pestic Sci 4:385–395
Sanders FE, Mosse B, Tinker PB (1975) Endomycorrhizas. Academic Press, London New York
Sanders FE, Tinker PB, Black RLB, Palmerley SM (1977) The development of endomycorrhizal root systems. I. Spread of infection and growth-promoting effects with four species of vesicular-arbuscular endophyte. New Phytol 78:257–268
Schank SC, Smith RL, Weiser GC, Bouton JH, Quesenberry KH, Tyler ME, Milam JR, Littell RC (1979) Fluorescent antibody technique to identify *Azospirillum brasilense* associated with roots of grasses. Soil Biol Biochem 11:287–296
Schenk NC, Kinloch RA, Dickson DW (1975) Interaction of endomycorrhizal fungi and root-knot nematode on soybean. In: Sanders FE, Mosse B, Tinker PB (eds) Endomycorrhizas. Academic Press, London New York
Schippers B, Vuurde van JWL (1978) Studies of microbial colonization of wheat roots and the manipulation of the rhizosphere microflora. In: Loutit MW, Miles JAR (eds) Microbial ecology. Springer, Berlin Heidelberg New York
Schönbeck F (1979) Endomycorrhiza in relation to plant diseases. In: Schippers B, Gams W (eds) Soil-borne plant pathogens. Academic Press, London New York
Schoknecht JD, Hattingh MT (1976) X-ray microanalysis of elements in cells of v.a. mycorrhizal and non-mycorrhizal onions. Mycologia 68:296–304
Schwartz F (1883) Die Wurzelhaare der Pflanzen. Untersuch Bot Inst Tübingen 1:135–188
Scott FM, Hamner KC, Baker E, Bowler F (1958) Electron microscope studies of the epidermis of *Allium cepa*. Am J Bot 45:449–461
Setterfield G, Bayley ST (1957) Studies on the mechanism of deposition and extension of primary cell walls. Can J Bot 35:435–444
Shamoot S, McDonald L, Bartholemew MV (1960) Rhizodeposition of organic debris in soil. Soil Sci Am Proc 32:817–820
Skinner MF, Bowen GD (1974a) The uptake and translocation of phosphate by mycelial strands of pine mycorrhizas. Soil Biol Biochem 6:53–56
Skinner MF, Bowen GD (1974b) The penetration of soil by mycelial strands of pine mycorrhizas. Soil Biol Biochem 6:57–61
Smiley RW (1974) Rhizosphere pH as influenced by plants, soils, and nitrogen fertilizers. Proc Soil Sci Soc Am 38:795–799
Smith SE, Nicholas DJD, Smith FA (1979) Effect of early mycorrhizal infection on nodulation and nitrogen fixation in *Trifolium subterraneum* L. Aust J Plant Physiol 6:305–316
Sperber JI (1958) The incidence of apatite-solubilizing organisms in the rhizosphere and soil. Aust J Agric Res 9:778–781
Starkey RL (1939) Some influence of the development of higher plants upon the microorganisms in the soil. VI. Microscopic examination of the rhizosphere. Soil Sci 45:207–209
Stribley DP, Read DJ (1974) The biology of mycorrhiza in the Ericaceae. IV. The effect of mycorrhizal infection on the uptake of $^{15}N$ from labelled soil by *Vaccinium macrocarpon* Ait. New Phytol 73:1149–1155
Stribley DP, Read DJ (1975) Some nutritional aspects of the biology of ericaceous mycorrhizae. In: Sanders FE, Mosse B, Tinker PB (eds) Endomycorrhizae. Academic Press, London New York

Stribley DP, Read DJ (1976) The biology of mycorrhiza in the Ericaceae. VI. The effects of mycorrhizal infection and concentration of ammonium nitrogen on growth of cranberry (*Vaccinium macrocarpon* Ait.) in sand culture. New Phytol 77:63–72

Stribley DP, Reid DJ, Hunt R (1975) The biology of mycorrhiza in the Ericaceae V. The effects of mycorrhizal infection, soil type and partial soil-sterilization (by gamma irradiation) on the growth of cranberry (*Vaccinium macrocarpon* Ait.). New Phytol 75:119–130

Sutton JC, Sheppard BR (1976) Aggregation of sand dune soil by endomycorrhizal fungi. Can J Bot 54:326–333

Theodorou C, Bowen GD (1970) Mycorrhizal responses of radiata pine in glasshouse and in field experiments with different fungi. Aust For 34:183–191

Theron JJ (1966) The mineralization of nitrogen in soils under grass. S Afr J Agric Sci 6:155–164

Timonin MI (1946) Microflora of the rhizosphere in relation to the manganese-deficiency disease in oats. Soil Sci Soc Am Proc 11:284–292

Tisdall JM, Oades JM (1979) Stabilization of soil aggregates by the root systems of rye grass. Aust J Soil Res 17:429–441

Trinick MJ (1970) *Rhizobium* interactions with soil microorganisms. PhD thesis Univ Washington WA

Trinick MJ (1977) Vesicular-arbuscular infection and soil phosphorus utilization in *Lupinus* spp. New Phytol 78:297–304

Trolldenier G (1965) Fluoreszenmikroskopische Untersuchung von Mikroorganismenkulturen in der Rhizosphäre. Z Bakteriol Parasitenkd Infekt Hyg III 119:256–259

Vancura V (1964) Root exudates of plants. I. Analysis of root exudates of barley and wheat in their initial phases of growth. Plant Soil 21:231–248

Vuurde van JWL, Elenbass PEM (1978) Use of fluorochromes for direct observation of microorganisms associated with wheat roots. Can J Microbiol 24:1272–1275

Warembourg FR (1975) Le dégagement de $CO_2$ dans la rhizosphère des plantes. Soc Bot Fr Colloq Rhizosphere 122:77–87

Warembourg FR, Billes G (1979) Estimating carbon transfers in the rhizosphere. In: Harley JL, Scott-Russell R (eds) The soil-root interface. Academic Press, London New York

Webley DM, Duff RB, Bacon JSD, Farmer VC (1965) A study of polysaccharide-producing organisms occurring in the root region of certain pasture grasses. J Soil Sci 16:149–157

Welte E, Trolldenier G (1965) Effect of soil microflora and seed microflora on the growth of plants. In: Macura J, Vancura V (eds) Plant microbe relationships. Czech Acad Sci, Prague

Woldendorp JW (1963) The influence of living plants on denitrification. Meded Landbhoogesch Wageningen 63(13):1–100

# I.3 Modern Solution Culture Techniques

C.J. ASHER and D.G. EDWARDS

It has been known since the very early days of plant physiology that land plants as well as water plants could be grown successfully without soil in aqueous culture media. Although the famous experiments of WOODWARD (1699) represent the earliest recorded use of solution culture[1] techniques to study the growth of land plants, the experiments of DE SAUSSURE (1804) are probably more significant in that they involved the first use of nutrient solutions of known initial composition, prepared by dissolving various salts in distilled water. Subsequently SACHS (1860) and KNOP (1865) showed that plants could be grown to maturity in simple solution culture systems not very different from those currently in use in many plant research laboratories throughout the world. The application of solution culture techniques has contributed much to current knowledge within the broad field of plant physiology, especially in the areas of mineral nutrition and water relations. However, with any set of techniques which have been in use for a long period of time there is the risk that excessive familiarity will cause research workers to lose sight of some inherent limitations of the techniques and use them in ways that are inappropriate.

In this chapter, emphasis will be placed on the uses and limitations of various types of solution culture techniques rather than on detailed descriptions of particular techniques. For the latter, readers are referred particularly to the excellent and comprehensive review by HEWITT (1966).

## 1 Major Differences Between Solution Culture and Soil Culture

Solution culture techniques are often used to simulate, under at least partially controlled conditions, situations which occur when plants are growing in the field. However, if solution culture is to be used as a model experimental system for soil-grown plants it is important that the differences between the two culture systems are kept firmly in mind.

### 1.1 Mechanical Support

Soil-grown plants receive their major mechanical support from the root system which ramifies through the soil matrix. This occurs also in sand culture and

1 Throughout the present chapter the general term *solution culture* has been taken to include a wide range of related plant culture techniques used for experimental purposes. These include conventional (non-flowing) water cultures, sand and gravel culture, mist culture and various types of flowing solution culture systems. The term *hydroponics* is used to denote crop production by a solution culture method

gravel culture, but in water culture systems an alternative means of mechanical support is required. Frequently the plants are supported by wrapping a short section of the stem with cotton or dacron wool, or a piece of soft plastic foam, and then wedging the stem into a hole in the lid of the culture vessel, or into a cork or rubber stopper inserted in a still larger hole in the lid. Where substantial radial expansion of the stem is expected during the course of an experiment, the cork should have a radial split so that it can be removed easily and replaced with a second cork with a larger hole, to prevent undue pressure on the base of the stem. This system works quite well with single-stemmed plants such as cotton or tomato, but is less convenient for plants which tiller from the base, e.g. most cereals and grasses.

For species which branch close to the ground or form a rosette of leaves at the soil surface, adequate support may be obtained by passing the tap root or the base of the stem through a slit in a piece of soft plastic film glued across a hole in the lid with contact adhesive (ASHER et al. 1965). Radial growth of the root or stem base is accommodated by stretching or tearing of the film. However, the method is unsuitable for supporting large plants or those which tiller from the base.

Basket supports provide a highly versatile plant support system suitable for species with widely differing growth habits (ASHER 1978). The use of these was advocated by GERICKE (1929) for intensive hydroponic crop production, but gravel culture systems proved to be more practical. For experimental purposes it is convenient to construct the baskets from chemically inert, non-phytotoxic plastics such as nylon, polyethylene, or polypropylene which are easily cleaned, a point of considerable importance in micronutrient studies. Usually the sides of the basket are solid, and the base is formed from plastic mesh through which the roots grow to reach the nutrient solution. Use of mesh in which the fibers are not bonded at points of intersection allows lateral displacement of the fibers to accommodate radial expansion of the roots. Usually the baskets are filled with a mulch of black polyethylene beads to exclude light from the nutrient solution and to provide additional support for the base of the stem. Spray from the bursting of bubbles at the surface of adequately aerated nutrient solutions keeps the lower 1 to 1.5 cm of the bead layer wet enough for seed germination and for the development of adventitious roots. The system is eminently suited for species which tiller or develop rhizomes. Baskets as small as 2 cm diameter (Fig. 1) and as large as 25 cm diameter have been found convenient in plant nutrient studies with various species (ASHER 1978). Two examples of basket supports are shown in Fig. 1.

### 1.2 Spatial Variation in Root Environment Parameters

Soil is an extremely complex and heterogeneous medium often varying substantially in physical, chemical or biological characteristics over comparatively small horizontal or vertical distances. A consequence of this variation is that different parts of a plant root system may be located in markedly different environments. In addition, the ionic environment experienced by any particular portion of a plant root system in the soil depends on a balance between rates of ion move-

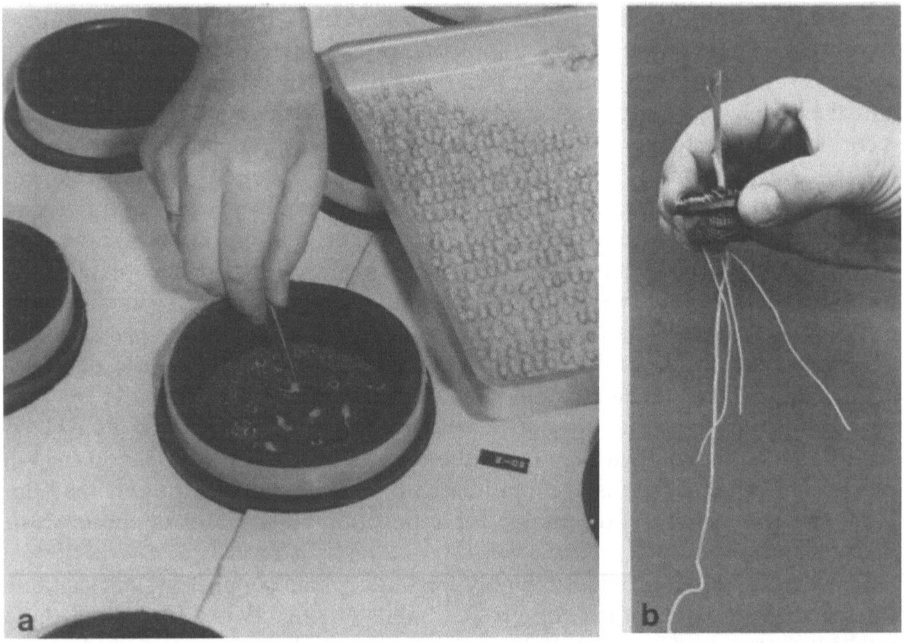


**Fig. 1a, b.** Basket supports for solution culture experiments. **a** 10 cm inside diameter basket being planted with sprouted wheat seeds for a flowing solution culture experiment (cf. Sect. 2.3). Mulch of black polyethylene beads in previously planted pots visible at *left* and in *background*. **b** 2 cm diameter basket used to support a maize seedling

ment to the root surface by mass flow and diffusion, and rates of absorption by that portion of root (BARBER 1962). Hence differential uptake of water and nutrient ions by different parts of the root system adds further to the heterogeneity of the environment to which the root system of a soil-grown plant is exposed.

Because of the complexity and heterogeneity of the soil it is extremely difficult to characterize the root environment of soil-grown plants precisely or to exercise adequate control over root environment parameters, the effects of which we may wish to study. However in well-stirred solution culture systems it is possible to provide a very nearly homogeneous root environment and to measure with acceptable accuracy such root environment parameters as pH, $pO_2$, $E_h$ (oxidation-reduction potential), nutrient ion concentrations, and temperature. Within wide limits these parameters can be varied independently of each other, something which frequently cannot be achieved in soil. Solution culture thus represents a preferred system for studying effects of root environment parameters on the physiology and growth of plants.

However, it must always be kept in mind that solution culture systems are much simpler than soil systems, and as a consequence, responses of soil-grown plants to various root environment parameters may be modified by factors not present in solution culture systems. Such factors include large numbers of soluble mineral elements not known to be essential for plant growth and therefore omitted from nutrient solutions, the wide variety of soluble organic compounds commonly occurring in the soil solutions, and the varied populations of microorganisms associated with the roots of soil-grown plants (see Sect. 1.4).

## 1.3 Temporal Variation in Root Environment Parameters

### 1.3.1 Nutrient Depletion

In soils, adsorption/desorption reactions, the slow dissolution of primary minerals, and the decomposition of soil organic matter serve to replenish the soil solution as mineral nutrients are withdrawn by the plant roots. Although both local depletion (BARBER 1962) and general depletion of the soil solution (BURD and MARTIN 1924) may occur over a period of time, in the majority of soils the solution in contact with the roots is buffered to a substantial degree against changes in mineral composition. Continued growth of roots into "new" soil further tends to stabilize the average ionic environment of the whole root system.

In simple water culture systems, the composition of the nutrient solution is essentially unbuffered and large changes in solution composition can occur within a relatively short space of time (Table 1). The rate of depletion depends on a number of factors including the rate of nutrient uptake by the plants, the initial concentrations of ions in the nutrient solution, and the volume of nutrient solution per plant. In sand culture the depletion problem tends to be more severe than in water culture because of the absence of stirring and the relatively small volumes of nutrient solution retained in the sand after drainage.

The need to supply adequate total amounts of mineral nutrients in conveniently small volumes of nutrient solution has led to the use of high initial concentrations of many ions. In the case of phosphorus, the initial concentrations frequently exceed those occurring in soil solutions by 3 to 4 orders of magnitude. Thus Hoagland's solution contains 1,000 μM phosphorus (HOAGLAND and ARNON 1950) and the Long Ashton solution, 1,330 μM (HEWITT

**Table 1.** Effects of initial potassium concentration and volume of nutrient solution per plant on changes with time in the concentration of potassium in nutrient solutions supporting the growth of young cassava plants. (Modified after SPEAR et al. 1978)

| Time from planting (days) | Volume of solution ($l\ plant^{-1}$) | | | | | | | | |
|---|---|---|---|---|---|---|---|---|---|
| | 1 | 2.5 | 8 | 1 | 2.5 | 8 | 1 | 2.5 | 8 |
| | Potassium concentrations in solution (μM)[a] | | | | | | | | |
| 0 | 20 | 20 | 20 | 300 | 300 | 300 | 1875 | 1875 | 1875 |
| 4 | 2 | 2 | 1 | 3 | 128 | 218 | 1065 | 1510 | 1590 |
| 10 | <1 | <1 | <1 | <1 | 1 | 95 | 130 | 950 | 1440 |
| 18 | <1 | <1 | <1 | <1 | <1 | 1 | <1 | 37 | 740 |
| 27 | <1 | <1 | <1 | <1 | <1 | <1 | <1 | 9 | 280 |

[a] Limit of detection (by atomic absorption spectroscopy) was 1 μM K

1966). However, the modal soil solution phosphorus concentration in 149 temperate soils was found to be less than 1 μM (REISENAUER 1966), and in six tropical soils GILLMAN and BELL (1978) found values below 0.2 μM.

### 1.3.2 pH Shifts

Unequal absorption of anions and cations from the root environment can lead to substantial changes in pH. Because of the large amounts in which nitrogen is required compared with the other mineral nutrients, the form in which nitrogen is supplied tends to exert a major influence on the direction of pH change. Thus absorption of nitrogen in anionic form ($NO_3^-$) generally leads to an increase in the pH of the root environment, whereas absorption of cationic nitrogen ($NH_4^+$) leads to a decrease in pH. TRELEASE and TRELEASE (1935) showed in water culture experiments with wheat that by varying the $NO_3^-/NH_4^+$ ratio they could cause the pH to increase, decrease, or remain approximately constant. However, species differences in nutrient absorption and assimilation may be important, since it has been consistently observed with non-nodulated jackbeans (*Canavalia ensiformis*) that the pH of the nutrient solution decreases markedly with time even when all the nitrogen is supplied in the $NO_3^-$ form (ASHER unpublished). With concentrated nutrient solutions some phosphate buffering of pH can be expected, particularly in near-neutral solutions ($pK_a = 7.21$). However, in general, pH shifts in nutrient solutions are likely to be greater than in soils because of the lack of an exchange complex to adsorb or desorb hydrogen ions.

## 1.4 Root–Microorganism Interactions

Roots of soil-grown plants exude substantial quantities of organic compounds into the soil. According to BARBER and MARTIN (1976) as much as 12%–18% of the total carbon fixed from the atmosphere by cereal plants may be lost in this manner. Accumulation of these organic substances in the soil adjacent to roots provides a rich substrate for the growth of a wide range of microorganisms. (For a review on factors affecting the colonization of root surfaces by microorganisms see BOWEN and ROVIRA 1976). In many cases the precise effects of these microorganisms on the growth and function of roots has not been elucidated and much remains to be learned in this area.

In solution culture roots appear to produce much smaller quantities of exudate than in soil (BOWEN and ROVIRA 1976) and the water-soluble components of the exudate will be dispersed through the bulk of a well-stirred nutrient solution. Hence the quantities of organic substrates available to support microbial growth on root surfaces of plants grown in solution culture are likely to be much less than for roots of soil-grown plants. However, despite this difference, it has proved possible to study some important root–microorganism interactions in solution culture and much has been learned from these experiments (see Sect. 2.1.7).

# 2 Uses and Limitations of Existing Solution Culture Methods

## 2.1 Non-Renewed or Intermittently Renewed Water Cultures and Sand Cultures

These are the simplest and most widely used solution culture methods in plant research laboratories today. Large variations exist among research workers in the initial composition of the nutrient solutions employed, the volume of solution provided per plant, and the frequency of solution renewal, if any (see HEWITT 1966). Frequently small volumes ($\leqslant 1$ l $plant^{-1}$) of rather concentrated nutrient solution ($\geqslant 20{,}000$ μM total nutrients) are employed for reasons of experimental convenience. In some cases the initial concentrations are so high that complete dissolution of the salts is not possible or precipitation subsequently occurs (e.g. BOND 1951; NORRIS and DATE 1976). In other cases thresholds for toxicity in sensitive species are exceeded (e.g. WILLIAMS and VLAMIS 1957) so that growth is restricted at least until nutrient depletion lowers the concentration of the toxic ion to a safe level. Since the initial concentrations commonly employed in nutrient solutions far exceed both the minimum concentrations necessary for healthy growth (ASHER 1978) and the concentrations commonly occurring in soil solutions, better results might often be obtained in solution culture experiments if the initial concentrations were reduced, and the volume of solution or frequency of renewal increased correspondingly. The composition of some well-known nutrient solutions with differing sources of nitrogen are shown in Table 2.

### 2.1.1 Use in Teaching, Demonstration, and Diagnosis

Water culture and sand culture provide convenient means of raising plant material in a standard manner for a wide variety of teaching and demonstration purposes. With water culture the study of root anatomy and morphology is facilitated by easy access to the root system. However, in soil-grown plants anatomical features may be modified by physical contact with the soil and by exposure to rhizosphere organisms not necessarily present in the nutrient solution. Hence care should be taken not to extrapolate too widely from observations made on the root systems of plants grown in water culture.

Symptoms of many nutritional disorders are readily induced by modifying the initial composition of the nutrient solution. For the majority of temperate species of economic importance, the visible symptoms of deficiency of at least most of the essential elements, and symptoms of the most common mineral toxicities, are well-known and adequately described in the literature. These descriptions represent an extremely valuable aid to the diagnosis of nutritional problems under both greenhouse and field conditions. However, for many tropical crop and pasture species, and for many native plants of both tropical and temperate regions this is not the case. Hence very considerable scope still exists for solution culture experiments aimed at the production and description of nutritional disorders in these species.

**Table 2.** Some well-known nutrient solutions having either no nitrogen (for studies on nitrogen fixation) or the nitrogen supplied as nitrate, ammonium, or nitrate plus ammonium. (Modified after ASHER 1978, all concentrations in µM)

| Nitrogen source | None | | Nitrate | Ammonium | | Nitrate plus ammonium | | |
|---|---|---|---|---|---|---|---|---|
| Reference | BOND (1951)[a] | NORRIS and DATE (1976)[a, b] | HOAGLAND and ARNON (1950) | SIDERIS et al. (1943)[c] | ADDOMS (1937)[d] | HOAGLAND and ARNON (1950) | MULDER (1948) | JOHNSON et al. (1957)[e] |
| Nitrogen | | | | | | | | |
| $NO_3^-$ | | | 15,000 | | | 14,000 | 3,460 | 14,000 |
| $NH_4^+$ | | | | 2,000 | 11,110 | 1,000 | 2,270 | 2,000 |
| Potassium | 10,060 | 3,000 | 6,000 | 1,500 | 6,390 | 6,000 | 6,370 | 6,000 |
| Calcium | 5,320 | 2,000 | 5,000 | 1,000 | 2,920 | 4,000 | 1,453 | 4,000 |
| Magnesium | 2,030 | 1,000 | 2,000 | 1,000 | 1,385 | 2,000 | 1,014 | 1,000 |
| Phosphorus | 2,610 | 1,000 | 1,000 | 500 | 6,390 | 1,000 | 2,190 | 2,000 |
| Sulfur | 4,930 | 3,005 | 2,000 | 1,500 | 6,950 | 2,000 | 3,620 | 1,000 |
| Chlorine | 10,075 | 1,000 | 18 | 2,000 | | 18 | 279 | 50 |
| Iron | 1,495 | 5.4 | 25[f] | 100 | 12 | 25[f] | 93 | 4[f] |
| Boron | 46 | 12 | 46 | 46 | 16 | 46 | 4 | 25 |
| Manganese | 9 | 4.6 | 9 | 9 | 7 | 9 | 4.5 | 5 |
| Zinc | 0.8 | 0.4 | 0.8 | 1.5 | | 0.8 | 0.9 | 2 |
| Copper | 0.3 | 0.16 | 0.3 | 1.6 | | 0.3 | 1 | 0.5 |
| Molybdenum | 0.1 | 0.03 | 0.1 | | | 0.1 | 0.4 | 0.1 |
| Other elements | | | | I, 0.8 | | | Na, 9 | |
| Total elements | 36,576 | 11,027 | 31,099 | 9,659 | 35,180 | 30,099 | 20,759 | 30,087 |

[a] Concentrations which would occur if all nutrient salts dissolved completely and no precipitation occurred
[b] A relatively dilute nutrient solution now commonly used in place of the more concentrated solution of NORRIS (1964) which has been found to inhibit the growth and nodulation of some legume species. Note however that precipitation (?$CaSO_4$) still occurs in this solution (NORRIS and DATE 1976)
[c] A relatively dilute nutrient solution used for experiments with pineapple
[d] Solution "B"
[e] Often used in place of Hoagland-Arnon solution, frequently at $^1/_{10}$th strength for young seedlings and $^1/_3$ to $^1/_2$ strength for older plants. $NH_4^+$:$NO_3^-$ of 1:7 maintains pH close to 5
[f] Renewed twice weekly

### 2.1.2 Production of Roots for Ion Transport Studies

Much of our existing knowledge concerning processes of ion uptake and long-distance transport has come from studies of whole plants or excised roots raised in simple water culture systems. Excised roots of low salt status are often produced by substituting a dilute calcium sulphate solution (ca. 200 μM) for the nutrient solution (see EPSTEIN 1972).

### 2.1.3 Nutrient Essentiality

The simplicity of water culture systems, and the availability of technology for reducing to very low levels contaminating elements in water and nutrient salts makes water culture a highly suitable technique for investigating the possible roles of elements not at present regarded as essential for plant growth. Sand culture is less suitable for this purpose, because of the need to purify the sand as well as the water and nutrient salts.

Less than half the elements known to occur in plant tissues have been shown to be essential for plant growth, and many of the remaining elements have not been studied carefully for possible roles using modern plant culture and purification techniques. Unfortunately this important area of plant nutrition research has been relatively neglected during the past two decades. Thus only two major discoveries have been made since 1959: the essential role of cobalt for the growth of nodulated legumes (AHMED and EVANS 1960; HALLSWORTH et al. 1960; REISENAUER 1960), and the essentiality of silicon for the growth of the sedge *Equisetum arvense* (CHEN and LEWIN 1969) and for tomato (MIYAKE and TAKAHASHI 1978). At present few laboratories appear to be directing significant resources to the search for additional essential elements for plant growth. However, following the demonstration by DIXON et al. (1975) that nickel is required for the functioning of the enzyme urease, a number of groups have recently commenced research on nickel as a possible essential element.

The present low level of research activity on nutrient essentiality for plants contrasts markedly with the high level of activity among animal science groups. Thus SCHWARZ (1974) drew attention to six discoveries of new trace elements for animals during the 15-year period 1959 to 1974 (chromium, tin, vanadium, fluorine, silicon, and nickel), and indicated that a further 14 elements were under intensive investigation at that time.

### 2.1.4 Effects of Root Environment Parameters

In general, simple water culture and sand culture systems are not very suitable for the study of effects of root environment parameters because of difficulty in maintaining constant pH and nutrient ion concentrations (see Sects. 1.3.1, 1.3.2 and 2.1.4.1). The preferred method for such studies is by means of continuously flowing solution cultures (see Sect. 2.3). However, since the majority of laboratories are not yet equipped with flowing culture equipment, it is important to examine critically the limitations of conventional solution culture systems and to determine the purposes for which they may legitimately be used, and those for which their use is inappropriate.

#### 2.1.4.1 Temperature Studies

Active ion uptake processes are markedly temperature-sensitive, but relatively little attention has been paid to effects of root temperature on growth and nutrient uptake by intact plants. Satisfactory control of root temperature can be obtained in water culture experiments by standing the pots in large constant temperature waterbaths. At the authors' laboratory four such waterbaths measuring approximately 2 m × 1 m × 20 cm deep have been in use for some years. The baths are constructed from stainless steel and insulated on the sides and bottom with polyurethane foam. The heat load from above is reduced by covering the baths with sheets of 1.5-cm-thick waterproof plywood carrying a pattern of holes of a suitable size to fit snugly around the tops of the pots. Aeration of the nutrient solutions provides sufficient stirring to prevent the development of appreciable temperature gradients in the nutrient solution. Using this equipment FORNO et al. (1979) showed that when five cassava cultivars were grown in full strength Hoagland's solution (HOAGLAND and ARNON 1950) all developed severe boron deficiency at a root temperature of 18° C, but absorbed sufficient boron for healthy growth at 28° C. Further experiments on effects of root temperature on mineral nutrition are needed.

#### 2.1.4.2 pH Studies

ARNON and JOHNSON (1942) successfully used water culture techniques to study effects of pH on tomato, lettuce and Bermuda grass. However, precise pH control was difficult to achieve even with a large volume of solution (10–20 l $plant^{-1}$), continuous slow addition of acid or alkali to the pots according to treatment, and twice-daily checks of pH on an individual pot basis. In some treatments deviations from the desired pH values exceeded 0.2 pH units, despite the precautions taken.

Where it is necessary only to maintain a favourable pH for plant growth, addition of solid $CaCO_3$ to the nutrient solution may be effective. Thus in experiments on the role of cobalt in symbiotic nitrogen fixation in lucerne, WILSON and REISENAUER (1963) were able to maintain a pH of between 6.5 and 7.0 by the addition of $CaCO_3$ at a rate of 1 g $l^{-1}$. KNYPL (1976) compared $CaCO_3$ and $CaHPO_4$ as pH buffering agents in culture media for *Spirodela oligorrhiza*. Although use of $CaCO_3$ gave markedly better control of pH, growth was superior with $CaHPO_4$. However, as a saturated solution of $CaHPO_4$ corresponds with a phosphate concentration of approximately 1,800 μM, use of this salt could lead to the maintenance of phosphate concentrations in the toxic range for susceptible species (cf. ASHER 1978).

#### 2.1.4.3 Effects of Nutrient Ion Concentrations

The usefulness of conventional solution culture methods for studying effects of nutrient ion concentrations on nutrient uptake and plant growth is severely restricted by the problem of nutrient depletion. Attention was drawn to this problem long ago by STILES (1916) but it is clear from a study of the modern plant nutrition literature that many research workers still do not take adequate account of nutrient depletion when designing solution culture experiments or interpreting their results.

In any particular situation the magnitude of the depletion problem will depend on a number of factors, important amongst which are (1) the accuracy with which concentration needs to be controlled to achieve the objectives of the experiment, (2) the range of concentrations over which the experiment is to be conducted, (3) the volume of solution provided per unit weight of roots, and (4) the mean rate of uptake per unit root weight at the specified concentrations. Since the root weight of young plants will often increase many-fold during the course of an experiment, the maximum probable root weight should be used when designing measures to restrict solution depletion to acceptable levels.

Probable levels of depletion can be characterized either in terms of the ratio of anticipated daily uptake to the total nutrient supply (ASHER 1981) or in terms of the frequency of renewal which would be required to limit depletion to a specified percentage of the initial concentration of an ion. In the latter case the maximum permissible time interval in hours between solution renewals (T) can be calculated approximately from the following equation:

$$T=\frac{D}{100}\cdot\frac{V}{R}\cdot\frac{C}{U} \tag{1}$$

where D = maximum acceptable depletion (%), V = volume of solution per pot (or per plant) (litres), C = initial concentration of an ion in the solution (μM), R = root weight per pot (or per plant) (g fresh wt.), and U = uptake rate per unit root weight (μmol $g^{-1}$ fresh wt. $h^{-1}$) at concentration C. (Strictly speaking, the value of U should be for the mean concentration between solution renewals, i.e. $\frac{C(200\text{-}D)}{200}$. However, for small values of D the slight underestimation of T resulting from using the U value corresponding with C can be neglected).

Values of U taken from short-term uptake experiments with low salt roots may seriously overestimate the average uptake rates that would occur in longer-term experiments (JOHANSEN et al. 1968) and hence yield misleadingly low values of T. Probably the best source of U values for experiments of several days duration or longer are flowing culture experiments in which the plant root systems have been exposed to a constant external concentration of the ion for an extended period of time.

In Table 3 maximum acceptable times between solution renewals have been calculated for solution to root $\left(\frac{V}{R}\right)$ ratios of 0.1, 0.5 and 2.5 l $g^{-1}$ fresh root wt., and maximum depletions (D) of 5% or 10%, using published uptake data from flowing culture experiments on zinc (CARROLL and LONERAGAN 1969) and phosphorus (LONERAGAN and ASHER 1967) as an example of a micronutrient and a macronutrient element. For most purposes $\frac{V}{R}$ values in excess of 0.5 l $g^{-1}$ are probably not practical. In Carroll and Loneragan's study, root weights after 46 days at 0.25 μM zinc ranged from 85 g $pot^{-1}$ (lucerne) to 154 g $pot^{-1}$

**Table 3.** Maximum acceptable times between solution renewals calculated for various solution: root ratios and maximum depletion levels, using published information on rates of zinc and phosphorus uptake

| C (μM) | Adequacy for plant growth | U ($\mu mol\ g^{-1}\ h^{-1}$) | $\frac{V}{R}$ ($l\ g^{-1}$) | | | | | |
|---|---|---|---|---|---|---|---|---|
| | | | D=5% | | | D=10% | | |
| | | | 0.1 | 0.5 | 2.5 | 0.1 | 0.5 | 2.5 |
| Zinc[a] | | | Solution renewal times, T (h) | | | | | |
| 0.01 | Deficient | 0.00013 | 0.4 | 2.0 | 9.8 | 0.8 | 3.9 | 20 |
| 0.25 | Adequate | 0.00076 | 1.6 | 8.2 | 41 | 3.3 | 16.4 | 82 |
| 6.25 | Adequate-toxic | 0.0134[b] | 2.3 | 11.7 | 58 | 4.5 | 23 | 117 |
| Phosphorus[c] | | | | | | | | |
| 0.04 | Deficient | 0.013 | 0.02 | 0.08 | 0.4 | 0.03 | 0.15 | 0.8 |
| 5 | Adequate | 0.437 | 0.06 | 0.3 | 1.4 | 0.1 | 0.6 | 2.9 |
| 24 | Adequate-toxic | 0.512[d] | 0.2 | 1.2 | 5.9 | 0.5 | 2.3 | 11.7 |

[a] Carroll and Loneragan (1969)
[b] Toxic for 3 out of 5 species studied
[c] Loneragan and Asher (1967)
[d] Toxic for 3 out of 8 species studied

(wheat). At $\frac{V}{R}=0.5$ these root weights would require pot sizes of 42.5 and 77 l respectively. Again, for experiments of more than a few days' duration it is probably impractical to renew nutrient solutions more often than once or at most twice per day because of the amount of labour involved.

The data in Table 3 show that for zinc concentrations in the adequate to toxic range it may be practical to use a simple, large volume water culture system particularly if 10% depletion is acceptable between solution renewals. However, the data indicate that adequate control of zinc concentration in the deficiency range is not likely to be achieved at practical values of T and $\frac{V}{R}$ (Table 3). For phosphorus, the data show that, because of the much higher rates of uptake at equivalent concentrations, adequate control of solution phosphorus concentration could not be obtained even at levels toxic to some species. The situation with other mineral elements will vary depending on the magnitude of the values to be inserted in Eq. (1). However, the limited evidence at present available suggests that effective control of concentration in simple solution culture systems is likely to be practical only with respect to studies of toxicity of the more slowly absorbed mineral elements, e.g. aluminum, micronutrients, and some environmental contaminants.

In sand culture systems the depletion problem tends to be intensified by the following factors:

1. relatively small volumes of solution are retained in the pots after drainage;
2. the solution retained in the pots is unstirred, hence zones of local depletion may occur adjacent to root surfaces;

3. in sub-irrigation systems, a substantial part of the total solution volume is stored outside the pots between flushing cycles, and consequently is not available for uptake for much of the time.

The impact of these factors can be lessened by increasing the frequency of flushing, but cannot be eliminated entirely. Sand cultures are thus unsuitable for quantitative studies on effects of external nutrient ion concentrations. As pointed out by ASHER (1980), nutrient depletion in solution culture experiments leads to overestimation of critical external concentrations for both deficiency and toxicity.

### 2.1.5 Establishment of Critical Tissue Concentrations

Critical tissue concentrations for the diagnosis of nutrient deficiencies and toxicities are frequently established from water culture or sand culture experiments. Although many plant and environmental factors have been shown to affect measured critical concentrations (BATES 1971), it has been widely assumed that critical tissue concentrations are a comparatively stable plant characteristic unlikely to be affected by temporal variation in the external supply of the element concerned. However, recent work of SPEAR et al. (1978, 1979) raises serious doubts about this assumption. Thus critical potassium concentrations in index leaves and whole tops of cassava plants grown in small pots (1 l solution plant$^{-1}$) in which depletion of solution potassium was rapid, were much lower than in plants grown in larger pots (2.5 or 8 l plant$^{-1}$) in which depletion was slower, or in flowing solution culture in which depletion was prevented. Similar results were obtained in a flowing culture experiment in which depletion was simulated by transferring plants from various constant potassium concentrations to a nutrient solution containing no added potassium.

These results suggest that arbitrary decisions concerning details of the solution culture technique to be employed can have large effects on measured critical concentrations. The results also call into question the validity of using critical concentrations established in solution culture experiments for diagnosis of nutritional problems in the field where temporal variations in nutrient supply may be quite different from those in a particular solution culture experiment.

### 2.1.6 Control of Plant Nutrient Status

One of the consequences of nutrient depletion in conventional solution culture systems is that large changes in the nutrient status of plants can occur during the course of an experiment. Thus plants which ultimately show severe deficiency of an element may have passed through an earlier period of luxury consumption or even toxicity of the same element (ASHER and COWIE 1970). Hence effects of nutrient status on the physiology and growth of plants are difficult to examine. Even in sand cultures watered at regular intervals with fresh nutrient solution the nutrient status of plants is likely to decline, since the total supply will be fixed by the characteristics of the system (initial concentration, volume retained, frequency of application) whereas growth and absolute demand for nutrients will tend to increase exponentially with time. Although these problems can be avoided by means of flowing solution culture techniques (Sect. 2.3), most plant science laboratories do not yet have the necessary equipment. Hence an alternative approach to the experimental control of plant nutrient status is required.

ASHER and COWIE (1970) devised a method by which the nitrogen status of grain sorghum plants could be controlled in water culture from an early seedling stage to grain maturity. Frequent small additions of nitrogen were made to each nutrient solution. In the control treatment the size and frequency of these were calculated from a previously established growth curve, and data on average concentrations of nitrogen in the tissues of healthy plants, to provide adequate nitrogen at all times. In the deficient treatments fixed percentages of the amounts of nitrogen applied in the control treatment were applied on each occasion. Since the success of the method depends on the careful preparation of a program of nutrient additions ASHER and COWIE (1970) suggested that the technique be called *Programmed Nutrient Addition*.

ASHER and COWIE (1970) showed that nutrient addition programs could be prepared for deliberate variations in nutrient status during the course of an experiment so that effects of nutrient deficiency at particular stages of growth could be studied. Subsequently LEE et al. (1981) used the technique to control the nitrogen status of ginger in the field (Fig. 2a). MALIK et al. (1978) used a variation of the technique to control the nitrogen status of cotton plants under greenhouse conditions. As no growth curve was available, a nitrogen concentration of approximately 4,000 μM was maintained in the control pots and uptake in this treatment computed every second day from solution analyses. Additions of nitrogen to the nitrogen-deficient treatments were then made on the basis of fixed percentages of the uptake in the control treatment. Where experiments are of short enough duration to be confined to the period of exponential growth, nutrient addition programs can be devised from theoretical growth curves based on the initial plant weight and assumptions concerning the probable relative growth rate of the control plants (Fig. 2b). A somewhat similar approach was adopted in mist culture experiments of INGESTAD and LUND (1979) who proposed the concept of a relative nutrient addition rate analogous to relative growth rate.

The technique has the advantages that no special equipment is required and that there is no restriction to the number of treatments which can be employed. In view of the success obtained with it so far under both greenhouse and field conditions the method appears to have wide applicability.

## 2.1.7 Study of Symbiotic Associations with Microorganisms

### *2.1.7.1 Nitrogen-Fixing Systems*

Sand culture and water culture methods have been used extensively in the study of symbiotic nitrogen fixation in legumes and other higher plants. The Leonard jar technique (LEONARD 1944) is a specialized sand culture system developed specifically for such studies and has become a standard technique in many laboratories. An automatic sub-irrigated sand culture system devised by

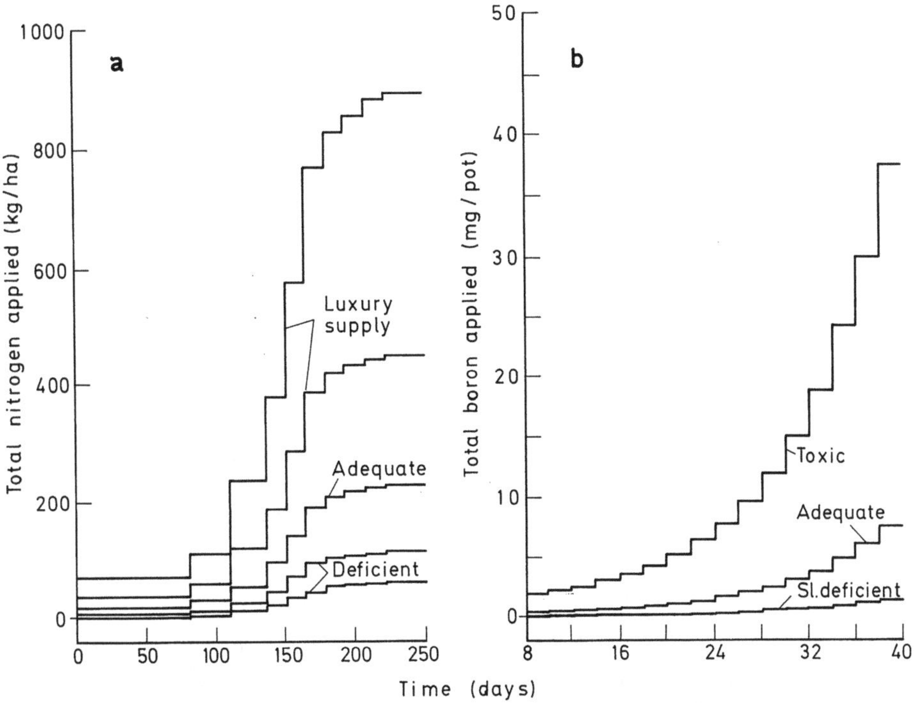


**Fig. 2a, b.** Examples of nutrient addition programs. **a** Nitrogen addition programs used to control the nitrogen status of ginger in the field. (Modified after LEE et al. 1981). **b** Boron addition programs used by FORNO et al. (1979) to control the boron status of cassava in water culture. (Two additional programs causing moderate and severe boron deficiency not shown)

ANDREW (1974) has proved effective in the study of soil acidity factors (pH, Ca, Al) affecting nodulation and nitrogen fixation in legumes (ANDREW 1976; CARVALHO et al. 1981). Water culture studies have also contributed much valuable information on effects of soil acidity factors (LONERAGAN and DOWLING 1958; MUNNS 1968).

Considerable scope exists for the further application of solution culture methods to the study of factors affecting nodulation in legumes. Unfortunately, however, some legume species do not nodulate readily in simple water culture systems and there is an urgent need for information on the reasons why this is so. One possibility which has not yet been examined in detail is that high initial nutrient concentrations may cause morphological changes in the roots which hinder nodulation. Thus BREWSTER et al. (1976) showed that in the non-leguminous species rape, the growth of root hairs was completely inhibited at phosphorus concentrations of 100 μM and higher. If a similar situation exists in some legumes in which infection by *Rhizobium* is via root hairs, the common use of initial phosphate concentrations of the order of 1,000 μM would effectively remove the potential sites of infection. Another possibility worthy of investigation is that inadequate production or retention of root exudates at root surfaces results in conditions unfavourable for multiplication of free-living *Rhizobium* prior to infection.

### *2.1.7.2 Mycorrhizal Associations*

The importance of ectomycorrhizal associations in the growth of some woody perennial species has long been recognized, and during the past two decades there has been growing recognition of the role of endomycorrhizas in the growth of many plant species, particularly under conditions of limiting mineral nutrient supply. So far only limited use of solution culture methods has been made for such studies. Recent research suggests that it may be important to use rather dilute culture solutions. Thus OWUSU-BENNOAH and MOSSE (1978) showed that infection of lettuce roots by a v.a. mycorrhizal fungus was reduced to zero by a mineral nitrogen concentration of 2,850 μM supplied as nitrate or an equimolar mixture of nitrate and ammonium. Similarly HOWELER et al. (1981) showed that formation of vesicular arbuscular mycorrhizas in cassava occurred at 0.1 μM and 1 μM phosphorus but not at 10 or 100 μM phosphorus.

### 2.1.8 Commercial Crop Production (see also Sect. 2.3.5)

The technical feasibility of crop production in solution culture systems was demonstrated in the U.S.A. in the late 1920's. Since then substantial use of these techniques, particularly gravel culture, has been made in the production of high-value greenhouse crops (DOUGLAS 1976). The main constraints to expansion of hydroponic crop production are economic rather than technical. This form of crop production becomes particularly attractive where centres of population exist in highly arid areas in which insufficient water is available for irrigation, or in situations where soil suitable for horticultural production is unavailable within a reasonable distance of the population centre. It could be that increasing costs of transporting produce associated with rising prices of liquid fuel, will lead to greater use of hydroponic production methods close to major cities and perhaps within them, flat roof tops being quite suitable for this purpose.

## 2.2 Mist Culture

Plants may be grown successfully with their roots suspended in a fine mist of nutrient solution generated pneumatically or by allowing the solution to drip onto a rapidly rotating disk. HEWITT (1966) discusses a number of variations of the technique. INGESTAD and LUND (1979) have described a sophisticated mist culture apparatus incorporating control of solution temperature, pH, and electrical conductivity, with provision for automatic nutrient additions. The main advantage of mist culture over other culture methods is the ease with which the root systems can be studied. A disadvantage of the method is that even brief interruptions to mist production due to mechanical problems, e.g. power failure or blocked atomizer nozzle, can lead to plant damage.

Usually concentrated nutrient solutions are supplied at rather low volumes per unit time. For example MARTIN and HENDRIX (1966) supplied Hoagland's solution to wheat plants at rates of the order of 65 ml $pot^{-1}$ $h^{-1}$. At such low solution supply rates substantial depletion of the solution in contact with the roots must occur (see Sects. 1.3.1, 2.3.1). Hence the method appears unsuited for critical studies on effects of solution composition on plant growth.

## 2.3 Flowing Solution Culture

Continuously flowing solution cultures provide the most practical means so far devised for experimental control of root environment parameters. Flowing culture techniques allow plants to be grown for extended periods of time with precise control of root temperature, pH and nutrient ion concentrations in dilute solutions comparable in composition with soil solutions. Such control is essential for quantitative studies on effects of various root environment parameters on nutrient uptake and growth in higher plants.

EDWARDS and ASHER (1974) made a distinction between those flowing culture systems in which the nutrient solution enters the pots at the required concentration of each ion under investigation ["Type (a)" systems] and those in which treatment concentrations are held constant by careful balancing of the rate of addition of one or more concentrated stock solutions against the rate of removal of ions by the test plants ["Type (b)" systems]. A well-known example of a Type (b) system is that devised by VAN DEN HONERT and subsequently described in detail by BECKING (1956). Type (b) flowing culture systems employ low flow rates and require only relatively small volumes of nutrient solution. However, the need to make frequent adjustments to stock solution flow rates *on an individual pot basis* means that a single operator can effectively manage only a small number or pots. It is perhaps for this reason that the majority of laboratories which have installed flowing culture equipment in recent years have opted for Type (a) systems.

### 2.3.1 The Flow Rate Problem

In Type (a) flowing culture systems it is essential that the flow rate through the pots is high enough to prevent appreciable alteration of solution composition as a result of contact with the roots. EDWARDS and ASHER (1974) showed that (F) the lowest acceptable flow rate in l $pot^{-1}$ $min^{-1}$ could be calculated approximately from the following equation:

$$F = \frac{100R}{D} \cdot \frac{U}{C} \tag{2}$$

where R is the weight of roots per pot (g fresh wt.), U is the rate of uptake of the test element (μmol $g^{-1}$ fresh wt. $min^{-1}$) at treatment concentration C (μM), and D is the maximum acceptable depletion of the test ion (%). Relationships between U and C are such that higher flow rates are required in the

**Table 4.** Nutrient solution flow rates used in experiments with Type (a)[a] flowing culture systems

| Reference | Flow rate (ml $pot^{-1}$ $min^{-1}$) | Reference | Flow rate (ml $pot^{-1}$ $min^{-1}$) |
|---|---|---|---|
| RORISON (personal communication) | 0.5 | CLEMENT et al. (1974) | 729 |
| FAGERIA (1976) | 2 | ASHER et al. (1965) | 695–900 |
| SABET et al. (1964) | 1.4–16.7 | BHAT (1980) | Approx 1000 |
| REISENAUER (1969) | 0–17.4 | ASHER and EDWARDS (1978) | 1,620 |
| VAN HAI and LAUDELOUT (1966) | 3.3–30 | EDWARDS and ASHER (1974) | 3–3000 |
| TIBBITTS et al. (1978) | 200 | VAN LUNE and VAN GOOR (1975) | 32,000 |
| TEMPLE-SMITH and MENARY (1977) | 704 | | |

[a] For explanation see text

deficiency range than in the adequate to toxic range of external concentrations (EDWARDS and ASHER 1974; ASHER 1981). Limited data suggest that higher rates are required for studies with macronutrients than with micronutrients (ASHER 1981). Both actual experimentation, and substitution into Eq. (2) of published data on uptake rates at particular concentrations, suggest that flow rates of the order of hundreds of ml $min^{-1}$ will often be required to control treatment concentrations to within 5% or 10% of their nominal values and that still higher flow rates will sometimes be needed.

Increases in R with time due to root growth mean that minimum acceptable flow rates also increase with time. REISENAUER (1969) sought to overcome this problem by constructing a system with continuously variable flow rates. However, the highest flow rate used (17.4 ml $min^{-1}$) was insufficient to prevent severe depletion of his nutrient solution by young wheat seedlings. Most modern flowing culture systems have incorporated constant high flow rates ($\gg$700 ml $min^{-1}$) (Table 4) to ensure adequate control of concentration at the highest R values likely to be encountered. All of these have been based on the recirculating principle introduced by ASHER et al. (1965). In most cases provision has been made for monitoring of the composition of the recirculating nutrient solution and the slow addition of reagents to maintain control of pH and test ion concentrations.

### 2.3.2 Composition of Flowing Culture Solutions

Although much remains to be learned about critical external concentrations for plant growth (cf. review by ASHER 1978), it is already clear that vigorous growth can be obtained in quite dilute flowing culture solutions (Table 5, Fig. 3). However, not all users of the technique have taken full advantage of the opportunity to work at such low concentrations. Thus in a study of nitrate uptake by *Lolium perenne*, CLEMENT et al. (1978) added additional calcium to all treatments to ensure a minimum concentration of 4,000 μM despite the observation

**Table 5.** Concentrations of elements in nutrient solutions employed in flowing culture experiments (μM)

| Reference | ISLAM et al. (1980) | WILD et al. (1974) | ASHER and LONERAGAN (1967) | TEMPLE-SMITH and MENARY (1977) | CLEMENT et al. (1978) |
|---|---|---|---|---|---|
| Element (or Ion) | | | | | |
| Nitrogen | | | | | |
| $NO_3^-$ | 250 | 700 | 750 | 3000 | 1.4–143,000 |
| $NH_4^+$ | – | – | 100 | – | – |
| Potassium | 250 | 1–33 | 250 | 1000 | 26 |
| Calcium | 250 | 420–470 | 250 | 1000 | 4,000–75,500 |
| Magnesium | 10[a] | 100 | 100 | 400 | 100 |
| Phosphorus | 15 | 50 | 0.04–25 | 0.06–7.7 | 50 |
| Sulfur | 260–560 | 100 | 100 | 400 | 4125 |
| Chlorine | 5 | 125 | 100 | 7.2 | 1.7 |
| Sodium | 40–63 | – | – | 3.6 | – |
| Silicon | 10 | – | – | – | – |
| Iron | 20 | 5 | 2 | 3.6 | 1.1 |
| Boron | 3 | 2.5 | 3 | 18 | 5 |
| Manganese | 0.25 | 0.5 | 1 | 3.6 | 1 |
| Zinc | 0.5 | 0.05 | 0.5 | 0.3 | 0.08 |
| Copper | 0.1 | 0.02 | 0.1 | 0.12 | 0.03 |
| Molybdenum | 0.02 | 0.005 | 0.02 | 0.04 | 0.05 |
| Cobalt | 0.04 | – | 0.04 | – | – |
| Total nutrients | 1114–1437 | 1454–1536 | 1658–1682 | 5837–5844 | 8311–222,810 |

[a] Adequate for plant growth at pH $\geqslant 5.5$, but inadequate at pH $\leqslant 4.0$

of LONERAGAN et al. (1968) that this species requires only 2.5 μM calcium for maximum growth.

### 2.3.3 Research Applications

Flowing solution culture represents a powerful tool for research into certain aspects of plant physiology and soil–plant relationships. Much greater use of the technique is likely in the future as increasing numbers of research workers come to recognize the importance of maintaining accurate control of the root environment in plant nutrition studies. The recent description of two flowing culture systems suitable for use in growth cabinets (TIBBITTS et al. 1978; BHAT 1980) raises exciting prospects for simultaneous control of root and shoot environments in plant physiological research.

Flowing solution culture at present offers the only practical means of obtaining quantitative data on the effects on plant growth of external nutrient ion concentrations ranging from deficiency to adequacy. The technique is also highly suitable for similar studies in the toxicity range, although in the case of some slowly absorbed elements, comparable results could probably be obtained in conventional water culture systems (cf. Sect. 2.1.4.3). Much work remains to be done on these concentration × growth relationships. Thus for many plant species no reliable information is available concerning limiting external concen-

**Fig. 3.** Vigorous plant growth obtained in a flowing solution culture experiment at the University of Queensland, using a culture solution with a total nutrient concentration of 3780 μM

trations for the deficiency or toxicity of even a single element. Similarly for a number of biologically important elements no reliable limiting concentrations have been established even for a single genotype. In addition, little is known about interactions amongst ions at concentrations likely to be encountered in soil solution. In studies with legumes, practically all the existing information concerns non-nodulated plants. However, there is good evidence that nodulation is more sensitive to adverse root environments than is the growth of the non-nodulated host plant (LONERAGAN and DOWLING 1958; LOWTHER and LONERAGAN 1968; CARVALHO et al. 1981). Hence more research is required on limiting external concentrations for the nodulation and growth of legumes.

Flowing solution culture equipment incorporating efficient temperature control is ideally suited for studying effects of root temperature on ion uptake and plant growth. In a recent experiment at the University of Queensland, WARRINER (unpublished) showed that in some species, above-optimum root temperature can increase markedly the susceptibility of plants to phosphorus toxicity (subterranean clover) and manganese toxicity (soybean, pigeon pea).

Further research is needed to establish the extent to which critical external concentrations may be influenced by root temperature.

ASHER and OZANNE (1977) used flowing solution culture to show that individual members of apparently uniform sets of *Arctotheca calendula* seedlings differed widely in susceptibility to potassium deficiency when grown at a constant low potassium concentration (2.6 μM). Hence flowing solution culture provides a potential method of identifying genotypes with resistance to specific nutritional disorders (ASHER and EDWARDS 1978). Again, JOHANSEN et al. (1968) showed that mean rates of potassium uptake by roots of intact plants grown in flowing culture at constant potassium concentrations were much lower than uptake rates by potassium-starved roots at similar concentrations in short-term uptake experiments. Data from flowing culture studies are thus highly relevant to the important question of the mechanisms by which higher plants regulate mineral uptake and mineral composition (cf. review by PITMAN and CRAM 1977). A further potential use for flowing culture technology is in producing plant material at known and constantly maintained levels of nutrient stress for studies on the biochemical role of specific mineral nutrients.

Recently, HOWELER et al. (1981) demonstrated the feasibility of using flowing solution culture to study effects of mycorrhizas on the mineral nutrition of higher plants. They showed that growth and phosphorus uptake by cassava from a dilute nutrient solution containing 1 μM phosphorus was markedly stimulated by v.a. mycorrhizas. Since no mycorrhizas formed at phosphorus concentrations $\geqslant 10$ μM, study of the root–fungus association would have been extremely difficult by any other solution culture technique. Further research is needed to see to what extent the technique is applicable to the study of mycorrhizas in other plant species.

### 2.3.4 Likely Future Developments

Considerable scope exists for improving the performance and ease of operation of existing flowing culture equipment. Although high flow rates can be used to ensure that only small changes in solution composition occur with each passage of the solution through the pots, a large volume of solution may be required to provide an adequate reserve of both treatment and basal nutrient ions in dilute solutions, e.g. up to 2,400 l $unit^{-1}$ (ASHER and EDWARDS 1978). The need for such large volumes of solution would be reduced greatly if a system could be devised for equilibrating the nutrient solutions with suitable cation and anion exchangers. Again, the concentrations of ions corresponding with nutrient deficiency are in a number of cases too low to be measured directly, so that time-consuming concentration procedures are needed prior to analysis of the nutrient solutions (e.g., potassium, ASHER and OZANNE 1967; zinc, CARROLL and LONERAGAN 1968; calcium, LONERAGAN et al. 1968; phosphorus, JINTAKANON et al. 1975).

The long-term solution to the problem of monitoring and controlling solution composition may come from developments in the field of electrochemistry. Glass electrode-autotitrator systems have long been in use for controlling solution pH and have been incorporated into some flowing culture installations

(ASHER et al. 1965; CLEMENT et al. 1974; ASHER and EDWARDS 1978; MORITSUGU and KAWASAKI 1979). The control of ions other than $H^+$ using appropriate ion-specific electrodes is foreshadowed by the description of such a system for controlling potassium and nitrate concentrations in flowing culture (CLEMENT et al. 1974). However, for some elements of biological interest no electrode is yet available and for many others the electrodes at present available lack the necessary sensitivity, selectivity, or speed of response at low concentrations to be effective in controlling solution composition in the low concentration range.

### 2.3.5 Commercial Crop Production

Although close control of the root environment is not essential for hydroponic crop production, there has been growing interest in continuously flowing culture methods for this purpose during the last decade.

#### *2.3.5.1 Use of Sewage Effluent*

Recent research at the University of California has demonstrated the feasibility of using secondary treated sewage effluent as a substrate for hydroponic vegetable production (BERRY et al. 1977; WALLACE et al. 1978). With troughs 6.2 m long and holding approximately 1,000 l of effluent, some gradients in plant growth were observed along the length of the bed. Thus with lettuce at a flow rate of $2.8 \times 10^3$ l $day^{-1}$ (approx. 2 ml $min^{-1}$) and with tomato at a flow rate of $2.8 \times 10^4$ l $day^{-1}$ (approx. 20 ml $min^{-1}$) there was some growth reduction at the outlet end of the bed. The causes of these reductions were not determined. Clearly further research is needed on the technique but it opens up exciting prospects for reducing hydroponic production costs by recycling mineral nutrients which might otherwise have contributed to the pollution of natural waters adjacent to centres of population.

#### *2.3.5.2 Nutrient Film Technique (N.F.T.)*

N.F.T. is a Type (a) recirculating flowing culture system (for definition see Sect. 2.3) in which the nutrient solution flows down shallow, gently sloping plastic gullies (slopes 1:30 to 1:100) in which are placed bare rooted or in some cases potted seedlings (for review see COOPER 1976). The pH of the circulating nutrient solution is controlled by additions of $H_3PO_4$ or $HNO_3$ and periodic additions of a nutrient mixture are made to keep the electrical conductivity of the solution between 2 and 3 mS $cm^{-1}$. The technique has gained substantial commercial acceptance, but it is clear that some technical problems remain and further research is needed. Thus RESH (1976) compared ten variations of N.F.T. for tomato production and found all to be inferior to standard commercial gravel culture.

In particular, research is needed on effects of flow rate. COOPER (1976) stresses the need to use flow rates sufficiently low to ensure that at least portion of the root system is exposed to air to ensure an adequate oxygen supply.

However MAHER (1976) favours higher flow rates which result in complete submergence of the root mat in the bottom of the gully and has shown that for even short gullies (7 m long) an oxygen deficit develops towards the outlet of the gully at flow rates below $4\,l\,min^{-1}$. Higher flow rates such as those advocated by Maher should also reduce gradients along the gully in mineral nutrient concentration and solution temperature.

## 3 Summary and Conclusions

The wide range of solution culture techniques available to plant scientists today carries with it a responsibility to select techniques consistent with the objectives of each particular study. In general the simplest acceptable method should be used.

Flowing solution culture is the only practical means available at present for studying relationships between plant growth and external nutrient ion concentrations in the deficiency range, and will usually be the best method also for studies in the toxicity range. However the large volumes of solution and large amounts of surface (pots, piping, tanks, pumps) in contact with the solution make the method unsuitable for studies on nutrient essentiality. Such studies are usually best done in simple non-renewed water culture. Simple water culture and sand culture methods are well suited to the study of symptoms of nutritional disorders. However, recent research raises doubts concerning the suitability of any water culture method for establishing critical tissue concentrations that may be applied to diagnosis in the field. Flowing solution culture has the advantage over other water culture methods that an adequate total supply of nutrients can be provided at low and controlled concentrations comparable with those commonly occurring in soil solutions. Programmed nutrient addition (cf. Sect. 2.1.6) offers a means by which an adequate total supply of nutrients can be provided without exposure of the roots to the high initial concentrations characteristic of conventional water culture and sand culture systems. However, the method is not suitable for studies in which control of external concentrations is required.

Despite the long period of time over which existing solution culture methods have been evolving, substantial scope still exists for improvement in these methods. Considerable opportunities exist also for novel applications of present techniques to problems in plant physiology and soil–plant relationships.

## References

Addoms RM (1937) Nutritional studies on loblolly pine. Plant Physiol 12:119–205

Ahmed S, Evans HJ (1960) Cobalt – a micronutrient element for the growth of soybean plants under symbiotic conditions. Soil Sci 90:205–210

Andrew CS (1974) Automatic sub-irrigation sand culture technique for comparative studies in plant nutrition. Lab Pract 23:20–21

Andrew CS (1976) Effect of calcium, pH and nitrogen on the growth and chemical composition of some tropical and temperate pasture legumes. I Nodulation and growth. Aust J Agric Res 27:611–623

Arnon DI, Johnson CM (1942) Influence of hydrogen ion concentration on the growth of higher plants under controlled conditions. Plant Physiol 17:525–539

Asher CJ (1978) Natural and synthetic culture media for spermatophytes. In: Rechcigl M Jr (ed) CRC handbook series in nutrition and food, section G: diets, culture media, food supplements, vol III. CRC Press, Cleveland, pp 575–609

Asher CJ (1981) Limiting external concentrations of trace elements for plant growth: use of flowing solution culture techniques. J Plant Nutr 3:163–180

Asher CJ, Cowie AM (1970) Programmed nutrient addition – a simple method for controlling the nutrient status of plants. Proc Aust Plant Nutr Conf, Mt Gambier, S Aust Sect 1(b), pp 28–32

Asher CJ, Edwards DG (1978) Relevance of dilute solution culture studies to problems of low fertility tropical soils. In: Andrew CS, Kamprath EJ (eds) Mineral nutrition of legumes in tropical and subtropical soils. CSIRO, Melbourne

Asher CJ, Loneragan JF (1967) Response of plants to phosphate concentration in solution culture. I. Growth and phosphorus content. Soil Sci 103:225–233

Asher CJ, Ozanne PG (1967) Growth and potassium content of plants in solution cultures maintained at constant potassium concentrations. Soil Sci 103:155–161

Asher CJ, Ozanne PG (1977) Individual plant variability in susceptibility to potassium deficiency: some observations on capeweed [*Arctotheca calendula* (L.), Levyns]. Aust J Plant Physiol 4:499–503

Asher CJ, Ozanne PG, Loneragan JF (1965) A method for controlling the ionic environment of plant roots. Soil Sci 100:149–156

Barber DA, Martin JK (1976) The release of organic substances by cereal roots into soil. New Phytol 76:69–80

Barber SA (1962) A diffusion and mass-flow concept of soil nutrient availability. Soil Sci 93:39–49

Bates TE (1971) Factors affecting critical nutrient concentrations in plants and their evaluation: a review. Soil Sci 112:116–130

Becking JH (1956) On the mechanism of ammonium ion uptake by maize roots. Acta Bot Neerl 5:1–79

Berry WL, Wallace A, Lunt OR (1977) Recycling municipal wastewater for hydroponic culture. Hort Sci 12:186

Bhat KKS (1980) A low-cost, easy-to-install flow culture system for use in a constant environment cabinet. J Exp Bot 31:1435–1440

Bond G (1951) The fixation of nitrogen associated with the roots of *Myrica gale* L. with special reference to its pH relation and ecological significance. Ann Bot 15:447–459

Bowen GD, Rovira AD (1976) Microbial colonization of plant roots. Ann Rev Phytopathol 14:121–144

Brewster JL, Bhat KKS, Nye PH (1976) The possibility of predicting solute uptake and plant growth response from independently measured soil and plant characteristics. IV The growth and uptake of rape in solutions of different phosphorus concentration. Plant Soil 44:279–293

Burd JS, Martin JC (1924) Secular and seasonal changes in the soil solution. Soil Sci 18:151–167

Carroll MD, Loneragan JF (1968) Response of plant species to concentrations of zinc in solution. I Growth and zinc content of plants. Aust J Agric Res 19:859–868

Carroll MD, Loneragan JF (1969) Response of plant species to concentrations of zinc in solution. II Rates of zinc absorption and their relation to growth. Aust J Agric Res 20:457–463

Carvalho de MM, Edwards DG, Andrew CS, Asher CJ (1981) Aluminum toxicity, nodulation and growth of *Stylosanthes* species. Agron J 73:261–265

Chen CH, Lewin JC (1969) Silicon as a nutrient element for *Equisetum arvense*. Can J Bot 47:125–131

Clement CR, Hopper MJ, Canaway RJ, Jones LHP (1974) A system for measuring the uptake of ions by plants from flowing solutions of controlled composition. J Exp Bot 25:81–99
Clement CR, Hopper MJ, Jones LHP (1978) The uptake of nitrate by *Lolium perenne* from flowing nutrient solution. I. Effect of $NO_3^-$ concentration. J Exp Bot 29:453–464
Cooper AJ (1976) Crop production with nutrient-film technique. Proc 4th Int Congr Soilless Cult (IWOSC), Las Palmas
Dixon NE, Gazzola C, Blakely RL, Zerner B (1975) Jackbean urease (EC 3.5.1.5). A metalloenzyme. A simple biological role for nickel? J Am Chem Soc 97:4131–4133
Douglas JS (1976) Advanced guide to hydroponics. Drake, New York
Edwards DG, Asher CJ (1974) The significance of solution flow rate in flowing culture experiments. Plant Soil 41:161–175
Epstein E (1972) Mineral nutrition of plants: principles and perspectives. Wiley and Sons, New York
Fageria NK (1976) Effect of P, Ca, and Mg concentrations in solution culture on growth and uptake of these ions by rice. Agron J 68:726–732
Forno DA, Asher CJ, Edwards DG (1979) Boron nutrition of cassava and the boron × temperature interaction. Field Crops Res 2:265–279
Gericke WF (1929) Aquaculture – a means of crop-production. Am J Bot 16:862
Gillman GP, Bell LC (1978) Soil solution studies on weathered soils from tropical North Queensland. Aust J Soil Res 16:1–11
Hai van T, Laudelout H (1966) Phosphate uptake by intact rice plants by the continuous flow method at low phosphate concentrations. Soil Sci 101:408–417
Hallsworth EG, Wilson SB, Greenwood EAN (1960) Copper and cobalt in nitrogen fixation. Nature 187:79–80
Hewitt EJ (1966) Sand and water culture methods used in the study of plant nutrition. Commonw Bur Hortic Plant Crops Tech Commun 22 (revised)
Hoagland DR, Arnon DI (1950) The water culture method for growing plants without soil. Coll Agric UC Berkeley, USA. Cal Agric Exp Stn Circ 347:1–32
Howeler RH, Edwards DG, Asher CJ (1981) Application of the flowing solution culture techniques to studies involving mycorrhizas. Plant Soil 59:179–183
Ingestad T, Lund AB (1979) Nitrogen stress in birch seedlings. I. Growth technique and growth. Physiol Plant 45:137–148
Islam AKMS, Edwards DG, Asher CJ (1980) pH optima for crop growth. Results of a flowing solution culture experiment with six species. Plant Soil 54:339–357
Jintakanon S, Kerven GL, Edwards DG, Asher CJ (1975) Measurement of low phosphorus concentrations in nutrient solutions containing silicon. Analyst 100:408–414
Johansen C, Edwards DG, Loneragan JF (1968) Interactions between potassium and calcium in their absorption by intact barley plants. II. Effects of calcium and potassium concentration on potassium absorption. Plant Physiol 43:1722–1726
Johnson CM, Stout PR, Broyer TC, Carlton AB (1957) Comparative chlorine requirements of different plant species. Plant Soil 8:337–353
Knop W (1865) Quantitative Untersuchungen über die Ernährungsprozesse der Pflanzen. Landwirtsch Vers Stn 7:93–107 (cited by Hewitt 1966)
Knypl JS (1976) Culture of *Spirodela oligorrhiza* in ammonium-media buffered with calcium carbonate or calcium phosphate. Biochem Physiol Pflanz 170:243–252
Lee MT, Asher CJ, Whiley AW (1981) Nitrogen nutrition of ginger (*Zingiber officinale*). I. Effects of nitrogen supply on growth and development. Field Crops Res 4:55–68
Leonard LT (1944) Method of testing legume bacteria cultures and results of tests of commercial inoculants in 1943. US Dep Agric Circ 703
Loneragan JF, Asher CJ (1967) Response of plants to phosphate concentration in solution culture: II. Rate of phosphate absorption and its relation to growth. Soil Sci 103:311–318
Loneragan JF, Dowling EJ (1958) The interaction of calcium and hydrogen ions in the nodulation of subterranean clover. Aust J Agric Res 9:464–472
Loneragan JF, Snowball K, Simmons WJ (1968) Response of plants to calcium concentration in solution culture. Aust J Agric Res 19:845–857

Lowther WL, Loneragan JF (1968) Calcium and nodulation in subterranean clover (*Trifolium subterraneum* L.). Plant Physiol 43:1362–1366
Lune van P, Goor van BJ (1975) A method of growing young fruiting apple trees in water culture. J Hortic Sci 50:129–133
Maher MJ (1976) The use of hydroponics for the production of greenhouse tomatoes in Ireland. Proc 4th Int Congr Soilless Cult (IWOSC). Las Palmas
Malik MNA, Evenson JP, Edwards DG (1978) The effect of level of nitrogen nutrition on earliness in upland cotton (*Gossypium hirsutum* L.). Aust J Agric Res 29:1213–1221
Martin NE, Hendrix JW (1966) An apparatus for the mist culture of wheat. Plant Disease Rep 50:369–371
Miyake Y, Takahashi E (1978) Silicon deficiency of tomato plant. Soil Sci Plant Nutr 24:175–189
Moritsugu M, Kawasaki T (1979) A new system of automatic pH regulation in solution culture. Ber Ohara Inst Landw Biol 17:171–178
Mulder EG (1948) Importance of molybdenum in the nitrogen metabolism of microorganisms and higher plants. Plant Soil 1:94–119
Munns DN (1968) Nodulation of *Medicago sativa* in solution culture. I. Acid sensitive steps. Plant Soil 28:129–146
Norris DO (1964) Techniques used in work with *Rhizobium*. In: Some concepts and methods in sub-tropical pasture research, Bull 47. Commonw Bur Pastures Field Crops
Norris DO, Date RA (1976) Legume bacteriology. In: Shaw NH, Bryan WW (eds) Tropical pasture research principles and methods, Bull 51. Commonw Bur Pastures Field Crops
Owusu-Bennoah E, Mosse B (1978) In: Development of VA mycorrhiza ($E_3$ and YU) in plants fed with nutrient solution in sand and nutrient film culture. Rothamsted experimental station annual report, Part 1
Pitman MG, Cram WJ (1977) Regulation of ion content in whole plants. In: Jennings DH (ed) Integration of activity in the higher plant. Soc Exp Biol Symp 31, Cambridge
Reisenauer HM (1960) Cobalt in nitrogen fixation by a legume. Nature 186:375–376
Reisenauer HM (1966) Mineral nutrients in soil solution. In: Altman PL, Dittmer DS (eds) Environmental biology. Fed Am Soc Exp Biol Bethesda Md USA
Reisenauer HM (1969) A technique for growing plants at controlled levels of all nutrients. Soil Sci 108:350–353
Resh HM (1976) A comparison of tomato yields, using several hydroponic methods. Proc 4th Int Congr Soilless Cult (IWOSC). Las Palmas
Sabet SA, Abdel Salam MA, Lagerwerff JV (1964) Growth and ion uptake by maize seedlings on solutions variable in phosphate and flow rate. Plant Soil 21:94–100
Sachs J (1860) Berichte über die physiologische Tätigkeit an der Versuchsstation in Tharandt. IV. Vegetationsversuche mit Ausschluss des Bodens über die Nährstoffe und sonstigen Ernährungsbedingungen von Mais, Bohnen und anderen Pflanzen. Landwirtsch Vers Stn 2:219–268 (cited by Hewitt 1966)
Saussure de T (1804) Recherches chimiques sur la végétation. Paris (cited by Hewitt 1966)
Schwarz K (1974) Recent dietary trace element research, exemplified by tin, fluorine and silicon. Fed Proc 33:1748–1757
Sideris CP, Young HY, Krauss BH (1943) Effects of iron on the growth and ash constituents of *Ananas comosus* (L) Merr. Plant Physiol 18:608–632
Spear SN, Edwards DG, Asher CJ (1978) Effects of nutrient supply on critical nutrient concentrations in cassava plants. In: Ferguson AR, Bieleski RL, Ferguson IB (eds) Plant nutrition 1978. Proc 8th Int Colloq Plant Anal Fert Probl, Auckland, NZ. NZ DSIR Inf Ser 134. Gov Printer, Wellington
Spear SN, Edwards DG, Asher CJ (1979) Response of cassava (*Manihot esculenta* Crantz) to potassium concentration in solution: critical potassium concentrations in plants grown with a constant or variable potassium supply. Field Crops Res 2:153–168
Stiles W (1916) On the interpretation of the results of water culture experiments. Ann Bot 30:427–436

Temple-Smith MG, Menary RC (1977) Growth and phosphate absorption in lettuce and cabbage plants in dilute solution culture. Aust J Plant Physiol 4:505–513
Tibbits TW, Palzkill DA, Frank HM (1978) Constructing a continuous circulation system for plant solution culture. Univ Wisconsin Coll Agric Life Sci, Res Bull R 2963
Trelease SF, Trelease HM (1935) Physiologically balanced culture solutions with stable hydrogen ion concentration. Science 78:438–439
Wallace A, Patel PM, Berry WL, Lunt OR (1978) Reclaimed sewage water: A hydroponic growth medium for plants. Res Rec Conserv 3:191–199
Wild A, Skarlou V, Clement CR, Snaydon RW (1974) Comparison of potassium uptake by four plant species grown in sand and flowing culture solutions. J Appl Ecol 11:801–812
Williams DE, Vlamis J (1957) Manganese and boron toxicities in standard culture solutions. Soil Sci Soc Am Proc 21:205–209
Wilson DO, Reisenauer HM (1963) Cobalt requirement of symbiotically grown alfalfa. Plant Soil 19:364–373
Woodward J (1699) Thoughts and experiments on vegetation. Philos Trans R Soc London 21:193 (cited by Hewitt 1966)

# I.4 Diagnosis of Mineral Deficiencies Using Plant Tests

D. BOUMA

## 1 Introduction

When agricultural scientists in the last century began to realize that mineral elements in a plant were taken up from the soil in which the plants grow, it was a logical step to suggest that chemical analysis of plants could be used as a means of assessing the nutrient supply of the soil. At the time it also appeared reasonable to suggest, as von Liebig did in the last century in his Law of Restitution (GOODALL and GREGORY 1947), that plant analysis could be used to determine the quantities of nutrients removed from the soil by a crop and, therefore, the amounts needed to maintain the supplying power of the soil.

Much of the early work on chemical analysis of plants was prompted by the desire to develop techniques that would supplement, or even replace, soil analysis and thus provide a biological method of soil analysis. A review containing many references on the history of plant analysis has been presented by GOODALL and GREGORY (1947). The idea of using plant analysis to determine the nutrient requirements of the soil has dominated this area of plant nutrition for many years. The previous edition of the Encyclopedia of Plant Physiology contains a section of six pages on plant and leaf analysis as part of a chapter on the determination of nutrient requirements of the soil (BERGMANN 1958). However, in the past 20 years or so a great deal of experimental work has caused a gradual but fundamental change in thinking, as it became progressively apparent that in many situations, and for a range of agricultural and horticultural plant species, the chemical composition of plants or plant tissue can directly reflect the nutrient status or the nutrient requirement of plants themselves.

In recent years, probably as a result of advances in knowledge and understanding of the role and function of nutrient elements, new approaches to diagnosis are being developed which differ in principle from plant analytical techniques. These are based on specific physiological or biochemical changes caused by deficiencies or, alternatively, on specific responses that can be induced in plants or plant tissue by the addition of a deficient element.

This chapter discusses the main features of plant analysis as a diagnostic technique and then discusses some of the newer approaches.

## 2 Plant Analysis

### 2.1 Physiological Basis

Within the limits imposed by genetic potential, plant growth depends on environmental factors including light, temperature and water supply, but also on the supply of essential nutrient elements. If all environmental factors except the supply of one essential element were optimal, then plant growth would be a function of the supply of this element. Increased plant growth resulting from increased supply of the element is usually accompanied by an increase in uptake of that element into the plant. Depending on many factors, some of which are discussed below, this increase may in turn be reflected in a changed concentration of the element in the plant dry matter. In plant analysis, our aim is to establish the relationship between nutrient concentration and growth (or yield), and then use this relationship in comparable situations to establish the nutrient status of a plant or crop. In this way, the nutrient requirement may be assessed (BATES 1971, BOULD 1968, GOODALL and GREGORY 1947, SMITH 1962, ULRICH 1952).

A relationship between plant nutrient concentration and yield may be established in pot experiments, in field experiments in which varied levels of one or more nutrient elements are applied, or in surveys of commercial fields. An example of the latter approach is the use of a relationship between root weight and petiolar nitrate content of sugar beet to determine the nitrogen requirements of the crop (ULRICH 1952).

The relationship is often curvilinear, but may also be linear or a combination of the two. An idealized example of the relationship between yield and nutrient supply is shown in Fig. 1a, where the full line represents the yield response to a nutrient element with all other elements present at optimal supplies, and the broken line the response when another less deficient element limits the full response. A relationship between yield and nutrient concentration in the dry matter is shown in Fig. 1b. Central to the diagnosis of nutrient deficiencies by plant analysis is the concept of a "critical concentration" for each element in the tissue. This concentration is the one separating the regions on the curve of sufficiency or "luxury consumption", and deficiency or "poverty adjustment" (arrow Fig. 1b; MACY 1936). A similar definition is as follows: that concentration which is just sufficient or just deficient for maximum growth or yield. In practice it is not often possible to define the critical concentration as sharply as implied above, and it is therefore often defined as that concentration where yield is 5% or 10% below the optimum (ULRICH and HILLS 1967).

Under conditions of extreme deficiency, small increases in yield that result from the application of the deficient element may sometimes be accompanied by a decrease in the concentration of the element in the dry matter (Fig. 1b, region I on the curve). PIPER (1942) first drew attention to this effect. It was later described in detail by STEENBJERG (1951) for the Cu concentration in barley grown in a pot experiment, and it is often referred to as the Piper-Steenbjerg effect. Several explanations have been given for this effect: BATES (1971) sug-

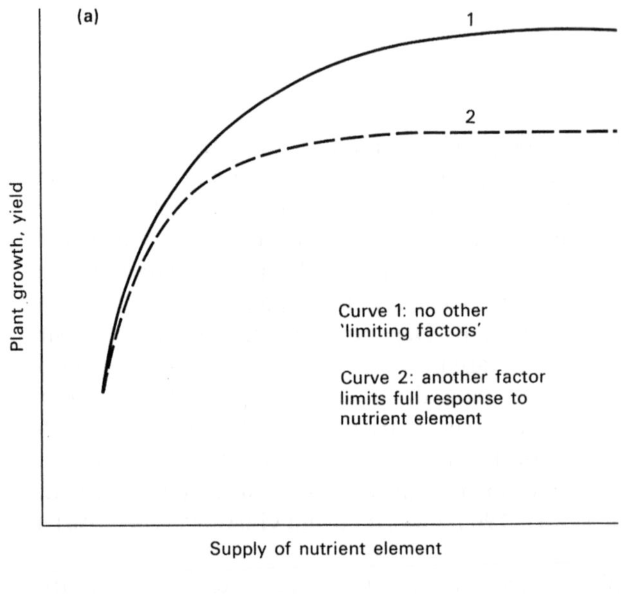


**Fig. 1 a, b.** Idealized curves representing the relationship between plant growth or yield and nutrient supply without and with a "limiting factor" (**a**). **b** shows relationship between nutrient concentration and growth or yield. See text for explanation

gested variations in physiological age caused by the selection of tissue. In wheat Loneragan (1978) has shown that Cu is readily lost from ageing leaves of plants grown with adequate Cu but not in Cu-deficient plants. Deficiency of Cu may delay the loss of Cu from leaves to such an extent that Cu concentrations in the corresponding leaves of non-deficient plants may fall below them, thus giving a marked Piper-Steenbjerg effect.

As the nutrient supply is further increased, there may be a region where there is increased growth (dry matter) without a marked change in the concentration of the deficient element (region II, Fig. 1b). With still higher levels of supply, growth as well as concentration of the element increase (region III, Fig. 1b) until the critical concentration is reached, when further additions cease to affect growth. However, in this range uptake of the element usually continues to increase, resulting in a further increase in the concentration of the element in the dry matter. This is the zone of "luxury consumption". Finally, with still higher nutrient applications growth may decline due to toxicity effects, while the concentration continues to increase (region V, Fig. 1b).

In most cases it is the total concentration of each element in the dry matter which has been used for diagnosis. However, there are many cases where the concentration of a specific fraction, either in fresh or in dried tissue, has been been found more useful, e.g. nitrate-N, sulfate-S, soluble K, to mention a few (FRENEY et al. 1978, SPENCER et al. 1977, ULRICH 1952). Occasionally, the ratio of one element to another has been used. In subterranean clover the ratio total N/total S has been suggested for the detection of S deficiency (SPENCER et al. 1977). In wheat, the ratio $SO_4^{2-}$/total S was least affected by N supply or plant age and was suggested as a suitable index of S status (FRENEY et al. 1978).

## 2.2 Choice of Tissue

Many different plant organs and tissues have been used for diagnosis (see GOODALL and GREGORY 1947 for a table of these). The tissues include roots, stems, bark, leaves of different age and position, laminae, petioles, midribs, seed, fruit and grain. In general, changes in nutrient concentration, caused by varying the nutrient supply, are greater in leaves than in other organs (GOODALL and GREGORY 1947). Although the entire leaf is usually sampled, the concentration in the petiole dry matter may occasionally be very different to that of the laminae, so that the inclusion of variable lengths of petiole will give variable results (BOULD 1961). Occasionally the petiole will give a better indication of the nutrient status than the lamina – this, however, may be true of one element but not of another. Thus in grapes the petioles gave a better index of K status than the laminae (SHAULIS 1961). This was also found for the K status of raspberry, while for N the reverse was the case (BOULD 1964). In sugar beet, nitrate-N in the petioles provided the best indication of the N status while sulfate-S in the laminae was preferred for the S status (ULRICH 1961).

In choosing an organ for assay, several criteria may be used (EMMERT 1959), but generally the two most important ones are the sensitivity of the response and its stability towards factors other than the supply of the particular nutrient in question. Change in the supply of nutrients can markedly affect plant morphology, thereby altering the proportion of dry matter distributed to different organs. It is important that this does not distort the correlation between nutrient content and yield or growth (GOODALL and GREGORY 1947). It will become apparent from the following pages that there are no general recommendations and that each crop, element and situation requires its own extensive study in order to develop a reliable diagnostic system based on plant analysis.

## 2.3 Factors Affecting the Relationship Between Nutrient Concentration and Yield

### 2.3.1 Plant Development

Within a given environment, the nutrient uptake by a plant depends on the supply to the roots and on the demands for the nutrient set up within the plant by the growth and normal functions of plant organs, including the roots (WILLIAMS 1955). From his studies of the growth of gramineous plants WILLIAMS (1948) concluded that each vegetative organ of a plant passes through phases of intake, a relatively constant content, and then export of many of its mineral nutrients. Each organ therefore has a certain capacity to accumulate nutrients, and may subsequently during senescence constitute a potential source of nutrients for younger plant parts. The demand of each organ for nutrients is usually met in part, if not entirely, by uptake from the medium, but the rate of uptake depends to a certain extent on the availability of the nutrient element within the plant. The importance of this principle was clearly illustrated for P, which is a mobile element, in an experiment with oats. Plants were grown at three levels of P supply. The plants grown with the lowest level (deficient) derived only 30% of their inflorescence P from other plant parts whereas those grown at the highest (=non-deficient) level derived no less than 93% from these sources. Thus in the latter case, though the external supply was plentiful, an abundant and more accessible supply was made available by senescent breakdown in leaves and roots, and later in the stems. For the P-deficient plants, remobilization was less significant, and P was preferentially obtained from the deficient medium. It should be pointed out that the pattern of redistribution of P in the non-deficient plants was caused by the demand of the inflorescences for N rather than for P. By contrast, the P-deficient plants had a much higher N status and any senescent breakdown in the leaves was probably initiated by a demand for P rather than for N (WILLIAMS 1948).

Substantial retranslocation, even over short periods, has also been shown for K (GREENWAY and PITMAN 1965). In young barley plants retranslocation was most evident in the mature leaf, which showed little K-uptake. Rapid rises in K concentration were evident in the developing youngest leaf in which about 50% of the K taken up during the 3-day experimental period had been derived from within the plant.

When one considers the effects of plant development on the mineral composition of plant organs, a distinction should be made between annual and perennial crops. In annuals, with their shorter period of growth and often with a more rapid development, changes in nutrient content tend to be more rapid and extensive. With them, it is generally difficult to use plant analysis to remedy current nutritional disorders, unless the diagnosis is made early enough in the life of the plant. Sometimes diagnosis of annual crops can help to regulate the quality of the harvested product. An example is the use of the nitrate content of petioles of sugar beet to monitor the N nutrition of the plant and through it to modify the yield of sugar by changing the supply of N, thereby altering the balance between top and root growth (ULRICH and HILLS 1967).

GREGORY (1953) made the point, worth repeating in the present context, that in a developing cereal plant over 90% of the nitrogen and phosphorus can be taken up during the time when dry matter is attaining only 25% of its final value. Subsequent development, which is accompanied by marked patterns of retranslocation, depends on this store of accumulated nutrients and determines final yields. Therefore, in the early stages of growth of annual plants the tissue concentration of nutrient elements tends to be high. As the plant develops, nutrient uptake continues, but for some elements, particularly mobile ones like N and P, the concentration in the plant tops or in the whole plant as a percentage of dry matter usually declines. The process is usually referred to as dilution and is partly due to an increase in the proportion of structural and storage materials relative to protoplasm (ULRICH 1952). Other factors can be a decline in the nutrient supply to the roots, in an absolute sense or relative to the rapidly increasing demands for growth. In the case of mobile elements there is also the redistribution between organs initiated by internal competition for nutrients or by senescence, even at adequate levels of external supply, referred to earlier.

Not all elements are mobile in the plant: for example, Ca is not (MILLIKAN and HANGER 1964). Some of the complications that may arise in the diagnosis of a deficiency of an immobile element are clearly illustrated in the work of LONERAGAN and SNOWBALL (1969), who studied the Ca nutrition of a number of grasses, herbs and legumes. Plants transferred from solutions containing luxury Ca levels to Ca-deficient solutions developed Ca deficiency symptoms, even though the plant tops had Ca concentrations three to ten times higher than the tops of healthy plants grown in a low but constant Ca supply. Even at a low and constant supply, older leaves may accumulate sufficient Ca to mask a deficiency in younger parts if analysis of whole plant tops is used as a criterion of Ca requirement. These considerations apply to any element with a restricted mobility in the plant. The difficulties in diagnosing deficiencies of these elements can be lessened by avoidance of old leaves and a careful selection of young tissues, e.g. leaves or even roots (LONERAGAN and SNOWBALL 1969).

In tree crops, differences in age are relatively unimportant, judging by the similarities in the nutrient concentration of comparable tissues from trees of different ages (SMITH 1962). In citrus, which can have two or more growth flushes in one season, the composition of leaves sampled from different flushes at comparable stages of development varied little, although the summer flush reached a relatively stable nutrient concentration sooner than the spring flush leaves (SMITH and REUTHER 1950).

### 2.3.2 Effects of Changes in Age of Tissue

The composition of each organ or tissue changes during its development under the influence of the same factors discussed before for the whole plant. Each element often shows a characteristic pattern of change in a given tissue as it develops, matures and finally senesces. Data that have been tabulated by SMITH (1962) indicate that concentrations of N, P, and K in the dry matter always declined with age, while others, like Ca and Mg, usually increased. These differences probably reflect differences in mobility, as discussed before. Such developmental changes with respect to leaves have been extensively studied and recorded for many crops, particularly horticultural ones, e.g. blackcurrants (BOULD 1961), peaches (EPSTEIN and LILLELAND 1942), oranges (JONES and PARKER 1950, SMITH and REUTHER 1950), and apples (REUTHER and BOYNTON 1939). EMMERT (1959) concluded from a survey of the literature that a knowledge of such trends may help in selecting the best sampling time for a tissue analysis test. In general, there are three more or less distinct growth periods. In the first period, when leaves are rapidly expanding, changes in concentration and day-to-day variations may be relatively large. Significant changes can also occur towards the end of the growing season, when senescence is often accompanied by retranslocation of mobile elements to woody tissues. The middle period is usually one of relative stability and may last from 3 to 6 months. Most of the diagnostic standards for fruit trees have been developed for leaves of this age (SMITH 1962). In some cases it is possible to reduce variation due to differences in leaf age by sampling leaves of the same physiological age, for example the youngest fully mature or expanded leaf. This is particularly useful in comparisons of nutritional treatments or when samplings during the growing season have to be done.

The selection criteria are usually governed by considerations of variability, reproducibility and the obvious need to obtain a reliable indication of the nutrient status of the plant, crop or tree. Requirements often conflict and a compromise is often necessary. Thus MASON (1958) concluded that for apples, the most suitable leaf position and sampling time were those for which variations in leaf composition were least. The best tissues are not necessarily those which show the biggest differences in composition. In addition, optimum tissues and conditions vary among elements and need to be determined separately.

### 2.3.3 Plant Age and Critical Levels

Although critical concentrations generally depend on the age of the selected tissue used for establishing the relationship between yield and nutrient concentration, the age of the plant itself is not always important. As pointed out before, the difficulty is one primarily confined to annuals and possibly biennial plants. ULRICH (1952, 1961) established a critical value of 1,000 ppm nitrate-N in the dry matter of petioles of the youngest mature leaves of sugar beets. This value did not vary during the growing season. Similar results were obtained for P in sugar beets (ULRICH 1961) and in Ladino clover (ULRICH 1948). SPENCER et al. (1977), in pot experiments on S requirements of subterranean clover, found a decrease in critical total S concentrations in young leaves from 0.19% 33 days from sowing to 0.15% and 0.10% for responses 61 and 133 days from sowing respectively. Similar results were obtained for sulfate-S. GREENWOOD (1966) found a very marked decrease in total N, soluble N and free ninhydrin-N content of the youngest fully expanded leaves of young wheat plants, sampled at 3 and 5 weeks from emergence, particularly in the "critical" range of the yield response curve to nitrogen. In a review, BATES (1971) concluded that the critical concentration in the whole plant changes with age, but that it is an open question whether a tissue selected to maintain a common physiological age at different sampling times does have a changing critical concentration. This again emphasizes the need for careful evaluation of the effects of tissue or plant age in the nutritional relationships of diagnosis.

### 2.3.4 Interactions Between Nutrient Elements

The statement was made in the beginning of this section that plant growth is a function of that nutrient present in suboptimal amounts, if other nutrient elements and environmental conditions are non-limiting. This is often referred to as the Law of the Minimum, first proposed by von Liebig (GOODALL and GREGORY 1947). Although true in principle, in practice there are few occasions where nutrients other than the limiting one are present at optimal levels. Nutrient interactions, involving a change in the concentration of one element in plant tissue caused by another, are quite common. They are often difficult to unravel, but need to be taken into account in diagnosis based on plant analysis. A shift in the content of one element invariably is accompanied by secondary changes in tissue content of other elements, even though there is no change in the availability of the interacting nutrient(s) (EMMERT 1961). Nutrient interactions may be positive or negative (PRÉVOT and OLLAGNIER 1961). Alternative expressions for the same phenomena are synergism and antagonism (SMITH 1962). Antagonism can occur during ion uptake, during translocation and accumulation in the tissues, or in metabolism [e.g. in enzyme systems, Fe versus Mn (PRÉVOT and OLLAGNIER 1961)]. It can involve competition between two or more elements, but also precipitation of nutrients or other phenomena.

Antagonisms during uptake may occur between cations but also between anions. Well-known examples of cation antagonism are those between K and Ca, K and Mg, Fe and Mn, to mention only a few (BOULD 1964, SMITH 1962). N supplied as the ammonium ion is antagonistic to other cations, but when supplied as nitrate it may compete with other anions, e.g. phosphate. Increasing the level of K to fruit trees has been shown to induce Mg deficiencies, accompanied by increases in the K concentration of the leaves (REUTHER et al. 1958). Conversely, changes in Mg concentration of leaf tissue are usually accompanied by opposite changes in leaf K (EMMERT 1961). Antagonistic effects have also been extensively documented for micronutrients. High levels of Cu in the soil can induce Fe deficiency. Pot cultures have shown that high levels of Cu, Zn or Mn can induce marked antagonistic effects between pairs of these ions on concentrations in leaves (REUTHER et al. 1958).

Antagonism during translocation is often caused by precipitation in the root tissues or elsewhere in the plant. Excess phosphate in the root environment causes precipitation of Zn and also of Fe along the veins, and this may lead to iron chlorosis or to "mottle leaf" as an expression of Zn deficiency (WEST 1938). In both cases a precipitation of the trace elements as phosphates appears to be involved. Another case of induced Zn deficiency was reported for subterranean clover growing in a soil low in Zn. An increase in N supply, regardless of N source, decreased the Zn content of the tops. This was attributed to a retention of Zn in the roots as a result of the formation of immobile Zn–protein complexes (OZANNE 1955). In water cultures, the critical Zn concentration of a number of subterranean clover cultivars showed a wide range of values, strongly dependent on the age of the plants and the phosphate supply. Plants in some treatments showed Zn deficiency symptoms, but had a higher Zn concentration than in others without these symptoms, again depending on phosphate supply. The severity of the Zn deficiency was related to the P/Zn ratio and not to the Zn concentration in the tissues (MILLIKAN 1963).

There is another type of interaction referred to as pseudo-antagonism by LUNDEGÅRDH (1966) or induced deficiency (EMMERT 1961), which does not involve direct ionic competion, and which can have various causes. A common interaction of this kind exists when two elements are deficient but different in severity, so that the deficiency of one element is masked by the more deficient element. When the supply of the latter is increased, a stage is reached where the first element becomes more deficient and then begins to determine plant response. This principle is of considerable importance in diagnosis based on plant analysis and is often referred to as the Law of the Minimum (see also Fig. 1A). The interaction between N and P is a well-known example (BOUMA 1956, 1959a, b, 1961, WILLIAMS 1955). In field experiments designed to develop criteria for the detection of S deficiency in subterranean clover pastures, BOUMA et al. (1969) found that correlations between pasture yield and total S or sulfate concentration in clover tops were much higher when plots had received phosphate than without applied phosphate. Similarly, correlations between yield responses to phosphate applications and the total P concentration in clover tops were higher when S deficiencies had been corrected. The implication of these examples is, as CHAPMAN (1966) and others have pointed out, that plant analysis can only detect a deficiency of one nutrient at a time. It also appears that, in general, analysis of plant tissue for a single element is of limited value in diagnosis, unless it is known that the supply of other elements is adequate. An example of an integrated approach to diagnosis is the one developed by BEAUFILS, referred to as Diagnosis and Recommendation Integrated System – DRIS (SUMNER 1974). This system attempts to take into account as many factors affecting yield as can be expressed quantitatively, including plant composition with respect to as many elements as are likely to affect crop performance. Composition is expressed in as many ways as possible, e.g. concentration (%) in the dry matter, or as ratios or products, in order to obtain the most appropriate comparison between different yield populations. By selecting the appropriate sets of expressions all elements can be related in the form of a statement listing them in decreasing order of requirement by the plant.

### 2.3.5 Environmental Factors

Maximum rates of plant growth depend on environmental factors such as temperature, light and moisture supply. When present at levels below the optimum for maximum growth, or yield, any of these factors may become "limiting" and thus cause a reduced nutrient requirement by the plant. As a result the nutrient concentration in the dry matter tends to increase. In subterranean clover plants grown at three temperatures (15 °C day/10 °C night, 21°/16 °C, 27°/22 °C), and at two P levels the response in growth to P was least at the lowest temperature. The P concentration in all plant parts (leaves, petioles and roots) was highest at the lowest temperature and decreased with rising temperature. This occurred at both P levels (BOUMA and DOWLING 1969a, b). Similar results were obtained for S (BOUMA unpublished).

There is potentially a limit where further decreases in temperature will not cause a further rise in nutrient concentration and may even cause a decrease. This will depend on the relative extent to which growth and nutrient uptake are reduced by lowering the temperature. ZURBUCKI (1961), for example, showed that the N, P and K concentrations of tomato plants grown at 12 °C were lower than of plants grown at 20 °C. At 50 days from sowing there was a fourfold difference in dry weight. In a case like this, where the temperature of 12 °C was much lower than the optimum, nutrient uptake may be reduced to an even greater extent than growth, resulting in a decrease in nutrient concentration.

Light influences nutrient concentrations in a similar manner. In an experiment (BOUMA unpublished) with subterranean clover plants grown at 4 P levels, concentrations of P in all plant parts (leaves, petioles and roots) were higher under a light intensity of 1300 ft.c. than under 2600 ft.c. These differences were relatively small at the lower P levels, but increased with the P supply due to the greatly reduced growth response to P at the low light level compared with that at the high level. This suggests a reduced P requirement at lower light levels. In cherry trees critical values for N were higher in sun leaves than in shade leaves (PROEBSTING and KENWORTHY 1954).

The effects of changes in soil moisture on nutrient concentrations in plant tissues are complex. Some obvious effects of excessive moisture are a loss of nutrients by leaching, erosion, and a decreased availability for some elements due to poor aeration (WADLEIGH and RICHARDS 1951). Some elements may become more available due to reduction processes in a poorly aerated soil, e.g. Fe and Mn. Uptake of nutrients, particularly anions, may also be reduced by a lack of oxygen in saturated soils. Changes in aeration may also affect root morphology. In general, all these factors can influence the nutrient concentration in plant tissues, depending on the extent to which they affect nutrient uptake, utilization and growth.

Reports on the effects of low soil moisture supplies on nutrient concentrations in the plant are conflicting. WADLEIGH and RICHARDS (1951) concluded that, for a given level of fertility, decreasing soil moisture causes an increase in N concentration in plant tissue, a decrease in K concentration, and a variable effect on P, Ca and Mg contents. On the other hand, WILLIAMS and SHAPTER (1955) concluded that in barley and rye, moisture stress gave increased leaf concentrations of K, N, Mg, Ca and Mn, although treatment effects on rye were relatively small initially.

It has often been suggested that drying of the soil causes an increase in concentration in the soil solution of at least some nutrient elements, and that a resulting increase in uptake by the plant may cause a corresponding increase in the tissues (WADLEIGH and RICHARDS 1951). However, GATES (1957, 1974) concluded that there was no evidence for this in his experiments, and that wilting both restricted growth and gave lower values for concentrations of a wide range of nutrient elements, including the principal determinants of protein synthesis, N and P. Moisture stress was considered to cause hydrolysis of organic compounds in older tissues, resulting in an increase in levels of soluble nutrients in the plant, so that the plant reduced uptake of these elements in response to a decline in internal demand, rather than because of changes in nutrient availability (GATES 1974). Upon rewatering, a complete reversal of these trends was found, in which the main factor controlling the absorption was suggested to be the internal demand caused by the renewal of synthetic processes (GATES 1957).

### 2.3.6 Other Factors Affecting Nutrient Composition

The nutrient concentration of plant tissue is strongly influenced by the fruit or crop load (EMMERT 1959, SMITH 1962). In general, the concentrations of N, Ca and Mg increase with the fruit load, while the K concentration usually decreases. The effects on P are variable. Leaf scorch caused by K deficiency has been reported to increase with the crop load by a number of workers. There are also marked interactions between the effects of fertilizer additions on crop load and the concentrations of other elements in leaf tissue. BOUMA (1959a) found a marked increase in crop load of navel oranges due to N applications, which was accompanied by a decrease in K concentration of the leaves, even with adequate K supplies in the soil (GROENEWEGEN and BOUMA 1960). This was attributed, at least in part, to a dilution effect following yield and growth responses to N. SATO (1961) considered a decline in leaf K with increasing fruit load in Satsuma oranges to be largely due to translocation of K to the fruit, possibly because the fruit had a relatively high K requirement. Related to this may be the finding in Florida that there was an almost linear increase in fruit drop when leaf K was less than 0.8% of the dry matter (WILSON 1961).

The nutrient composition of the leaves also depends on their position relative to the fruit. In oranges, for example, N and P concentrations are, as a rule, lower in leaves from fruiting shoots than in leaves from vegetative shoots, probably because the fruit competes for available nutrient elements (BOUMA 1959b, 1961). For this reason leaf sampling is usually standardized, some workers preferring leaves from fruiting shoots and others those from vegetative shoots (see CHAPMAN 1966 for a tabulation of tissues sampled).

Other factors which can affect the nutrient composition of plant tissues are pests and diseases, and these need to be taken into account to avoid misleading interpretations of tissue composition (BERGMANN et al. 1974, KIRKPATRICK et al. 1964). Genetic factors can be important (see Chap. IV.1, this Vol.). A special case is the horticultural one where one genotype (scion) is grafted on to another (stock) and where the nutrient-absorbing and -utilizing parts of the plant are different in origin and character (EPSTEIN 1972, KENWORTHY 1967, REUTHER and SMITH 1954).

# 3 Physiological and Biochemical Approaches to Diagnosis

## 3.1 Introductory Remarks

Significant and impressive though the past progress in the field of plant analysis has been, there are indications that research along traditional lines in this area and its application in diagnosis have reached a point of diminishing returns. BAR-AKIVA (1971) expressed the view that many of the difficulties of plant analysis stem from its empirical nature. GREENWOOD (1976) concluded that nutrient response curves based on plant analysis as outlined above give very little information about the current intensity of deficiency. Because of the many factors that can affect response curves, they may also be quite misleading.

It is not surprising, therefore, that in recent years there has been a search for other approaches based on specific metabolic, enzymatic or physiological changes caused by deficiencies, or on responses that can be induced in plants or in plant tissue by the addition of a deficient element. They can be broadly grouped, for ease of presentation, as physiological or biochemical, although the division is an arbitrary one in some cases.

## 3.2 Physiological Approaches

### 3.2.1 Physiological Assessment

In a series of papers BOUMA and DOWLING (1962, 1966a, b, 1967) and BOUMA et al., (1969) presented a diagnostic method, to which they referred as "the physiological assessment of the nutrient status of plants". It is based on the differential leaf area responses induced by transfer of deficient young subterranean clover plants to complete nutrient solutions and to other solutions each without a different element. Leaf areas of plants were measured by comparing individual leaves with a set of photographic standards. This process is quick and simple. Responses in leaf area were relatively greater than those in dry weight, thus enabling an earlier identification of induced responses. For example, P-deficient plants transferred to solutions without P showed smaller increases in leaf area than corresponding plants transferred to complete solutions or to solutions lacking an element other than P. When plants not deficient in P were transferred, there were no significant differences in leaf area increase between any of the solutions. In this way it was possible to make a clear distinction between P-deficient and non-deficient plants. Similar results were obtained with plants raised at different levels of S, K or B. With an unknown deficiency, involving any one of these elements, therefore, the deficient element can be identified by the element omitted from that solution in which the transferred plants showed the smallest leaf area increases during the test period. In most cases the relative differences in leaf area between plants in complete solutions and in solutions without the deficient element increased with the severity of the deficiency. This was therefore suggested as a measure of the degree of the deficiency in the field.

Obviously the approach only lends itself to those plants which can be transferred to test solutions and which can subsequently continue or resume growth.

The approach has been further developed by JANSSEN (1974) and by MULLER (1974), and used to advantage in several countries for a number of crops, including wheat and maize, and also for perennials such as *Pinus caribea* and cacao. The emphasis has been somewhat different in that the authors were primarily assessing the likelihood of finding nutrient deficiencies in soils to be used for agricultural development and testing the suitability of those soils for certain crops. They therefore grew seedlings in small pots containing the soil concerned. These pots had a gauze bottom so that the roots could extend into nutrient solutions on which the pots were placed. These solutions included, as above, a complete nutrient solution and others each without a different element. A deficiency was identified by a difference in growth, using a suitable parameter such as leaf area or leaf length, between plants on a solution without a certain element and on the other solutions. JANSSEN (1974) used a "sufficiency quotient" as an index of nutrient supply in the soil, this being defined as the ratio of the relative growth rate of plants in pots on the deficient solution to that on the complete solution.

### 3.2.2 Nutrient Stress

In a review summarizing the development of an approach having features in common with the one discussed above, GREENWOOD (1976) suggested the concept of "nutrient stress" to quantify the current nutrient status of plants. He defined this as the proportion by which the growth rate of the plant or crop is limited by a particular nutrient element under the prevailing conditions. GREENWOOD et al. (1965) developed this concept to estimate N deficiency in a grass sward, and expressed N stress ($S_N$) as the percentage reduction in the relative growth rate of deficient plants compared with that of plants growing with non-limiting N. To measure $S_N$ they used a split-plot technique, applying sufficient N for maximum growth to one half and leaving the other half untreated. Dry matter on these plots was measured at intervals of 18–14 days. Since there are some obvious disadvantages in the use of dry matter as an index of growth, particularly under field conditions, they also investigated the concentration of total and ninhydrin-N in the youngest fully expanded leaf of the tillers of annual ryegrass and wheat, and also leaf elongation as parameters of N status that could be used to determine $S_N$. $S_N$ could be estimated reliably by all three indices over the whole range of experimental treatments employed, but leaf elongation was found to be the best. In wheat and ryegrass, plant symptoms of N deficiency appeared when $S_N$ was greater than 40% (GREENWOOD 1966, GREENWOOD and TITMANIS 1966).

### 3.2.3 Approaches Based on Photosynthesis

In a further development of physiological assessment, BOUMA aimed at inducing responses in detached leaves rather than in intact plants. This would have some obvious advantages, one being the possible simplification of any routine diag-

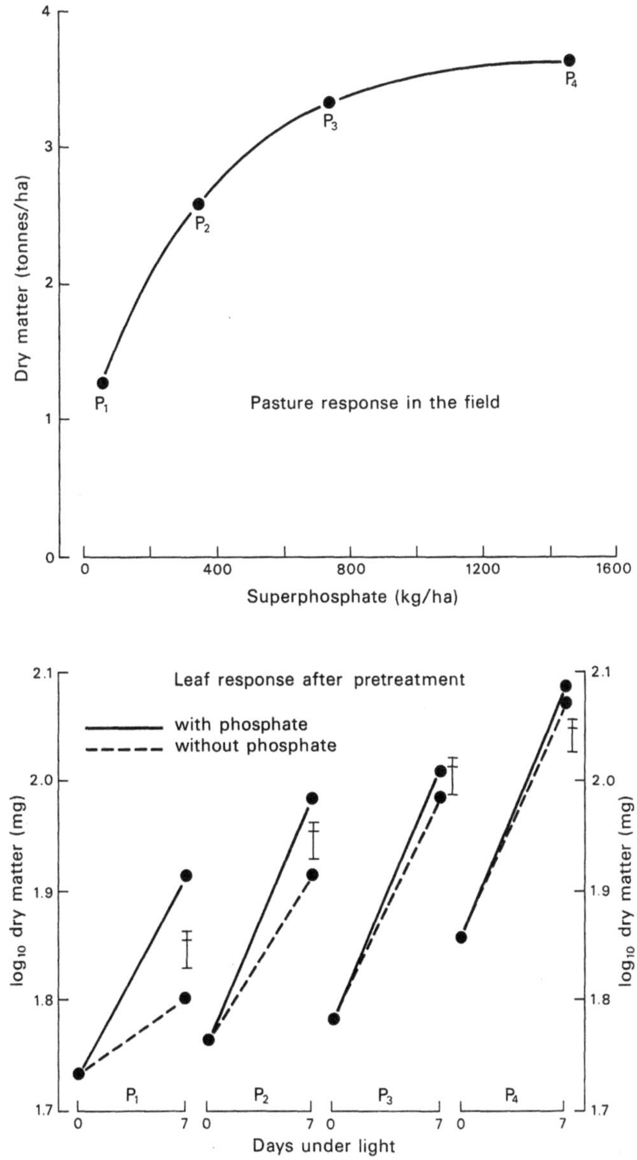


**Fig. 2.** Effect of phosphate supply on yield of a subterranean clover-dominant pasture (*top*). *Bottom graph* shows differences in dry matter production by leaves sampled in the four phosphate levels when placed under fluorescent lights after a brief pretreatment with and without inorganic phosphate (8 h; 6 mM). Significant differences (P=0.05 and 0.01) shown by *vertical lines* on day 7. (BOUMA and DOWLING 1980)

nostic application. Use was made of a previous finding that P deficiency causes a marked reduction in photosynthesis of subterranean clover plants, and that a transfer of plants to complete nutrient solutions is followed by a return to normal levels of photosynthesis within 24–48 h (BOUMA 1967). A similar response was found in leaves detached from P-deficient plants by placing them for about 8 h in a 6 mM inorganic phosphate solution (approximately 40 times the concentration required for intact plants). The leaves were then returned to water and photosynthesis was measured the following day (Bouma 1975). Subsequently it was shown (Bouma and Dowling 1976) that a corresponding response in dry matter increase could be obtained when leaves were placed under fluorescent lights for a period of 5–6 days after the initial phosphate uptake period. By comparing dry matter increases of leaves treated with phosphate after detachment with those of comparable untreated leaves, a simple and direct method for the detection of P deficiency is being developed. In a variety of pot experiments (Bouma and Dowling 1976) and field studies (Bouma unpublished), it has been found that dry matter differences between treated and untreated leaves depend on the P status of the plants from which the leaves were detached. These differences were not significant in the case of plants of an adequate P status, but increased with a decline in the P status of the plant. This is illustrated in Fig. 2, which shows the response of a subterranean clover dominant pasture to superphosphate (top portion). Dry matter differences induced in clover leaves as outlined above were closely related to the phosphate status of the pasture plots from which the leaves were sampled (bottom portion).

The measurement of photosynthesis to determine the critical concentrations of K and Mg of corn plants was suggested by Peaslee and Moss (1966). They plotted net rates of photosynthesis against extractable leaf K or Mg and obtained the usual response curves for both elements, with critical concentrations close to those found by other workers for field-grown plants. They suggest that this approach is likely to give a more accurate and more easily obtained estimate of the critical concentration than the usual correlative approach. Photosynthesis reflects the instantaneous performance of the leaf or plant that can be related to its concomitant nutrient concentration. However, not all nutrient elements affect photosynthesis in the same manner. There may even be differences in the effect of an individual element between plant species. Bouma et al. (1979), for example, investigated the effects of potassium and magnesium on net photosynthesis of intact subterranean clover plants and found that K had little effect on net photosynthesis in early stages of a deficiency, while Mg deficiency caused an immediate decline. This is somewhat at variance with the results of Peaslee and Moss (1966), who found an early decline in photosynthesis with a K deficiency while Mg caused a decrease at a more advanced stage of deficiency, as judged by plant size. Nevertheless, their suggestion remains valid as a principle, but would require investigation in each case.

### 3.2.4 Other Approaches

Randall (1969a) used detached leaves for the estimation of the S status of subterranean clover. He incubated young leaves from plants grown at five S levels, in solutions with and without sulfate. Over a period of 4 days alcohol-insoluble N of leaves in solutions without sulfate declined, irrespective of S status. The presence of sulfate in the treating solution arrested or reversed the decline in leaves from S-deficient plants, but had no such effect on leaves from non-deficient plants.

GREENHAM et al. (1972) investigated the possible use of electrical impedance methods in the detection of Ca and P deficiency in subterranean clover. This is based on the fact that some deficiencies can cause injuries of plant tissue. They impaled petioles with electrodes and studied the changes in capacitive impedance of an alternating current bridge. They were able to establish some clear and consistent differences in impedance characteristics between deficient and non-deficient plants.

### 3.3 Biochemical Approaches

#### 3.3.1 Nitrogen and Molybdenum

Mo is a constituent of nitrate reductase, an adaptive enzyme which catalyzes the reduction of nitrate to nitrite (EVANS and NASON 1953, NASON and EVANS 1953, NICHOLAS and NASON 1955). The activity of the enzyme can be measured colorimetrically from nitrite formed after incubation of leaf or root tissue with a buffer containing nitrate. Very low nitrate reductase activities are found in Mo-deficient plants. When Mo is added to the deficient leaf tissue, induction of the enzyme is rapid, maximum activity values occurring within 4 h (MULDER et al. 1959). N-deficient plants also show a low nitrate reductase activity. In this case, addition of nitrate to the rooting medium or to detached leaves causes a rapid rise in activity. It is not surprising, therefore, that there are several reports on the use of changes in nitrate reductase activity for the diagnosis of N and Mo deficiencies.

BAR-AKIVA and STERNBAUM (1965) grew young grapefruit trees in soil at five levels of ammonium sulfate and found that nitrate reductase activity in newly detached leaves, incubated for 2 h with nitrate ("initial activity"), increased approximately 4.5 times from the lowest to the highest level of N supply. They also measured the enzyme activity of leaf samples incubated with nitrate for 24 h prior to the enzyme assay, calling it inducible activity. This treatment caused considerable increases in activity of the deficient leaves, but not of the N-sufficient leaves. As a result, the gap between initial and inducible activity increased from nearly zero at the highest level to quite high values at the lowest level. The procedure was therefore suggested as a useful tool to assess the N status of citrus trees.

Citrus usually does not accumulate nitrate. In subsequent work BAR-AKIVA et al. (1970) investigated this approach for annual ryegrass, which can accumulate considerable amounts of nitrate. They found that, in spite of high endogenous nitrate levels in normal tissues, activity could be induced further, and that the response was negatively correlated with the N supply to the plant. They calculated the ratio of induced to endogenous enzyme activity, and suggested that a value of 1.5 or higher indicated N deficiency. This was similar to the value found in citrus. The results of the ryegrass work are shown in Fig. 3.

WITT and JUNGK (1974), however, found this ratio unsatisfactory for cauliflower and poinsettia plants without modifying the approach. Instead they made use of the fact that additional nitrate reductase activity can be induced by incubating leaf tissue with nitrate in the light as compared with the dark. This

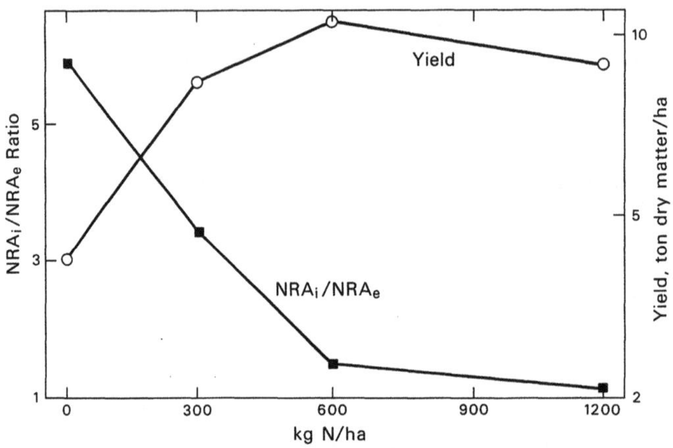


**Fig. 3.** Ratio of inducible/endogenous nitrate reductase and yield as affected by nitrogen supply. (BAR-AKIVA et al. 1970)

"additional" nitrate reductase activity was negatively correlated with nitrate supply to the plants. It was suggested that this relationship provided a good indication of the N status of the plant.

The diagnosis of Mo deficiency on the basis of nitrate reductase activity has been suggested by several workers. SHAKED and BAR-AKIVA (1967) found a direct relationship between activity and Mo supply to sweet lime seedlings growing in solution cultures. They also compared enzyme activity before and after treatment of leaf tissue with Mo. There was no response by leaves from plants grown at the highest Mo level. The differences in activity between leaves incubated with and without Mo increased as the Mo supply to the plants decreased, and reached quite a high value for the plants grown without Mo.

RANDALL (1969b) found that light stimulated the induction of nitrate reductase in Mo-treated leaf tissue from Mo-deficient wheat plants. To assess the Mo status of plants he incubated leaf fragments with buffered $KNO_3$ with or without Mo for 4 h in the light to induce the reductase, followed by 1 h in darkness to allow accumulation of nitrite, and permit in vivo assay of the enzyme. There was a good quantitative relationship between the magnitude of the response in enzyme activity of the incubated leaf tissue to Mo and the yield responses to Mo of the plants from which the leaf fragments came.

### 3.3.2 Phosphorus

In tissues of many plant species the activity of acid phosphatase is enhanced markedly by phosphate deficiency (HEWITT and TATHAM 1960, BESFORD 1978b, 1979a, b, c). The enzyme catalyzes the hydrolysis of phosphate esters. Assay rests on the hydrolysis of p-nitrophenyl phosphate by tissue extracts, the yellow colour of the reaction product, p-nitrophenol, being proportional to enzyme activity. BESFORD has shown that in tomatoes (1979a) increased enzyme activity was specific for P deficiency. In this and further work with seven other plant species (BESFORD 1979b) enzyme assay provided a rapid and sensitive index of the plant's P status. After subsequent experiments BESFORD (1980) concluded that visual comparison of the yellow colour formed during incubation of tomato leaf discs may eliminate the need for laboratory facilities in the diagnosis of P deficiency.

### 3.3.3 Potassium and Magnesium

In the glycolytic pathway the enzyme pyruvate kinase (PK) catalyses the transfer of phosphate from phosphoenolpyruvate (PEP) to adenosine diphosphate (ADP), yielding adenosine triphosphate (ATP) and pyruvate. The enzyme requires K as well as Mg ions (BONNER and VARNER 1965) although BESFORD and MAW (1975) were unable to demonstrate an absolute requirement for Mg by PK in tomato plants. In enzyme preparations there is also present a complicating reaction caused by phosphatase which hydrolyzes PEP in the absence of ADP. The latter enzyme is non-selectively inhibited by monovalent cations but stimulated by Mg. Activity of PK can be found by measuring the amount of pyruvate formed in presence and absence of ADP, then subtracting pyruvate found in the absence of ADP (=phosphatase activity) from the activity produced in the presence of ADP. In tomato plants PK activity was greatest in the leaves, and the activity increased with maturity of the tissues (BESFORD and MAW 1975).

In experiments with K in tomatoes, and K and Mg in cucumbers, BESFORD (1975) measured the PK activity of leaf extracts in the presence of optimum concentrations of Mg when assessing the K status of the plants, and in the presence of optimum K in the assessment of the Mg status. These PK activities were then compared with the PK activities of corresponding leaf extracts assayed in the presence of optimum concentrations of both K and Mg (controls). The assay clearly distinguished between plants receiving adequate or inadequate K supplies. The test also detected sub-optimal K levels that existed in the leaves before deficiency symptoms had developed. The assay of phosphatase, as well as of PK activity, reflected the Mg status of the cucumber plants. Subsequent more detailed work with K and Mg in tomato plants fully confirmed previous results and showed that a simultaneous assessment of the K and Mg status of plants can be completed within 2 h of receiving the leaf samples (BESFORD 1978a). The relationships between leaf K or leaf Mg and PK activities in tomatoes are shown in Fig. 4a and b respectively.

Potassium deficiency leads to characteristic deficiency symptoms in many crops. In barley the well-defined necrosis which accompanies K deficiency appears to be caused by the accumulation of the amine putrescine (RICHARDS and COLEMAN 1952). COLEMAN and RICHARDS (1956) were able to induce a leaf necrosis characteristic of K-deficiency in high-K barley plants by feeding putrescine. They were also able to cause a disappearance of putrescine accumulated in K-deficient barley by supplying K. BAR-AKIVA et al. (1974) suggested that the knowledge of biochemical effects of nutrient excesses or deficiencies may help in alleviating harmful effects on crop yields. They were able to overcome at least partly the undesirable effects of high K on fruit quality of citrus

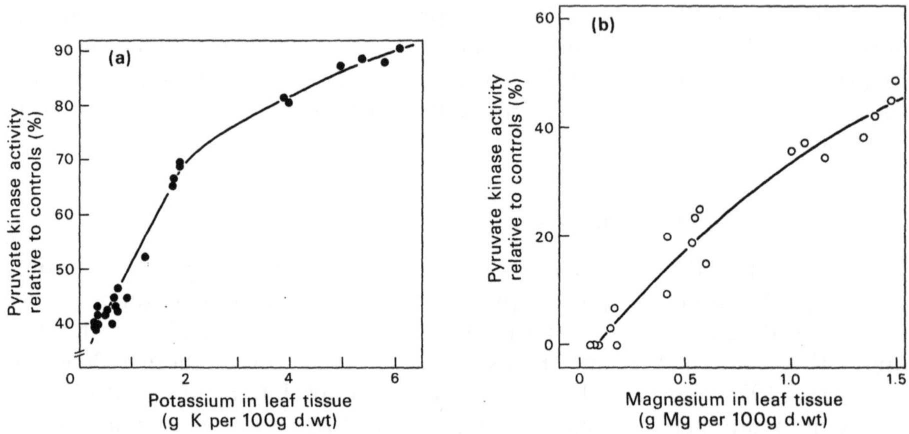


**Fig. 4.** Relationship between pyruvate kinase activity and leaf potassium concentration (**a**) and leaf magnesium concentration (**b**). The enzyme activity was measured in the presence of added Mg (for K) or added K (for Mg) and expressed as a percentage of the activity in the presence of K + Mg. (BESFORD 1978a)

(skin coarseness) by supplying putrescine to the leaves in orchard trees, or to the media in tissue culture. Although there is no published information available, putrescine may be a useful index of K-deficiency.

### 3.3.4 Iron and Manganese

Fe is possibly the classical example of an element for which it is often difficult, if not impossible, to identify deficiency on the basis of plant or leaf analysis. REUTHER et al. (1949) found no relationship between visible deficiency symptoms and Fe concentration of citrus leaves. AGARWALA et al. (1965) found a poor correlation between Fe supply and concentration in the tops of corn plants. An additional difficulty is that symptoms of Fe and Mn deficiencies are sometimes difficult to differentiate (BAR-AKIVA 1961, O'SULLIVAN et al. 1969).

Is is not surprising, therefore, that alternative methods are being investigated for the detection of Fe deficiency. Fe is a constituent of several metallo-enzymes in which it is chelated by a porphyrin group. These enzymes include several oxidases, peroxidase being one, which catalyze oxidation reactions in respiration (DAVIES et al. 1964). Peroxidase activity is measured by the oxidation of pyrogallol by the leaf extract, in the presence of hydrogen peroxide (BAR-AKIVA et al. 1967). The method is simple and close relationships have been established between enzyme activity and Fe status of species as diverse as citrus (BAR-AKIVA 1961, 1964, BAR-AKIVA and LAVON 1968) and Yorkshire fog, oats and tomatoes (O'SULLIVAN et al. 1969).

Estimation of peroxidase activity has also been proposed as a means of distinguishing between Fe and Mn deficiency in citrus (BAR-AKIVA 1961, 1964), peroxidase activity showing a marked decrease when Fe is deficient, but an increase with Mn deficiency. This has also been established for timothy, oats

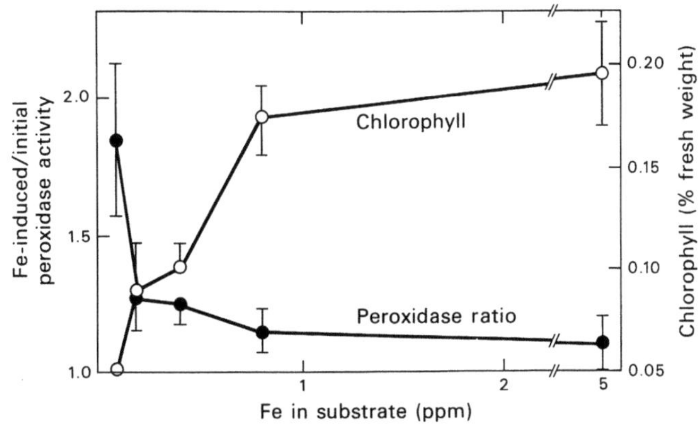


**Fig. 5.** Ratio Fe-induced/initial peroxidase activity and chlorophyll content as affected by Fe supply in substrate. (BAR-AKIVA and LAVON 1968)

and tomatoes, but not for Yorkshire fog (O'SULLIVAN et al. 1969). As an additional test for Mn deficieny, pentose (xylose) accumulation, which is specific for this deficiency and is simple to measure colorimetrically, has been suggested (BAR-AKIVA 1961, 1964).

Enzyme activation by incubation of leaves with ferrous sulphate has been used as a test for Fe deficiency in lemon trees (BAR-AKIVA and LAVON 1968). A close relationship was found between the level of Fe supply to the trees and the peroxidase activity of leaves ("initial activity"), and a close negative relationship between Fe supply and the ratio induced/initial peroxidase activity. This is illustrated in Fig. 5.

### 3.3.5 Copper

Cu is a constituent of the enzyme ascorbic acid oxidase which catalyzes the oxidation of ascorbic acid. The enzyme is widely distributed in plants (DAVIES et al. 1964). BROWN (1953) compared a range of crop plants and found that ascorbic acid oxidase was a good index of available Cu in most plants, whether they showed visual Cu deficiency symptoms or not. An increased enzyme activity was found in all of the plants studied after addition of Cu to the soil. Similar results were obtained with apple seedlings grown in solutions at different copper levels (PERUMAL and BEATTIE 1966). In lemon trees the relationship between growth and Cu concentration in the leaves was poor, but between growth and enzyme activity quite close (BAR-AKIVA et al. 1969). In the same work it was demonstrated that the enzyme could be induced by infiltration of Cu-déficient leaf tissue with Cu, and that the response was greater the lower the Cu status of the plants from which the leaves had been detached. The difference in response between leaf tissue treated with and without Cu was, therefore, inversely related to Cu status.

Such an approach overcomes many of the difficulties of leaf analysis in the diagnosis of Cu deficiency. Enzyme activity is relatively easy to measure, either manometrically using ascorbic acid as substrate (BROWN 1953) or colorimetrically as the residual ascorbic acid after incubating leaf tissue with a known amount of ascorbic acid (BAR-AKIVA et al. 1969).

### 3.3.6 Zinc

Zn is a component of the enzyme carbonic anhydrase (CA; EVANS and SORGER 1966), which catalyzes the reaction: $CO_2 + H_2O \rightarrow H^+ + HCO_3^-$, and is thus thought to function in the transfer of $CO_2$ through the liquid phase of the cell to the chloroplast surface (HATCH and SLACK 1970). In a wide range of crops, including tomatoes, oats, spinach, citrus and beans, there is a close relationship between Zn concentration in leaf tissue and CA (BAR-AKIVA and LAVON 1969, EDWARDS and MOHAMED 1973, OHKI 1976, RANDALL and BOUMA 1973, WOOD and SILBY 1952). While for some crops there is also a reasonably close relationship between photosynthesis and CA, in others photosynthesis does not decline except at very low CA levels, e.g. spinach (RANDALL and BOUMA 1973).

CA activity has been proposed as a diagnostic test for Zn deficiency in a number of agricultural species, including citrus, wheat, mustard and maize (BAR-AKIVA and LAVON 1969, DWEVIDI and RANDHAWA 1974). In Zn-deficient citrus leaves, CA activities decreased to values only 25%–30% of those for non-deficient leaves. Although deficiencies of some other minor elements also caused reductions in CA activity, these were relatively small (BAR-AKIVA and LAVON 1969). DWEVIDI and RANDHAWA (1974) used CA activity as a measure of the Zn status of several crops at progressive stages of Zn deficiency. They compared several indices of CA activity, referred to as "quick tests", with a standard extraction procedure for CA and found correlation coefficients higher than 0.9. The "quick tests" involved the estimation of $CO_2$ liberated by leaf tissue incubated in the presence of $NaHCO_3$. Although the activity values determined by these "quick tests" were lower than those for the extraction of the enzyme, the relationship between these values was close enough to permit the use of "quick tests" in the detection of Zn deficiency before visual symptoms appeared, and earlier than could be detected by the Zn concentrations of leaf tissue.

## 4 Prospects for the Future

The successful diagnosis of the nutrient status of a crop or a plant based on plant analysis depends on the closeness of the relationship between nutrient concentration at a certain stage of development and growth or yield responses to nutrients at that time or subsequently, established in prior field or pot experiments. It also depends on the ability to recognize, evaluate and take into account the many factors, discussed in preceding pages, which can modify these relationships.

Although there is no doubt that plant analysis has made significant, and often spectacular, progress in the past, further progress appears likely to remain slow and painstaking. The plant physiological and nutritional principles underlying this approach have been known for many years and there appear to be no reasons to expect, or hope for, significant changes in the trends which have characterized progress in the more recent past. These trends, which can be summarized as the further extension and refinement of plant analysis as a diagnostic technique, are likely to continue. The best prospects for progress here appear to lie in the further development of the nutritional relationships in diagnosis, in understanding the factors that can affect these, and finally in the integration of these factors in diagnostic functions that can cope with the routine application in providing reliable fertilizer advice for the farmer.

The physiological and biochemical techniques discussed in the second part of this chapter are less empirical and, because they are based on well-known physiological or biochemical functions or activities in the plant, they should provide a more direct approach to diagnosis. This in itself could be a significant step forwards.

Unfortunately, in much of the work reported above, the induced responses, or the physiological or biochemical indices used to assess the nutrient status, were correlated with nutrient supply, often in pot experiments, rather than with crop or plant responses. To assess the value of these approaches, which are often claimed to be more reliable and more rapid, they need to be tested more rigorously than in the past under conditions applying in the real world. It is difficult to escape the impression that at least some of this work is little more than superficial applied biochemistry. In spite of the fact that work in this area has been in progress for 10–20 years, as far as the author is aware, relatively little is being applied in routine diagnosis of deficiencies.

It is suggested that the greatest challenge for future progress in this area lies in the field testing and evaluation of these approaches to establish beyond doubt their advantages as alternative diagnostic techniques.

*Acknowledgements.* Thanks are due to Dr. E.A.N. Greenwood, Division of Land Resources Management, CSIRO, Wembley, W.A., and Mr K. Spencer, Division of Plant Industry, CSIRO, Canberra, A.C.T., for their criticism of the manuscript.

## References

Agarwala SC, Sharma CP, Farooq S (1965) Effect of iron supply on growth, chlorophyll, tissue iron and activity of certain enzymes in maize and radish. Plant Physiol 40:493–499

Bar-Akiva A (1961) Biochemical indications as a means of distinguishing between iron and manganese deficiency symptoms in *Citrus* plants. Nature 190:647–648

Bar-Akiva A (1964) Visible symptoms and chemical analysis versus biochemical indicators as a means of diagnosing iron and manganese deficiencies in *Citrus* plants. In: Bould C, Prévot P, Magness JR (eds) Plant analysis and fertilizer problems, vol IV. Am Soc Hortic Sci, E Lansing, Mich

Bar-Akiva A (1971) Functional aspects of mineral nutrients in use for the evaluation of plant nutrient requirement. In: Samish RM (ed) Recent advances in plant nutrition, Proc 6th Int Colloq Plant Anal Fert Probl, vol I. Gordon & Breach New York

Bar-Akiva A, Lavon R (1968) Peroxidase activity as an indicator of the iron requirement of *Citrus* plants. Isr J Agric Res 18:145–153
Bar-Akiva A, Lavon R (1969) Carbonic anhydrase activity as an indicator of zinc deficiency in *Citrus* leaves. J Hortic Sci 44:359–362
Bar-Akiva A, Sternbaum J (1965) Possible use of the nitrate reductase activity of leaves as a measure of the nitrogen requirement of *Citrus* trees. Plant Cell Physiol 6:575–577
Bar-Akiva A, Kaplan M, Lavon R (1967) The use of a biochemical indicator for diagnosing micronutrient deficiencies of grapefruit trees under field conditions. Agrochim 11:283–288
Bar-Akiva A, Lavon R, Sagiv J (1969) Ascorbic acid oxidase activity as a measure of the copper nutrition requirement of Citrus trees. Agrochim 14:47–54
Bar-Akiva A, Sagiv J, Leshem J (1970) Nitrate reductase activity as an indicator for assessing the nitrogen requirement of grass crops. J Sci Food Agric 21:405–407
Bar-Akiva A, Sagiv J, Renveni O (1974) Physiological approaches to plant nutritional problems. In: Wehrmann J (ed) Proc 7th Int Colloq Plant Anal Fert Probl, vol I. Germ Soc Plant Nutr, Hannover
Bates TE (1971) Factors affecting critical nutrient concentrations in plants and their evaluation: A review. Soil Sci 112:116–130
Bergmann EL, Boyle JS, Fries RE, Ferretti PA (1974) Influence of virus infection on elemental content of selected vegetable crops. In: Wehrmann W (ed) Proc 7th Int Colloq Plant Anal Fert Probl, vol I. Germ Soc Plant Nutr, Hannover
Bergmann W (1958) Die Ermittlung der Nährstoffbedürftigkeit des Bodens. In: Ruhland W (ed) Encyclopedia of plant physiology, vol IV. Springer, Berlin Göttingen Heidelberg
Besford RT (1975) Pyruvate kinase and a phosphatase as potential indicators of potassium and magnesium status of tomato and cucumber plants. J Sci Food Agric 26:125–133
Besford RT (1978a) Use of pyruvate kinase activity of leaf extracts for the quantitative assessment of potassium and magnesium status of tomato plants. Ann Bot 42:317–324
Besford RT (1978b) A phosphatase as a potential indicator of the phosphorus status of the glasshouse cucumber (*Cucumis sativus*). J Sci Food Agric 29:87–91
Besford RT (1979a) Nutrient imbalances in tomato plants and acid phosphatase activity in the leaves. J Sci Food Agric 30:275–280
Besford RT (1979b) Phosphorus nutrition and acid phosphatase activity in the leaves of seven plant species. J Sci Food Agric 30:281–285
Besford RT (1979c) Quantitative aspects of leaf acid phosphatase activity and the phosphorus status of tomato plants. Ann Bot 44:153–161
Besford RT (1980) A rapid tissue test for diagnosing phosphorus deficiency in the tomato plant. Ann Bot 45:225–227
Besford RT, Maw GA (1975) Some properties of pyruvate kinase extracted from *Lycopersicon esculentum*. Phytochemistry 14:677–682
Bonner J, Varner JE (1965) The path of carbon in respiratory metabolism. In: Bonner J, Varner JE (eds) Plant biochemistry. Academic Press, London New York
Bould C (1961) Leaf analysis as a guide to the nutritional status of soft fruit crops. In: Reuther W (ed) Plant analysis and fertilizer problems. Am Inst Biol Sci, Washington DC
Bould C (1964) Leaf analysis in relation to raspberry nutrition. In: Bould C, Prévot P, Magness JR (eds) Plant analysis and fertilizer problems, vol IV. Am Soc Hortic Sci, E Lansing, Mich
Bould C (1968) Leaf analysis as a diagnostic method and advisory aid in crop nutrition. Exp Agric 4:17–27
Bouma D (1956) Studies in citrus nutrition. II. Phosphorus deficiency and fruit quality. Aust J Agric Res 7:261–271
Bouma D (1959a) Growth, yield, and fruit quality in a factorial field experiment with citrus in relation to changes in phosphorus nutrition. Aust J Agric Res 10:41–51
Bouma D (1959b) The development of the fruit of the Washington Navel orange. Aust J Agric Res 10:804–817

Bouma D (1961) The development of cuttings of the Washington Navel orange to the stage of fruit set. IV. The effect of different nitrogen and phosphorus levels on fruiting cuttings. Aust J Agric Res 12:1089–1099
Bouma D (1967) Growth changes of subterranean clover during recovery from phosphorus and sulphur stresses. Aust J Biol Sci 20:51–66
Bouma D (1975) Effects of some metabolic phosphorus compounds on rates of photosynthesis of detached phosphorus-deficient subterranean clover leaves. J Exp Bot 26:52–59
Bouma D, Dowling EJ (1962) The physiological assessment of the nutrient status of plants. I. Preliminary experiments with phosphorus. Aust J Agric Res 13:791–800
Bouma D, Dowling EJ (1966a) The physiological assessment of the nutrient status of plants. II. The effect of the nutrient status of the plant with respect to phosphorus, sulphur, potassium, calcium, or boron on the pattern of leaf area response following the transfer to different nutrient solutions. Aust J Agric Res 17:633–646
Bouma D, Dowling EJ (1966b) The physiological assessment of the nutrient status of plants. III. Experiments with plants raised at different nitrogen levels. Austr J Agric Res 17:647–655
Bouma D, Dowling EJ (1967) The physiological assessment of the nutrient status of plants. IV. The effect of the interaction between nutrient elements on leaf area responses. Aust J Agric Res 18:223–233
Bouma D, Dowling EJ (1969a) Effects of temperature on growth and nutrient uptake in subterranean clover during recovery from phosphorus stress. I. Growth changes. Aust. J Biol Sci 22:505–514
Bouma D, Dowling EJ (1969b) Effects of temperature on growth and nutrient uptake in subterranean clover during recovery from phosphorus stress. II. Phosphorus uptake and distribution. Aust J Biol Sci 22:515–522
Bouma D, Dowling EJ (1976) The relationship between the phosphorus status of subterranean clover plants and the dry weight responses of detached leaves in solutions with and without phosphate. Aust J Agric Res 27:53–62
Bouma D, Dowling EJ (1980) Field evaluation of a test for phosphorus deficiency in pastures based on dry matter responses, induced in detached subterranean clover leaves. Commun Soil Sci Plant Anal 11:861–872
Bouma D, Dowling EJ, Wahjoedi H (1979) Some effects of potassium and magnesium on the growth of subterranean clover (*Trifolium subterraneum* L.). Ann Bot 43:529–538
Bouma D, Spencer K, Dowling EJ (1969) Assessment of the phosphorus and sulphur status of subterranean clover pastures. 3. Plant tests. Aust J Exp Agric Anim Husb 9:329–340
Brown JC (1953) The effect of the dominance of a metabolic system requiring iron or copper on the development of lime-induced chlorosis. Plant Physiol 28:495–502
Chapman HD (ed) (1966) Diagnostic criteria for plants and soils. Univ Cal, Div Agric Sci, pp 1–793
Coleman RG, Richards FJ (1956) Physiological studies in plant nutrition. XVIII. Some aspects of nitrogen metabolism in barley and other plants in relation to potassium deficiency. Ann Bot 20:393–409
Davies DD, Giovanelli J, Ap Rees T (1964) Plant biochemistry. Blackwell, Oxford
Dwevidi RS, Randhawa NS (1974) Evaluation of a rapid test for the hidden hunger of zinc in plants. Plant Soil 40:446–450
Edwards GE, Mohamed AK (1973) Reduction in carbonic anhydrase activity in zinc-deficient leaves of *Phaseolus vulgaris* L. Crop Sci 13:351–354
Emmert FH (1959) Chemical analysis of tissue as a means of determining nutrient requirements of deciduous fruit plants. Proc Am Soc Hort Sci 73:521–547
Emmert FH (1961) The bearing of ion interactions on tissue analysis results. In: Reuther W (ed) Plant analysis and fertilizer problems. Washington, D.C.: Am Inst Biol Sci, Washington DC
Epstein E (1972) Mineral nutrition of plants: Principles and perspectives. Wiley and Sons, New York

Epstein E, Lilleland O (1942) A preliminary study of the manganese content of the leaves of some deciduous fruit trees. Proc Am Soc Hort Sci 41:11–18
Evans HJ, Nason A (1953) Pyridine nucleotide-nitrate reductase from extracts of higher plants. Plant Physiol 28:233–254
Evans HJ, Sorger GJ (1966) Role of mineral elements with emphasis on the univalent cations. Annu Rev Plant Physiol 17:47–76
Freney JR, Spencer K, Jones MB (1978) The diagnosis of sulphur deficiency in wheat. Aust J Agric Res 29:727–738
Gates CT (1957) The response of the young tomato plant to a brief period of water shortage. III. Drifts in nitrogen and phosphorus. Aust J Biol Sci 10:125–146
Gates CT (1974) Water shortage and agriculture: Some responses. J. Aust Inst Agric Sci 40:121–142
Goodall DW, Gregory FG (1947) Chemical composition of plants as an index of their nutritional status. Tech Commun 17, Imp Bur Hortic Plant Crops, E Malling
Greenham CG, Randall PJ, Ward MM (1972) Impedance parameters in relation to phosphorus and calcium deficiencies in subterranean clover (*Trifolium subterraneum* L.). J Exp Bot 23:197–209
Greenway H, Pitman MG (1965) Potassium retranslocation in seedlings of *Hordeum vulgare*. Aust J Biol Sci 18:235–247
Greenwood EAN (1966) Nitrogen stress in wheat – its measurement and relation to leaf nitrogen. Plant Soil 24:279–288
Greenwood EAN (1976) Nitrogen stress in plants. Adv Agron 28:1–35
Greenwood EAN, Titmanis ZV (1966) The effect of age on nitrogen stress and its relation to leaf nitrogen and leaf elongation in a grass. Plant Soil 24:379–389
Greenwood EAN, Goodall DW, Titmanis ZV (1965) The measurement of nitrogen deficiency in grass swards. Plant Soil 23:97–116
Gregory FG (1953) The control of growth and reproduction by external factors. In: Rep 13th Int Hortic Congr London, R Hortic Soc
Groenewegen H, Bouma D (1960) The chemical composition of the soil in a factorial experiment with Citrus. I. Exchangeable metal cations and their effect on the cation content of citrus leaves. Aust J Agric Res 11:208–222
Hatch MD, Slack CR (1970) Photosynthetic $CO_2$-fixation pathways. Annu Rev Plant Physiol 21:141–162
Hewitt EJ, Tatham P (1960) Interaction of mineral deficiency and nitrogen source on acid phosphatase activity in leaf extracts. J Exp Bot 11:367–375
Janssen BH (1974) A double pot technique for rapid soil testing. Trop Agric (Trinidad) 51:161–166
Jones WW, Parker ER (1950) Seasonal variations in mineral composition of orange leaves as influenced by fertilizer practices. Proc Am Soc Hort Sci 55:92–100
Kenworthy AL (1967) Plant analysis and interpretation of analysis for horticultural crops. In: Hardy GW et al. (eds) Soil testing and plant analysis, part II, plant analysis. Soil Sci Soc Am Madison, Wisc
Kirkpatrick JD, Gundy van SD, Mai WF (1964) Interrelationships of plant nutrition, growth, and parasitic nematodes. In: Bould C, Prévot P, Magness JR (eds) Plant analysis and fertilizer problems, vol IV. Am Soc Hortic Sci, E Lansing, Mich
Loneragan JF (1978) Anomalies in the relationship of nutrient concentrations to plant yield. In: Ferguson AR, Bieleski RL, Ferguson IB (eds) Plant nutrition 1978. DSIR Inf Ser 134. Government Printer, Wellington
Loneragan JF, Snowball K (1969) Rate of calcium absorption by plant roots and its relation to growth. Aust J Agric Res 20:479–490
Lundegårdh H (1966) Plant Physiology. Edinburgh and London: Oliver and Boyd, Edinburgh London
Macy P (1936) The quantitative mineral nutrient requirements of plants. Plant Physiol 11:749–764
Mason AC (1958) The concentration of certain nutrient elements in apple leaves taken from different positions on the shoot and at different dates through the growing season. J Hortic Sci 33:128–138

Millikan CR (1963) Effects of different levels of zinc and phosphorus on the growth of subterranean clover (*Trifolium subterraneum* L.). Aust J Agric Res 14:180–205
Millikan CR, Hanger BC (1964) Effect of calcium level in the substrate on the distribution of $^{45}Ca$ in subterranean clover (*Trifolium subterraneum* L.). Aust J Biol Sci 17: 823–844
Mulder EG, Boxma R, Veen van WL (1959) The effect of molybdenum and nitrogen deficiencies on nitrate reduction in plant tissues. Plant Soil 10:335–355
Muller A (1974) Nutrient deficiencies in a volcanic ash soil from Ecuador. In: Wehrmann J (ed) Proc 7th Int Colloq Plant Anal Fert Probl, vol II. Hannover, W. Germ.: Germ Soc Plant Nutr Hannover
Nason A, Evans HJ (1953) Triphosphopyridine nucleotide-nitrate reductase in *Neurospora*. J Biol Chem 202:655–673
Nicholas DJD, Nason A (1955) The role of molybdenum as a constituent of nitrate reductase from soybean leaves. Plant Physiol 30:135–138
Ohki K (1976) Effect of zinc nutrition on photosynthesis and carbonic anhydrase activity in cotton. Physiol Plant 38:300–304
O'Sullivan M, Flynn MJ, Codd FJ (1969) A biochemical method for diagnosing micronutrient deficiencies in plants. Ir J Agric Res 8:111–119
Ozanne PG (1955) The effect of nitrogen on zinc deficiency in subterranean clover. Aust J Biol Sic 8:47–55
Peaslee DE, Moss DN (1966) Photosynthesis in K- and Mg-deficient maize (*Zea mays* L.) leaves. Soil Sci Soc Am Proc 30:220–223
Perumal A, Beattie JM (1966) Effect of different levels of copper on the activity of certain enzymes in leaves of apple. Proc Am Soc Hortic Sci 88:41–47
Piper CS (1942) Investigations on copper deficiency in plants. J Agric Sci 32:143–178
Prévot P, Ollagnier M (1961) Law of the minimum and balanced mineral nutrition. In: Reuther W (ed) Plant analysis and fertilizer problems. Am Inst Biol Sci, Washington DC
Proebsting EL, Kenworthy AL (1954) Growth and leaf analysis of Montmorency cherry trees as influenced by solar radiation and intensity of nutrition. Proc Am Soc Hortic Sci 63:41–48
Randall PJ (1969a) Restoration of protein level in detached leaves by supply of sulphate. Plant Soil 31:385–388
Randall PJ (1969b) Changes in nitrate and nitrate reductase levels on restoration of molybdenum to molybdenum-deficient plants. Aust J Agric Res 20:635–642
Randall PJ, Bouma D (1973) Zinc deficiency, carbonic anhydrase, and photosynthesis in leaves of spinach. Plant Physiol 52:229–232
Reuther W, Boynton D (1939) Variations in potassium content of foliage from certain New York apple orchards. Proc Am Soc Hortic Sci 37:32–38
Reuther W, Smith PF (1954) Leaf analysis of *Citrus*. In: Childers NF (ed) Fruit nutrition. Hortic Publ Rutgers Univ, New Brunswick, New Jersey
Reuther W, Embleton TW, Jones WW (1958) Mineral nutrition of tree crops. Annu Rev Plant Physiol 9:175–206
Reuther W, Smith PF, Specht A (1949) A comparison of the mineral composition of Valencia orange leaves from the major producing areas of the United States. Proc Fla Hortic Soc 60:38–45
Richards FJ, Coleman RG (1952) Occurrence of putrescine in potassium-deficient barley. Nature 170:460
Sato K (1961) The effects of growth and fruiting on leaf composition of citrus trees. In: Reuther W (ed) Plant analysis and fertilizer problems. Am Inst Biol Sci, Washington DC
Shaked A, Bar-Akiva A (1967) Nitrate reductase activity as an indication of molybdenum level and requirement of citrus plants. Phytochemistry 6:347–350
Shaulis NJ (1961) Associations between symptoms of potassium deficiency, plant analysis, growth and yield of Concord grapes. In: Reuther W (ed) Plant analysis and fertilizer problems. Am Inst Biol Sci, Washington DC
Smith PF (1962) Mineral analysis of plant tissues. Annu Rev Plant Physiol 13:81–108

Smith PF, Reuther W (1950) Seasonal changes in Valencia orange trees. I. Changes in leaf dry weight, ash, and macro-nutrient elements. Proc Am Soc Hortic Sci 55:61–72

Spencer K, Jones MB, Freney JR (1977) Diagnostic indices for sulphur status of subterranean clover. Aust J Agric Res 28:401–412

Steenbjerg F (1951) Yield curves and chemical plant analyses. Plant Soil 3:97–109

Sumner ME (1974) An evaluation of Beaufils' physiological diagnosis technique for determining the nutrient requirements of crops. In: Wehrmann J (ed) Proc 7th Int Colloq Plant Anal Fert Probl, vol II. Hannover, W. Germ. Germ Soc Plant Nutr, Hannover

Ulrich A (1948) Plant analysis – methods and interpretation of results. In: Kitchen HB (ed) Diagnostic techniques for soils and crops. Am Pot Inst, Washington DC

Ulrich A (1952) Physiological bases for assessing the nutritional requirements of plants. Annu Rev Plant Physiol 3:207–228

Ulrich A (1961) Plant analysis in sugarbeet nutrition. In: Reuther W (ed) Plant analysis and fertilizer problems. Am Inst Biol Sci, Washington DC

Ulrich A, Hills FJ (1967) Principles and practices of plant analysis. In: Hardy GW et al. (eds) Soil testing and plant analysis, part II, plant analysis. Madison, Wisc

Wadleigh CH, Richards LA (1951) Soil moisture and the mineral nutrition of plants. In: Truog E (ed) Mineral nutrition of plants. Univ Wisc Press, Madison

West ES (1938) Zinc-cured mottle leaf in citrus induced by excess phosphate. J Counc Sci Ind Res Aust 11:182–184

Williams RF (1948) The effects of phosphorus supply on the rates of intake of phosphorus and nitrogen and upon certain aspects of phosphorus metabolism in gramineous plants. Aust J Sci Res Ser B 1:336:361

Williams RF (1955) Redistribution of mineral elements during development. Annu Rev Plant Physiol 6:81–108

Williams RF, Shapter RE (1955) A comparative study of growth and nutrition in barley and rye as affected by low water treatment. Aust J Biol Sci 8:435–466

Wilson AE (1961) Leaf analysis and fertility control in commercial citrus production. In: Reuther W (ed) Plant analysis and fertilizer problems. Am Inst Biol Sci, Washington DC

Witt HH, Jungk A (1974) The nitrate inducible nitrate reductase activity in relation to nitrogen nutritional status of plants. In: Wehrmann J (ed) Proc 7th Int Colloq Plant Anal Fert Probl, vol II. Hannover, W. Germ. Germ Soc Plant Nutr Hannover

Wood JG, Silby PM (1952) Carbonic anhydrase activity in plants in relation to zinc content. Aust J Sci Res Ser B 5:244–255

Zurbucki ZI (1961) Dependence of mineral composition of plants on environmental conditions. In: Reuther W (ed) Plant analysis and fertilizer problems. Am Inst Biol Sci Washington DC

# I.5 Interactions Between Nutrients in Higher Plants

A.D. Robson and M.G. Pitman

## 1 Introduction

Interactions between nutrients in higher plants occur when the supply of one nutrient affects the absorption, distribution or function of another nutrient. Thus, depending on nutrient supply, interactions between nutrients can either induce deficiencies or toxicities and can modify growth response. Interactions between nutrients can be assessed by examining both the relationship between nutrient supply and growth and the relationship between nutrient concentrations in plants and growth. Additionally where nutrient supply is neither deficient nor toxic for plant growth, interactions can be assessed by considering nutrient concentrations and contents within the plant.

There are several ways in which nutrients can interact either within the soil or plant to affect nutrient absorption or utilization. However, these interactions can be classified in two major categories. *Firstly there are interactions which occur between ions because the ions are able to form a chemical bond* (either ionic or covalent). Hence interactions can occur due to the formation of precipitates or complexes. This form of interaction is generally most marked when the interacting ions have very different chemical properties [for example anions interacting with cations, a small cation with a high charge density (electron acceptor or hard acid) with a small anion with a high charge density (electron donor or hard base), or a large cation with a low charge density (soft acid) with a large anion with a low charge density (soft base)].

Interactions between nutrients relating to the formation of precipitates or complexes (either in the soil or the plant) can be predicted from solubility products and binding or stability constants. However, to use this chemical data to define the role of these processes in interactions among nutrients, concentrations and forms of nutrients (oxidation state, complexed anion or cation) in each of the compartments (soil solution, cell wall of root, cytoplasm of leaf, vacuole in leaf and phloem) must be known. Our knowledge of the concentrations and chemical states of nutrients in many of these compartments is very limited. Many micronutrients do not exist as divalent cations in either the soil solution or the xylem (Table 1). The formation of complex anions between micronutrients and organic ligands may have several important consequences. For example, from a consideration of the solubility of Cu and Zn compounds alone it would be predicted that liming soils would decrease the concentrations of both Cu and Zn in soil solutions to a similar extent and hence decrease the uptake of Cu and Zn (Lindsay 1978). However, because Cu is more complexed by soluble organic matter than Zn (Hodgson et al. 1966), effects of

**Table 1.** Concentrations and forms of mineral nutrients in soil solutions, xylem and phloem

| Nutrient | Soil solution[a] | | | Xylem[b] | | Phloem[c] | |
|---|---|---|---|---|---|---|---|
| | Conc (μM) | | Form | Conc (μM) | Form | Conc (μM) | Form |
| | Range | Median values | | | | | |
| Calcium | 0–5,000 | 1,500 | $Ca^{2+}$ | 400 to 4,500 | $Ca^{2+}$ | 250 to 2,650 | nd |
| Potassium | 0–5,000 | 1,300 | $K^+$ | 500 to 11,500 | $K^+$ | 20,000 to 85,000 | nd |
| Magnesium | 0–6,000 | 4,000 | $Mg^{2+}$ | 80–1,125 | $Mg^{2+}$ | 2,100 to 23,000 | nd |
| Sodium | nd[d] | 125 | $Na^+$ | 1,600 to 2,600 | $Na^+$ | 60 to 19,000 | nd |
| Phosphorus | 0–2 | 1 | $H_2PO_4^-$, $HPO_4^{2-}$ | 30–1,500 | $H_2PO_4^-$ | 10,000 | nd |
| Sulphur | 0–2,000 | 500 | $SO_4^{2-}$ | 1,400 | $SO_4^{2-}$ | 900 to 1,200 | no $SO_4^{2-}$ |
| Nitrogen | 0–3,000 | 1,500 | $NO_3^-$, $NH_4^+$ | 700 to 30,000 | $NO_3^-$, amino-N | | no $NO_3^-$ |
| Chlorine | nd | 200 | $Cl^-$ | 1,300 to 3,300 | $Cl^-$ | 7,900 to 11,900 | nd |
| Iron | <0.03–3 | | Complexed | 15–30 | Fe-complex (anion) | 75–220 | nd |
| Copper | <0.01 to 0.6 | | Complexed (89%–100%) | 1.5–2.0 | Cu-complex (anion) | 3–140 | nd |
| Zinc | <0.03–5 | | Complexed (28%–99%) | 5–10 | Zn-complex (anion) | 3–170 | Zn-complex (anion) |
| Manganese | <0.02 | | Complexed (84%–99%) | 7–11 | Cation | 9–25 | Mainly cation |
| Cobalt | <0.007 to 0.2 | | Complexed (8%–50%) | 0.4–14 | nd | nd | Co-complex (anion) |
| Boron | 3–1,000 | | $H_3BO_3$ | 60–75 | $H_3BO_3$(?) | nd | nd |
| Molybdenum | 0.02 to 0.08 | | $MoO_4^{2-}$ | 0.5–4 | $MoO_4^{2-}$(?) | 0.1 | nd |
| pH | 4–8 | | | 4.9–6.3 | | 7.9–8.2 | |

[a] Values are from REISENAUER (1966) for macronutrients and from LONERAGAN (1975) for micronutrients. Values for sodium and chloride are those cited by RUSSELL (1973) for typical soils

[b] Values are collated from BOLLARD (1960), HUSA and MCILRATH (1965), TIFFIN (1967), JONES and ROWE (1968), PATE (1975) and KIRKBY and ARMSTRONG (1980)

[c] Values are collated from TAMMES and VAN DIE (1964), MACROBBIE (1971), PATE (1975) and VAN GOOR and WIERSMA (1974)

[d] nd no data available

increasing soil pH are more marked on Zn uptake than on Cu uptake by plants (PIPER and BECKWITH 1949, WILLIAMS 1977). An understanding of the forms and concentrations of nutrients in soil solution is also required if studies of interactions in solution culture are to be applicable to interactions occurring within soil. Many experiments examining interactions involving micronutrients are conducted using concentrations rarely, if ever, found in soil solutions and using cations even though many of the micronutrients occur as complex anions in soil solutions (Table 1).

To avoid both precipitation either in the soil or within plant cells and immobilization by adsorption to cell walls, ions may be absorbed and transported in different forms to that required for metabolism within the cell. Similarly nutrients may be stored in cells in forms (including as insoluble salts) other than as simple anions or cations to avoid interactions between nutrients affecting metabolic roles.

*The second form of interactions is between ions with chemical properties sufficiently similar to compete for sites of adsorption* (on clay minerals and oxides in soil and in cell walls of plants), *absorption* (both influx and efflux), *transport* (within xylem and phloem) *and function* (at active sites). For these interactions, competition will be greatest for nutrients with similar size, charge, geometry of co-ordination and electronic configuration. If competition occurs between ions in absorption and transport but at the site of function, there is more marked specificity, the effect of one ion in reducing the concentration of the required ion at the site of action may decrease the rate of that reaction. However, some roles of nutrients are sufficiently non-specific (for example maintenance of electroneutrality or osmotic regulation) that one nutrient can substitute for another (for example, Na for K).

Interactions between nutrients related to competition between ions occur because there is insufficient specificity. The degree of specificity or selectivity varies widely for different processes. For adsorption of cations onto surfaces of clay minerals in soils, competition is related largely to the concentration in solution and the ratio of charge to radius (see RUSSELL 1973). Similarly the adsorption of ions by fixed carboxylates in the cell walls of plants is a relatively non-selective process. Root surfaces are thought to act as cation exchangers with little selectivity for particular cations other than that related to charge density (LEDIN and WIKLANDER 1974); however, their selectivity has been examined for relatively few cations (mainly alkalis and alkaline earth cations). However, for active absorption of ions across the plasmalemma only ions with extremely similar chemical properties compete at low concentrations ($SeO_4^{2-}/SO_4^{2-}$, $K^+/Rb^+$ but not $K^+/Na^+$, $Ca^{2+}/Sr^{2+}$).

For incorporation into metalloproteins and for the functioning of those metalloproteins, selectivity is generally more marked. Indeed for some metalloproteins only one metal ion is incorporated into their structure. Additionally, if more than one ion can be incorporated into a metalloprotein, biological activity may only be retained with one ion. It is interesting that the activity of metal ions with enzymes does not correlate with strength of binding or stability of metal chelates (PRICE 1970, SUELTER 1974).

In competitive binding the amount of any ion bound will be related to the concentration of that ion relative to its competitors and the affinity constants for the ion-ligand pairs (DA SILVA and WILLIAMS 1977). The affinity constants depend upon both properties of the ions (charge, size, electronic configuration) and of the ligand (size of cavity, charge, extent of chelation, co-ordination geometry). Additionally the stability of complexes formed by ions and ligands will vary with the solution pH (affect dissociation of the ion and ligand), the ionic strength of the solution, the effective redox potential (affect oxidation state of the ion) and the nature of the solvent (affect the hydration of the ion). Hence since these chemical properties differ within different compartments within the plant, selectivity of ligands could be different in different compartments (for example comparison of pH in xylem and phloem (Table 1).

DA SILVA and WILLIAMS (1977) also point out that incorporation of metal ions into specific sites may be achieved by different rates of reactions (kinetic factors) as well as by thermodynamic considerations. They suggest that the extremely high specificity of incorporation of Mg into chlorin, Cu into uroporphyrin, Fe into protoporhyrin and Co into corrin (in vivo which is difficult to achieve in vitro) is due to differences in rates of reactions involving intermediate products.

Cations differ in their affinity for ligands due to differences in their electronic configuration (electropositivity and electronegativity). Sodium and K have high affinities for ether, alcohol or carbonyl-O-ligands; Ca and Mg have high affinities for carboxylate or phosphate-O-ligands, while the transition metals have high affinities for -N- and -S-ligands (DA SILVA and WILLIAMS 1977). Biological systems must have a wide variety of ligands separated in space to concentrate and to separate the inorganic elements. The degree of strength of complexing differs markedly between the cations. Potassium, Na, Mg, Ca and Mn exist mainly as free or reversibly bound ions, whereas Fe, Cu, Zn and Mo are present mainly as structural chelates or metalloproteins (MENGEL and KIRKBY 1978).

Anions also differ in their affinity for ligands. Ions like $Cl^-$ are only weakly bound at hydrophobic centres, whereas ions like $H_2PO_4^-$ may form moderately strong complexes through H bonding and condensation. Of the anions, P, B and Si are present largely as oxyanions or undissociated molecules, N and S are covalently bonded, whereas Cl is present as free or reversibly bound $Cl^-$ (MENGEL and KIRKBY 1978). Molybdenum which exists largely as $MoO_4^{2-}$ (oxidation state VI) within the soil solution is reduced within the cell to Mo (V) and behaves like a metal with affinity for both N and S ligands.

As well as interactions which occur because of effects of one element on the absorption, transport or utilization of another there may be interactions relating to growth responses. Where two nutrients correct separate deficiencies, the interaction between them on growth is positive, i.e. the effect of one essential element on growth is greatest when all other nutrients are in optimum supply. There are many examples of positive interactions between essential nutrients on growth (see ANDERSON 1956). Where two nutrients correct the same deficiency the interaction between them on growth is negative (for example see Sect. 2.2.2.3).

Many interactions between nutrients have been noted in the literature (Table 2). As well as the interactions listed, competitive interactions can occur

**Table 2.** Interactions occurring between elements which affect the absorption and utilization of nutrients by plants. (+, interaction recorded; o, no interaction observed after investigation; no satisfactory data for other combinations)

| | $NH_4^+$ | $K^+$ | $Na^+$ | $Ca^{2+}$ | $Mg^{2+}$ | $Mn^{2+}$ | $Cu^{2+}$ | $Zn^{2+}$ | $Fe^{2+}$ | $Al^{3+}$ | $Cl^-$ | $NO_3^-$ | $H_2PO_4^-$ | $SeO_4^{2-}$ | $SO_4^{2-}$ | $MoO_4^{2-}$ | $H_3BO_3$ | $H_4SiO_4$ |
|---|---|---|---|---|---|---|---|---|---|---|---|---|---|---|---|---|---|---|
| $NH_4^+$ | | | | | | | | | | | | | | | | | | |
| $K^+$ | + | | | | | | | | | | | | | | | | | |
| $Na^+$ | | + | | | | | | | | | | | | | | | | |
| $Ca^{2+}$ | | + | + | | | | | | | | | | | | | | | |
| $Mg^{2+}$ | | + | | + | | | | | | | | | | | | | | |
| $Mn^{2+}$ | | | | + | | | | | | | | | | | | | | |
| $Cu^{2+}$ | | | | | | | | | | | | | | | | | | |
| $Zn^{2+}$ | | | | + | + | | + | | | | | | | | | | | |
| $Fe^{3+}/Fe^{2+}$ | | | | + | | + | | | | | | | | | | | | |
| $Al^{3+}$ | | | | + | | | + | | | | | | | | | | | |
| $Cl^-$ | | | | + | | | | | | | | | | | | | | |
| $NO_3^-$ | | | | | | | | | | | + | | | | | | | |
| $H_2PO_4^-/HPO_4^{2-}$ | | | | + | | | + | + | + | + | o | o | | | | | | |
| $SeO_4^{2-}$ | | | | | | | | | | | | | | | | | | |
| $SO_4^{2-}$ | | | | | | | | | | | o | o | | + | | | | |
| $MoO_4^{2-}$ | | | | | | | + | | + | | | | + | | + | | | |
| $H_3BO_3$ | | | | + | | | | | | | | | + | | | | | |
| $H_4SiO_4$ | | | | | | + | | | | | | | + | | | | | |

between essential nutrients and other elements in the environment (particularly trace elements with other heavy metals, for example Zn-Cd interactions). Because of the chemical basis of the interaction between ions, many interactions which affect the nutrition of plants also affect the nutrition of animals (see review by UNDERWOOD 1977).

In this review we have decided not to consider all of the interactions that occur between elements which affect the absorption and utilization of nutrients. Instead we have chosen specific examples that show the mechanisms that give rise to the interaction. Interaction between nutrients is covered also in Chap. IV.2 and in Chap. V.3, this Volume.

# 2 Interactions Between Nutrients in Monoculture

## 2.1 Interactions Between Nutrients Affecting the Absorption of Nutrients

There are many interactions between nutrients which change the relationship between nutrient supply and growth without changing the relationship between nutrient concentrations in the plant and growth. Additionally, where nutrients are not deficient for plant growth, the application of one nutrient may depress both the concentration and content of another nutrient within the plant. Interactions of this nature operate either in the soil or at the absorbing sites within the root.

### 2.1.1 Interactions Occurring in the Soil

Plants absorb nutrients from the soil solution. However, the concentration of nutrients in the soil solution depends upon reactions occurring within the soil (Fig. 1). Interactions may arise from the complex equilibria between the solution and other components of the soil. The formation of complexes, the precipitation of insoluble salts and the adsorption by soil colloids can all be affected by the concentration of other ions. We will consider here the nature of interactions between ions related to either adsorption or precipitation.

#### *2.1.1.1 Adsorption*

Concentrations of both anions and cations in solution can be controlled by adsorption onto soil colloids (see reviews by RUSSELL 1973, NYE and TINKER 1977). Ion adsorption by soil components is termed either non-specific or specific (chemisorption). Non-specific adsorption occurs for ions adsorbed onto charged surfaces of opposite signs (for example cation adsorption on negatively charged surfaces of clay minerals). Ions of equal valency are adsorbed in simple proportion to their equilibrium activity in solution. In contrast to non-specific adsorption, specific adsorption can occur on uncharged surfaces or even on surfaces of like sign, as well as on surfaces of unlike sign (adsorption of cations and anions on surface of iron oxides). Specifically adsorbed ions are held by chemical bonds as well as by electrostatic attraction (see BOWDEN et al. 1973 for a fuller discussion). Competition between ions which are specifically adsorbed is more complex than competition between non-specifically adsorbed ions because the adsorption of one ion can reverse the sign of the surface charge.

Effects of soil acidity on nutrient uptake can arise through effects of acidity on the adsorption of ions. For example, the application of $CaCO_3$ to acid soils can markedly increase the growth of legumes on acid soils by effects on the uptake of Mo by plants (ANDERSON and OERTEL 1946). The adsorption of $MoO_4^{2-}$ by soils and by oxides of Al and Fe decreased as pH was increased above 4 (JONES 1957). The adsorption of both $SO_4^{2-}$ and $H_2PO_4^-$ was also decreased by increasing soil pH although the effects are much less marked than those observed for $MoO_4^{2-}$ (BARROW 1970). The effect of pH on the adsorption of anions can be related to the variable surface charge of Fe and Al oxides

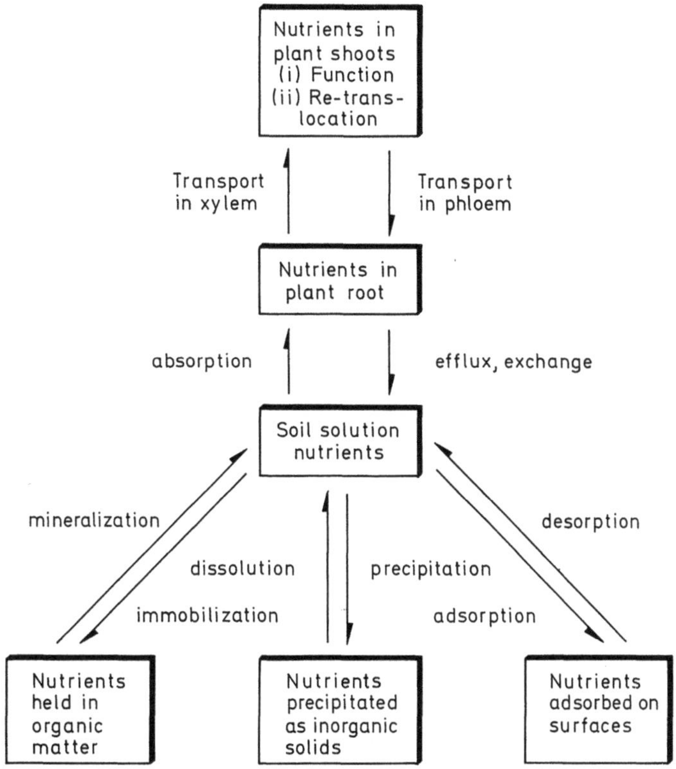


**Fig. 1.** Processes in soils and plants where interactions between nutrients may occur

(BOWDEN et al. 1973). As the solution pH nears the $pK_a$ of the appropriate acid the proportion of negatively charged ions in solution increases rapidly. Hence adsorption will increase because only the negatively charged ions are adsorbed. However, as the pH increases beyond the $pK_a$ the tendency for the surface to carry a positive charge decreases and adsorption will decrease. Hence maximum adsorption for particular anions tends to occur when solution pH equals the $pK_a$ of their acid (for example at pH 4.0 for molybdate, at pH 9.2 for silicate). For polybasic acids with large differences in $pK_a$ (for example $H_3PO_4$) there are discontinuities at the $pK_a$ values rather than a maximum. The anions of fully dissociated acids are not adsorbed unless the surface is positive. Thus $SO_4^{2-}$ is adsorbed only on the acid side of the zero point of charge (the point at which equal amounts of $H^+$ and $OH^-$ are adsorbed).

Interactions between elements affecting nutrient uptake by plants can arise through effects of one ion on the adsorption of another. Here we will consider (1) competition in the adsorption of anions (silicate-phosphate, phosphate-molybdate), (2) adsorption and incorporation of a cation into an oxide surface (Co-Mn) and (3) the effect of anion adsorption on the adsorption of a cation (P-Zn).

The application of $CaSiO_3$ may increase the growth of plants by increasing P uptake. For example, on highly weathered soils in Hawaii, $CaSiO_3$ increased the growth of soybeans only when P supply was limited (FOX 1978). In these situations where the P concentrations in the soil solution are low, the added soluble silicates may displace $H_2PO_4^-$ adsorbed onto Al and Fe oxides particularly in neutral or alkaline soils. The competition between phosphate and silicate for adsorption onto the surfaces of goethite depends upon solution pH (HINGSTON et al. 1971, 1972). Silicate can only be adsorbed in the presence of $H_2PO_4^-$ when it can increase the negative charge on the surface, that is above pH 7. Competition between other anions in adsorption also occurs. For example adding $H_2PO_4^-$ to an acid soil very low in $MoO_4^{2-}$ may increase Mo concentrations in soil solutions and the uptake of Mo by plants by displacing $MoO_4^{2-}$ from the surface of soil colloids (STOUT et al. 1951).

The adsorption of cations onto the surfaces of oxides can also lead to interactions affecting nutrient uptake. One particularly striking example is the close association between Co and Mn in soils. Differences among soils in the availability of applied Co to plants was inversely related to their total Mn content (ADAMS et al. 1969). On soils not containing high contents of montmorillonite, adsorption capacities for Co were positively correlated with their Mn content (TILLER et al. 1969). Waterlogging soils results in the dissolution of Mn oxides [Mn (IV) reduced to Mn (II)] and markedly increases the uptake of Co by plants (ADAMS and HONEYSETT 1964). These observations are related to the ability of Mn oxides to adsorb Co from solution and to hold it in a form unavailable to plants. MCKENZIE, on the basis of an elegant series of experiments (MCKENZIE 1967, 1970, 1971, 1972), has concluded that $Co^{2+}$ adsorbed onto Mn oxides can be slowly oxidized to $Co^{3+}$. The trivalent ion is incorporated into surface layers of the crystal lattice. While other ions are adsorbed by Mn oxides (for example $Cu^{2+}$, $Ni^{2+}$) only Co is incorporated into the surface layers of the oxide. This specificity appears to be due to the high crystal field stabilization energy of the low spin $Co^{3+}$ ion (MCKENZIE 1970, 1972). Only $Co^{2+}$ could be oxidized by $Mn^{3+}$ and only the low-spin $Co^{3+}$ ion has a higher crystal field stabilization energy than $Mn^{3+}$.

The adsorption of anions by oxides and hydroxides of Fe and Al can enhance the adsorption of cations. For example the adsorption of $H_2PO_4^-$ by both amorphous hydrous oxides of Fe and Al and goethite enhanced the adsorption of Zn (STANTON and DU BURGER 1967, 1970, BOLLAND et al. 1977) and resulted in decreased absorption of Zn by plant roots (STANTON and DU BURGER 1967). This phenomenon could arise through the formation of complexes between Zn and $H_2PO_4^-$ on the surface of the oxides.

#### *2.1.1.2 Precipitation*

Interactions between ions can occur because the addition of one ion either enhances the precipitation or dissolution of sparingly soluble compounds. Three examples will be considered here: (1) the effect of soil acidity on the toxicity of Al to plants (2) the effect of soil acidity on the availability of Mn to plants and (3) the dissolution of calcium phosphates in alkaline soils.

Aluminium toxicity may decrease the growth of plants on acid soils (see reviews by FOY 1974 and HELYAR 1978). The application of compounds which increase soil pH ($CaCO_3$, $K_2CO_3$) can eliminate Al toxicity (for example see MUNNS 1965a). In soils the solubility product of $Al(OH)_3$ (pAl + 3pOH) varied from 32 for some freshly limed soils to 35 for other soils (MARION et al. 1976). In solution the solubility products of gibbsite and amorphous $Al(OH)_3$ were 34 and 32.3 respectively. At pH 5 the activity of Al in the soil solution would therefore be 0.1 μM if gibbsite was the predominant form and 5.0 μM if amorphous $Al(OH)_3$ predominated. The growth of cotton roots in both soil and nutrient solutions was depressed when the activity of the Al ion in solution exceeded 2 μM (ADAMS and LUND 1966). Hence while plants differ greatly in their tolerance of high concentrations of Al in solution (see review by FOY 1974), it is likely that toxic concentrations of Al will only occur at soil pH values around 5 and less.

Avoidance of Al toxicity by tolerant species is in many cases related to maintenance of rhizosphere pH above 4.0 (FOY 1974). When anion absorption exceeds cation absorption, the rhizosphere will become less acid (KIRKBY 1968). Indeed differences among tropical pasture legumes in tolerance to Al toxicity (ANDREW et al. 1973) appeared to be related to differences among them in the excess of anion absorption over cation absorption (HELYAR 1978). HELYAR further predicted that legumes dependent on symbiotic nitrogen fixation – where rhizosphere pH falls (NYATSANGA and PIERRE 1973) – will be more susceptible to Al toxicity than legumes supplied with $NO_3^-$. The effect of high concentrations of Al in solution on the growth of plants can be modified by both P and Ca supply (MUNNS 1965a, b). These interactions, which may involve several processes, are considered in Section 2.3.1.

The concentration of $Mn^{2+}$ in the soil solution is dependent on the soil pH and on its oxidation-reduction potential because of the reaction $MnO_2 + 2H^+ \rightleftharpoons Mn^{2+} + \frac{1}{2}O_2 + H_2O$. Manganese deficiency in plants occurs mainly on alkaline soils and Mn toxicity occurs mainly on acid soils because plants probably only take up $Mn^{2+}$ (HALSTEAD and BARBER 1968). Reactions occurring within the rhizosphere may markedly affect the dissolution of insoluble and unavailable Mn oxides. Firstly exudates from plant roots may solubilize amorphous Mn oxides (JONES and LEEPER 1951). Secondly bacterial activity in the rhizosphere may also affect the solubility of Mn oxides. Thus varieties of oats susceptible to Mn deficiency can have more Mn-oxidizing bacteria around their roots than those less susceptible (TIMONIN 1946). Thirdly the acidity of the rhizosphere may vary particularly in relation to the form of N supplied. Plants fed $NH_4^+$ (which leads to acidification of the rhizosphere) are less likely to be Mn-deficient than those fed $NO_3^-$ (which leads to increased pH in the rhizosphere) (HOYT and MYOVELLA 1979).

The availability of P for plants on neutral and alkaline soils may be controlled by the dissolution and precipitation of sparingly soluble Ca phosphates such as hydroxy-apatite (LARSEN 1967). For example if the concentration of $H_2PO_4^-$ in the soil solution is controlled by the solubility of hydroxy-apatite [$Ca_5(PO_4)_3OH$] the relationship $10(pH - \frac{1}{2}pCa) - 3(pH + pH_2PO_4)$ must be constant [$pH_2PO_4 = constant + 2pH + pPO_4$ and $p(OH) = constant - pH$].

Hence it can be calculated that for hydroxy-apatite increasing the pH from 6 to 7 in the presence of $5 \times 10^{-3}$ M Ca will decrease the P concentration in solution from $3.1 \times 10^{-3}$ M to $2.5 \times 10^{-7}$ M (RUSSELL 1973).

However, it is difficult to use solubility products to predict the concentrations of $H_2PO_4^-$ in solution (see discussions by LARSEN 1967, RUSSELL 1973). Firstly solubility products predict concentrations at equilibrium which may be slowly reached in soils. Secondly the solubility constant of hydroxy-apatite can depend on whether the hydroxy-apatite is being precipitated or dissolved (BJERRUM 1949). Thirdly impurities in the hydroxy-apatite such as F and $CaCO_3$ may modify its solubility (LARSEN 1967, RUSSELL 1973). Finally the neutral molecule $(CaHPO_4)^0$ may exist in the soil solution (LARSEN 1967) and further complicate the use of solubility constants.

Notwithstanding these difficulties, solubility products of various Fe, Al and Ca phosphates have been used to construct a unified solubility diagram (for example LINDSAY and VLEK 1977) for P in soils with a wide range of soil pH. The value of such an approach is not certain.

### 2.1.2 Absorption from Solution at the Root Surface

Interactions among ions can arise through the effect of one ion on the absorption of another by plant roots from solution. Absorption of ions from solution may be affected by other ions through effects due to (1) decreased or enhanced access to the sites of absorption, to (2) competition at the site of absorption and (3) interaction with metabolic controls of absorption (such as an ATPase or an electrogenic $H^+$ transport system). Each of these three types of interaction that occur within the root will be considered in turn. Unfortunately in many instances it is not possible to define which of these three mechanisms is involved in the enhancement or depression of the absorption of one ion by another.

#### *2.1.2.1 Access to the Sites of Absorption*

One ion may affect the absorption of another ion by affecting its concentration at the site of absorption in the plasmalemma. There are fixed negative charges in the cell walls of plants associated with carboxylate groups on which cations are adsorbed. While there is little relationship between cation adsorption and cation absorption, and adsorption is not a necessary prerequisite to absorption (see review by HAYNES 1980), adsorbed cations may influence the absorption of both anions and cations.

For example, Ca concentrations in solution may markedly affect the rate of absorption of both anions and other cations (see review by HAYNES 1980). Two effects are probably involved. Firstly, low concentrations of Ca in solution (about 50 μM) are required to maintain the integrity of membranes (EPSTEIN 1962). This effect is specific for Ca (EPSTEIN 1962). Secondly, higher concentrations of Ca in solution may depress cation absorption [for example $Mn^{2+}$ (ROBSON and LONERAGAN 1970) and $Zn^{2+}$ (CHAUDHRY and LONERAGAN 1972a)] and enhance anion absorption [for example, $Br^-$ (HOOYMANS 1964); $SO_4^{2-}$ (FRANKLIN 1971) and $H_2PO_4^-$ (EDWARDS 1968, FRANKLIN 1969, 1970, ROBSON et al. 1970)], perhaps by affecting the access of these ions to the sites of absorption. Effects of Ca on anion absorption are established almost instantaneously (ROBSON et al. 1970).

The effect of Ca is not specific. Trivalent cations enhance the absorption of $H_2PO_4^-$ more than divalent cations which have a larger effect than monovalent cations (FRANKLIN 1969, 1970, HYDE 1966). Calcium and other cations may enhance anion absorption by contracting the electrical double layer associated with fixed negative charges in the cell wall and thus allowing greater access to sites of absorption. This effect may partly explain the observation that $Cl^-$ uptake by *Beta* tissue was increased by an increased ratio of $Ca^{2+}$ to $K^+$ in solutions of equal total cation concentrations (PITMAN 1964). Increasing the solution pH generally increases the rate of absorption of cations and decreases the rate of absorption of anions (see for example ISLAM et al. 1980). These effects may at least partly reflect effects of $H^+$ ions on the access of ions to sites of absorption.

The effect of Ca on the absorption of anions and cations varies with plant species. Increasing Ca concentration in solution depressed Mn absorption more and enhanced P absorption more for *Medicago truncatula* than for *Trifolium subterraneum* (ROBSON and LONERAGAN 1970, ROBSON et al. 1970). Perhaps these species differ in the distribution of negative charges in the pores of their cell wall.

*2.1.2.2 Competition at Site of Absorption*

Ion absorption (i.e. movement across the plasmalemma) by plant roots is a very selective process at low concentration in solution (see EPSTEIN 1972). For the univalent cations $Na^+$ does not compete with $K^+$ for uptake at low concentrations (<0.2 mM) in solution (see EPSTEIN 1972). However, ions more similar to $K^+$ can compete with it and decrease its absorption. Both $Rb^+$ and $NH_4^+$ can depress the absorption of $K^+$ by plant roots. Indeed the effect of $NH_4^+$ on $K^+$ absorption from solution (TROMP 1962) may change markedly the relationship between potassium concentrations in solutions and growth (see SPEAR et al. 1978a). For the alkaline earth cations only $Sr^{2+}$ and $Ca^{2+}$ compete in their absorption from solution (EPSTEIN 1962).

Increases in the external K concentration often reduce the absorption of Ca (LAZAROFF and PITMAN 1966, OSMOND 1966, JOHANSEN et al. 1968, SPEAR et al. 1978b) and Mg (YOSHIDA 1966, LEGGETT and GILBERT 1969, FAGERIA 1974). In some instances long-distance transport of Ca is reduced at high concentrations of monovalent cations due to a reduction in transpiration rate (LAZAROFF and PITMAN 1966). In other instances other explanations are required. Effects of Ca on K absorption are complex. Apart from effects of Ca in maintaining membrane integrity and affecting the access of ions to sites of absorption (Sect. 2.1.2.1), Ca may affect K absorption through competition for absorption for the low affinity mechanism operating at high concentrations of ions in solution.

The absorption of the transition metal cations from concentrations of the order of 1 μM is strongly inhibited by the alkali and alkaline earth cations (Fig. 2) and by other transition metal cations. The effect of alkali and alkaline

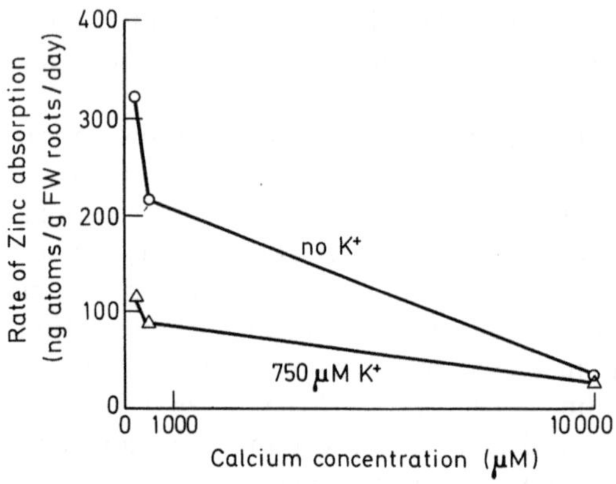


**Fig. 2.** The effect of other cations on the absorption of zinc by wheat roots. (CHAUDHRY and LONERAGAN 1972a). Zinc added as zinc chloride at a concentration of 1 μM

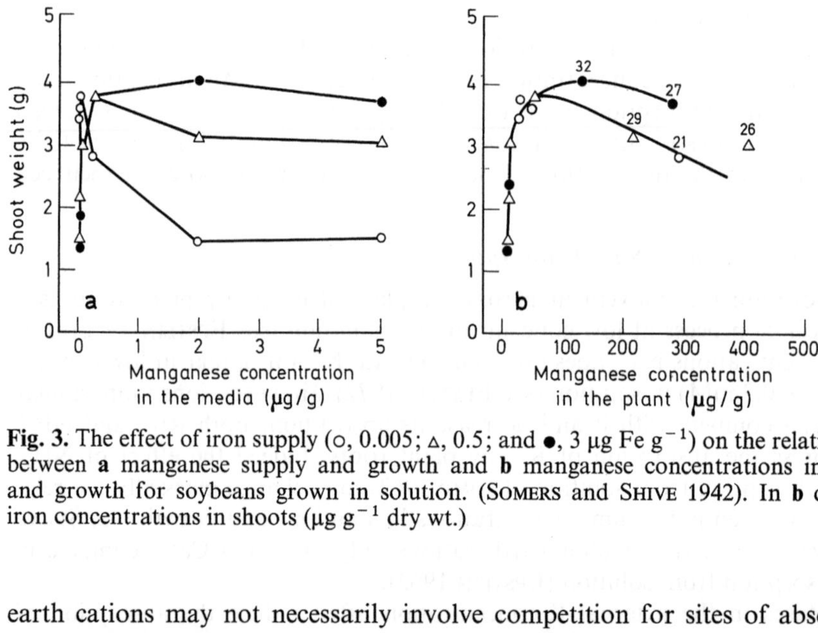


**Fig. 3.** The effect of iron supply (o, 0.005; △, 0.5; and •, 3 μg Fe $g^{-1}$) on the relationships between **a** manganese supply and growth and **b** manganese concentrations in shoots and growth for soybeans grown in solution. (SOMERS and SHIVE 1942). In **b** data are iron concentrations in shoots (μg $g^{-1}$ dry wt.)

earth cations may not necessarily involve competition for sites of absorption across the plasmalemma (see Sect. 2.1.2.2). Nevertheless there may be competition between some transition metal cations. The absorption of $Zn^{2+}$ was strongly inhibited by $Cu^{2+}$, weakly by $Co^{2+}$ and not at all by $Mn^{2+}$ or $Fe^{2+}$ (HAWF and SCHMID 1967, BOWEN 1969, CHAUDHRY and LONERAGAN 1972b). At low concentrations of $Zn^{2+}$ and $Cu^{2+}$, and in the presence of $Ca^{2+}$ at concentrations commonly present in soil solutions, the effect of $Cu^{2+}$ was competitive, suggesting that it would be additive to that of $Ca^{2+}$. Both $Zn^{2+}$ and $Mn^{2+}$ inhibited Fe absorption from $Fe^{2+}$ by tomato roots (TIFFIN 1967) and by tobacco leaf cells (KANNAN 1969). Iron added either as an inorganic salt (SOMERS and SHIVE 1942) or as a chelate (MORAGHAN 1980) depressed Mn absorption by plants. Indeed the much-cited interaction between Fe and Mn in plants may result from effects on absorption rather than antagonisms within the plant (Fig. 3).

For the anions the absorption of $H_2PO_4^-$ by plant roots is not affected by the concentrations of $Cl^-$ (CARTER AND LATHWELL 1967), $NO_3^-$ (ALBERDA 1948), or $SO_4^{2-}$ (TANADA 1955) in the solution. Similarly the absorption of $NO_3^-$ by low salt roots of barley was not affected by $Cl^-$, $Br^-$ or $SO_4^{2-}$ concentrations in solution (RAO and RAINS 1976). For the halides, $Cl^-$ and $Br^-$ compete for the same transport site where $Br^-$ replaces $Cl^-$ with nearly the same efficiency (ELZAM and EPSTEIN 1965, SMITH and FOX 1977). By contrast, $F^-$ and $I^-$ compete only weakly with $Cl^-$ (ELZAM and EPSTEIN 1965), possibly due to interference with metabolism. In other plant systems however, there is clear ability of $I^-$ to substitute for $Cl^-$ (e.g. in beet root slices or certain algae). The uptake of $SO_4^{2-}$ is unaffected by the concentration of other anions (LEGGETT and EPSTEIN 1956) except $SeO_4^{2-}$ (EPSTEIN 1962). For $MoO_4^{2-}$, absorption by tomato roots was strongly inhibited by $SO_4^{2-}$ (STOUT et al. 1951). A particularly puzzling interaction is the marked stimulation of the absorption of $MoO_4^{2-}$ by tomato roots associated with increasing the $H_2PO_4^-$ concentration in solution (STOUT et al. 1951). BARSHAD (1951) suggested that P may stimulate Mo uptake because of the formation of a phosphomolybdate complex absorbed more readily by the plant.

#### *2.1.2.3 Metabolic Controls of Ion Absorption*

Interactions between ions may be due to activation or deactivation of the driving forces such as an ATPase or an electrogenic $H^+$ transport system. Interaction may also occur due to suppression of a transporting compound.

The deficiency or toxicity of one ion may affect the control of absorption of another. For example Zn-deficient plants appear to be more susceptible to P toxicity. Zinc-deficient plants may have higher rates of P absorption than plants adequately supplied with Zn (PARIBOK and ALEKSEEVA-POPOVA 1965, LONERAGAN et al. 1979). The mechanism of this effect is not understood. Another example is the apparent lack of control of $H_2PO_4^-$ absorption (leading to concentration of $H_2PO_4^-$ within the plant high enough to cause P toxicity) in corn plants grown in saline media (NIEMAN and CLARK 1976). High P concentrations may have accumulated within the plant because salinity impaired growth but not P uptake.

Apart from metabolic changes that result from deficiency and toxicity, other controls of the absorption of one ion may be modified by another. The activity of anion transport may be regulated by vacuolar contents of the ion concerned, by $OH^-/H^+$ in the cytoplasm or by products of metabolism such as organic acids. For example the addition of $NO_3^-$ may decrease $Cl^-$ influx (and vice versa) by carrot tissue (CRAM 1973), excised barley roots (SMITH 1973) and citrus leaf slices (SMITH and FOX 1977). However in wheat roots there is an inducible $NO_3^-$ uptake system that is separate from the $Cl^-$ uptake process (JACKSON et al. 1973). Similarly $Cl^-$ did not affect the uptake of $NO_3^-$ by low salt roots of barley (RAO and RAINS 1976). It is therefore likely that the interaction between $Cl^-$ and $NO_3^-$ is related to some general aspect of ion absorption.

High intra-cellular $NO_3^-$ concentrations greatly decreased $Cl^-$ influx (and vice versa), suggesting that the internal $Cl^-$ plus $NO_3^-$ concentration regulated the influx of these ions (CRAM 1973, SMITH 1973, SMITH and FOX 1977). Organic anions within roots (for example malate) did not affect the influx of either $NO_3^-$ or $Cl^-$ (CRAM 1973, SMITH 1973). It is therefore necessary to distinguish between situations where reduced $NO_3^-$ can be removed from the system, as in transfer from roots to shoots, and studies where the vacuole was a sink for $Cl^-$ and $NO_3^-$. Additionally it may be important to differentiate between species where $NO_3^-$ is reduced in roots and those in which $NO_3^-$ is reduced in shoots.

For avocados grown in soil, the N concentration in leaves was unaffected by the addition of 20 mM NaCl which produced symptoms of $Cl^-$ toxicity and reduced growth by nearly 60% (DOWNTON 1978). Chloride concentrations in leaves increased more than tenfold. Even though the ratio of $Cl^-$ to $NO_3^-$ added in solution was about 100:1 in saline soils there was effective discrimination in uptake between $NO_3^-$ and $Cl^-$.

Another source of interaction among ions occurs due to the regulation of cation/anion balance in cells. This process is discussed in detail by SMITH and RAVEN (1976). Uptake of more equivalents of N as $NO_3^-$ than equivalents of cation must be accomplished by release of $OH^-$ to the solution while uptake of N as $NH_4^+$ or of more $K^+$ than $NO_3^-$ results in $H^+$ excretion. Changes in rhizosphere pH can affect ion uptake both by affecting the concentration of ions in solution (see Sect. 2.1.1) and by affecting the absorption of ions from solution by plant roots (see Sect. 2.1.2.1).

When cation absorption exceeds anion absorption organic acids may be synthesized. Synthesis of organic acids is also the regular way of eliminating $OH^-$ formed in reduction of $NO_3^-$ or $SO_4^{2-}$. Organic acids are important in the transport of some divalent cations across the root. In particular citrate appears to chelate Fe and possibly Zn (TIFFIN 1967). There have been various observations that organic acid synthesis is affected by deficiencies. DEKOCK and MORRISON (1958) found that Fe-induced chlorosis reduced concentrations of malic and citric acids by 50%. By contrast, chlorosis associated with Ni toxicity was accompanied by increased citrate and decreased malate concentrations. DEKOCK has suggested that the ratio of citrate to malate is related to the balance of P to Fe and K to Ca, but there are other examples where the ratio of citrate to malate changed, that could not be associated with these particular ratios (see HEWITT 1963). The increase in citrate could be a response to low Fe supply to facilitate transport in the plant.

## 2.2 Interactions Between Nutrients Affecting the Utilization of Nutrients Within the Plant

Interactions involving effects on nutrient utilization within the plant may involve either of two distinct components. First the transport of the nutrient (in either xylem or phloem) to the site of function may be impaired. Second at this site the function of the nutrient may be impaired or enhanced. Effects of one ion on the utilization of another is usually assessed by considering its effect on the relationship between nutrient concentration in whole shoots and growth (see BATES 1971). Changed relationships may, however, reflect either of the two components of nutrient utilization. Antagonism or synergism between nutrients in their function can only be assessed if concentrations at the site of function (functional nutrient requirement; LONERAGAN 1968) can be assessed. In this section of this review we will consider interactions that are due to effects on the distribution and function of nutrients.

### 2.2.1 Distribution

One nutrient may affect the distribution of another in at least three distinct ways. Firstly, transport in either xylem or phloem may be impaired by precipitation. Secondly, nutrients may not be able to enter the phloem for retranslocation from old to young leaves because they are immobilized in old leaves. Finally, one nutrient may modify the distribution of another within the leaf of the plant. Each of these types of interaction will be considered.

#### *2.2.1.1 Precipitation Within the Plant*

High concentrations of P in solution may increase the incidence of chlorosis associated with Fe deficiency. While these effects may partly relate to effects of P on the concentrations of Fe available for absorption, there are also interactions between Fe and P within the plant. OLSEN (1935) first suggested that when Fe is taken up from neutral or alkaline solutions it can be precipitated as ferric phosphate in the vascular bundles along the veins of a leaf. Autoradiographs of leaves of plants grown with an initial external $H_2PO_4^-$ concentration of 100 μM and an initial Fe concentration (as labelled $FeCl_3$) of 18 μM showed a heavy deposition of Fe along the veins, while the rest of the leaf blade contained little Fe (REDISKE and BIDDULPH 1953).

Under certain conditions it appears likely that immobilization of Fe, perhaps as Fe phosphates, can occur in the vascular system to such an extent that the adjacent tissues become markedly deficient in Fe. In some species Fe is transported in the xylem as an anion complexed with citrate (TIFFIN 1967). Soybean varieties susceptible to Fe chlorosis appeared less able to keep the Fe in their tissues mobile than tolerant varieties when supplied with high $H_2PO_4^-$ concentrations (BROWN et al. 1959). Perhaps varieties differ in the concentrations of citrate available to compete with $H_2PO_4^-$ for the Fe.

The formation of precipitates may also limit the transport of ions within the phloem. When Mn and Ca chlorides were added to phloem sap from *Yucca* and *Ricinus*, precipitation occurred when the Ca concentration in the sap exceeded 0.2–0.5 mM (depending on species and Ca/K ratio in growth media) and when the Mn concentration exceeded 0.004–0.18 mM (depending on species) (VAN GOOR and WIERSMA 1974). By contrast, precipitation did not occur until much higher concentrations of K and Mg chlorides were added. VAN GOOR and WIERSMA did not identify the compounds precipitated but considered that Ca phosphates would be precipitated when Ca concentrations in sap exceeded 0.6 mM.

### 2.2.1.2 *Immobilization in Old Leaves*

One nutrient may affect the retranslocation of another from old to young leaves. When plants are grown with an ample N supply, S deficiency leads to chlorosis of new leaves (ERGLE 1954). When N supply is low or when plants are relying on symbiotic activity for the N supply, symptoms of S deficiency may be indistinguishable from N deficiency, developing most strongly in older leaves (ANDERSON and SPENCER 1950a, EATON 1966). The effect of N supply is probably due to an effect on the rate of hydrolysis of proteins. Sulphur present as $SO_4^{2-}$ in old leaves moves readily to new growth. However, when cotton plants were transferred from plus-S to minus-S solutions old leaves lost none of their protein-S or soluble organic S when amply supplied with N (ERGLE 1954).

Similarly, N supply may influence the retranslocation of Cu from old to new leaves. The movement of Cu out of old leaves of wheat closely paralleled the movement of N (Fig. 4; LONERAGAN et al. 1976, 1980b, HILL et al. 1978, 1979b). Copper deficiency, which delayed the retranslocation of Cu from the oldest leaf of wheat, also delayed the retranslocation of N and leaf senescence. Moreover when the oldest leaves of Cu-deficient and Cu-sufficient plants were shaded and induced to senesce, their Cu and N contents declined simultaneously (HILL et al. 1979a). In Cu-adequate plants N and Cu concentrations in the oldest leaf of wheat decreased slowly until the leaf senesced (HILL et al. 1979b). After senescence N and Cu concentrations fell rapidly. In Cu-deficient plants which did not senesce, Cu and N concentrations decreased slowly. KANNANGARA and WOOLHOUSE (1968) have concluded that the initial gradual decline in total protein content after full expansion of leaves of *Perilla frutescens* was associated with a decline in the content of Fraction 1 proteins, whereas the Fraction 2 proteins declined only during the sharp decline in total protein content, just prior to abscission. The strong binding of Cu by nitrogen ligands may thus be involved in the relationship of Cu movement out of old leaves to N movement out of these leaves.

In the study of HILL et al. (1979b), the movement of Zn out of the oldest leaf of wheat supplied with adequate Zn also paralleled the movement of N out of these leaves. The Zn content of these leaves did not decline until the leaf senesced.

### 2.2.1.3 *Distribution of Nutrients Within the Leaf*

Increasing the supply of Si can alleviate Mn toxicity in cereals without affecting the uptake of Mn by the plant or the concentration of Mn in the entire leaf (Fig. 5; WILLIAMS and VLAMIS 1957, VLAMIS and WILLIAMS 1967). With high Mn supply, necrotic, brown spots occurred on the old leaves. The addition of sodium silicate but not sodium sulphate eliminated these symptoms. Autoradiography using $^{54}Mn$ indicated that the brown spots were localized high concentrations of Mn. When Si was added the $^{54}Mn$ was distributed uniformly throughout the leaf. These effects of Si on Mn toxicity and the distribution of Mn within the leaf were also observed for a legume (HORST and MARSCHNER 1978a) even though it is likely that concentrations of Si in the shoots of legumes

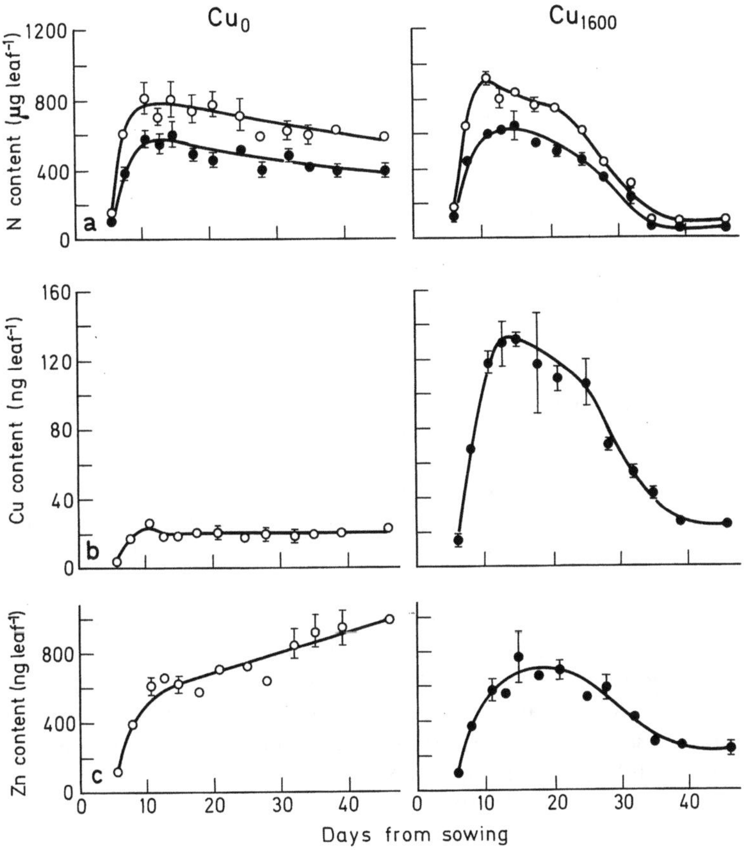


**Fig. 4.** The effect of copper supply on the changes in contents of **a** nitrogen (total nitrogen, o; alcohol insoluble nitrogen •), **b** copper and **c** zinc in the oldest leaf of wheat (HILL et al. 1979b). Chlorophyll contents for copper-adequate plants ($Cu_{1600}$) fell sharply after day 28, whereas those of copper-deficient plants ($Cu_0$) did not. At each harvest the mean and the limits of two replicates are shown

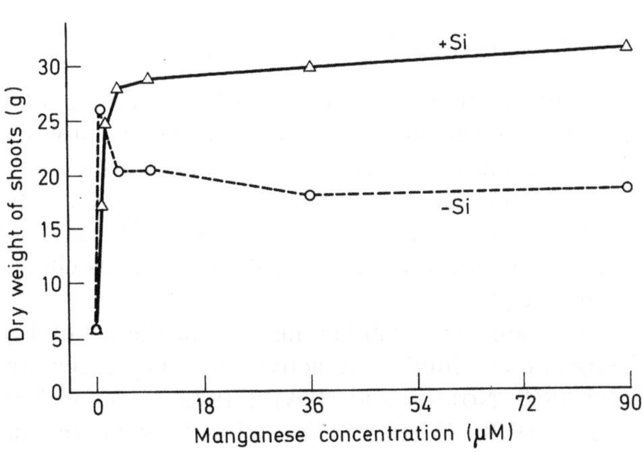


**Fig. 5.** The effect of silicon (o no silicon, Δ silicon added) in alleviating manganese toxicity in wheat. (VLAMIS and WILLIAMS 1967)

would be much lower than in the shoots of cereals (see JONES and HANDRECK 1967).

The mechanism by which Si alters the distribution of Mn within the leaf is not understood. HORST and MARSCHNER (1978b) consider that Si facilitates the movement of Mn from the xylem vessels into the surrounding tissue and thereby prevents local accumulation of Mn near the vessels. The addition of Si to culture solutions increased the proportion of Mn in shoots of ryegrass associated with the cell walls (JARVIS and JONES, personal communication). Silicon may also affect the distribution of Mn within the cell. The addition of Si increased the Mn concentration in the press sap from bean leaves suggesting that Si may favour the transport of Mn into the vacuoles (HORST and MARSCHNER 1978b). Direct interactions between Si and Mn in the press sap could not be detected by thin layer chromatography or electrophoresis (HORST 1976). However, Si may affect the form of Mn within the plant. The addition of Si to the culture solution doubled the amount of Mn extracted from the leaves of bean plants by methanol but did not affect the amount of Mn extracted by water, KCl or EDTA (HORST and MARSCHNER 1978c). Water extracted 80% of the total Mn content of leaves, whereas methanol extracted less than 10%.

There may be interactions between Si and Mn on plant growth which are not related to effects of Si on Mn distribution within leaves. In studies with Sudan grass, increasing the Si in solution decreased Mn uptake and either induced Mn deficiency or alleviated Mn toxicity depending on the level of Mn supply (BOWEN 1972). Similarly, the application of silicic acid to an acid soil depressed the concentration of Mn in leaves and total uptake of Mn by tomatoes (PEASLEE and FRINK 1969).

### 2.2.2 Function

Interactions between nutrients may occur because (1) one nutrient competes with another at the site of its function for incorporation into active sites, (2) one nutrient can substitute for another and (3) one nutrient is required for the assimilation or metabolism of another.

#### *2.2.2.1 Competition at Site of Action*

Although antagonisms at the site of function are often invoked to explain interactions between nutrients, there are few examples which clearly demonstrate this phenomenon in vivo. To demonstrate such an antagonism one ion must inhibit a physiological process or a metabolic reaction at low but not at high concentrations of the nutrient at the site of its action. Otherwise the antagonism affecting plant growth may result from effects either on uptake or distribution within the plant.

An example of an antagonism is the interaction between $WO_4^{2-}$ and $MoO_4^{2-}$. Tungstate can inhibit the activity of $NO_3^-$ reductase in higher plants (HEIMER et al. 1969, NOTTON and HEWITT 1971, RAO and RAINS 1976, HARPER and NICHOLAS 1978). When Mo is replaced by W an analogue of $NO_3^-$ reductase

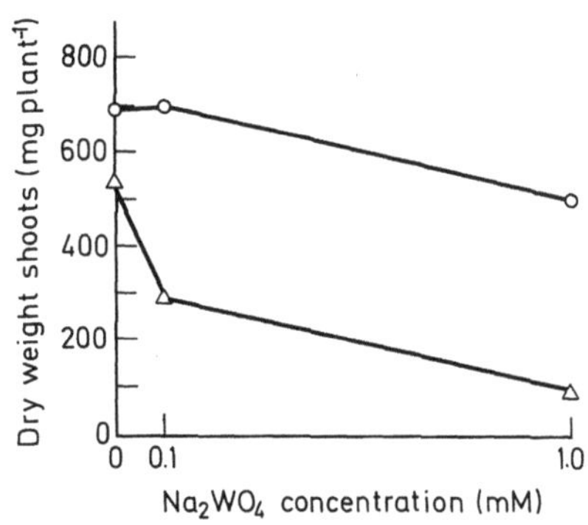


**Fig. 6.** The effect of tungstate on the growth of *Alnus glutinosa* supplied with either nitrate (△) or urea (○). (PIZELLE and THIERY 1979)

is formed (NOTTON and HEWITT 1971). This enzyme, while still displaying dehydrogenase activity, is inactive in $NO_3^-$ reduction. Molybdenum and W were not interchangeable in vivo.

The effect of this interaction on plant growth depends both on the source of N and the ratio of $WO_4^{2-}$ to $MoO_4^{2-}$ in the external media. Tungstate to $MoO_4^{2-}$ ratios of 1,500:1 did not decrease growth of white clover reliant either on symbiotically fixed N or $NO_3^-$ (QUIN and HOGLUND 1976). However, $WO_4^{2-}$ inhibited the growth of shoots and roots of *Alnus glutinosa* (Fig. 6; PIZELLE and THIERY 1979); the effect was more marked when the source of N was $NO_3^-$ than when it was urea. For *Alnus*, inhibiting $NO_3^-$ reduction with $WO_4^{2-}$ increased nodule number without affecting the weight of each nodule. Hence with $NO_3^-$-fed plants, acetylene reduction activity per plant but not per g nodule was increased by adding 0.01 mM $Na_2WO_4$. Further increasing the concentration of $Na_2WO_4$ to 0.1 mM depressed acetylene reduction activity perhaps due to incorporation of W in place of Mo in nitrogenase.

The effects of $WO_4^{2-}$ on plant growth may also result from competition between $WO_4^{2-}$ and $MoO_4^{2-}$ for absorption, as well as for incorporation into $NO_3^-$ reductase or nitrogenase. In many studies with $WO_4^{2-}$, Mo concentrations within the plant are not presented.

### *2.2.2.2 Sparing Effect*

Sodium can substitute for part of the requirement for K for cereals (WEHUNT and COLLINS 1953, DE WIT et al. 1963), grasses (GAMMON 1953, HYLTON et al. 1967, SMITH 1974) and dicotyledons (HOLT and VOLK 1945). The addition of Na to soil can decrease both the external and internal requirements for K (for example SMITH 1974), indicating that some species can use Na in place of K for some but not all physiological or metabolic functions. Plant growth may decrease if a balance between inorganic cations and anions is not maintained (DE WIT et al. 1963). Sodium may have a sparing effect on K requirements by contributing to the cation content of the cell vacuole in place of K.

Some plant species have a specific requirement for Na (BROWNELL and WOOD 1957, BROWNELL 1965). Increasing the K supply did not eliminate effects of Na on the growth of several species of *Atriplex* (BROWNELL 1968). Species with the $C_4$ dicarboxylic acid pathways for $CO_2$ generally have a requirement for Na (BROWNELL and CROSSLAND 1972).

### *2.2.2.3 Involvement of One Nutrient in the Assimilation and Metabolism of Another*

Three examples will be considered here to demonstrate this type of interaction.

The form of N supplied can influence the response in growth by plants to nutrients. For example the growth of tobacco, rice and soybean cell cultures was enhanced by Ni when urea was the source of N (POLACIO 1977). Substituting $NH_4^+$ for urea eliminated any requirement for Ni. Nickel is a component of jackbean urease (DIXON et al. 1975) which hydrolyzes urea. Another example is the greater effect of Mo on algal and plant growth when $NO_3^-$ rather than $NH_4^+$ is the source of N (ISHIOKA and ARNON 1955, HEWITT and MCCREADY 1956). Molybdenum is a constituent of nitrate reductase (NICHOLAS and NASON 1955).

The third example relates to the involvement of mineral nutrients in symbiotic N fixation by both legumes and non-legumes. Mineral nutrients may limit symbiotic N fixation by legumes by affecting growth and survival of rhizobia, infection and nodule development and nodule function as well as by affect-

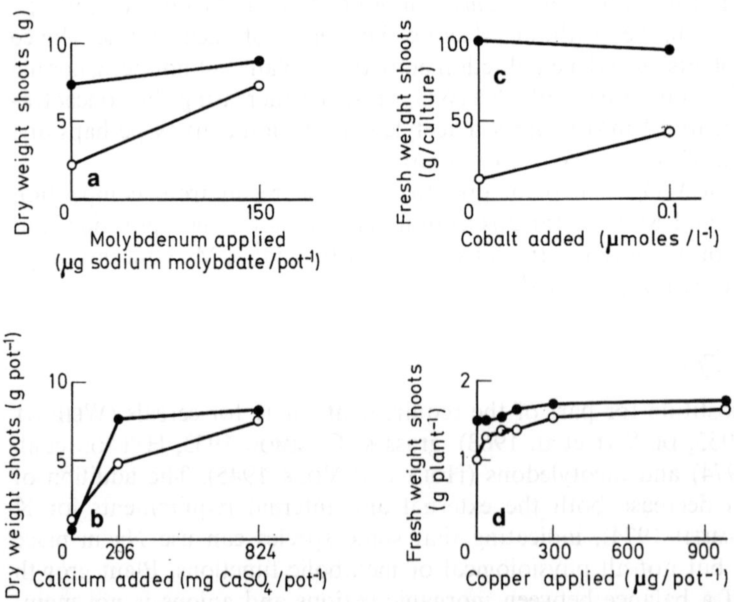


**Fig. 7.** The interaction between the application of nitrogen (o, no nitrogen; •, nitrogen applied) and **a** molybdenum (ANDERSON and SPENCER 1950a), **b** calcium (LONERAGAN 1959), **c** cobalt (DELWICHE et al. 1961) and **d** copper (SNOWBALL et al. 1980) on the growth of legumes. For calcium there is a positive interaction for severely calcium-deficient plants; for calcium when moderately deficient and for the other nutrients the interaction is negative

ing host plant growth. For particular nutrients where the requirements (either external or internal) for symbiotic N fixation are greater than for growth of the host legume, effects of nutrient deficiency on growth can be alleviated by supplying mineral N (see ROBSON 1978). There are several examples. Firstly the growth of legumes responds to Co only when reliant on symbiotically fixed N (DELWICHE et al. 1961). For other nutrients – Mo (ANDERSON and SPENCER 1950b), Ca (LONERAGAN 1959) and Cu (SNOWBALL et al. 1980) – the interaction between nutrient supply and N on growth may also be negative (i.e. correct the same deficiency) but less complete than that for Co (Fig. 7). For all other nutrients the interaction between their supply and N on the growth of legumes is positive (see summary in ROBSON 1978), indicating that host plant growth has a higher requirement than symbiotic N fixation for these nutrients.

For Mo, Ca and Cu, nodule function appears to be the component of the symbiosis which has the highest requirement. Molybdenum is a component of nitrogenase (see review by DILWORTH 1974). The role of Cu and Ca in nodule function is less clear (see review by ROBSON 1978).

## 2.3 Complex Interactions Between Nutrients Involving Several Processes

Some interactions between nutrients involve effects of one nutrient on both the absorption and utilization of another nutrient. These interactions may operate through effects both in the external media and within the plant. Two such interactions will be considered.

### 2.3.1 Calcium/Aluminium/Phosphate

Plants grown in external media with high concentrations of Al may display symptoms associated with either P deficiency (stunting, small dark green leaves, purpling of stems and leaves) or Ca deficiency (cupping or rolling of young leaves). The interactions between Al and P and between Al and Ca arise through effects on both absorption and utilization.

The growth of roots is markedly affected by Al concentrations in solution (see review by JACKSON 1967). At low concentrations of Al in solution the primary root may elongate normally but lateral growth is inhibited. At severely toxic concentrations of Al even primary roots are stunted and thickened. The effects of Al on the uptake of nutrients including Ca and P may reflect effects on root growth as well as interactions at the site of absorption.

In soil $H_2PO_4^-$ is adsorbed onto the surfaces of Al oxides and may also be precipitated as Al phosphates (see RUSSELL 1973). Both the application of $CaCO_3$ and $H_2PO_4^-$ to soil can decrease Al concentrations in solution and alleviate Al toxicity by precipitating either Al hydroxides or Al phosphates (Fig. 8; MUNNS 1965c). Effects of $H_2PO_4^-$ in alleviating Al toxicity probably result from these reactions in the external media rather than from effects within the plant. Under conditions where precipitation did not occur, increasing the $H_2PO_4^-$ concentration in solution from 1 to 20 μM (both adequate concentrations) did not alleviate Al toxicity (MUNNS 1965b). Moreover, high concentra-

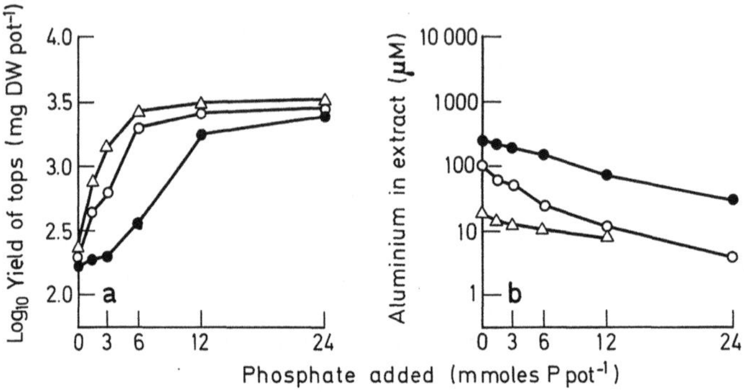


**Fig. 8.** The interaction between the application of $CaCO_3$ (●, nil; ○, 15 mmol; △, 30 mmol) and $KH_2PO_4$ on **a** the growth of *Medicago sativa* on an acid soil and **b** the concentration of aluminium in $CaCl_2$ extracts. (MUNNS 1965c)

tions of Al in solution can depress plant growth even though there are normally adequate concentrations of P in tops (MUNNS 1965b, ANDREW et al. 1973).

Phosphate and Al can form neutral aluminophosphate complexes (WHITE et al. 1976). Solution pH markedly affects this reaction. The optimum pH for the formation and polymerization of aluminophosphate complexes in dilute solutions was about 5 and no complexed P was detected in solution at pH 4.3 or less.

In solutions where Al and $H_2PO_4^-$ would not precipitate, increasing the concentration of Al in solution to 20 μM depressed the growth of lucerne more at pH 4.5 than at pH 5 particularly at high $H_2PO_4^-$ concentrations (20 μM) (WHITE 1976). WHITE (1976) suggested that the greater accumulation yet lower toxicity of Al in plants grown at pH 5 may have been due to passive influx and transport of the uncharged aluminophosphate polymers into vacuoles.

There are also other effects of Al on P uptake by plant roots. Pre-treatment of roots with $Al_2(SO_4)_3$ increased the amount of P in roots (CLARKSON 1966). However, the accumulation of this additional P, which was readily exchangeable with non-radioactive P, was unaffected by dinitrophenol and low temperatures. CLARKSON (1966) suggested that reactions between Al and P occurred in the free space of the root external to the plasmalemma. Aluminium and P were also co-precipitated in the same zones of roots of maize grown in nutrient solutions containing 3,000 μM Al (RASMUSSEN 1968). The precipitation of Al in the free space of the root as either hydroxide or phosphate may regulate the transport of Al and in some instances P to shoots.

Plant species tolerant of Al toxicity differ from sensitive species in the effects of Al on P transport from roots to shoots. High Al concentrations in solution markedly depressed P concentrations in the shoots of a sensitive species, *Medicago sativa,* but did not affect P concentrations in the shoots of a tolerant species, *Trifolium subterraneum* (MUNNS 1965b). Similarly for several tropical and temperate legumes, increasing Al concentrations in solution decreased P transport from roots to shoots considerably more in sensitive than in tolerant species (ANDREW and VANDENBERG 1973).

Aluminium can also interfere with the metabolism of P within the plant. Aluminium inhibits the activity of hexokinases (RORISON 1965, CLARKSON 1966), acid phosphatases (WOOLHOUSE 1969) and ATPases (WOOLHOUSE 1969) leading to reduced incorporation of P into phosphorylated hexose sugars (CLARKSON 1966).

The interactions between Al and Ca appear to be less complex than those between Al and P. Increasing the Ca concentration in solution can often mitigate the effects of low concentrations of Al on plant growth (see for example MUNNS 1965b, JACKSON 1967). Part of this effect can be attributed to effects of increasing the Ca concentration in solution in reducing the activity but not the concentration of Al ions in solution (see HELYAR 1978). Increased Ca supply can also depress Al uptake by plants (MUNNS 1965b). Aluminium can also decrease Ca uptake (see review by JACKSON 1967) perhaps through effects on root growth as well as interactions at the root surface. The marked effects of $CaCO_3$ in alleviating Al toxicity are not however associated in many instances with effects of Ca on Al toxicity because $K_2CO_3$ may be equally effective (see for example MUNNS 1965a).

### 2.3.2 Zinc/Phosphate

The interactions between Zn and P have been extensively studied with several mechanisms proposed to explain the effect of P application in inducing Zn deficiency. The effects of P fertilizers on plant response may involve a number of discrete phenomena. The importance of a particular phenomenon will depend upon plant species and environmental conditions. It is likely to be simplistic to consider that all interactions between P and Zn result from a single phenomenon.

Interactions between P and Zn can be divided into three major categories. First, there are interactions because of effects of P and Zn on plant growth. Second, there are interactions between P and Zn which result from both effects of P on Zn uptake and Zn on P uptake. Finally there may be interactions which result from interactions between P and Zn within the plant.

Phosphorus applications which increase plant growth may decrease Zn concentrations in tops to deficient levels by dilution (BOAWN et al. 1954, LAMBERT et al. 1979) without decreasing Zn uptake. Phosphorus applications which correct P deficiencies may also decrease the proportion of total plant weight which is roots (for example LONERAGAN and ASHER 1967). Zinc deficiency may induce P toxicity by decreasing growth without decreasing the accumulation of P (LONERAGAN et al. 1979). Each of these three effects of either P or Zn on plant growth may lead to effects of high P applications in "inducing" symptoms attributed to Zn deficiency.

A complicating factor in the study of interactions between P and Zn is the high concentrations of Zn in some phosphatic fertilizers. Superphosphate contains 250–750 $\mu g\ g^{-1}$ (WILLIAMS 1977), sufficient to alleviate Zn deficiency, and leading to negative interactions between the application of superphosphate and Zn in some instances (ANDERSON 1946).

Phosphate applications may also reduce Zn contents and concentrations within plants. Phosphate enhances Zn adsorption by oxides and hydroxides of Fe and Al and decreases Zn absorption by plant roots (see Sect. 2.1.1.1). It is unlikely that the precipitation of Zn phosphates is involved in phosphate-induced Zn deficiency because Zn phosphate is a readily available source of Zn for plant growth (BOAWN et al. 1957).

The application of phosphates to soil may decrease Zn uptake by plants due to effects of the associated cations in decreasing Zn absorption (CHAUDHRY and LONERAGAN 1972a). Another mechanism by which P applications may decrease Zn absorption is through effects of P supply in decreasing the development of vesicular arbuscular mycorrhizas (LAMBERT et al. 1979). Increasing the rate of P application (as $CaH_2PO_4$) from 75 to 200 ppm in the media, which increased the growth of non-mycorrhizal corn and soybean but did not affect the growth of mycorrhizal plants, decreased the percentage of root volume infected by mycorrhizal fungi. This increase in P supply decreased Zn uptake more for plants inoculated with mycorrhizal fungi than for non-mycorrhizal plants. Increasing P supply to plants may decrease both percentage infection and the weight of mycorrhizal roots (ABBOTT and ROBSON 1977). Mycorrhizal roots absorb more Zn than non-mycorrhizal roots both from solution (BOWEN et al. 1974) and from soil (GILMORE 1971, PAIRUNAN et al. 1980).

As well as effects of P application on Zn uptake, Zn deficiency may affect P absorption. For sugar beets (ROSELL and ULRICH 1964), peas (PARIBOK and ALEKSEEVA-POPOVA 1965) and for okra (*Hibiscus esculentus*) (LONERAGAN et al. 1979) Zn deficiency enhanced the accumulation of P by leaves. Coupled with the reduced growth due to Zn deficiency, this effect may lead to marked P toxicity in Zn-deficient plants (LONERAGAN et al. 1979).

Interactions between Zn and P within the plant (for example MILLIKAN 1951, 1963, BOAWN and LEGGETT 1964) have often been invoked to explain the effects of P supply on Zn deficiency. Effects of P supply on Zn concentrations in whole tops were poorly related to effects on symptoms and growth. Apart from the possible involvement of P toxicity in these experiments, Zn concentrations in whole tops may be a poor guide to the Zn status of plants (REUTER 1980). Increasing the P supply from that deficient for plant growth to that toxic for plant growth did not affect the critical concentrations of Zn in young tissues of subterranean clover (REUTER 1980).

# 3 Interactions Between Nutrients in Mixed Communities

Interactions between nutrients in mixed communities are more complex than those in monoculture because of genetic variation in absorption and utilization of nutrients. Additionally, the ability of legumes to symbiotically fix N produces marked interactions between nutrients in grass–legume swards.

Fertilizer application may change the botanical composition of a legume–grass sward in several ways. A limiting N supply may prevent grasses but not

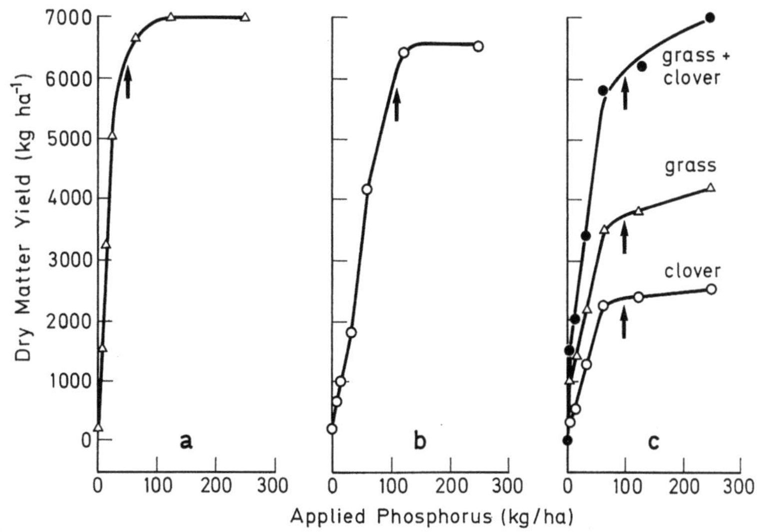


**Fig. 9.** The effect of phosphate supply on the growth of monocultures of **a** rye grass **b** subterranean clover and **c** a mixed sward of rye grass and subterranean clover. (OZANNE et al. 1976). *Arrows* indicate the amount of phosphate required for 90% of maximum yield

legumes from responding to applications of other nutrients. Fertilizers may thus initially increase legume growth only. Subsequently the increased N supply due to N fixation may enable grasses to respond to such an extent that legume growth is reduced by shading. Hence the effects of fertilizer on the botanical composition of a pasture can change markedly with time. An example is the effect of P fertilizers. Continued P applications to pastures generally leads to non-legume (usually grass) dominance (for example see TRUMBLE and DONALD 1938 and review by ROSSITER 1966). The initial application of P may, however, increase the proportion of legume in the pasture (WILLOUGHBY 1954, WILLIAMS et al. 1956). Grass dominance with continued P application probably reflects an improved N supply associated with an initial response by legumes to P (see review by LONERAGAN 1964).

Grass grown with clover responded to a much higher level of applied P than when it was grown as a single species with high rates of $NH_4NO_3$ (Fig. 9; OZANNE et al. 1976). The response of clover in the mixed sward to applied P was very similar to its response in a pure sward. The effect of clover on the response of grass to applied P was attributed to increased availability of N from symbiotically fixed N due to effects of P in stimulating clover growth.

Potassium-deficient pastures are generally grass-dominant, and K application markedly increases the legume content (for example ROSSITER 1947, FITZPATRICK and DUNNE 1956, JONES 1966). In monocultures the growth of most grasses appears less affected by low K supply than the growth of legumes and other herbs (ASHER and OZANNE 1967). Both when growing singly (ASHER 1964, JACKMAN 1965) and when growing in association with a legume (WHELAN and

**Table 3.** The influence of an associated grass species on the response of *Macroptilium atropurpureum* to equivalent amounts of either sodium or potassium sulphate (HALL 1978)

| Treatment | Mono-culture | With *Chloris gayana* | With *Setaria anceps* |
|---|---|---|---|
| | Yield (g $plant^{-1}$) | | |
| Nil | 1.22 | 1.09 | 1.21 |
| $+Na_2SO_4$ | 1.20 | 1.49 | 1.21 |
| $+K_2SO_4$ | 1.53 | 1.76 | 1.79 |

EDWARDS 1975) grasses absorbed K at a faster rate per g roots than did legumes. When a tropical legume and tropical grass were grown with both their tops and roots competing, the growth of the legume was markedly depressed by the grass at low but not at high K supply (HALL 1971). When only the tops were competing, the competitive abilities of the grass and legume were unaffected by K supply. The differential effects of K on the growth of legume and grasses may lead to complex interactions between K and N on the growth of mixed swards.

HALL (1978) has examined the effect of Na and K supply on the growth of a legume either in monoculture or in association with either of two grass species which differ in their ability to use Na in cation balance. Potassium application increased the growth of the legume in all three situations (Table 3). Sodium application did not increase the growth of the legume in monoculture or in association with a grass unable to use Na in cation balance. However, when grown in association with a grass able to substitute Na for K for some functions, Na application increased its growth.

Sulphur-deficient pastures are generally also grass-dominant (for example CONRAD 1950, HILDER and SPENCER 1954, WALKER et al. 1956). Sulphur application to these pastures stimulated legume growth and increased the contribution of the legume to total production. In all these situations response of the grass but not the legume may have been limited by N deficiency. In monoculture there is little evidence for differences between grasses and legumes in response to S (see review by ROBSON and LONERAGAN 1978).

The effect of Cu and Zn application on botanical composition and growth of pastures sown on newly cleared land (ANDERSON 1946) provides another example. Both the application of Cu and Zn increased total yield. When neither nutrient was supplied *Phalaris tuberosa* was the dominant species. Addition of Cu alone gave dominance of *Medicago sativa,* whereas addition of Zn alone increased the competitive advantage of *P. tuberosa.* Addition of both nutrients increased the growth of *P. tuberosa* less than the growth of *M. sativa* so that *M. sativa* became dominant. It is not clear whether these responses are produced by differences in the absorption or the utilization of Cu and Zn by these species.

# 4 Conclusion

Interactions between nutrients can arise through several distinct mechanisms; the importance of a particular mechanism will depend upon environmental conditions. It is important in the study of interactions to identify whether an interaction occurs due to effects on growth or interactions in the media, or to interactions within the plant. Moreover interactions occurring within the plant may occur due to effects of one nutrient on the absorption, transport or function of another. A major limitation in the study of interactions is the lack of knowledge of the concentrations and forms of nutrients in different compartments of the soil and the plant.

# References


Abbott LK, Robson AD (1977) Growth stimulation of subterranean clover with vesicular arbuscular mycorrhizas. Aust J Agric Res 28:639–649

Adams F, Lund ZF (1966) Effect of chemical activity of soil solution aluminium on cotton root penetration of acid subsoils. Soil Sci 101:193–198

Adams SN, Honeysett JL (1964) Some effects of soil waterlogging on the cobalt and copper status of pasture plants grown in pots. Aust J Agric Res 15:357–367

Adams SN, Honeysett JL, Tiller KG, Norrish K (1969) Factors controlling the increase of cobalt in plants following the addition of a cobalt fertilizer. Aust J Soil Res 7:29–42

Alberda T (1948) The influence of some external factors on growth and phosphate uptake of maize plants of different salt conditions. Rec Trav Bot Neerl 41:541–561

Anderson AJ (1946) Fertilizers in pasture development on peat soils in the lower south east of South Australia. J Counc Sci Ind Res Aust 19:394–403

Anderson AJ (1956) Effects of fertilizer treatments on pasture growth. Proc 7th Int Grassl Congr, pp 323–333

Anderson AJ, Oertel AC (1946) Plant responses to molybdenum as a fertilizer. Counc Sci Ind Res Bull 198:25–44

Anderson AJ, Spencer D (1950a) Molybdenum in nitrogen metabolism of legumes and non-legumes. Aust J Sci Res Ser B 3:414–430

Anderson AJ, Spencer D (1950b) Sulphur in nitrogen metabolism of legumes and non-legumes. Aust J Sci Res Ser B 3:431–449

Andrew CS, Vandenberg PJ (1973) The influence of aluminium on phosphate sorption by whole plants and excised roots of some pasture legumes. Aust J Agric Res 24:341–351

Andrew CS, Johnson AD, Sandland RL (1973) Effect of aluminium on the growth and chemical composition of some tropical and temperate pasture legumes. Aust J Agric Res 24:325–339

Asher CJ (1964) Potassium uptake by pasture plants as affected by root development and the potassium concentration in the root environment. PhD thesis, Univ West Aust

Asher CJ, Ozanne PG (1967) Growth and potassium content of plants in solution cultures maintained at constant potassium concentration. Soil Sci 103:155–161

Barrow NJ (1970) Comparison of the adsorption of molybdate, sulphate and phosphate by soils. Soil Sci 109:282–289

Barshad I (1951) Factors affecting the molybdenum content of pasture plants. 2. Effect of soluble phosphates, available nitrogen and soluble sulphates. Soil Sci 71:387–398

Bates TE (1971) Factors affecting critical nutrient concentrations in plants and their evaluation – a review. Soil Sci 112:116–130
Bjerrum N (1949) Selected papers. Munksgaard, Copenhagen
Boawn LC, Leggett GE (1964) Phosphorus and zinc concentrations in Russet Burbank potato tissues in relation to development of zinc deficiency symptoms. Soil Sci Soc Am Proc 28:229–232
Boawn LC, Viets FG, Crawford CL (1954) Effect of phosphate fertilizer on zinc nutrition of field beans. Soil Sci 78:1–7
Boawn LC, Viets FG, Crawford CL (1957) Plant utilization of zinc from various types of zinc compounds and fertilizer materials. Soil Sci 83:219–227
Bolland MDA, Posner AM, Quirk JP (1977) Zinc adsorption by goethite in the absence and presence of phosphate. Aust J Soil Res 15:279–286
Bollard EG (1960) Transport in the xylem. Annu Rev Plant Physiol 11:141–166
Bowden JW, Bolland MDA, Posner AM, Quirk JP (1973) Generalized model for anion and cation adsorption at oxide surfaces. Nature Phys Sci 245:81–83
Bowen GD, Skinner MF, Bevege DI (1974) Zinc uptake by mycorrhizal and uninfected roots of *Pinus radiata* and *Araucaria cunninghamii*. Soil Biol Biochem 6:141–144
Bowen JE (1969) Absorption of copper, zinc and manganese by sugar cane tissue. Plant Physiol 44:255–261
Bowen JE (1972) Manganese-silicon interaction and its effect on growth of sudan grass. Plant Soil 37:577–588
Brown JC, Holmes RS, Tiffin LO (1959) Hypothesis concerning iron chlorosis. Soil Sci Soc Am Proc 23:231–234
Brownell PF (1965) Sodium as an essential micronutrient element for a higher plant (*Atriplex vesicaria*). Plant Physiol 40:460–468
Brownell PF (1968) Sodium as an essential micronutrient for some higher plants. Plant Soil 28:161–164
Brownell PF, Crossland CJ (1972) The requirement for sodium as a micronutrient by species having the $C_4$ dicarboxylic photosynthetic pathway. Plant Physiol 49:794–797
Brownell PF, Wood JG (1957) Sodium as an essential micronutrient element for *Atriplex vesicaria* Howard. Nature 179:635–636
Carter OG, Lathwell DJ (1967) Effect of chloride on phosphorus uptake by corn roots. Agron J 59:250–253
Chaudhry FM, Loneragan JF (1972a) Zinc absorption by wheat seedlings. I. Inhibition by macronutrient ions in short term studies and its relevance to long term zinc nutrition. Soil Sci Soc Am Proc 36:323–327
Chaudhry FM, Loneragan JF (1972b) Zinc absorption by wheat seedlings. II. Inhibition by hydrogen ions and by micronutrient cations. Soil Sci Soc Am Proc 36:327–331
Clarkson DT (1966) Effect of aluminium on the uptake and metabolism of phosphorus by barley seedlings. Plant Physiol 41:165–172
Conrad JP (1950) Sulfur fertilization in California and some related factors. Soil Sci 70:43–54
Cram WJ (1973) Internal factors regulating nitrate and chloride influx in plant cells. J Exp Bot 24:328–341
DeKock PC, Morrison RI (1958) The metabolism of chlorotic leaves. 2. Organic acids. Biochem J 70:272–277
Delwiche CC, Johnson CM, Reisenauer HM (1961) Influence of cobalt on nitrogen fixation by *Medicago*. Plant Physiol 36:73–78
Dilworth MJ (1974) Dinitrogen fixation. Annu Rev Plant Physiol 25:81–114
Dixon NE, Gazzola C, Blakeley RL, Zerner B (1975) Jackbean urease (EC 3.5.1.5), a metalloenzyme. A simple biological role for nickel? J Am Chem Soc 97:4131–4133
Downton WJS (1978) Growth and flowering in salt-stressed avocado trees. Aust J Agric Res 29:523–534
Eaton FM (1966) Sulfur. In: Chapman HD (ed) Diagnostic criteria for plants and soils. Univ Cal Div Agric Sci, Berkeley

Edwards DG (1968) Cation effects on phosphate absorption from solution by *Trifolium subterraneum*. Aust J Biol Sci 21:1–11

Elzam OE, Epstein E (1965) Absorption of chloride by barley roots: kinetics and selectivity. Plant Physiol 40:620–624

Epstein E (1962) Mutual effects of ions in their absorption by plants. Agrochimica 6:293–322

Epstein E (1972) Mineral nutrition of plants: Principles and perspectives. Wiley and Sons, New York

Ergle DR (1954) The utilization of storage sulphur by cotton and the effect on growth and chloroplast pigments. Bot Gaz 115:225–234

Fageria NK (1974) Uptake of potassium and its influence on growth and magnesium uptake by groundnut (*Arachis hypogaea* L) plants. Biol Plant 16:210–213

Fitzpatrick EN, Dunne TC (1956) Potassium for subterranean clover. J Dep Agric West Aust Ser 3:321–326

Fox RL (1978) Studies on phosphorus nutrition in the tropics. In: Andrew CS, Kamprath EJ (eds) Mineral nutrition of legumes in tropical and sub-tropical soils. CSIRO, Melbourne

Foy CD (1974) Effect of aluminium on plant growth. In: Carson EW (ed) The plant root and its environment. Univ Press Virg Charlottesville

Franklin RE (1969) Effect of adsorbed cations on phosphorus uptake by excised roots. Plant Physiol 44:687–700

Franklin RE (1970) Effect of adsorbed cations on phosphorus absorption by various plant species. Agron J 62:214–216

Franklin RE (1971) Cation effects on chloride, sulphate and phosphate uptake by excised roots. Soil Sci 112:343–347

Gammon N (1953) Sodium and potassium requirements of pangola and other pasture grasses. Soil Sci 76:81–90

Gilmore AE (1971) The influence of endotrophic mycorrhiza on the growth of peach seedlings. J Am Soc Hort Sci 96:35–38

Goor van BJ, Wiersma D (1974) Redistribution of potassium, calcium, magnesium and manganese in plants. Physiol Plant 31:163–168

Hall RL (1971) The influence of potassium supply on competition between Nandi *Setaria* and Greenleaf *Desmodium*. Aust J Exp Agric Anim Husb 11:415–419

Hall RL (1974) Analysis of the nature of the interference between plants of different species. II. Nutrient relations in a Nandi *Setaria* and Greenleaf *Desmodium* association with particular reference to potassium. Aust J Agric Res 25:749–756

Hall RL (1978) The analysis and significance of competitive and noncompetitive interference between species. In: Wilson JR (ed) Plant relations in pastures. CSIRO, Melbourne

Halstead EH, Barber SA (1968) Manganese uptake attributed to diffusion from soil. Proc Soil Sci Soc Am 32:540–542

Harper JE, Nicholas JC (1978) Nitrogen metabolism of soybeans. I. Effect of tungstate on nitrate utilization, nodulation and growth. Plant Physiol 62:662–664

Hawf LR, Schmid WE (1967) Uptake and translocation of zinc by intact plants. Plant Soil 27:249–260

Haynes RJ (1980) Ion exchange properties of roots and ionic interactions within the root apoplasm: their role in ion accumulation by plants. Bot Rev 46:75–99

Heimer YM, Wray JL, Filner P (1969) The effect of tungstate on nitrate assimilation in higher plant tissues. Plant Physiol 44:1197–1199

Helyar KR (1978) Effects of aluminium and manganese toxicities on legume growth. In: Andrew CS, Kamprath EJ (eds) Mineral nutrition of legumes in tropical and sub-tropical soils. CSIRO, Melbourne

Hewitt EJ (1963) The essential nutrient elements: requirements and interactions in plants. In: Steward FC (ed) Plant physiology: A treatise, vol III. Academic Press, London New York

Hewitt EJ, McCready CC (1956) Molybdenum as a plant nutrient, VII. The effects

of different molybdenum and nitrogen supplies on yields and composition of tomato plants grown in sand culture. J Hortic Sci 31:284–290
Hilder EJ, Spencer K (1954) The influence of sulphur on a natural *Medicago* pasture. J Aust Inst Agric Sci 20:171–176
Hill J, Robson AD, Loneragan JF (1978) The effect of copper and nitrogen supply on the retranslocation of copper in four cultivars of wheat. Aust J Agric Res 29:925–939
Hill J, Robson AD, Loneragan JF (1979a) The effect of copper supply and shading on Cu retranslocation from old wheat leaves. Ann Bot 43:449–457
Hill J, Robson AD, Loneragan JF (1979b) The effect of copper supply on the senescence and the retranslocation of nutrients of the oldest leaf of wheat. Ann Bot 44:279–287
Hingston FJ, Posner AM, Quirk JP (1971) Competitive adsorption of negatively changed ligands on oxide surfaces. Faraday Soc. Disc 52:334–342
Hingston FJ, Posner AM, Quirk JP (1972) Anion adsorption by goethite and gibbsite. I. The role of the proton in determining adsorption envelopes. J Soil Sci 23:177–192
Hodgson JF, Lindsay WL, Trierweiler JF (1966) Micronutrient cation complexing in soil solution. II. Complexing of zinc and copper in displaced solution from calcareous soils. Soil Sci Soc Am Proc 30:723–726
Holt ME, Volk NJ (1945) Sodium as a plant nutrient and substitute for potassium. J Am Soc Agron 37:821–827
Hooymans JJM (1964) The role of calcium in the absorption of anions and cations by excised barley roots. Acta Bot Neerl 13:507–540
Horst WJ (1976) Einfluss von Silizium auf die Mangantoleranz von Buschbohnen (*Phaseolus vulgaris* L). Diss Nr 59 TU, Berlin
Horst WJ, Marschner H (1978a) Effect of silicon on manganese tolerance of bean plants (*Phaseolus vulgaris* L). Plant Soil 50:287–303
Horst WJ, Marschner H (1978b) Symptome von Mangan-Überschuss bei Bohnen (*Phaseolus vulgaris* L). Z Pflanz Bodenkd 141:129–142
Horst WJ, Marschner H (1978c) Einfluss von Silizium auf den Bindungszustand von Mangan im Blattgewebe von Bohnen (*Phaseolus vulgaris*). Z Pflanz Bodenkd 141:487–497
Hoyt PB, Myovella GGS (1979) Correction of severe manganese deficiency in wheat with chemical fertilizers. Plant Soil 52:437–444
Husa JG, McIlrath WJ (1965) Absorption and translocation of boron by sunflower plants. Bot Gaz 126:186–194
Hyde AH (1966) Nature of the calcium effect in phosphate uptake by barley roots. Plant Soil 24:328–332
Hylton LO, Ulrich A, Cornelius DR (1967) Potassium and sodium interrelations in growth and mineral content of Italian ryegrass. Agron J 59:311–314
Ishioka PS, Arnon DI (1955) Molybdenum in relation to nitrogen metabolism II. Assimilation of ammonia and urea without molybdenum by *Scenedesmus*. Physiol Plant 8:552–560
Islam AKMS, Edwards DG, Asher CJ (1980) pH optima for crop growth. Results of a flowing solution culture experiment with six species. Plant Soil 54:339–357
Jackman RH (1965) The uptake of rubidium by the roots of some graminaceous and leguminous plants. NZJ Agric Res 8:763–777
Jackson WA (1967) Physiological effects of soil acidity. In: Pearson RW, Adams F (eds) Soil acidity and liming. Am Soc Agron, Madison, Wisc
Jackson WA, Flesher D, Hageman RH (1973) Nitrate uptake by dark-grown corn seedlings: some characteristics of apparent induction. Plant Physiol 51:120–127
Johansen C, Edwards DG, Loneragan JF (1968) Interactions between potassium and calcium in their absorption by intact barley plants. 1. Effect of potassium on calcium absorption. Plant Physiol 43:1717–1721
Johansen C, Edwards DG, Loneragan JF (1969) Interactions between potassium and calcium in their absorption by intact barley plants. II. Effects of calcium and potassium concentration on potassium absorption. Plant Physiol 43:1722–1726
Jones LHP (1957) The solubility of molybdenum in simplified systems and aqueous soil suspensions. J Soil Sci 8:312–327

Jones LHP, Handreck KA (1967) Silica in soils, plants and animals. Adv Agron 19:107–149

Jones LHP, Leeper GW (1951) The availability of various manganese oxides to plants. Plant Soil 3:141–153

Jones OP, Rowe RW (1968) Sampling the transpiration stream in woody plants. Nature 219:403

Jones RJ (1966) Nutrient requirements of improved pasture on podzolic soils developed in phyllite at North Deep Creek. Proc Trop Grassl Soc Aust 6:23–30

Kannan S (1969) Factors related to iron absorption by enzymically isolated leaf cells. Plant Physiol 44:1457–1460

Kannangara CG, Woolhouse HW (1968) Changes in the enzyme activity of soluble protein fractions in the course of foliar senescence in *Perilla frutescens* L. Butt. New Phytol 67:533–542

Kirkby EA (1968) Influence of ammonium and nitrate nutrition on the cation anion balance and nitrogen and carbohydrate metabolism of white mustard plants grown in dilute nutrient solutions. Soil Sci 105:133–141

Kirkby EA, Armstrong MJ (1980) Nitrate uptake by roots as regulated by nitrate assimilation in the shoot of castor oil plants. Plant Physiol 65:286–290

Lambert HD, Baker DE, Cole H (1979) The role of mycorrhizae in the interactions of phosphorus with zinc, copper and other elements. Soil Sci Soc Am J 43:976–980

Larsen S (1967) Soil phosphorus. Adv Agron 19:151–210

Lazaroff N, Pitman MG (1966) Calcium and magnesium uptake by barley seedlings. Aust J Biol Sci 19:991–1005

Ledin S, Wiklander L (1974) Exchange acidity of wheat and pea roots in salt solutions. Plant Soil 41:403–413

Leggett JE, Epstein E (1956) Kinetics of sulfate absorption by barley roots. Plant Physiol 31:222–226

Leggett JE, Gilbert WA (1969) Magnesium uptake by soybeans. Plant Physiol 44:1182–1186

Lindsay WL (1978) Chemical reactions affecting the availability of micronutrients in soil. In: Andrew CS, Kamprath EJ (eds) Mineral nutrition of legumes in tropical and sub-tropical soils. CSIRO, Melbourne

Lindsay WL, Vlek PLG (1977) Phosphate minerals. In: Dixon JB, Weed SB (eds) Minerals in the soil environment. Soil Sci Soc Am, Madison, Wisconsin

Loneragan JF (1959) Calcium in the nitrogen metabolism of subterranean clover. Aust J Biol Sci 12:26–39

Loneragan JF (1964) The nutrition of grasslands. In: Barnard C (ed) Grasses and grasslands. MacMillan, London Melbourne

Loneragan JF (1968) Nutrient requirements of plants. Nature 188:1307–1308

Loneragan JF (1975) The availability and absorption of trace elements in soil-plant systems and their relation to movement and concentrations of trace elements in plants. In: Nicholas DJD, Egan AR (eds) Trace elements in soil-plant-animal systems. Academic Press, London New York

Loneragan JF, Asher CJ (1967) Response of plants to phosphate concentration in solution culture. II. Rate of phosphate absorption and its relation to growth. Soil Sci 103:311–318

Loneragan JF, Snowball K, Robson AD (1976) Remobilization of nutrients and its significance in plant nutrition. In: Wardlaw IF, Passioura JB (eds) Transport and transfer processes in plants. Academic Press, London New York

Loneragan JF, Grunes DL, Welch RM, Aduayi EA, Tengah A, Lazar DA, Cary EE (1979) P accumulation and toxicity in leaves in relation to zinc supply. Annu Meet, Fort Collins. Agron Abstr, pp 175–176

Loneragan JF, Grove TS, Robson AD, Snowball K (1980a) Phosphorus toxicity as a factor in zinc-phosphorus interactions in plants. Soil Sci Soc Am J 43:966–972

Loneragan JF, Snowball K, Robson AD (1980b) Copper supply in relation to content and redistribution of copper among organs of the wheat plant. Ann Bot 45:621–632

Macrobbie EAC (1971) Phloem translocation, facts and mechanisms: A comparative study. Biol Rev 46:429–481

Marion GM, Hendricks DM, Dutt GR, Fuller WH (1976) Aluminium and silica solubility in soils. Soil Sci 121:76–85
Mengel K, Kirkby EA (1978) Principles of plant nutrition. Berne Int Potash Inst
Millikan CR (1951) Diseases of flax and linseed. Dep Agric Vic Aust Tech Bull 9:1–120
Millikan CR (1963) Effects of different levels of zinc and phosphorus on the growth of subterranean clover (*Trifolium subterraneum*). Aust J Agric Res 14:180–205
McKenzie RM (1967) The sorption of cobalt by manganese minerals in soils. Aust J Soil Res 5:235–246
McKenzie RM (1970) The reaction of cobalt with manganese dioxide minerals. Aust J Soil Res 8:97–106
McKenzie RM (1971) The influence of cobalt on the reactivity of manganese dioxide. Aust J Soil Res 9:55–58
McKenzie RM (1972) The sorption of some heavy metals by the lower oxides of manganese. Geoderma 8:29–35
Moraghan JT (1980) Distribution of selected elements within flax plants as affected by Fe EDDHA. Plant Soil 54:153–158
Munns DN (1965a) Soil acidity and growth of a legume. I. Interactions of lime with nitrogen and phosphorus on growth of *Medicago sativa* L. and *Trifolium subterraneum* L. Aust J Agric Res 16:733–741
Munns DN (1965b) Soil acidity and growth of a legume. II. Reactions of aluminium and phosphate in solution and effects of aluminium, phosphate, calcium and pH on *Medicago sativa* L and *Trifolium subterraneum* in solution culture. Aust J Agric Res 16:743–755
Munns DN (1965c) Soil acidity and growth of a legume. III. Interaction of lime and phosphate on growth of *Medicago sativa* L in relation to aluminium toxicity and phosphate fixation. Aust J Agric Res 16:757–766
Nicholas DJD, Nason A (1955) Role of molybdenum as a constituent of nitrate reductase from soybean leaves. Plant Physiol 30:135–138
Nieman RH, Clark RA (1976) Interactive effects of salinity and phosphorus nutrition on the concentration of phosphate and phosphate esters in mature photosynthesising corn leaves. Plant Physiol 57:157–161
Notton BA, Hewitt EJ (1971) The role of tungsten in the inhibition of nitrate reductase in spinach (*Spinacia oleracea* L) leaves. Biochem Biophys Res Commun 44:702–710
Nyatsanga T, Pierre WH (1973) Effects of nitrogen fixation by legumes on soil acidity. Agron J 65:936–940
Nye PH, Tinker PB (1977) Solute movement in the soil-root system. Blackwell, Oxford
Olsen C (1935) Iron absorption and chlorosis in green plants. CR Trav Lab Carlsberg Ser Chim 21:15–63
Osmond CB (1966) Divalent cation absorption and interaction in *Atriplex*. Aust J Biol Sci 19:37–48
Ozanne PG, Howes KMW, Petch A (1976) The comparative phosphate requirements of four annual pastures and two crops. Aust J Agric Res 27:479–488
Pairunan AK, Robson AD, Abbott LK (1980) The effectiveness of vesicular-arbuscular mycorrhizas in increasing growth and phosphorus uptake of subterranean clover from phosphorus sources of different solubilities. New Phytol 84:327–338
Paribok TA, Alekseeva-Popova NV (1965) Effect of zinc on absorption and utilization of phosphorus by plants. Sov Plant Physiol 12:514–518
Pate JS (1975) Exchange of solutes between phloem and xylem and circulation in the whole plant. In: Zimmermann MH, Milburn JA (eds) Transport in plants I. Encyclopedia of plant physiology new ser vol 1. Springer, Berlin Heidelberg New York
Peaslee DE, Frink CR (1969) Influence of silicon acid on uptake of Mn, Al, Zn, Cu by tomatoes. Soil Sci Soc Am Proc 33:569–571
Piper CS, Beckwith RS (1949) The uptake of copper and molybdenum by plants. In: Plant and animal nutrition in relation to soil and climatic factors. Br Comm Spec Conf Agric HMSO, London
Pitman MG (1964) The effect of divalent cations on the uptake of salt by beetroot tissue. J Exp Bot 15:444–456

Pizelle G, Thiery G (1979) Influence du tungstate sur la nutrition azotée nitrique et symbiotique d'*Alnus glutinosa*. Physiol Veg 17:769–775

Polacio JC (1977) Nitrogen metabolism in soybean tissue culture. II. Urea utilization and urease synthesis require $Ni^{2+}$. Plant Physiol 59:827–830

Price CA (1970) Molecular approaches to plant physiology. McGraw-Hill, New York

Quin BF, Hoglund JH (1976) The effects of tungstate and nitrogen source on the dry weight and nitrogen yields and molybdenum and tungsten content of white clover. Plant Soil 45:201–212

Rao KP, Rains DW (1976) Nitrate absorption by barley. II. Influence of nitrate reductase activity. Plant Physiol 57:59–62

Rasmussen HP (1968) Entry and distribution of aluminium in *Zea mays*: Electron microprobe X-ray analysis. Planta 81:28–37

Rediske JH, Biddulph O (1953) The absorption and translocation of iron. Plant Physiol 28:576–593

Reisenauer HM (1966) Mineral nutrients in soil solution. In: Environmental biology. Fed Am Soc Exp Biol Biol Handb, Bethesda Md

Reuter DJ (1980) Distribution of copper and zinc in subterranean clover in relation to deficiency diagnosis. PhD thesis, Murdoch Univ West Aust

Robson AD (1978) Mineral nutrients limiting nitrogen fixation in legumes. In: Andrew CS, Kamprath EJ (eds) Mineral nutrition of legumes on tropical and sub-tropical soils. CSIRO, Melbourne

Robson AD, Loneragan JF (1970) Sensitivity of annual *Medicago* species to manganese toxicity as affected by calcium and pH. Aust J Agric Res 21:223–232

Robson AD, Loneragan JF (1978) Responses of pasture plants to soil chemical factors other than nitrogen and phosphorus, with particular emphasis on the legume symbiosis. In: Wilson JR (ed) Plant relations in pastures. CSIRO, Melbourne

Robson AD, Edwards DG, Loneragan JF (1970) Calcium stimulation of phosphate absorption by annual legumes. Aust J Agric Res 21:601–612

Rorison IH (1965) The effect of aluminium on the uptake and incorporation of phosphate by excised sainfoin roots. New Phytol 64:23–27

Rosell RA, Ulrich A (1964) Critical zinc concentrations and leaf minerals of sugar beet leaves. Soil Sci 97:152–167

Rossiter RC (1947) Effect of potassium on the growth of subterranean clover and other pasture plants on Crawley sand. 2. Field plot experiments. J Counc Sci Ind Res 20:389–401

Rossiter RC (1966) Ecology of the Mediterranean annual type pasture. Adv Agron 18:1–56

Russell EW (1973) Soil conditions and plant growth. Longman, London

Silva da JJR Fr, Williams RJP (1977) The uptake of elements by biological systems. Struct Bond 29:67–121

Smith FA (1973) The internal control of nitrate uptake into barley roots with differing salt content. New Phytol 72:769–782

Smith FA, Fox AL (1977) Interactions between chloride and nitrate uptake in *Citrus* leaf slices. Aust J Plant Physiol 4:177–182

Smith FA, Raven JA (1976) Nitrogen assimilation and transport in vascular land plants in relation to intracellular pH regulation. New Phytol 76:415–431

Smith FW (1974) The effect of sodium on potassium nutrition and ionic relations in rhodes grass. Aust J Agric Res 25:407–414

Snowball K, Robson AD, Loneragan JF (1980) The effect of copper on nitrogen fixation in subterranean clover (*Trifolium subterraneum*). New Phytol 85:63–72

Somers JJ, Shive JW (1942) The iron-manganese relation in plant metabolism. Plant Physiol 17:582–602

Spear SN, Asher CJ, Edwards DG (1978a) Response of cassava, sunflower and maize to potassium concentration in solution. I. Growth and plant potassium concentration. Field Crop Res 1:347–361

Spear SN, Edwards DG, Asher CJ (1978b) Response of cassava, sunflower and maize to potassium concentration in solution. II. Interaction between potassium, calcium and magnesium. Field Crop Res 1:375–389

Stanton DA, Burger du TR (1967) Availability to plants of zinc sorbed by soil and hydrous iron oxides. Geoderma 1:13–17
Stanton DA, Burger du TR (1970) Studies on zinc in selected Orange Free State soils. V. Mechanics for the reactions of zinc with iron and aluminium oxides. Agrochemophysica 2:65–76
Stout PR, Meagher WR, Pearson GA, Johnson CM (1951) Molybdenum nutrition of crop plants. I. The influence of phosphate and sulphate on the absorption of molybdenum from soils and solution cultures. Plant Soil 3:51–87
Suelter CH (1974) Monovalent cations in enzyme-catalyzed reactions. In: Sigel H (eds) Metal ions in biological systems. Decker, New York
Tammes PML, Die van J (1964) Studies on phloem exudation from *Yucca flaccida* flaw. I. Some observations on the phenomenon of bleeding and the composition of the exudate. Acta Bot Neerl 13:76–83
Tanada T (1955) Effect of ultra violet radiation and calcium and their interaction on salt absorption by excised mung bean roots. Plant Physiol 30:221–225
Tiffin LO (1967) Translocation of manganese, iron, cobalt and zinc in tomato. Plant Physiol 42:1427–1432
Tiller KG, Honeysett JL, Hallsworth EG (1969) The isotopically exchangeable form of native and applied cobalt in soils. Aust J Soil Res 7:43–56
Timonin MI (1946) Microflora of the rhizosphere in relation to the manganese deficient disease of oats. Soil Sci Soc Am Proc 11:284–292
Tromp J (1962) Interactions in the absorption of ammonium, potassium and sodium by wheat roots. Acta Bot Neerl 11:147–192
Trumble HC, Donald CM (1938) The relation of phosphate to the development of seeded pasture on a podsolized sand. Counc Sci Ind Res Aust Bull 116:1–47
Underwood EJ (1977) Trace elements in human and animal nutrition, 4th edn. Academic Press, London New York
Vlamis J, Williams DE (1967) Manganese and silicon interaction in the Gramineae. Plant Soil 27:131–140
Walker TW, Adams AFR, Orchiston HD (1956) The effect of levels of calcium sulphate on the yield and composition of a grass and clover pasture. Plant Soil 7:290–300
Wehunt RL, Collins WD (1953) Response of oats to Na and K on Norfolk sandy loam at two residual K levels. Soil Sci 76:91–96
Whelan BR, Edwards DG (1975) Uptake of potassium by *Setaria anceps* and *Macroptilium atropurpureum* from the same standard solution culture. Aust J Agric Res 26:819–829
White RE (1976) Studies on mineral ion absorption by plants. III. The interaction of aluminium, phosphate and pH on the growth of *Medicago sativa*. Plant Soil 46:195–208
White RE, Tiffin LO, Taylor AW (1976) The existence of polymeric complexes in dilute solutions of aluminium and orthophosphate. Plant Soil 45:521–529
Williams CH (1977) Trace metals and superphosphate: toxicity problems. J Aust Inst Agric Sci 43:99–109
Williams DE, Vlamis J (1957) The effect of silicon on yield and manganese 54 uptake and distribution in the leaves of barley plants grown in culture solutions. Plant Physiol 32:404–409
Williams WA, Love RM, Conrad JP (1956) Range improvement in California by seeding annual clovers, fertilization and grazing management. J Range Manage 9:28–33
Willoughby WM (1954) Some factors affecting grass-clover relationships. Aust J Agric Res 5:157–180
Wit de CT, Dijkshoorn W, Noggle JC (1963) Ionic balance and growth of plants. Versl Landbouwkd Onderz 69:15
Woolhouse HW (1969) Differences in the properties of the acid phosphatases of plant roots and their significance in the evolution of edaphic ecotypes. In: Rorison IH (ed) Ecological aspects of the mineral nutrition of plants. Blackwell, Oxford
Yoshida F (1966) Inter-relationships between potassium and magnesium absorption by oats (*Avena sativa* L). Thesis State Agric Univ Wageningen

# I.6 Import and Export of Mineral Nutrients in Plant Roots [1]

U. LÜTTGE

## 1 Introduction: The Dual Role of Roots in the Evolution of Higher Land Plants

The aqueous medium of water plants provides two essentials; firstly, *support,* which is obtained by various buoyancy mechanisms, and secondly, a *nutrient supply* where the solution is in contact with the whole plant. With the evolution of higher land plants roots acquired the dual role of anchoring the plant in the ground and serving to exchange material with the soil medium. Root-like structures, rhizoids, are frequently observed in lower plants. They are often lengthy single cells or may be multicellular such as in kelp. In benthic algae these rhizoids normally only function for attachment. In liverworts and mosses the unicellular rhizoids may serve both anchoring and transport functions. But only in higher land plants has there been an evolution of complex structures to satisfy both requirements.

These structures, the roots, are the major absorptive organs of higher plants, and it is no surprise that they have been the material most frequently used for studies of general mechanisms of ion uptake by plant tissues (reviews, e.g., PITMAN 1976, EPSTEIN 1976). From the nutritional point of view, however, uptake and accumulation of ions in the root tissue itself are often less pertinent than the transport through roots, followed by export and translocation of ions to the shoot system (reviews, e.g. PITMAN et al. 1976, ANDERSON 1976).

The dual role of roots needs to be stressed even in a review that focuses only on the physiological function. Root structure must serve both functions. Although highly optimized in the evolutionary process, the structural features that have evolved for one function may interfere with the other function and vice versa. A typical example is the occurrence of secondary growth of roots, which is particularly developed in larger dicotyledonous plants and trees, and necessary in providing their special anchoring requirements. The associated lignification, periderm formation and suberization of the major part of the root system prevents uptake of water and solutes and restricts it to the very tips of roots, i.e. to a quite small fraction of the entire root system. It is conceivable that ectotrophic mycorrhizas of trees have evolved as a consequence of this. The hyphae of mycorrhizal fungi extend several centimeters from the root surface into the soil, and intrude the cell walls of the root cortex up to the endodermis. In this way the surface area which is in contact with the soil and which acts as a pathway for transfer of nutrients from the soil to the root is much increased (see PITMAN 1976, CLARKSON and HANSON 1980).

1 Dedicated to the memory of NOE HIGINBOTHAM, distinguished promoter of this field

# 2 Relations Between Structure and Transport Functions Along the Length of Roots

## 2.1 The Phenomenon of Variations in Transport Functions Along the Length of Roots

Variations in transport functions along the length of roots have long been documented experimentally, in particular by the use of radioactive tracers administered to different root zones (e.g. WIEBE and KRAMER 1954, SHONE et al. 1969, BOWEN 1969, 1970, ROVIRA and BOWEN 1970, SMITH 1970, MARSCHNER and RICHTER 1973). A general simple picture of the most effective root zone(s) in transport has not emerged, because there is considerable diversity of behav-

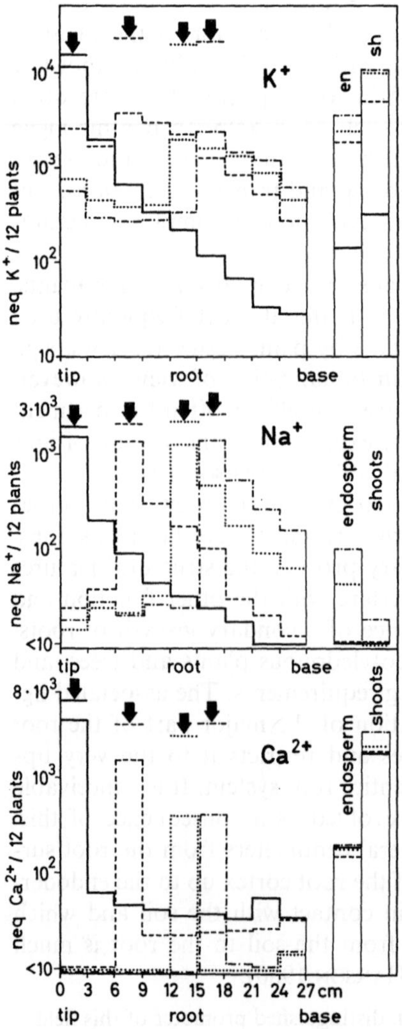


**Fig. 1.** Absorption and translocation of $K^+$, $Na^+$ and $Ca^{2+}$ in various root zones of 9 day old maize seedlings (*Zea mays* L. cv. Velox). 1 mEq $l^{-1}$ solutions of these ions (labelled with $^{42}K$, $^{22}Na$, and $^{45}Ca$) were administered to various root zones as indicated by the *arrows,* and the remainder of the roots were bathed with non-labelled nutrient solution of otherwise identical composition. After 24 h plants were harvested, roots were rinsed for 15 min in non-labelled solution and then separated in 3-cm segments; root segments, grains (endosperm) and shoots were analyzed separately. The amounts of labelled $K^+$, $Na^+$ or $Ca^{2+}$ found after 24 h in the various zones are given in the histograms as nEq per 12 plants calculated on the basis of the specific activity of the labelling solution. The lines touching the arrows give the sums, i.e. the total amount of label taken up. Note the logarithmic scale of the ordinate. The reader can make the various comparisons allowed by the instructive experiment by himself. Some interesting aspects are as follows: $K^+$ and $Na^+$ absorption are slightly smaller in the root zones closer to the tip than further up; conversely $Ca^{2+}$ absorption is larger in the tip region. $K^+$ and $Na^+$ are translocated towards the root tip and the root base, grain and shoot but the profiles are different. $Na^+$ transport towards the tip seems to be less than $K^+$ transport. $Na^+$ export seems to be similar in all root zones, whereas $K^+$ transport out of the tip region is smaller than export from the other zones. The latter is also observed with $Ca^{2+}$, but $Ca^{2+}$ is only transported upwards to the grains and shoots and not at all downwards to the tip. (Data from MARSCHNER and RICHTER 1973)

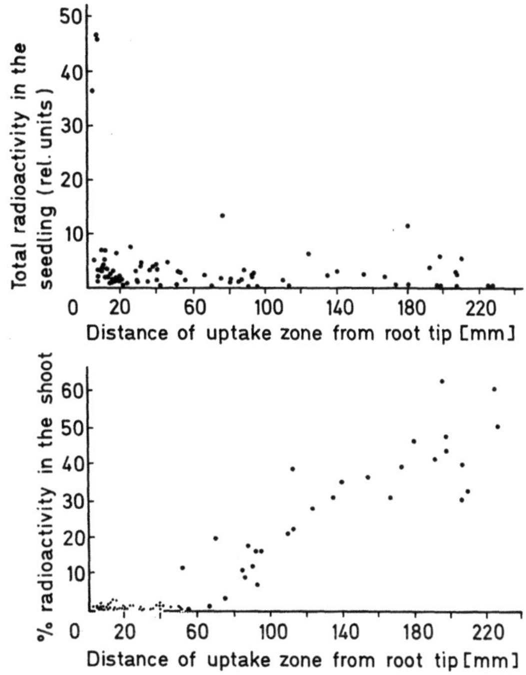


**Fig. 2.** Phosphate uptake and translocation in various root zones of 6-day-old maize seedlings (*Zea mays* cv. Golden Cross Bantam). (BURLEY et al. 1970)

iour (summarized in the introduction of MARSCHNER and RICHTER 1973, CLARKSON and HANSON 1980). This is due to wide differences between plant species, ionic species (Fig. 1), environmental and nutritional conditions (BOWEN 1970, CLARKSON et al. 1978c) and positions in the root-branching system, i.e. primary root, lateral root or adventitious root (RUSSELL and SANDERSON 1967). Distinction also must be made between uptake and accumulation in the root tissue on the one hand and translocation on the other (Figs. 1 and 2, CLARKSON et al. 1968, SHONE et al. 1969). Clearly both structural and physiological variations along the length of the root must be responsible for the observed diversity.

## 2.2 Structure-Function Relations in Various Root Zones

A three-dimensional view of the root is required in order to understand the complex structural relations along and across the root in respect to the transport functions of uptake and translocation and also with regard to the role of the roots in anchoring the plant and exploiting the soil (Fig. 3).

### 2.2.1 The Root Surface

#### *2.2.1.1 The Mucigel of the Tip*

Glycoprotein secretion (GREEN and NORTHCOTE 1979) and degeneration of the root cap cells form a slimy material at the surface of the very tip of roots.

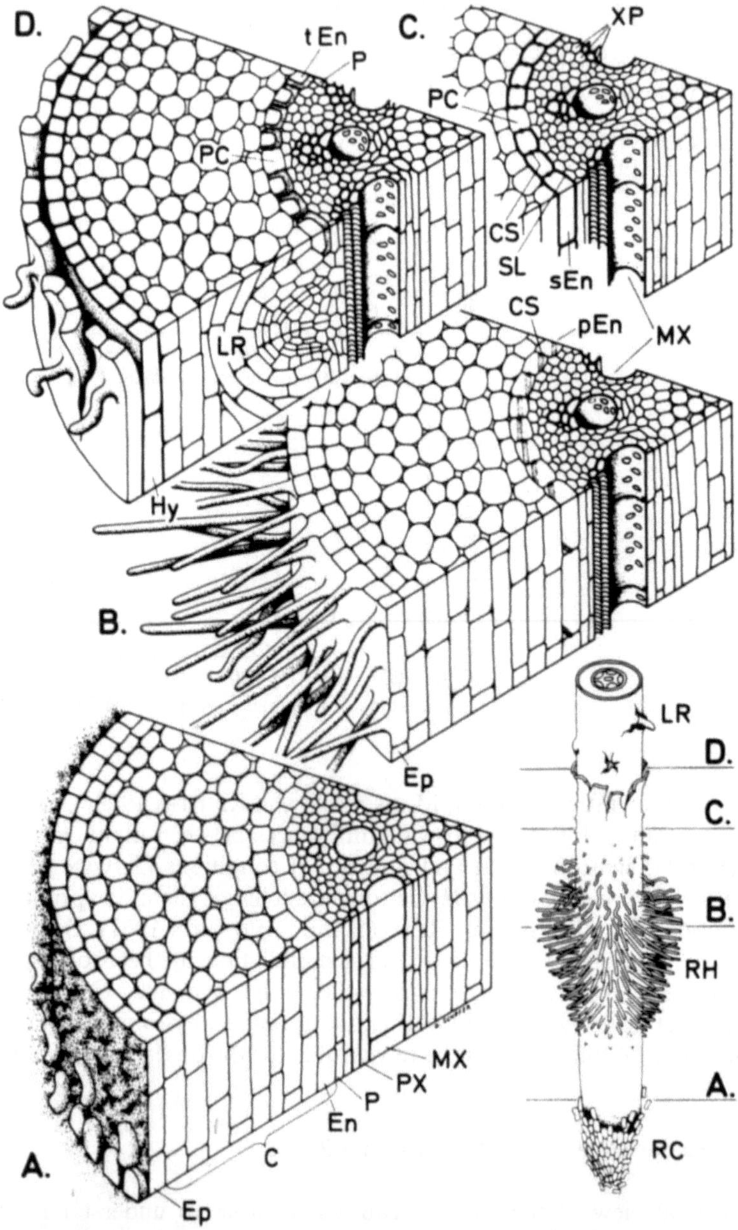


**Fig. 3A–D.** Three dimensional scheme of a root (*lower right*) with details of four successive root zones, as discussed in Sections 2.2.1–2.2.4, describing transport across the root in the various zones and along the length of roots. Anatomy of grass roots was taken as a basis for the scheme; distances along the length of the root are not indicated because the differentiation can vary greatly, depending on species and growth conditions. **A** Zone of primary root differentiation, **B** Root hair zone with primary endodermis, **C** Zone of secondary endodermis, **D** Zone of lateral root formation, hypodermis and tertiary endodermis. *C* cortex; *CS* Casparian strip; *En* endodermis (*pEn* primary, *sEn* secondary, *tEn* tertiary endodermis); *Ep* epidermis; *Hy* hypodermis; *LR* lateral root; *MX* metaxylem; *P* pericycle; *PC* passage cell; *Ph* phloem; *PX* protoxylem; *RC* root cap; *RH* root hair; *SL* suberin lamella; *XP* xylem parenchyma

But also further up the primary root the epidermis is covered by an amorphous material, a mucigel, containing free carboxyl groups and other material of hydrophilic nature (see Chap. I.2, this Vol. for a full description). This peripheral mucigel considerably facilitates the uptake of ions by ion exchange with solid soil particles (see LÄUCHLI 1976a, PITMAN et al. 1976). Autoradiographs demonstrate that there can be a preferential adsorption of polyvalent ions in the mucigel, a process which is passive and thus independent of temperature (CLARKSON and SANDERSON 1969). Presumably this adsorption is also responsible for the relatively high uptake and retention of $Ca^{2+}$ in the apical 3 cm of maize roots (Fig. 1).

### *2.2.1.2 The Primary Root Epidermis*

The dominating role of the outermost cell layer of the root, the primary root epidermis, in ion adsorption, absorption and uptake can be documented by autoradiographs (Fig. 4) and X-ray microanalysis (LÄUCHLI 1967, LÄUCHLI et al. 1971). Pectic substances in the relatively thin cell walls again may be important for adsorption of cations prior to absorption. The formation of root hairs is essential in providing a much increased external surface for these processes.

Recently it was discovered that, in response to various situations of stress, the root epidermis also can increase its inner surface. The primary roots thicken in a zone which contains the root hairs. The cell walls of the root hairs and the outer tangential walls of the epidermal cells form protuberances or ingrowths, leading to a considerable increase of the plasmalemma surface. Thus, these cells develop to transfer cells, which are generally characterized by such cell wall labyrinths frequently occurring at apoplast–symplast boundaries in sites where there is considerable transport across the plasmalemma (GUNNING 1977). One example is salt stress of *Atriplex hastata* plants which develop root thickenings and peripheral transfer cells 1–3 mm behind the tips when grown in media with 100–600 mM NaCl (Fig. 5, KRAMER et al. 1978). Similarly, *Helianthus annuus* plants form root thickenings and transfer cells 3–15 mm behind the tips when Fe is absent from the nutrient solution for 24 h to 48 h, i.e. under conditions of stress due to Fe deficiency (Fig. 6, KRAMER et al. 1980).

*Helianthus* has been characterized as belonging to the "Fe-efficient" plants which respond to Fe deficiency by $H^+$ extrusion and production of reducing substances. These two processes considerably increase the capacity of the plants for uptake of Fe, since Fe is largely transported in the form of $Fe^{2+}$ (MARSCHNER et al. 1974, BROWN 1978). This behaviour is related to adaptation to calcareous soils, i.e. "Fe-efficient" plants are calcicole, whereas "Fe-inefficient" ones are calcifuge (BROWN 1978, CLARKSON and HANSON 1980).

Conceivably, the increased plasmalemma surface of root epidermis transfer cells formed under conditions of Fe deficiency serves increased $H^+$ pumping out of the cells. The ubiquitous occurrence of $H^+$ extrusion pumps in higher plant cells including roots has been stressed repeatedly (PITMAN et al. 1975, ANDERSON et al. 1977, reviews MARRÈ 1979, SMITH and RAVEN 1979, see Sect. 2.2.4.3). Additional experiments with *A. hastata* (KRAMER unpublished) show that NaCl stress induces symptoms of Fe deficiency, although Fe is present

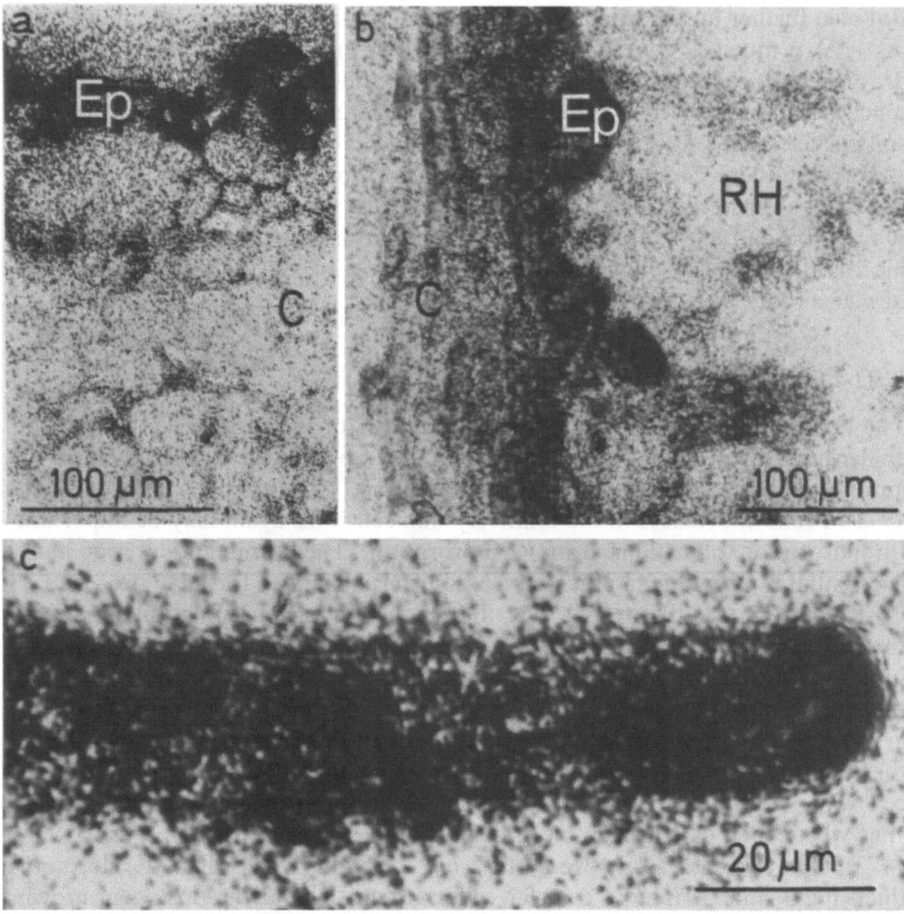


**Fig. 4.** Adsorption and absorption of sulphate by the peripheral cell layers of *Zea mays* (**a**) and *Acer platanoides* (**b**) seminal roots, and by a single root hair of *A. platanoides* (**c**). 20 mM $^{35}SO_4^{2-}$ was applied for 1 h (**a**) and 2 h (**b, c**), respectively. The root in (**a**) was inhibited with 1 mM $NaN_3$ (see Sect. 2.2.3). **a** from WEIGL and LÜTTGE (1962), **b** from LÜTTGE (1964). *C* cortex; *Ep* epidermis; *RH* root hairs

in the medium. This can be due to effects of ion competition. The formation of an increased plasmalemma surface in the transfer cells may be a response to particular passive and active transport problems in the presence of high NaCl concentrations.

Aerial roots of orchids often have a special *velamen radicum,* which morphogenetically corresponds to a multilayered root epidermis. The cells of the functional *velamen* are dead; they have complex cell wall structures with pores and cracks creating capillary forces, which allow the *velamen* to rapidly imbibe any water sprayed on the plant (rain) (CAPESIUS and BARTHLOTT 1975). The outermost layer of the cortex, i.e. the hypodermis forms a sheath which resembles endodermal differentiations having Casparian strips and secondary and

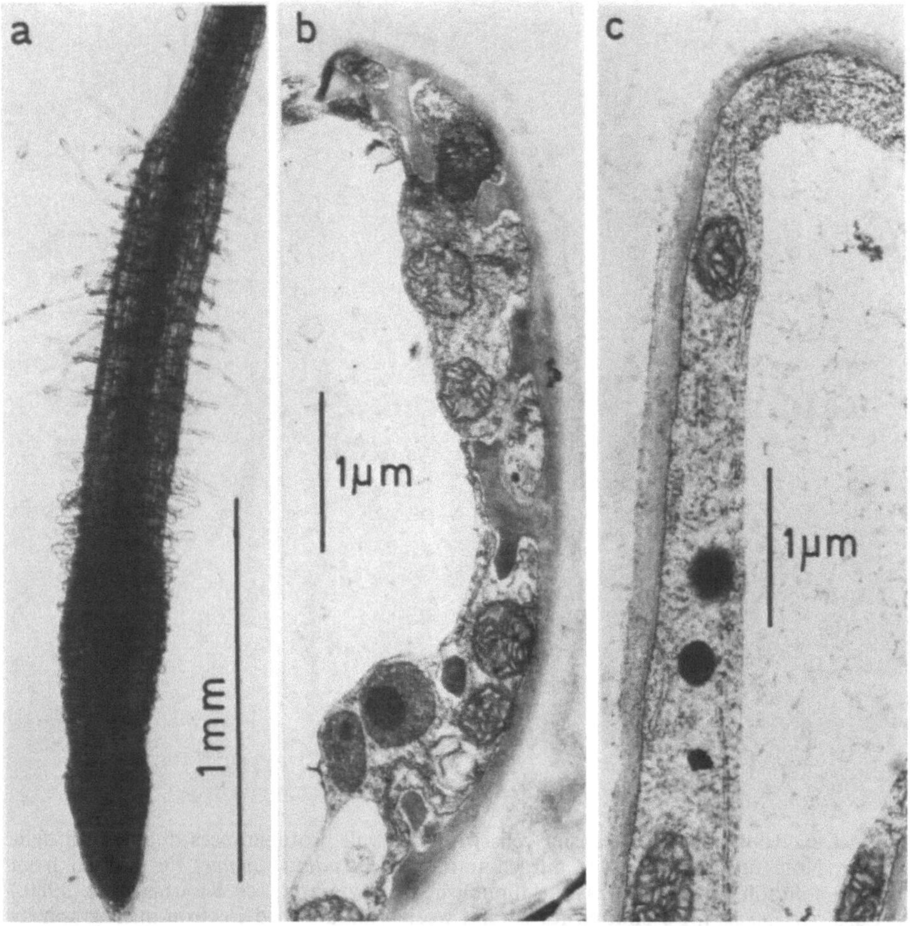


**Fig. 5.** **a** *Atriplex hastata* root grown in 400 mM NaCl with root thickening behind the tip and with numerous root hairs. **b** Cell wall ingrowths of a root hair from a root grown in 400 mM NaCl. **c** Cell wall of a root epidermis cell of the same ontogenetic level as in **b** but from a root grown in the absence of NaCl. (KRAMER et al. 1978)

tertiary developmental stages including formation of passage cells (see Sect. 2.2.3). Tritiated water and $^{32}$P-labelled phosphate are readily taken up from the *velamen* by the underlying living root tissues (CAPESIUS and BARTHLOTT 1975).

#### *2.2.1.3 Root Hypodermis and Periderm*

Above the zone of the root hairs, the epidermis often deteriorates and the outer cells of the cortex begin to suberize, forming a hypodermis (or exodermis) on the root surface. Suberization of cell walls may also occur in the epidermal cells and even in the root hairs. In *Zea mays* it was shown that the deposition

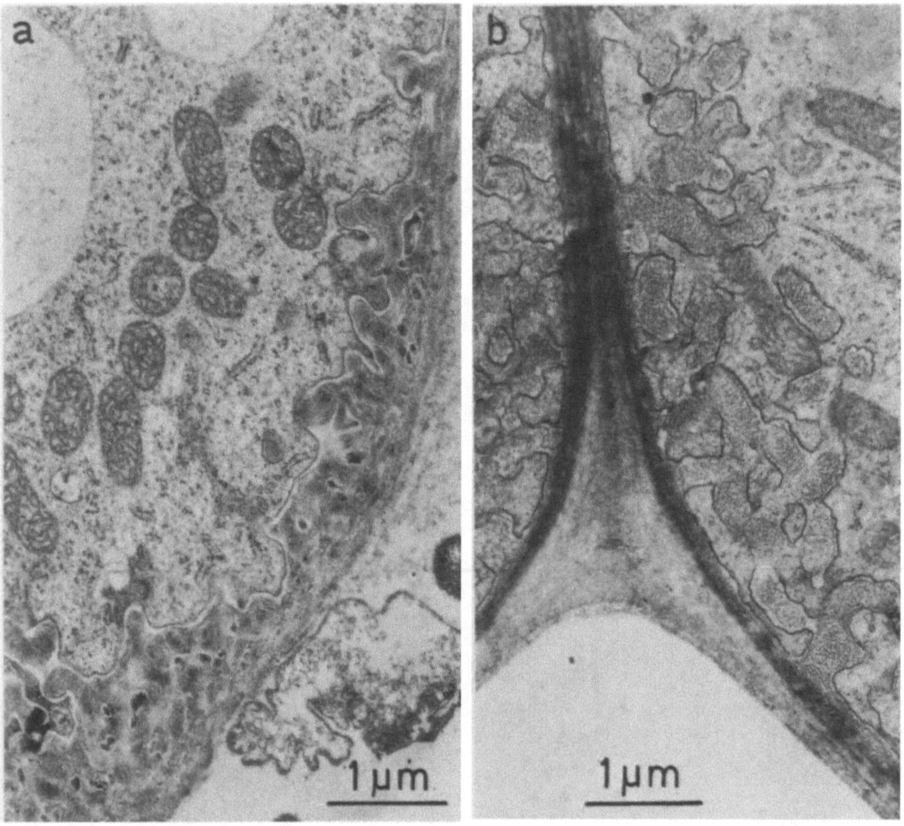


**Fig. 6a, b.** Root-epidermal transfer cells with cell wall protuberances due to iron deficiency. Note the abundance of mitochondria. **a** *Helianthus annuus,* Fe omitted from culture solution for several days. (Unpublished micrograph, see KRAMER et al. 1980.) **b** *Capsicum annuum,* Fe omitted for several weeks (unpublished electron-micrograph by courtesy of Dr. D. KRAMER)

of suberin lamellae in hypodermal walls led to a marked reduction of phosphate uptake (FERGUSON and CLARKSON 1976b). Similar results were obtained with *Carex arenaria,* where several suberized cell layers in the outer cortex largely prevent the passage of $H_2O$, $Ca^{2+}$ and phosphate (CLARKSON et al. 1978a). But this conclusion cannot be generalized. In another species, *Allium cepa,* suberized cells walls of the hypodermal cell layers seem to be of a porous nature remaining highly permeable for $H_2O$, ions and small organic molecules. In this species, obviously, hydrophilic channels pass across the suberin lamellae, and solutes are selected on the basis of their molecular weight (CLARKSON et al. 1978b). In salt-stressed *Atriplex hastata* roots and in Fe-stressed *Helianthus* roots, hypodermal cells even form cell wall labyrinths on their outer tangential walls when epidermal cells die, and thus seem to serve special transport functions (see Sect. 2.2.1.2., KRAMER et al. 1978, KRAMER et al. 1980).

**Fig. 7.** Electrical analogue depicting alternative ion uptake pathways across the root. *RH* root hair; *Ep* epidermis; *C* cortex; *En* endodermis; *XP* xylem parenchyma cells; *CS* Casparian strip

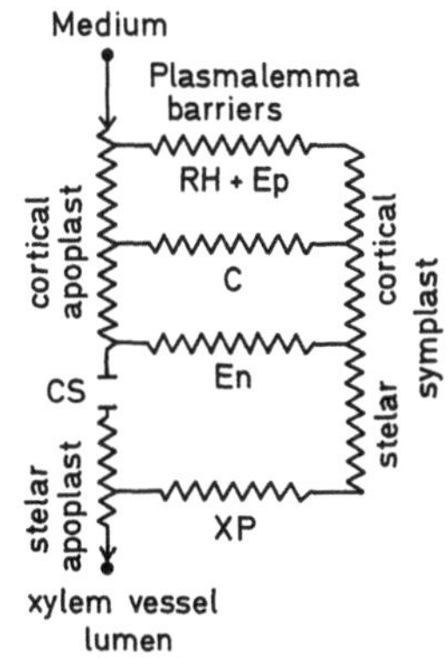


In *Iris* roots, where the cortical cells have storage functions, a periderm develops at the root surface. This periderm has been found to be entirely impermeable to sulphate (ZIEGLER et al. 1963).

Thus in these root zones uptake is often much reduced or negligible. Lateral roots, initiated in the stele and breaking through the cortex and root surface, begin to exploit the soil in these zones. It has often been suspected that rupture of the outer root tissues by the lateral roots opens routes for uncontrolled passive entry of solutes into the stele, but quantitatively this seems to be not important (see also Sect. 2.2.3).

### 2.2.2 The Cortex

The cortex usually comprises several layers of large, rather non-differentiated and highly vacuolated parenchyma cells. When a root fulfils storage functions, the cortex is particularly thick. In other cases, such as in the roots of *Calluna vulgaris,* the cortex can be very much reduced and is only represented by the endodermis, which morphogenetically corresponds to the innermost layer of the cortex (pp. 379–380 in TROLL and HÖHN 1973).

The cortex must be traversed by water and solutes during their transport from the medium to the stele where they can then be exported from the root. In principle this passage can occur via either apoplastic or symplastic pathways, and the situation can be depicted by an electrical analogue (e.g. GLINKA 1977, Fig. 7). Apoplast and symplast constitute two parallel pathways of different resistances. The apoplastic pathway is blocked at the endodermis by the Casparian strip (see Sect. 2.2.1.4). The symplastic pathway can be entered by passage through the plasmalemma in the root hair and epidermal cells, in the cortical cells and in the endodermal cells. The relative importance of the various pathways for transport across the cortex depends on their relative resistances, which may vary for plant species, solute species, environmental conditions (e.g. solute concentration in the medium), and intrinsic factors (e.g. hormonal balance, state of ionic relations, etc.). Perhaps this consideration can help to resolve a long-lasting controversy on whether or not the cortical apoplast is an important and rather freely accessible pathway ("free space", see LÄUCHLI 1976a) for entry of solutes deeper into the root. On the one hand, for example, in the Crafts-Broyer hypothesis of ion transport across the root, the entire cortical apoplast is taken as supplying a considerably increased surface for membrane-controlled ion uptake (CRAFTS and BROYER 1938, LATIES 1967, 1969). Conversely, VAKHMISTROV (1967), BANGE (1973), and VAN IREN and VAN DER SLUIJS (1980) maintain that uptake is exclusively limited to the root surface and thus "the epidermis is the only actively absorbing part of the root, the salt solution in the free space representing in a sense a ballast volume". Histochemical localization of ATPase activity in corn roots suggests that the epidermis and outer cortex must have a primary energy-linked role in ion absorption by the root (MALONE et al. 1977).

There is much evidence that, as the electrical analogue suggests, both the apoplastic pathway up to the endodermis and the symplastic pathway are in fact used. Cytological labelling techniques have demonstrated contributions of both the apoplastic (see LÄUCHLI 1976a) and the symplastic pathways of transport across the root. The endoplasmic reticulum extending from cell to cell through plasmodesmata is used as a route in symplastic transport (STELZER et al. 1975, VAN IREN and VAN DER SPIEGEL 1975, ROBARDS and CLARKSON 1976). Visible cytoplasmic streaming in the root cortex apparently is not important for radial transport (GLASS and PERLEY 1979). GRUNDWALDT et al. (1979) suggest that at low ion concentrations in the external medium uptake is largely restricted to the root surface, but that uptake from the cortical apoplast begins to play a role as ion concentrations are increased. PITMAN (1965) argued that in a solution of 0.5 mM $K^+$ +9.5 mM $Na^+$ the transport of $K^+$ and presumably also of $Na^+$ across the cortex of barley roots was too large to be explained solely by diffusion in the apoplast and that the symplastic pathway must play an important role. Since the cell wall carries a negative charge, the apoplastic pathway would be even less efficient for anions than for cations.

In conclusion it seems that, while it is not always superfluous, the cortex is possibly not essential for ion absorption and translocation. This is also demonstrated by *Calluna* roots in which the cortex is lacking, and is reflected in the fact that only the root hairs and the *outer* epidermal or hypodermal cell walls form protuberances in response to salt stress and iron deficiency (Sect. 2.2.1.2).

### 2.2.3 The Endodermis

Very much has been written on the role of changes of endodermal structure along the length of the root in relation to transport across the root (e.g. LÄUCHLI 1976a). The endodermis is morphogenetically the innermost layer of the cortex. It usually deteriorates during secondary growth of dicotyledonous roots before it can develop further; but in the monocotyledons the primary endodermis goes through further developmental stages of secondary and tertiary endodermis (Fig. 3). In the primary endodermis, usually in the zone of the root hairs, the anticlinal walls of the endodermal cells are incrusted with suberin-like material, forming the so-called Casparian strip. The lipid composition of suberin compounds of endodermis has recently been analyzed (ESPELIE and KOLATTUKUDY 1979, SCOTT and PETERSON 1979a, b).

In the secondary endodermis, suberin lamellae are layered on the walls all around the endodermal cells; in the tertiary endodermis, further cellulose layers are deposited, which may alternate with bands of lignin (SCOTT and PETERSON 1979a, b). Only a few so-called passage cells in the endodermis are exempt from these secondary and tertiary changes.

Electron microscopical localization studies using electronopaque tracers clearly show that the apoplastic pathway in the cortex can only be used up to the primary endodermis (reviewed by LÄUCHLI 1976a). Transport across the endodermis is dependent on metabolic energy; i.e. at least one of the plasmalemma barriers depicted in Fig. 7 must be passed by energy-dependent transport. It is seen on micro-autoradiographs of maize roots that in the presence of azide, $^{35}SO_4^{2-}$ accumulates outside the endodermis in the cortex and epidermis (WEIGL and LÜTTGE 1962). In some cases additional cortex cell layers develop sheaths around the stele with phi thickenings of their radial walls, e.g. in cycads (PLAUT 1910), *Taxus baccata* (p. 297 in STRASBURGER 1923) and some angiosperms (HAAS et al. 1976, MACKENZIE 1979). Phi thickenings are lignified but not suberized. In apple (*Pyrus malus*) and geranium (*Pelargonium hortorum*) roots, by contrast to the Casparian strip of the endodermis, the phi thickenings do not block apoplastic transport of a fluorescent dye (PETERSON et al. 1981). STELZER and LÄUCHLI (1977) observed that the roots of the flooding- and salt-tolerant grass *Puccinellia peisonis* develop a second endodermal cell layer with secondary (suberin lamellae) and tertiary stages (thickened walls). This may provide additional protection from passive and uncontrolled entry of salt into the stele under conditions of salinity.

An intriguing question is why passive diffusion into the stele of roots cannot bypass the apoplast barrier of the Casparian strip by using apoplastic routes in other root zones. Apoplastic routes exist in the youngest part of the root below the zone of the primary endodermis, i.e. where the endodermis is not yet differentiated from the rest of the periblem layers that will eventually form the cortex, or in the region where the endodermal layer is becoming discernible but has not yet developed a Casparian strip. Clearly translocation in this zone will be minimal because vessels have also not yet developed (but see Sect. 2.2.4). Physiological factors may however, also be involved. Micro-autoradiographs show that in the very tip region of the roots, $Ca^{2+}$ and sulphate only penetrate up to the border between periblem and plerome and not beyond, even though there is no visible structural barrier (LÜTTGE and WEIGL 1962, Fig. 8).

In the zone of a developed primary endodermis URSPRUNG and BLUM (1921) suggested the existence of an abrupt physiological gradient at the endodermis, the so-called "Endodermissprung" of water potential. Although the finding of these authors has never been confirmed, various approaches have been used to assess the possibility of physiological gradients existing at the endodermis between the cortex and the stele.

Carbohydrate metabolism is very active both in the cortex and the stele. In the stele there is a somewhat larger activity of pentosephosphate cycle as compared to glycolysis than in the cortex. Pentosephosphate cycle activity also

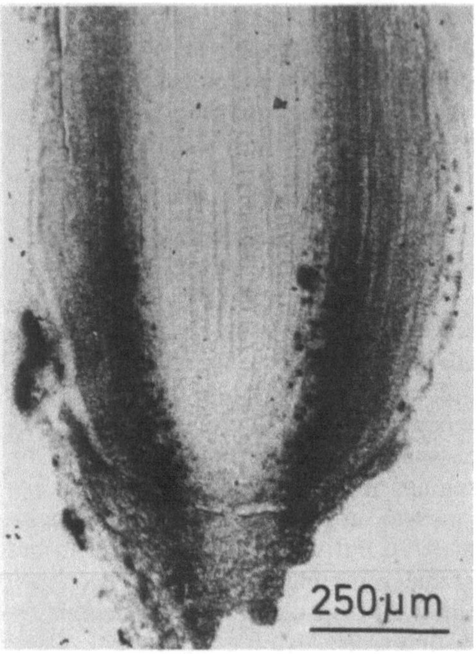


**Fig. 8.** Autoradiograph of a longitudinal section of a maize root tip to which 20 mM $^{35}SO_4^{2-}$ has been applied for 5 min. Label penetrated only up to the border between periblem and plerome. (LÜTTGE and WEIGL 1962)

increases as the stele matures; presumably it is required to provide NADPH + $H^+$ for lignin biosynthesis (WONG and AP REES 1971).

Cytological techniques like microautoradiography and X-ray microanalysis have suggested that ion concentrations are larger in the stele than in the cortex (WEIGL and LÜTTGE 1962, 1965, LÜTTGE and WEIGL 1964, LÄUCHLI 1967, 1972, LÄUCHLI et al. 1971). This conclusion can be criticized, however, because these methods did not allow resolution between vacuoles and cytoplasm. If the concentration of the ions investigated were higher but uniform in the whole root symplast as compared with the vacuoles, in view of the larger relative content of cytoplasm in the stelar parenchyma cells these cytological observations would only demonstrate a larger amount of ions in individual stelar cells but not larger concentrations, let alone electrochemical potentials (LÜTTGE and HIGINBOTHAM 1979). Indeed, BOWLING and coworkers observed no electrochemical gradients of ions like $K^+$, $Ca^{2+}$, $Mg^{2+}$, $Na^+$, $Cl^-$, $NO_3^-$, and $SO_4^{2-}$ within the cortex and stele; the $O_2$ gradient is very small; a gradual increase of vacuolar pH is as yet unexplained (DUNLOP and BOWLING 1971a, b, c, BOWLING and ANSARI 1972, BOWLING 1976). Their measurements have been also criticized, however, because blindly pushing a microelectrode across the root – as done in the experiments of BOWLING and co-workers – may cause artefacts due to the wound created by the electrode (ANDERSON and HIGINBOTHAM 1975). Notwithstanding these uncertainties, it appears that no sudden physiological gradients occur at the endodermis, and that the endodermis has no gland-like functions pumping ions across the roots, as suggested by older hypotheses (see SUTCLIFFE 1962). Cells of the primary endodermis are highly vacuolated with a thin layer of cytoplasm along their walls, just as the cortical cells, and their most conspicuous peculiarity is the apoplast barrier of the Casparian strip. Recently ROBARDS et al. (1980) have shown that density of protein or glycoprotein particles on the freeze-etching fracture faces of the plasmalemma of endodermal cells of *Zea mays* is about two to three times larger than in cortical cells. This is discussed as an indication of a special activity of the endodermal plasmalemma due to the particular necessity to take up ions from the apoplast. This interpretation, however, is only pertinent if the apoplastic pathway up to the endodermis is not of minor importance (see Sect. 2.2.2).

By the suberization and lignification in the secondary and tertiary endodermis, the plasmalemma of the endodermal cells becomes inaccessible to the apoplastic pathway of the cortex, and thus uptake into the symplast across the endodermal plasmalemma becomes impossible (see Fig. 7). Working with roots of *Hordeum vulgare, Zea mays,* and *Cucurbita pepo* CLARKSON and coworkers (CLARKSON et al. 1971, HARRISON-MURRAY and CLARKSON 1973, FERGUSON and CLARKSON 1975, 1976a) found that this much reduces the radial transport of $Ca^{2+}$ and $Mg^{2+}$ which are not transported symplastically (FERGUSON 1979) but not of phosphate and $K^+$. Presumably this also explains findings of RICHTER and MARSCHNER (1974) that in *Zea mays* and *Phaseolus vulgaris* $Ca^{2+}$ levels are higher in the cortex than in the stele; $K^+$ was equally distributed and $Na^+$ levels were larger in the stele.

Obviously the cells of secondary and tertiary endodermis studied by CLARKSON and co-workers were still alive and were traversed by plasmodesmata that could allow symplastic transport across them. The passage cells, which are exempt from suberisation and lignification of the secondary and tertiary endodermis, must have been of minor importance for the transport of $Ca^{2+}$ and $Mg^{2+}$. In *Iris* roots the walls of the tertiary endodermis are heavily thickened by apposition of cellulose layers and the cells eventually die. Radial transport then becomes restricted exclusively to the passage cells. Since in this zone of *Iris* roots a peripheral periderm also prevents ion uptake from the medium, radial transport through the passage cells is largely directed from the stele towards the cortex (ZIEGLER et al. 1963).

It could be argued that lateral roots breaking through the endodermis and cortex further up the root may provide pathways for passive movement of ions across the endodermis. A new endodermis with a Casparian strip enclosing steles of main root and lateral root is formed, however. In the stages of lateral root formation where formation of the Casparian strip lags behind division of endodermal cells, no xylem is developed and the cells of the lateral root tip are only slightly vacuolated. In this respect the situation much resembles that of the growing main root shown in Fig. 8 (DUMBROFF and PEIRSON 1971).

### 2.2.4 The Stele

The stele fulfils mechanical functions in that the central and radial arrangement of xylem confers considerable tensile strength on the root. It also fulfils physiological functions providing the pathways of long-distance transport, i.e. phloem and xylem, and mechanisms of loading and unloading of these pathways.

#### 2.2.4.1 Long-Distance Transport in Roots

Export of ions to the shoots obviously depends greatly on the extent of development of the xylem; it is very small from the tips of roots and increases in root zones further up (e.g. Figs. 1 and 2; BURLEY et al. 1970, MARSCHNER and RICHTER 1973). Comparative physiological and anatomical studies combined with radioactive or dye labelling suggest that the metaxylem vessels provide the major pathway for translocation to the shoots (LÜTTGE and WEIGL 1962, BURLEY et al. 1970, ROBARDS and CLARKSON 1976).

LUNDEGÅRDH's (1950) classical experiment, in which a cut cylinder of a wheat root produced exudate from both ends, seems to suggest that there is no absolute polarity of translocation in the root. In isolated onion root segments, $Cl^-$ appears to be transported in both directions, $K^+$ transport is more strongly upward than downward to the tip, $Ca^{2+}$ is transported solely upward, and $Na^+$ is largely transported downward (MACKLON 1975a, b). SINGH and JACOBSON (1977) observed strictly polar upward transport of $Na^+$, $Rb^+$, $Ca^{2+}$ and $Cl^-$ to the cut end of excised barley roots. Polarity of translocation in roots certainly needs more attention in relation to transport mechanisms (SINGH and JACOBSON 1977). EVANS and VAUGHAN (1966) think that the apparent polarity of $Ca^{2+}$ transport in isolated root segments of maize is due to wounding responses. Thus, experiments with *intact* maize seedlings performed by MARSCHNER and RICHTER (1973) appear to be crucial. $K^+$ and $Na^+$ are clearly transported in both directions, but $Ca^{2+}$ exclusively moves upward to the shoot (Fig. 1). This obviously suggests that phloem plays a considerable role in downward transport of ions, since the phloem-immobile ion $Ca^{2+}$ is not translocated in that direction, whereas phloem-mobile $K^+$ is moved.

#### 2.2.4.2 Loading of the Xylem Vessels

The mechanism of loading of ions into the xylem vessels has long been the most outstanding issue in research on ion transport of roots. Transport of many major nutrient ions from the soil or culture solution into the cells of the root and into the xylem fluid is against an electrochemical gradient (DUNLOP and BOWLING 1971a, b, c, BOWLING and ANSARI 1972, BOWLING 1976, DAVIS and HIGINBOTHAM 1976, reviews e.g., LÜTTGE 1973, CLARKSON 1974, ANDERSON 1976, LÜTTGE and HIGINBOTHAM 1979). Thermodynamically, a single active step in the system would be sufficient to transport ions against a gradient across the root. CRAFTS and BROYER (1938) envisaged such a single step being the uptake from the cell wall apoplast across the plasmalemma of root hair, epidermis and cortex cells into the symplast of these cells with subsequent passive transport through the root symplast and diffusive release of ions into the dead xylem vessels. This hypothesis has found experimental support by ARISZ (1956). Conversely, it had been suggested that the xylem parenchyma cells might actively excrete ions into the xylem vessels (SUTCLIFFE 1962 p 120). WEIGL and LÜTTGE (1965) and LÄUCHLI (LÄUCHLI 1967, LÄUCHLI et al. 1971) supported this by microautoradiographic and X-ray microprobe evidence showing that the labelling of the conducting metaxylem vessels was always much higher than in the surrounding xylem parenchyma. The drawbacks of these cytological methods have, however, already been mentioned above. Later the groups of LATIES (cf. LATIES 1967, 1969, LÜTTGE 1969) and EPSTEIN (LÄUCHLI and EPSTEIN 1971, LÄUCHLI et al. 1971, LÄUCHLI 1972) debated the existence of active transport only in the root tissue peripheral to the endodermis versus operation of ion pumps in the stelar parenchyma on the basis of conflicting interpretations of kinetic results. More recently, the work of PITMAN and coworkers has made it convincingly clear, however, that two pumps of distinctly different properties must be involved, one located at the outer face of the symplast taking up ions from the outer medium into the root cells, and one located at the inner face of the symplast actively excreting ions into the xylem vessels (reviewed by LÄUCHLI 1976b, PITMAN 1977, LÜTTGE and HIGINBOTHAM 1979).

In the overall process of ion transport from the medium across the root and into the vessels, differentiations of the root surface (Sect. 2.2.1) and the endodermis (Sect. 2.2.3) are naturally important. In discussions of the actual loading mechanism, however, variations in structure along the length of roots were not considered important. Both the passive loading mechanism of the "one-pump hypothesis" of CRAFTS and BROYER (1938) and the active loading mechanism of the "two-pump hypothesis" (PITMAN 1977) consider that loading occurs from the xylem parenchyma cells into fully mature, i.e. cytoplasm-free and dead xylem vessels.

The evaluation of another hypothesis first conceived by HYLMÖ (1953) and then reconsidered by SCOTT (1965), ANDERSON and HOUSE (1967), HIGINBOTHAM et al. (1973), DANILOVA and STAMBOLTSYAN (1975) and DAVIS and HIGINBOTHAM (1976), however, requires us to consider xylem element differentiation along the length of roots in relation to the loading process. In the tip region of the root the developing vessel elements still contain cytoplasm with organelles, plasmalemma and tonoplast. In HYLMÖ's hypothesis, ions are loaded by transport across the tonoplast into the vacuoles of the immature vessel cells in this region of the roots. HIGINBOTHAM et al. (1973) observed that in corn roots cytoplasm lined the early metaxylem cell walls up to 10 cm behind the root tip. In a zone 5–10 cm behind the tip the perforation plates of the cross walls between vessel elements were already formed and the vacuolar sap appeared to be continuous with the xylem sap in the dead and mature elements. Thus symplastic transport into the cytoplasm of young xylem elements followed by active transport across the tonoplast could constitute the loading mechanism of vessels. According to this hypothesis, transport across followed by translocation out of the roots should be small in the very tip region, where xylem elements are not yet discernible, then should reach a peak, where xylem cells develop but still have plasmalemma and tonoplast, and should decline again further behind the tip, where xylem is fully mature. This seems not to be born out by the experiments of Figs. 1 and 2. If such a phenomenon were observed, it could also be due to variation of physiological activities along the length of roots (Sect. 3) and not simply to the structural differentiations of the xylem. Quantitative aspects also argue against a loading only in the limited region of a vessel, where its elements are still immature. The development of the xylem, and hence root growth, would need to be unusually rapid (LATIES 1969), or fluxes across the tonoplast of living vessel cells must be extraordinarily large (ANDERSON et al. 1970, ANDERSON 1976), or both.

LÄUCHLI et al. (1978) have demonstrated that a catalyzed transport mechanism from living xylem parenchyma into dead xylem elements clearly must be a key feature of the loading process. They used the amino acid analogue p-fluorophenylalanine (FPA), whose presence in cells during protein synthesis may lead to formation of ineffective proteins. FPA appears to be a specific inhibitor of the xylem loading process, because it is incorporated in a protein of rapid turnover, which is involved in the loading mechanism (SCHAEFER et al. 1975, PITMAN 1977). This protein might be a transport-ATPase as suggested by the cytological demonstration of $K^+$-stimulated diethylstilbestrol-inhibited ATPase activity in xylem parenchyma cells adjacent to the dead vessels (WINTER-SLUITER et al. 1977). FPA inhibited $Cl^-$ translocation out of barley roots in zones where all the xylem vessels were mature and not just in those zones where some immature vessels were present.

The presence of half-bordered pits with relatively thin cell wall areas sited between the mature, dead vessels and the xylem parenchyma cells with their comparatively dense cytoplasm (LÄUCHLI et al. 1974b) seems to be a structural feature that would greatly facilitate metabolism-dependent transport.

The loading mechanisms which actively concentrate ions in the xylem are usually thought to be the driving force for osmotic water flow across the roots and into the vessels, thus powering root pressure exudation (see ANDERSON 1976) and guttation from leaves with passive hydathodes (DIEFFENBACH et al. 1980a, b). Organic molecules like malate and glutamine transported in the vessels, however, may also contribute to the driving force of exudation (BUTZ and LONG 1979).

### *2.2.4.3 Reabsorption from the Vessels During Translocation in the Root*

It seems to be trivial that the composition of the solution translocated in the xylem vessels will be modified en route not only by secretion into the vessels, but also by reabsorption from the vessels. This was already demonstrated by ZIEGLER et al. (1963; see Sect. 2.2.3). JACOBY (1964, 1965) and JACOBY and RATNER (1974) found that bean and maize roots accumulate $Na^+$ in their upper regions, which allows a certain control of the supply of $Na^+$ to the shoot under conditions of salt stress. LÄUCHLI and coworkers have much extended this comparing the behaviour of pairs or groups of varieties or related species with different responses to NaCl stress, e.g. the cereals *Zea mays* and *Hordeum vulgare,* and the soybean varieties *Glycine max* cv. Jackson, cv. Bragg and cv. Lee and bean (*Phaseolus coccineus*).

In *Zea mays* 50 mM NaCl inhibits growth by 30%–40% (LÄUCHLI et al. 1976), and concentrations above 75 mM cause poor yield and mortality (YEO et al. 1977a). Nevertheless maize obviously can still be grown or kept for a limited amount of time on 150 mM NaCl or 75 mM $Na_2SO_4$ (YEO et al. 1977b). As shown by X-ray microanalysis under such conditions of stress, the xylem parenchyma cells in the upper root zones of *Zea mays* absorb $Na^+$ and $Cl^-$ from the transpiration stream; xylem parenchyma cells contained high $Na^+$

levels and almost no $K^+$; and $Na^+$ levels in the stele relative to those in the cortex were also high (YEO et al. 1977a). In the mature parts of the roots, but not in young roots, the cell walls in the areas of the half-bordered pits between the vessels and the xylem parenchyma cells are strikingly increased in thickness but of a very loose fibrillar texture (Fig. 9; YEO et al. 1977a). Thus maize plants respond to salt stress by the attempt to exclude NaCl from the shoots. *Hordeum vulgare* can tolerate much higher NaCl levels than maize. Barley is a salt includer; its response to salt stress is based on mechanisms in the leaves being able to cope with higher NaCl levels (see WYN JONES et al. 1979). The xylem parenchyma cells of barley always have high $K^+$ contents (YEO et al. 1977b).

The *Glycine max* varieties Jackson and Bragg are damaged already by 10 mM NaCl, whereas growth of the variety Lee and of *Phaseolus coccineus* is only inhibited by NaCl concentrations around 50 mM (KRAMER et al. 1977, LÄUCHLI and WIENEKE 1979). X-ray microanalysis showed that the less sensitive variety and *P. coccineus* accumulate $Na^+$ in the xylem parenchyma cells of the upper root zones. These xylem parenchyma cells resemble transfer cells (GUNNING 1977) having cell wall protuberances particularly at their half-bordered pits towards the vessels (Fig. 9b, c). Unlike $Na^+$, $Cl^-$ is not reabsorbed from the transpiration stream but appears to be accumulated in the cortex of the apical regions of the root (WIENEKE and LÄUCHLI 1979).

It remains to be seen whether real salt tolerance in halophytes can be based not only on salt inclusion but also on salt exclusion. It is quite clear, however, from the work of LÄUCHLI and coworkers that in response to NaCl stress some glycophytes exchange $Na^+$ for $K^+$ at the level of the xylem parenchyma cells (LÄUCHLI 1976b). The capacity of the xylem parenchyma cells to accumulate $Na^+$ must be rapidly exhausted, however, during longer duration of NaCl stress. Furthermore, one may have to assume that, as in other metabolically active cells, high levels of $Na^+$ and $Cl^-$ will be incompatible with the cytoplasm of the xylem parenchyma cells and must be sequestered in the vacuoles, so that compatible organic solutes will have to be synthesized and accumulated in the cytoplasm to provide osmotic balance (WYN JONES et al. 1979). NaCl also could be compartmented in the ER cisternae, which may serve symplastic movement (STELZER et al. 1975, KRAMER et al. 1977). Therefore, more peripheral tissues (e.g. the cortex) must participate in the exhange. In fact LÄUCHLI (personal communication in relation to LÄUCHLI 1979, WIENEKE and LÄUCHLI 1980) has suggested a $Na^+$ circulation in the roots. This model assumes $Na^+$ uptake in the root zones closer to the tip and $Na^+$ release to the medium in those root zones where xylem parenchyma cells reabsorb $Na^+$ from the vessels. Questions regarding electrical charge balance have so far been left open. $Cl^-$ is taken up together with $Na^+$, but when $Na^+$ is exchanged for $K^+$ in the vessels more distant from the root tip (LÄUCHLI 1976b) and $Cl^-$ is also retained in these root zones (LÄUCHLI and WIENEKE 1979, WIENEKE and LÄUCHLI 1979) the stoichiometry of exchanges appears problematic. Although the work of LÄUCHLI and WIENEKE (1979) clearly shows smaller NaCl levels in the leaves of the variety Lee as compared with Jackson, i.e. partial salt exclusion in the less sensitive variety, there are some results which appear to contradict the

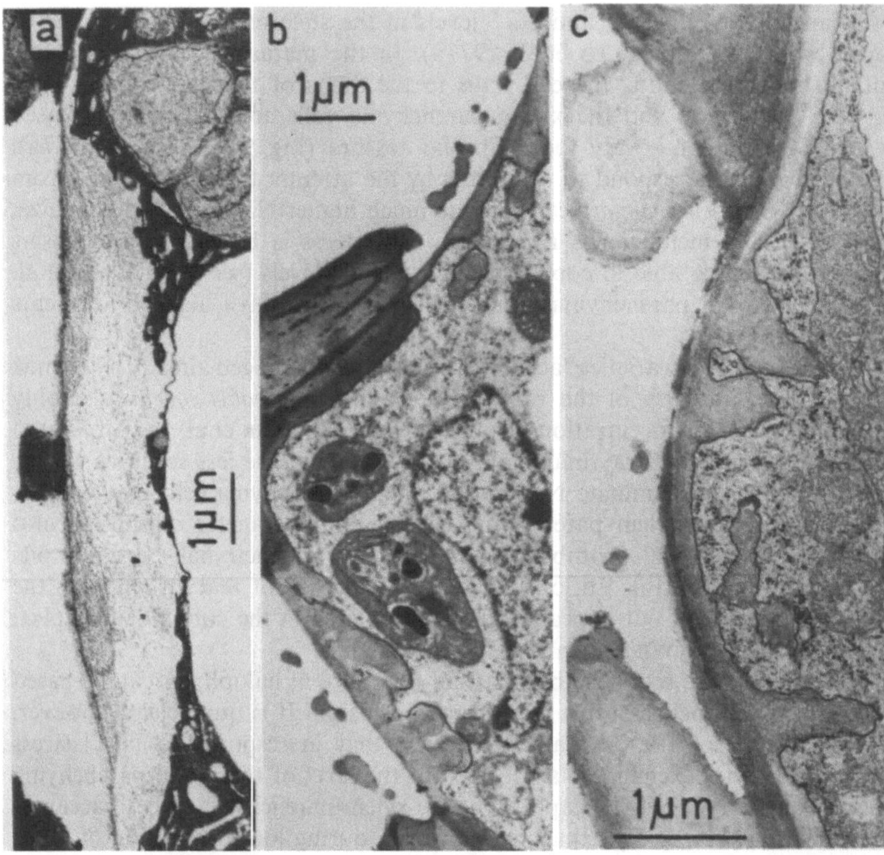


**Fig. 9a–c.** Cell wall modifications in the area of half-bordered pits between xylem vessels and xylem parenchyma cells. **a** Cell wall thickening and loose fibrillar texture, maize root stressed with 50 mM $Na_2SO_4$, 250 mm behind the tip. (YEO et al. 1977a). **b** Protuberances, root of the soybean cultivar Lee, 155 mm behind the tip. (Unpublished electron-micrograph; see LÄUCHLI et al. 1974a). **c** Protuberances, *Phaseolus coccineus* root. (Unpublished electron-micrograph; see KRAMER et al. 1977)

model; e.g. the upper root parts of variety Lee grown in 50 mM NaCl contain larger $K^+$ levels than plants grown without NaCl, and the upper root parts of varieties Lee and Jackson contain similar amounts of $Na^+$ when both are grown on 50 mM NaCl (Fig. 2 in LÄUCHLI and WIENEKE 1979).

Transfer cells of the xylem parenchyma and of the pericycle are also involved in the regulation of transport functions of $N_2$-fixing root nodules (PATE 1976, NEWCOMB and PETERSON 1979). It will continue to be stimulating to follow up the work described above and in Section 2.2.1.2 and obtain an integrated picture of the distribution and role of transfer cells in angiosperm roots (LETVENUK and PETERSON 1976, NEWCOMB and PETERSON 1979).

The quantitative questions need assessment on the basis of exchange mechanisms discussed for the two pumps involved in transport across the root (Sect.

2.2.4.2). JESCHKE (1970, 1972, 1977a, b, 1979, JESCHKE and STELTER 1973) has demonstrated a $Na^+/K^+$ exchange mechanism operating between barley roots and the external medium, which is of similar activity in root tips and differentiated root tissue. COLOMBO et al. (1979) showed that this could be a complex system of two exchange mechanisms, i.e. an active ATP-consuming and electrogenic $H^+$ extrusion mechanism capable of exchanging $H^+$ for $K^+$, plus a $Na^+/H^+$ exchange, combining together to give a net release of $Na^+$ and uptake of $K^+$. Since most workers now seem to accept the two-pump hypothesis (Sect. 2.2.4.2), the next step in our research must be to characterize in more detail both the outer and inner pumps. It has been suggested that proton pumping or charge separation at the site of both pumps may be the driving force for a number of uniport, symport (or co-transport) and antiport (or counter-transport) mechanisms whose differential activities or resistances at both faces of the symplast may be the basis for regulation of ion uptake, accumulation and export by roots (HANSON 1978, OKAMOTO et al. 1978, 1979). Together with observations of variation of pumping activities along the length of roots (Sect. 3.3), this may provide an avenue allowing the quantitative evaluation of cycling models explaining selectivity.

# 3 Variations of Physiological Activities Along the Length of Roots

## 3.1 Growth, Differentiation and Hormonal Gradients

Different physiological and biochemical activities are associated with cell division, extension growth and differentiation. This brings about variations of physiological features along the length of roots. The question here is to what extent and in which way this can influence transport physiologically, i.e. apart from the structural interactions discussed in Section 2.

Astonishingly little work has been devoted to this aspect. JESCHKE and STELTER (1976) have considered the different relative cytoplasm contents of cells in various root zones. Cells in the tip meristem and below the extension zone contain much more cytoplasm relatively to the cell volume than cells further behind the tip. It is known that cytoplasmic reaction systems require certain $K^+$ levels, but that $Na^+$ levels must be kept low, and $Na^+$ is sequestered in the vacuoles (WYN JONES et al. 1979). $K^+/Na^+$ ratios in the root tip are always very high, even in $K^+$-free media. Presumably $K^+$ can be transported in the phloem to the root tips from reserves in other parts of the seedlings (RICHTER and MARSCHNER 1973). $Na^+/K^+$ exchange mechanisms are operative in the more distant root zones (see Sect. 2.2.4.3). On the basis of cytometric considerations for the various root zones JESCHKE and STELTER (1976) then arrive at cytoplasmic and vacuolar $K^+$ concentrations of 110 and 20 mM, respectively, i.e. $K^+$ levels in the cytoplasm high enough for biochemical reaction systems.

An outstanding physiological factor is the existence of hormonal gradients in roots, because growth and development of roots is regulated by phytohormones transported from the shoots into the roots and also in the opposite direction out of the tips towards more distant root zones. Indoleacetic acid (IAA) is transported from the tops of plants downwards. Other hormones may be transported in both directions although their synthesis may be restricted to either roots or shoots. Abscisic acid (ABA) is largely formed in shoots but can also be synthesized in roots (WALTON et al. 1976), and transport upwards and downwards can occur both in the phloem and xylem of plants (BELLANDI and DÖRFFLING 1974, ZEEVAART 1977). Cytokinins, gibberellic acid, and unknown inhibitors are exported from the root tips (ATKIN et al. 1973). Cytokinin synthesis is restricted to the meristematic initials in the very tip of roots (VAN STADEN and DAVEY 1979). Cytokinins are transported, however, both from roots to shoots and from shoots to roots. Transport from the shoot is probably based on cytokinins originally synthesized in the root tips but stored in the form of glycosides in the leaves. Kinetins can be translocated in the xylem and phloem, and retranslocated in the xylem and phloem, and retranslocation and cycling in the whole plant is important (VONK 1979, VAN STADEN and DAVEY 1979). The ethylene precursor 1-aminocyclopropane-1-carboxylic acid is synthesized in roots and translocated in the xylem to the shoot where conversion to ethylene takes place (BRADFORD and YANG 1980). Thus all classes of hormones (auxins, gibberellins, abscisic acid, cytokinins, and ethylene) can be transported in the xylem (BRADFORD and YANG 1980, KING 1976). Phytohormones and their effects on transport processes in plants are discussed in Volume 2.B.7, this Series, and so this can be only touched here. All the phytohormones mentioned above are known to be involved in control and regulation of membrane-electric phenomena and of transport in plant cells and tissues (VAN STEVENINCK 1976). It seems to be noteworthy though, that naturally occurring hormonal gradients have hardly been considered yet as a physiological basis of differences of bioelectric and transport features along the length of roots.

### 3.2 Bioelectrical Fields Along Roots

B.I.H. SCOTT and coworkers (review: SCOTT 1967) have described bioelectrical fields along growing roots, which have recently been reinvestigated by WEISENSEEL et al. (1979). Figure 10 shows that a current is directed inwards in the meristem and elongation zones. Current also enters at the tips of root hairs. Current leaves in the root hair zone across the epidermal surface beneath the root hairs. The root cap has either inward or outward current. Experiments with roots imbedded in bromocresol purple-stained agar reveal a colour-banding phenomenon corresponding to the current distribution (Fig. 10), which suggests that much of the current consists of hydrogen ions. Hydrogen ions leak into growing cells or cell parts and are pumped out of non-growing ones. The biological function of these currents may be related to growth (WEISENSEEL et al. 1979), but they also determine patterns of ion uptake along the length of roots (SCOTT 1967).

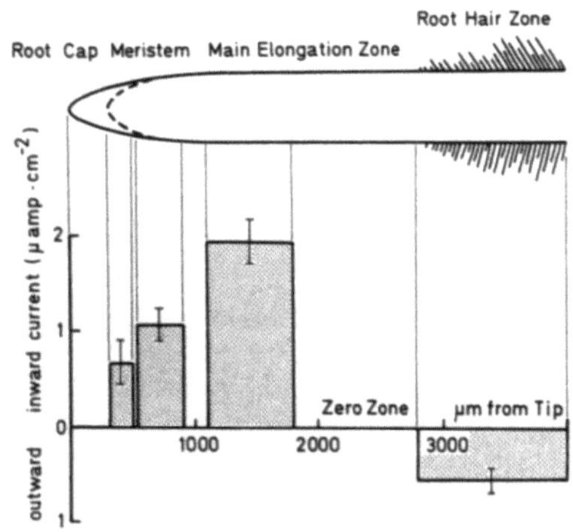


**Fig. 10.** Pattern and density of current traversing growing barley roots (growth rate $5.1 \pm 0.7$ μm $min^{-1}$) in solutions of 1 mM NaCl, 0.1 mM KCl and 0.1 mM $CaCl_2$ or in 0.5 mM $CaCl_2$, pH 5.2–5.7. *Width of the bars* represents the spread of the measuring positions; "zero zone" refers to current almost entirely flowing parallel to the surface. Errors are SEM. (WEISENSEEL et al. 1979)

### 3.3 Differences in Ion Transport Mechanisms Along Roots

Assimilate supply from the shoot to various root zones can vary and is temperature-dependent. This may be one reason for variations in ion transport along the roots (ROVIRA and BOWEN 1973). ESHEL and WAISEL (1972, 1973) observed variations in the uptake of sodium along roots of corn and barley seedlings, not only in terms of transport rates but also with regard to qualitative differences in the responses to temperature and ion concentration. Electrical measurements also show that different parts of roots have different physiological transport properties (HELMY et al. 1973).

Differences in Fe uptake capacity along roots have already been mentioned in relation to epidermal transfer cells formed in apical root zones in response to Fe stress (Sect. 2.2.1.2). In maize Fe translocation along the length of roots is rather uniform (KASHIRAD et al. 1973); in contrast, the rates of translocation in barley are very much higher in a zone 1–4 cm from the root tip than elsewhere in the root, and the differences seem to be related to metabolic factors and not to root anatomy (CLARKSON and SANDERSON 1978). Since Fe can only be absorbed in the divalent ferrous state, the release of reductants or in situ reduction could be involved and might be a phenomenon preferentially located in the younger root tissues (BROWN 1978).

In different zones of soybean roots, TRAVIS et al. (1979) have found different activities of the plasmalemma-associated $K^+$-stimulated ATPase, which is thought to be an integral part of ion uptake systems of higher plant roots (HODGES 1976). ATPase activity was low in the region of the root cap, increased to a maximum in the meristematic region, decreased to a minimum as cell elongation proceeded, and then increased as lateral root development began. This ATPase could be an $H^+$ extrusion pump exchanging $H^+$ for $K^+$ and driving an $H^+$-$Na^+$ antiport (COLOMBO et al. 1979, Sect. 2.2.4.3). However, maximum ATPase activity in soybean roots (TRAVIS et al. 1979) does not occur in zones similar to those of barley roots where a strong active $H^+$ extrusion was observed (WEISENSEEL et al. 1979, Fig. 10). Almost the inverse relation seems to apply. It would be very interesting to perform both kinds of study with the same material and under comparable conditions.

# 4 Root–Shoot Interactions and Circulation in the Whole Plant

## 4.1 Some Examples Illustrating General Aspects of Circulation

Transport from root to shoot is often investigated by analysis of exudation from the cut ends of isolated roots that are held either in air, with a capillary attached, or immersed in solution in a separate compartment of a transport chamber (e.g. LÄUCHLI et al. 1978). It can also be studied by analysis of guttation fluid, which is driven out of passive hydathodes by root pressure (e.g. DIEFFENBACH et al. 1980a, b), or by shoot analysis when the solutes involved can only arrive in the shoot via the root. One-way considerations of root–shoot interactions focus on supply of water and inorganic ions from root to shoot in the xylem and of assimilates from shoot to root in the phloem. In addition the transport of phytohormones (see Sect. 3.1) and of yet unknown factors is involved. Root factors can regulate photosynthetic activity in leaves (CARMI and KOLLER 1979). A shoot factor is needed for the development of nitrite uptake machinery in roots, in which protein synthesis is involved (JACKSON et al. 1974).

The shoot affects transport out of the root directly via passive driving forces and indirectly via metabolism-dependent driving forces. In addition to root pressure built up by ion pumping in the root, transpiration is the driving force for ascent of water and solutes in the plant. Only in smaller plants is root pressure by itself able to drive the ascent of sap, when the water potential gradient between the plant and the atmosphere is close to zero. Normally transpiration is the predominant driving force. This passive force is regulated by active stomatal control in the shoots. Somewhat less directly, the rate of transpiration may also affect ion transport across the root and xylem loading (ANDERSON 1976). It has been clearly shown that the supply of metabolic substrates from shoot to root is essential for the functions of root cells in ion uptake and translocation, including the maintenance of membrane potentials of root cells (PITMAN 1972, HATRICK and BOWLING 1973, GRAHAM and BOWLING 1977). When relative growth rate of barley seedlings is reduced by limited daily photoperiod (2 h illumination in 24 h) the metabolism-dependent ion transport machineries in the roots are much more affected than in the leaves (PITMAN 1972, PITMAN et al. 1974a).

An example of hormonal interaction has been given by selective wilting of shoots of barley seedlings, whose roots were sealed in a moist chamber that prevented water flow to the shoots but kept the roots turgid. When brief wilting of the shoots was reversed by submerging the roots again in solution, the physiological functions in the shoots recovered within a very short period but $^{86}Rb^{+}$ export from the roots was inhibited for up to 3 h after the onset of recovery (PITMAN et al. 1974b). It seems that some factor regulating ion export by the roots, presumably abscisic acid formed in the wilting leaves, has been transported from shoot to root. After water stress, abscisic acid is readily distributed throughout plants via phloem and xylem (HOAD 1978).

This already implies that the one-way considerations are over-simplifications. Many solutes are mobile both in the xylem and phloem. They can be retranslocated in the phloem and in the xylem, respectively, arriving in the shoot via the xylem and in the root via the phloem. Furthermore, the phloem supplies sinks not only downwards from the mature exporting leaves but also upwards in the growing parts of shoots. Shoot–root competition for nutrients, carbon skeletons and metabolites can be excellently documented by the study of utilization of seed reserves in germinating seedlings during the first week of development (SUTCLIFFE 1976).

In *Trifolium alexandrinum*, which is a pasture plant used in arid regions on water- and salt-stressed soils, a retranslocation of $Na^+$ seems to control $Na^+$ levels in the leaves. This plant can withstand NaCl-salinity up to 50 mM. $Na^+$ supplied from the roots in the transpiration stream is reabsorbed by special transfer cells in the phloem of the petioles of the trifoliate leaves and retranslocated to the roots and into the medium (WINTER 1982a, b; WINTER and LÄUCHLI 1982; WINTER and PRESTON 1982).

Interesting limitations to this are set by the immobility of some important solutes in the phloem. It has long been known that $Ca^{2+}$ is rather immobile (see ZIEGLER 1975). The pH of the phloem sap is always slightly alkaline (pH 7–8,5; ZIEGLER 1975). RAVEN (1977) has drawn attention to this in relation to the cytoplasmic nature of sieve tube contents which never allows large deviation from this pH value. Thus the export of surplus $H^+$ or $OH^-$ equivalents, generated during metabolism in the leaves, is not possible in the phloem. RAVEN has pointed out that in evolution this may have contributed to specialization and division of labour between plant organs.

## 4.2 Nitrogen, Sulphur and Phosphorus

At low $NO_3^-$ concentrations in the root medium, $NO_3^-$ is largely reduced in the roots, but at higher concentrations (0.8–8 mM) it needs to be transported to the shoots for reduction (WALLACE and PATE 1967, KIRKBY and KNIGHT 1977, ROBIN et al. 1979). DIJKSHOORN (1970) and BEN ZIONI et al. (1971) developed a cycling model for the cooperation of roots and shoots in this mechanism: $KNO_3$ is transported in the xylem to the shoots, where $NO_3^-$ is reduced to $NH_3$ and $OH^-$; $OH^-$ is neutralized by malic acid synthesis by $CO_2$ fixation via phosphoenolpyruvate carboxylase; $K^+$-malate is transported in the phloem to the roots, where malate is decarboxylated; the resulting $HCO_3^-$ then exchanges for the $NO_3^-$ taken up from the root medium and the resulting pyruvate is used as a substrate for root metabolism. For various reasons this model has been subject to criticism. KIRKBY and ARMSTRONG (1980) suggest that it does not operate in plants like tomato, sugar cane and tobacco, where the ratio of cation and anion uptake by roots is close to unity and little $HCO_3^-$ is released, but that it seems to work in castor bean.

Notwithstanding the criticisms, the DIJKSHOORN-BEN ZIONI hypothesis seems to continue to stimulate research. Sulphur is a case similar to that of nitrogen. Uptake at the root level is in the oxidized form as sulphate, but $SO_4^{2-}$ reduction in photoautotrophic plants is light-enhanced and largely confined to leaves (KYLIN 1960, SCHWENN and TREBST 1976, RENNENBERG et al. 1979). As in $NO_3^-$ reduction, $OH^-$ is formed in $SO_4^{2-}$ reduction. Reduced sulphur is exported from the leaves of tobacco plants in the phloem in the form of glutathione (67%–70%), methionine (27%–30%) and cysteine (2%–8%) (RENNENBERG et al. 1979).

The third major nutrient, phosphorus, which is available as an inorganic anion, is different, because it is used in metabolism in the same oxidized form, i.e. as phosphate, in which it is taken up by the root. Shoot–root interactions are also important, however, and phosphate-deficiency experiments suggest that phosphate transport in the root is under control of the shoot (CLARKSON et al. 1978c).

## 5 Conclusion

Among the higher plants there is a large morphological variety of roots, depending on special functions like anchoring, water and nutrient uptake, formation of symbiotic systems, storage, etc. Roots of a given plant species are, however, also extraordinarily complex with their three-dimensional pattern of structural and physiological differentiations. Variations in time, e.g. due to changing environmental conditions of nutrient and micronutrient supply, salinity and water stress, add to the complexity of responses. But only in such a way can roots be versatile enough to satisfy the many nutritional requirements of plants, as excellently and concisely reviewed by CLARKSON and HANSON (1980). Physiological activities of roots need more attention as possible bases for ecological adaptations.

*Acknowledgement.* I thank Dr. DETLEF KRAMER for reading the manuscript, critical discussions and for his help in evaluating some of the literature.

## References

Anderson WP (1976) Transport through roots. In: Lüttge U, Pitman MG (eds) Transport in plants II, part B, tissues and organs. Encyclopedia of plant physiology New Ser, vol II. Springer, Berlin Heidelberg New York

Anderson WP, Higinbotham N (1975) A cautionary note on plant root electrophysiology. J Exp Bot 26:533–536

Anderson WP, House CR (1967) A correlation between structure and function in the root of *Zea mays*. J Exp Bot 18:544–555

Anderson WP, Aikman DP, Meiri A (1970) Excised root exudation: a standing gradient osmotic flow. Proc R Soc London Ser B 174:445–458

Anderson WP, Robertson RN, Wright BJ (1977) Membrane potentials in carrot root cells. Aust J Plant Physiol 4:241–252

Arisz WH (1956) Significance of the symplasm theory for transport in the root. Protoplasma 46:5–62
Atkin RK, Burton GE Robinson DK (1973) Effect of root-growing temperature on growth substances in xylem exudate of *Zea mays*. J Exp Bot 24:475–487
Bange GG (1973) Diffusion and absorption of ions in plant tissue. III. The role of the root cortex cells in ion absorption. Acta Bot Neerl 22:529–542
Bellandi DM, Dörffling K (1974) Transport of abscisic acid-2-C-14 in intact pea seedlings. Physiol Plant 32:365–368
Ben Zioni A, Vaadia Y, Lips SH (1971) Nitrate uptake by roots as regulated by nitrate reduction products of the shoot. Physiol Plant 24:288–290
Bowen GD (1969) The uptake of orthophosphate and its incorporation into organic phosphates along roots of *Pinus radiata*. Aust J Biol Sci 22:1125–1135
Bowen GD (1970) Effects of soil temperature on root growth and on phosphate uptake along *Pinus radiata* roots. Aust J Soil Res 8:31–42
Bowling DJF (1976) Ionic gradients in higher plant tissues. Persp Exp Biol 2:391–399
Bowling DJF, Ansari AQ (1972) Control of sodium transport in sunflower roots. J Exp Bot 23:241–246
Bradford KJ, Yang SF (1980) Xylem transport of 1-aminocyclopropane-1-carboxylic acid, an ethylene precursor, in waterlogged tomato plants. Plant Physiol 65:322–326
Brown JC (1978) Mechanism of iron uptake by plants. Plant Cell Environ 1:249–257
Burley JWA, Nwoke FIO, Leister GL, Popham RA (1970) The relationship of xylem maturation to the absorption and translocation of $^{32}P$. Am J Bot 57:504–511
Butz RG, Long RC (1979) L-Malate as an essential component of the xylem fluid of corn seedling roots. Plant Physiol 64:684–689
Capesius I, Barthlott W (1975) Isotopen-Markierungen und rasterelektronenmikroskopische Untersuchungen des Velamen radicum der Orchideen. Z Pflanzenphysiol 75:436–448
Carmi A, Koller D (1979) Regulation of photosynthetic activity in the primary leaves of bean (*Phaseolus vulgaris* L.) by materials moving in the water-conducting system. Plant Physiol 64:285–288
Clarkson DT (1974) Ion transport and cell structure in plants. McGraw-Hill, London
Clarkson DT, Hanson JB (1980) The mineral nutrition of higher plants. Annu Rev Plant Physiol 31:239–298
Clarkson DT, Sanderson J (1969) The uptake of a polyvalent cation and its distribution in the root apices of *Allium cepa:* tracer and autoradiographic studies. Planta 89:136–154
Clarkson DT, Sanderson J (1978) Sites of absorption and translocation of iron in barley roots. Tracer and microautoradiographic studies. Plant Physiol 61:731–736
Clarkson DT, Sanderson J, Scott-Russell R (1968) Ion uptake and root age. Nature 220:805–806
Clarkson DT, Robards AW, Sanderson J (1971) The tertiary endodermis in barley roots: fine structure in relation to radial transport of ions and water. Planta 96:292–305
Clarkson DT, Sanderson J, Robards AW (1978a) Permeability studies on the epidermal/hypodermal layers of the sand sedge (*Carex arenaria* L.). Agric Res Counc Letcombe Lab, Annu Rep 1977:25
Clarkson DT, Robards AW, Sanderson J, Peterson CA (1978b) Permeability studies on epidermal-hypodermal sleeves isolated from roots of *Allium cepa* (onion). Can J Bot 56:1526–1532
Clarkson DT, Sanderson J, Scattergood CB (1978c) Influence of phosphate-stress on phosphate absorption and translocation by various parts of the root system of *Hordeum vulgare* L. (barley). Planta 139:47–53
Colombo R, Bonetti A, Lado P (1979) Promoting effect of fusicoccin on $Na^+$ efflux in barley roots: evidence for a $Na^+$-$H^+$ antiport. Plant Cell Environ 2:281–285
Crafts AS, Broyer TC (1938) Migration of salts and water into xylem of the roots of higher plants. Am J Bot 24:415–431
Danilova MF, Stamboltsyan EY (1975) The ultrastructure of differentiating primary xylem cells of the root in relation to hypothesis of transfer of solutes into xylem elements. Bot Zh 60:913–926

Davis RF, Higinbotham N (1976) Electrochemical gradients and $K^+$ and $Cl^-$ fluxes in excised corn roots. Plant Physiol 57:129–136
Dieffenbach H, Kramer D, Lüttge U (1980a) Release of guttation fluid from passive hydathodes of intact barley plants. I. Structural and cytological aspects. Ann Bot 45:397–401
Dieffenbach H, Lüttge U, Pitman MG (1980b) Release of guttation fluid from passive hydathodes of intact barley plants. II. The effects of abscisic acid and cytokinins. Ann Bot 45:703–712
Dijkshoorn W (1970) Partition of ionic constituents between organs. Proc 6th Int Coll Plant Anal Fert Probl, Tel Aviv
Dumbroff EB, Peirson DR (1971) Probable sites for passive movement of ions across the endodermis. Can J Bot 49:35–38
Dunlop J, Bowling DJF (1971a) The movement of ions to the xylem exudate of maize roots. I. Profiles of membrane potential and vacuolar potassium activity across the root. J Exp Bot 22:434–444
Dunlop J, Bowling DJF (1971b) The movement of ions to the xylem exudate of maize roots. II. A comparison of the electrical potential and electrochemical potential of ions in the exudate and in the root cells. J Exp Bot 22:445–452
Dunlop J, Bowling DJF (1971c) The movement of ions to the xylem exudate of maize roots. III. The location of the electrical and electrochemical potential differences between the exudate and the medium. J Exp Bot 22:453–464
Epstein E (1976) Kinetics of ion transport and the carrier concept. In: Lüttge U, Pitman MG (eds) Transport in plants II, part B, tissues and organs. Encyclopedia of plant physiology New Ser, vol II. Springer, Berlin Heidelberg New York
Eshel A, Waisel Y (1972) Variations in sodium uptake along primary roots of corn seedlings. Plant Physiol 49:585–589
Eshel A, Waisel Y (1973) Variations in uptake of sodium and rubidium along barley roots. Physiol Plant 28:557–560
Espelie KE, Kolattukudy PE (1979) Composition of the aliphatic components of suberin of the endodermal fraction from the first internode of etiolated *Sorghum* seedlings. Plant Physiol 63:433–435
Evans EC, Vaughan BE (1966) Wounding response in relation to polar transport of radiocalcium in isolated root segments of *Zea mays*. Plant Physiol 41:1145–1151
Ferguson IB (1979) The movement of calcium in non-vascular tissue of plants. Commun Soil Sci Plant Anal 10:217–224
Ferguson IB, Clarkson DT (1975) Ion transport and endodermal suberization in the roots of *Zea mays*. New Phytol 75:69–79
Ferguson IB, Clarkson DT (1976a) Simultaneous uptake and translocation of magnesium and calcium in barley (*Hordeum vulgare* L.) roots. Planta 128:267–269
Ferguson IB, Clarkson DT (1976b) Ion uptake in relation to the development of a root hypodermis. New Phytol 77:11–14
Glass ADM, Perley JE (1979) Cytoplasmic streaming in the root cortex and its role in the delivery of potassium to the shoot. Planta 145:399–401
Glinka Z (1977) Effects of abscisic acid and of hydrostatic pressure gradient on water movement through excised sunflower roots. Plant Physiol 59:933–935
Graham RD, Bowling DJF (1977) Effect of the shoot on the transmembrane potentials of root cortical cells of sunflower. J Exp Bot 28:886–893
Green JR, Northcote DH (1979) Location of fucosyl transferases in the membrane system of maize root cells. J Cell Sci 40:235–244
Grunwaldt G, Ehwald R, Pietzsch W, Göring H (1979) A special role of the rhizodermis in nutrient uptake by plant roots. Biochem Physiol Pflanzen 174:831–837
Gunning BES (1977) Transfer cells and their roles in transport of solutes in plants. Sci Prog Oxford 64:539–568
Haas DL, Carothers ZB, Robbins RR (1976) Observations on the phi-thickenings and casparian strips in *Pelargonium* roots. Am J Bot 63:863–867
Hanson JB (1978) Application of the chemiosmotic hypothesis to ion transport across the root. Plant Physiol 62:402–405

Harrison-Murray RS, Clarkson DT (1973) Relationship between structural development and the absorption of ions by the root system of *Cucurbita pepo*. Planta 114:1–16
Hatrick AA, Bowling DJF (1973) A study of the relationship between root and shoot metabolism. J Exp Bot 24:607–613
Helmy AK, Peinemann N, Ferreiro EA (1973) Effect of root position on measured root membrane potentials. J Exp Bot 24:29–32
Higinbotham N, Davis RF, Mertz SM, Shumway LK (1973) Some evidence that radial transport in maize roots is into living vessels. In: Anderson WP (ed) Ion transport in plants. Academic Press, London New York
Hoad GV (1978) Effect of water stress on abscisic acid levels in white lupin (*Lupinus albus* L.) fruit, leaves and phloem exudate. Planta 142:287–290
Hodges TK (1976) ATPases associated with membranes of plant cells. In: Lüttge U, Pitman MG (eds) Transport in plants II, part A, Cells. Encyclopedia of plant physiology, New Ser, vol II. Springer, Berlin Heidelberg New York
Hylmö B (1953) Transpiration and ion absorption. Physiol Plant 6:333–405
Iren van F, Sluijs van der BP (1980) Symplasmic and apoplasmic radial ion transport in plant roots. Cortical plasmalemmas lose absorption capacity during differentiation. Planta 148:130–137
Iren van F, Spiegel van der A (1975) Subcellular localization of inorganic ions in plant cells by in vivo precipitation. Science 187:1210–1211
Jackson WA, Johnson RE, Volk RJ (1974) Nitrite uptake by nitrogen-depleted wheat seedlings. Physiol Plant 32:37–42
Jacoby B (1964) Function of bean roots and stems in sodium retention. Plant Physiol 39:445–449
Jacoby B (1965) Sodium retention in excised bean stems. Physiol Plant 18:730–739
Jacoby B, Ratner A (1974) Mechanism of sodium exclusion in bean and corn plants – a reevaluation. In: Wehrmann J (ed) Plant analysis and fertiliser problems. Ger Soc Plant Nutr, Hannover
Jeschke WD (1970) Evidence for a $K^+$-stimulated $Na^+$ efflux at the plasmalemma of barley root cells. Planta 94:240–245
Jeschke WD (1972) Wirkung von $K^+$ auf die Fluxe und den Transport von $Na^+$ in Gerstenwurzeln, $K^+$-stimulierter $Na^+$-Efflux in der Wurzelrinde. Planta 106:73–90
Jeschke WD (1977a) $K^+$-$Na^+$ exchange and selectivity in barley root cells, effects of $K^+$, $Rb^+$, $Cs^+$, and $Li^+$ on the $Na^+$ fluxes. Z Pflanzenphysiol 84:247–264
Jeschke WD (1977b) $K^+$-$Na^+$ exchange and selectivity in barley root cells: effect of $Na^+$ on the $Na^+$ fluxes. J Exp Bot 28:1289–1305
Jeschke WD (1979) New $K^+$-$Na^+$ exchange across the plasmalemma of meristematic root tissues. Z Pflanzenphysiol 94:325–330
Jeschke WD, Stelter W (1973) $K^+$-dependent net $Na^+$ efflux in roots of barley plants. Planta 114:251–258
Jeschke WD, Stelter W (1976) Measurement of longitudinal ion profiles in single roots of *Hordeum* and *Atriplex* by use of flameless atomic absorption spectroscopy. Planta 128:107–112
Kashirad A, Marschner H, Richter C (1973) Absorption and translocation of $^{59}Fe$ from various parts of the corn root. Z Pflanzerernähr Bodenk 134:136–147
King RW (1976) Implications for plant growth of the transport of regulatory compounds in phloem and xylem. In: Wardlaw JF, Passioura JB (eds) Transport and transfer processes in plants. Academic Press, London New York
Kirkby EA, Armstrong MJ (1980) Nitrate uptake by roots as regulated by nitrate assimilation in the shoot of castor oil plants. Plant Physiol 65:286–290
Kirkby EA, Knight AH (1977) Influence of the level of nitrate nutrition on ion uptake and assimilation, organic acid accumulation and cation–anion balance in whole tomato plants. Plant Physiol 60:349–353
Kramer D, Läuchli A, Yeo AR, Gullasch J (1977) Transfer cells in roots of *Phaseolus coccineus*: ultrastructure and possible function in exclusion of sodium from the shoot. Ann Bot 41:1031–1040
Kramer D, Anderson WP, Preston J (1978) Transfer cells in the root epidermis of *Atriplex*

*hastata* L. as a response to salinity: a comparative cytological and X-ray microprobe investigation. Aust J Plant Physiol 5:739–747

Kramer D, Römheld V, Landsberg E, Marschner H (1980) Induction of transfer-cell formation by iron deficiency in the root epidermis of *Helianthus annuus* L. Planta 147:335–339

Kylin A (1960) The incorporation of radio-sulphur from external sulphate into different sulphur fractions of isolated leaves. Physiol Plant 13:366–379

Laties GG (1967) Metabolic and physiological development in plant tissues. Aust J Sci 30:193–203

Laties GG (1969) Dual mechanisms of salt uptake in relation to compartmentation and long-distance transport. Annu Rev Plant Physiol 20:89–116

Läuchli A (1967) Untersuchungen über Verteilung und Transport von Ionen in Pflanzengeweben mit der Röntgen-Mikrosonde. I. Versuche an vegetativen Organen von *Zea mays*. Planta 75:185–206

Läuchli A (1972) Translocation of inorganic solutes. Annu Rev Plant Physiol 23:197–218

Läuchli A (1976a) Apoplasmic transport in tissues. In: Lüttge U, Pitman MG (eds) Transport in plants II, part B. Tissues and organs. Encyclopedia of plant physiology New Ser, vol II. Springer, Berlin Heidelberg New York

Läuchli A (1976b) Symplasmic transport and ion release to the xylem. In: Wardlaw IF, Passioura JB (eds) Transport and transfer processes in plants. Academic Press, London New York

Läuchli A (1979) Regulation des Salztransportes und Salzausschließung in Glykophyten und Halophyten. Ber Dtsch Bot Ges 92:87–94

Läuchli A, Epstein E (1971) Lateral transport of ions into the xylem of corn roots. I. Kinetics and energetics. Plant Physiol 48:111–117

Läuchli A, Wieneke J (1979) Studies on growth and distribution of $Na^+$, $K^+$ and $Cl^-$ in soybean varieties differing in salt tolerance. Z Pflanzenernähr Bodenkd 142:3–13

Läuchli A, Spurr AR, Epstein E (1971) Lateral transport of ions into the xylem of corn roots. II. Evaluation of a stelar pump. Plant Physiol 48:118–124

Läuchli A, Kramer D, Stelzer R (1974a) Ultrastructure and ion localization in xylem parenchyma cells of roots. In: Zimmermann U, Dainty J (eds) Membrane transport in plants. Springer, Berlin Heidelberg New York

Läuchli A, Kramer D, Pitman MG, Lüttge U (1974b) Ultrastructure of xylem parenchyma cells of barley roots in relation to ion transport to the xylem. Planta 119:85–99

Läuchli A, Pitman MG, Lüttge U, Kramer D, Ball E (1978) Are developing xylem vessels the site of ion exudation from root to shoot? Plant Cell Environ 1:217–223

Läuchli A, Yeo T, Kramer D (1976) Distribution of soluble inorganic ions in maize roots; microprobe analysis of deep frozen specimens. 6th Eur Congr Electron Microsc, Jerusalem

Letvenuk LJ, Peterson RL (1976) Occurrence of transfer cells in vacuolar parenchyma of *Hieracium florentinum* roots. Can J Bot 54:1458–1471

Lundegårdh H (1950) The translocation of salts and water through wheat roots. Physiol Plant 3:103–151

Lüttge U (1964) Mikroautoradiographischer Nachweis der Aufnahme von $^{35}SO_4^{--}$ in die Wurzelhaare von Ahornkeimlingen. Naturwissenschaften 12:296–297

Lüttge U (1969) Aktiver Transport (Kurzstreckentransport bei Pflanzen). Protoplasmatologia VIII 7b; 1–146

Lüttge U (1973) Stofftransport der Pflanzen. Springer, Berlin Heidelberg New York

Lüttge U, Higinbotham N (1979) Transport in plants. Springer, Berlin Heidelberg New York

Lüttge U, Weigl J (1962) Mikroautoradiographische Untersuchungen der Aufnahme und des Transportes von $^{35}SO_4^{--}$ und $^{45}Ca^{++}$ in Keimwurzeln von *Zea mays* L. und *Pisum sativum* L. Planta 58:113–116

Lüttge U, Weigl J (1964) Der Ionentransport in intakten und entrindeten Wurzeln. Ber Dtsch Bot Ges 77:63–70

MacKenzie KAD (1979) The development of the endodermis and phi layer of apple roots. Protoplasma 100:21–32

Macklon AES (1975a) Cortical cell fluxes and transport to the stele in excised root segments of *Allium cepa* L. I. Potassium, sodium and chloride. Planta 122:109–113
Macklon AES (1975b) Cortical cell fluxes and transport to the stele in excised root segments of *Allium cepa* L. II. Calcium. Planta 122:131–141
Malone CP, Barke JJ, Hanson JB (1977) Histochemical evidence for the occurrence of oligomycin-sensitive ATPase in corn roots. Plant Physiol 60:916–922
Marrè E (1979) Fusicoccin: a tool in plant physiology. Ann Rev Plant Physiol 30:273–288
Marschner H, Richter C (1973) Akkumulation und Translokation von $K^+$, $Na^+$ und $Ca^{2+}$ bei Angebot zu einzelnen Wurzelzonen von Maiskeimpflanzen. Z Pflanzenernaehr Bodenkd 135:1–15
Marschner H, Kalisch A, Römheld V (1974) Mechanism of iron uptake in different plant species. Proc 7th Int Colloq Plant Anal Fert Probl, Hannover 2:274–281
Newcomb W, Peterson RL (1979) The occurrence and ontogeny of transfer cells associated with lateral roots and root nodules in Leguminosae. Can J Bot 57:2583–2602
Okamoto H, Ichino K, Katon K (1978) Radial electrogenic activity in the stem of *Vigna sesquipedalis:* involvement of spatially separate pumps. Plant Cell Environ 1:279–284
Okamoto H, Katon K, Ichino K (1979) Distribution of electric potential and ion transport in the hypocotyl of *Vigna sesquipedalis*. VI. The dual structure of radial electrogenic activity. Plant Cell Physiol 20:103–114
Pate JS (1976) Transport in symbiotic systems fixing nitrogen. In: Lüttge U, Pitman MG (eds) Transport in plants II, part B. Tissues and organs. Encyclopedia of plant physiology New Ser, vol II. Springer, Berlin Heidelberg New York
Peterson CA, Emanuel ME, Weerdenburg CA (1981) The permeability of phi thickenings in apple (*Pyrus malus*) and geranium (*Pelargonium hortorum*) roots to an apoplastic fluorescent dye tracer. Can J Bot 59:1107–1110
Pitman MG (1965) Sodium and potassium uptake by seedlings of *Hordeum vulgare*. Aust J Biol Sci 18:10–24
Pitman MG (1972) Uptake and transport of ions in barley seedlings. III. Correlation between transport to the shoot and relative growth rate. Aust J Biol Sci 25:905–919
Pitman MG (1976) Ion uptake by plant roots. In: Lüttge U, Pitman MG (eds) Transport in plants II, part B, Tissues and organs. Encyclopedia of plant physiology New Ser, vol II. Springer, Berlin Heidelberg New York
Pitman MG (1977) Ion transport into the xylem. Annu Rev Plant Physiol 28:71–88
Pitman MG, Lüttge U, Läuchli A, Ball E (1974a) Ion uptake to slices of barley leaves, and regulation of K content in cells of the leaves. Z Pflanzenphysiol 72:75–88
Pitman MG, Lüttge U, Läuchli A, Ball E (1974b) Effect of previous water stress on ion uptake and transport in barley seedlings. Aust J Plant Physiol 1:377–385
Pitman MG, Schaefer N, Wildes RA (1975) Relation between permeability to potassium and sodium ions and fusicoccin-stimulated hydrogen-ion efflux in barley roots. Planta 126:61–73
Pitman MG, Anderson WP, Lüttge U (1976) Transport processes in roots. In: Lüttge U, Pitman MG (eds) Transport in plants II, part B. Tissues and organs. Encyclopedia of plant physiology New Ser, vol II. Springer, Berlin Heidelberg New York
Plaut M (1910) Untersuchungen zur Kenntnis der physiologischen Scheiden bei Gymnospermen, Equiseten und Bryophyten. Jahrb Wiss Bot XLVII:121–185
Raven JA (1977) $H^+$ and $Ca^{2+}$ in phloem and symplast. Relation of relative immobility of the ions to the cytoplasmic nature of the transport paths. New Phytol 75:465–480
Rennenberg H, Schmitz K, Bergmann L (1979) Long-distance transport of sulfur in *Nicotiana tabacum*. Planta 147:57–62
Richter Ch, Marschner H (1973) Umtausch von Kalium in verschiedenen Wurzelzonen von Maiskeimpflanzen. Z Pflanzenphysiol 70:211–221
Richter Ch, Marschner H (1974) Verteilung von $K^+$, $Na^+$ und $Ca^{++}$ zwischen Wurzelrinde und Zentralzylinder. Z Pflanzenphysiol 71:95–100
Robards AW, Clarkson DT (1976) The role of plasmodesmata in the transport of water and nutrients across roots. In: Gunning BES, Robards AW (eds) Intracellular communication in plants: studies on plasmodesmata. Springer, Berlin Heidelberg New York
Robards AW, Newman TM, Clarkson DT (1980) The distinctive nature of the plasma-

membrane of the endodermis in roots as shown by freeze-fracture electron microscopy. Letcombe Lab Annu Rep 1979, Agric Res Counc, Wantage
Robin P, Blayac D, Salsac L (1979) Influence de l'alimentation nitrique sur la teneur en nitrate et l'activité nitrate réductase des racines et des feuilles de plantules de mais. Physiol Vég 17:55–56
Rovira AD, Bowen GD (1970) Translocation and loss of phosphate along roots of wheat seedlings. Planta 93:15–25
Rovira AD, Bowen GD (1973) The influence of root temperature on $^{14}C$ assimilate profiles in wheat roots. Planta 114:101–107
Russell RS, Sanderson J (1967) Nutrient uptake by different parts of the intact roots of plants. J Exp Bot 18:491–508
Schaefer N, Wildes RA, Pitman MG (1975) Inhibition by p-fluorophenylalanine of protein synthesis and of ion transport across the roots in barley seedlings. Aust J Plant Physiol 2:61–73
Schwenn JD, Trebst A (1976) Photosynthetic sulfate reduction by chloroplasts. In: Barber J (ed) The intact chloroplast. Elsevier, Amsterdam
Scott BIH (1967) Electric fields in plants. Annu Rev Plant Physiol 18:409–418
Scott FM (1965) The fine structure of xylem vessels. In: Baker KF, Snyder WC (eds) Ecology of soilborne pathogens. Univ Cal Press, Berkeley
Scott MG, Peterson RL (1979a) The root endodermis in *Ranunculus acris*. I. Structure and ontogeny. Can J Bot 57:1040–1062
Scott MG, Peterson RL (1979b) Histochemistry of the endodermis and the synthesis of phenolic compounds in roots. Can J Bot 57:1063–1077
Shone MGT, Clarkson DT, Sanderson J (1969) The absorption and translocation of sodium by maize seedlings. Planta 86:301–314
Singh C, Jacobson L (1977) Polar movement of ions in barley roots. Physiol Plant 39:37–78
Smith FA, Raven JA (1979) Intracellular pH and its regulation. Annu Rev Plant Physiol 30:289–311
Smith RC (1970) Time course of exudation from excised corn root segments of different stages of development. Plant Physiol 45:571–575
Staden van J, Davey JE (1979) The synthesis, transport and metabolism of endogenous cytokinins. Plant Cell Environ 2:93–106
Stelzer R, Läuchli A (1977) Salz- und Überflutungstoleranz von *Puccinellia peisonis*. II. Strukturelle Differenzierung der Wurzel in Beziehung zur Funktion. Z Pflanzenphysiol 84:95–108
Stelzer R, Läuchli A, Kramer D (1975) Intrazelluläre Transportwege in Wurzeln intakter Gerstenpflanzen. Cytobiologie 10:449–457
Steveninck van RFM (1976) Effect of hormones and related substances on ion transport. In: Lüttge U, Pitman MG, (eds) Transport in plants II, part B. Tissues and organs. Encyclopedia of plant physiology New Ser, vol II. Springer, Berlin Heidelberg New York
Strasburger E (1923) Das botanische Praktikum. Fischer, Jena
Sutcliffe JF (1962) Mineral absorption in plants. Pergamon, New York London
Sutcliffe JF (1976) Regulation in the whole plant. In: Lüttge U, Pitman MG (eds) Transport in plants II, part B. Tissues and organs. Encyclopedia of plant physiology New Ser, vol II. Springer, Berlin Heidelberg New York
Travis RL, Geng S, Berkowitz RL (1979) Analysis of the distribution of potassium-stimulated adenosine triphosphatase activity in soybean root. Plant Physiol 63:1187–1190
Troll W, Höhn K (1973) Allgemeine Botanik. Enke, Stuttgart
Ursprung A, Blum G (1921) Zur Kenntnis der Saugkraft. IV. Die Absorptionszone der Wurzel. Der Endodermissprung. Ber Dtsch Bot Ges 39:70–79
Vakhmistrov DB (1967) On the function of apparent free space in plant roots. A study of the absorbing power of epidermal and cortical cells in barley roots. Sov Plant Physiol 14:103–107

Vonk CR (1979) Origin of cytokinins transported in the phloem. Physiol Plant 46:235–240
Wallace W, Pate JS (1967) Nitrate assimilation in higher plants with special reference to the cocklebur (*Xanthium pennsylvanicum* Wallr.). Ann Bot 31:213–228
Walton DC, Harrison MA, Cote P (1976) The effects of water stress on abscisic-acid levels and metabolism in roots of *Phaseolus vulgaris* L. and other plants. Planta 131:141–144
Weigl J, Lüttge U (1962) Mikroautoradiographische Untersuchungen über die Aufnahme von $^{35}SO_4^{--}$ durch Wurzeln von *Zea mays* L. Die Funktion der primären Endodermis. Planta 59:15–28
Weigl J, Lüttge U (1965) Die Ionenaufnahme durch die Luftwurzeln von *Epidendrum*. Protoplasma 60:1–6
Weisenseel MH, Dorn A, Jaffe LF (1979) Natural $H^+$ currents traverse growing roots and root hairs of barley (*Hordeum vulgare* L.). Plant Physiol 64:512–518
Wiebe HH, Kramer PJ (1954) Translocation of radioactive isotopes from various regions of roots of barley seedlings. Plant Physiol 29:342–348
Wieneke J, Läuchli A (1979) Short-term studies on the uptake and transport of $Cl^-$ by soybean cultivars differing in salt tolerance. Z Pflanzenernaehr Bodenkd 142:799–814
Wieneke J, Läuchli A (1980) Effects of salt stress on distribution of $Na^+$ and some other cations in two soybean varieties differing in salt tolerance. Z Pflanzenernaehr Bodenkd 143:55–67
Winter E (1982a) Salt tolerance of *Trifolium alexandrinum* L. II. Ion balance in relation to its salt tolerance. Aust J Plant Physiol 9:227–237
Winter E (1982b) Salt tolerance of *Trifolium alexandrinum* L. III. Effects of salt on ultrastructure of phloem and xylem transfer cells in petioles and leaves. Aust J Plant Physiol 9:239–250
Winter E, Läuchli A (1982) Salt tolerance of *Trifolium alexandrinum* L. I. Comparison of the salt response of *T. alexandrinum* and *T. pratense*. Aust J Plant Physiol 9:221–226
Winter E, Preston J (1982) Salt tolerance of *Trifolium alexandrinum* L. IV. Ion measurements by X-ray microanalysis in unfixed, frozen hydrated leaf cells at various stages of salt treatment. Aust J Plant Physiol 9:251–259
Winter-Sluiter E, Läuchli A, Kramer D (1977) Cytochemical localization of $K^+$-stimulated adenosine triphosphatase activity in xylem parenchyma cells of barley roots. Plant Physiol 60:923–927
Wong WL, Ap Rees T (1971) Carbohydrate oxidation in the stele and cortex isolated from roots of *Pisum sativum*. Biochim Biophys Acta 252:296–304
Wyn Jones RG, Brady CJ, Speis J (1979) Ionic and osmotic relations in plant cells. In: Laidman DL, Wyn Jones RG (eds) Recent advances in the biochemistry of cereals. Academic Press, London New York
Yeo AR, Kramer D, Läuchli A, Gullasch J (1977a) Ion distribution in salt-stressed mature *Zea mays* roots in relation to ultrastructure and retention of sodium. J Exp Bot 28:17–29
Yeo AR, Läuchli A, Kramer D, Gullasch J (1977b) Ion measurements by x-ray microanalysis in unfixed, frozen, hydrated plant cells of species differing in salt tolerance. Planta 134:35–38
Zeevaart JAD (1977) Sites of abscisic acid synthesis and metabolism in *Ricinus communis* L. Plant Physiol 59:788–791
Ziegler H (1975) Nature of transported substances. In: Zimmermann MH, Milburn JA (eds) Transport in plants I, phloem transport. Encyclopedia of plant physiology New Ser vol I. Springer, Berlin Heidelberg New York
Ziegler H, Weigl J, Lüttge U (1963) Mikroautoradiographischer Nachweis der Wanderung von $^{35}SO_4^{--}$ durch die Tertiärendodermis der *Iris*-Wurzel. Protoplasma 56:362–370

# I.7 Cycling of Elements in the Biosphere

C.C. DELWICHE

## 1 The Sources of Plant Constituents

### 1.1 Soil and Atmospheric Sources

Of the 18 or more elements required by plants or beneficial to their development, only carbon has its immediate source in the atmosphere. Except in a few special cases such as lichens and some aerial plants, other required elements are obtained from the soil. One exception to this is the fixation of $N_2$ by some symbiotic associations (Chapt. II.2). Under most circumstances the availability of elements is limited, however, and so the death and decomposition of plant material is essential to the maintenance of the supply of necessary elements for new growth (Gosz et al. 1973).

The system is not a completely closed one, and there is a continual loss from the soil by volatilization, erosion and leaching. The replenishment of elements lost in this manner is required to maintain productivity. The atmosphere is the principal vehicle by which some of these elements are returned to the soil from the sea or other reservoirs where they have been concentrated. This process is not a complete one either. Some elements, such as phosphorus, which is present in seawater only at very low concentrations, must be provided by other means. Because phosphorus forms some compounds of exceedingly low solubility (explaining its low concentration in seawater), its mobility in soil is low and so the rate of its replenishment need not be as great as that of a more mobile element such as sulfur, even though the requirements of the two elements by plants are of the same order of magnitude (VAN WAZER 1961). Thus the weathering of soil materials although it takes place slowly under most circumstances suffices to resupply phosphorus as well as some other elements to the system.

### 1.2 The Weathering Process

Any effort to classify the mineral elements required by plants introduces some inconsistencies, but in considering the weathering process as a contributor to the nutrient requirements of plants it is helpful to distinguish between those elements which are volatile or form volatile compounds such as carbon, oxygen, hydrogen, nitrogen, sulfur and the halides (and which are of relatively low concentration in igneous rocks), as contrasted with the other elements which are igneous rock constituents. The sedimentary rocks that provide the parent material for many soils can have higher concentrations of the volatile elements.

**Fig. 1.** Relative quantities of various elements in plants, animals, igneous rocks and seawater on a dry weight basis. Note that the scale is logarithmic

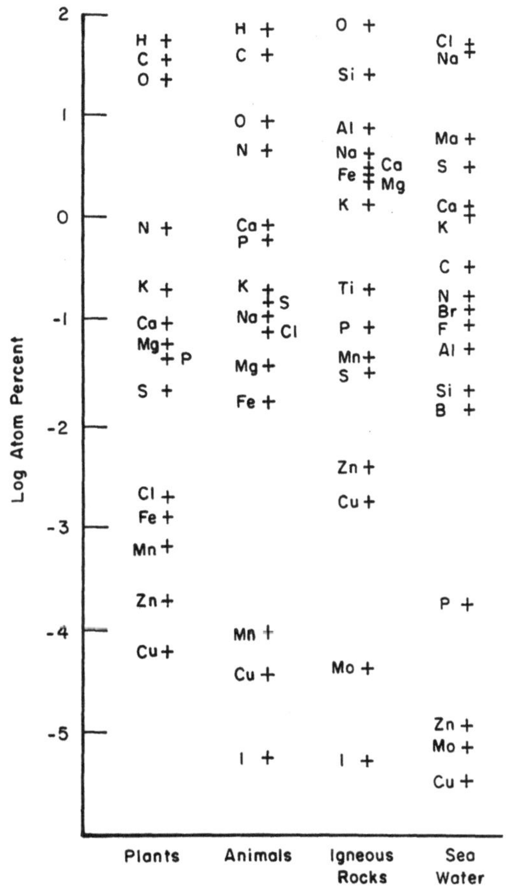


Carbon and sulfur are major constituents of some sedimentary materials (GARRELS and MACKENZIE 1971).

Figure 1 gives a comparison of the concentration of certain elements in biological materials, igneous rocks and seawater. It is not surprising that biological materials are made up of those elements readily available in the environment in which they evolved and roughly in proportion to their availability. What is of interest is the manner in which some elements commonly available in abundance have been shunned by the biological world and, if present in plants or animals, are there only as passive or apparently inadvertent constituents. An example of these is aluminum, which is quite ubiquitous but which serves no known biological role.

"Weathering" is a broad term including both the physical processes resulting from alternate heating and cooling, wind and water erosion, and other mechanical processes and the chemical processes of solution and reprecipitation. Elements liberated in the weathering process become available to plants, and plants are active participants in the weathering process. The organic constituents of plants and the products of microbial degradation of plant material participate in and accelerate the chemical weathering process. As a consequence, the weathering process is itself a function of the vegetation present. The vegetation is determined by the nature of the weathering products as well as other climatic and biotic factors. Some of these intricate interrelationships will be discussed later.

## 2 The Nature of Cycles

### 2.1 The Hydrologic Cycle

The concept of cyclic natural processes probably has always been a part of philosophical thought. As our understanding of natural processes has expanded, and as we have gained an appreciation for the significance of the approximately $5 \times 10^9$ years that the Earth has existed, we continue to refine our concepts of these cycles. Depending upon our interests or the degree of our myopia we can define any number of cycles based on elements, organisms or processes. Our primary interest here is to deal in detail with some of the elemental cycles of significance to plants. But before we can treat these effectively it is necessary to give brief consideration to the hydrologic cycle, the sedimentary cycle and the magmatic cycle. All of these cycles are related and each influences the others. The biological world (and therefore the biological cycles) are strongly dependent upon the hydrologic, sedimentary and magmatic cycles and in turn have strongly influenced these. Corollary to this concept is the realization that all of the cycles are now quite different than they were in early pre-Cambrian times prior to the evolution of life on the planet. Moreover, as the characteristics of life have changed through evolution so has the nature of these cycles been altered.

The hydrologic cycle, perhaps the first to have been recognized as such, not only dominates the biological world, but is one of the most important determinants of the character of the globe.

Most of the water of the Earth that we can recognize (about $1.4 \times 10^{18}$ m$^3$ or 80% of the total) is in the oceans. Pore water of sedimentary rocks, the next largest fraction, constitutes about $3.2 \times 10^{17}$ m$^3$ (18.8%). Another 1.2% is tied up in ice. Fresh-water lakes and rivers make up about 0.002% and the atmosphere contains a small 0.006% of the total. About 0.026% of the total falls in precipitation every year (and an equivalent amount is evaporated annually). Although annual evaporation (and rainfall) equals about 1 part in 30,000 of the water of the hydrosphere, it would be misleading to use this figure to estimate the time constant of the hydrologic cycle. Other superimposed cycles, such as the circulation of the oceans and the storage of water in glaciers, have periods of thousands of years, as does the storage of water underground in some places. The cycle involving pore water of sediments lasts millions of years. Thus, although the quantity of water which is in the atmosphere and that which falls as precipitation on land are small compared with the total, they are vital to plants not only because of the water that is supplied, but because of the weathering and erosional processes that result and the elements which are transported from the atmosphere to the soil by rainfall (PENMAN 1973).

The processes of evaporation and transport of water provide the driving force for circulation and mixing of the atmosphere; and the effect of clouds and precipitation on the reflectivity of the Earth are major moderating factors in determining the temperature of the continents. Water vapor in the atmosphere serves to limit radiative loss of heat, narrowing the temperature differentials between seasons and from day to night. Although one could visualize the adaptation of life to extremes of temperature, the unique characteristics of water as a solvent make it particularly suited as a medium for life, and by inference suggest an ambient temperature within its liquid range. Higher atmospheric pressures could expand this range but not indefinitely.

Thus, the hydrologic cycle can be looked upon as a steady state with minor fluctuation (on a geologic time-scale as well as the short-term fluctuation which we refer to as "weather").

## 2.2 The Sedimentary Cycle

Because of the length of time over which it has been operating, the sedimentary cycle also can be viewed as something near a steady-state condition with the weathering of old sediments and the formation of new sediments taking place at approximately equal rates. As new igneous rocks are exposed by weathering or brought to the surface by vulcanism, they are also subject to weathering and contribute to the sedimentary burden. Sedimentary rocks constitute about 75% of the total rock exposed on the continents and so their contribution to soil and plants is a major one (GARRELS and MACKENZIE 1971). The difference in composition of sedimentary and igneous rocks, as given in Table 1, would suggest a considerable difference in the composition of soils coming from these two sources. Differences are recognizable in soils and in the vegetation they support, but these are not as great as the composition figures alone would suggest. Climatic and biotic factors are also important and the plant population tends to regulate the characteristics of the soil and the mineral elements available in it. In some cases differences in elemental composition caused by processes of sedimentation and metamorphosis can result in deficiencies of some essential elements or toxic concentrations of others which in turn limit vegetation. Under these conditions the vegetation mosaic will reflect the differences in tolerance, deficiency or toxicity involved. The most familiar examples of this sort are those resulting from extremes of water availability or total salt concentration, but other more subtle manifestations are common.

The processes of weathering and sediment formation include the mechanical and chemical weathering of the parent material, the transfer of the products of weathering either laterally or downward, and the formation of secondary precipitates including secondary clay minerals. These small particulates are transferred to other locations, eventually precipitate and result in the formation of new sediments. The new sediments usually contain additional materials of

**Table 1.** A comparison of the elemental composition of various rock types. (Derived from the data of HORN and ADAMS 1966, TAYLOR 1964 and POLDERVAART 1955)

| Element | μg atoms $g^{-1}$ in | | | |
|---|---|---|---|---|
| | Igneous rocks | Shales | Sandstones | Limestones |
| O | 29,000 | 30,200 | 30,700 | 31,000 |
| Si | 10,000 | 2,600 | 13,000 | 850 |
| Al | 3,040 | 2,960 | 920 | 150 |
| H | 1,400 | 5,600 | 1,800 | 860 |
| Ca | 1,030 | 550 | 975 | 7,530 |
| Na | 1,030 | 420 | 140 | 17 |
| Fe | 1,000 | 850 | 180 | 70 |
| Mg | 960 | 600 | 290 | 1,900 |
| K | 540 | 680 | 270 | 70 |
| Ti | 120 | 96 | 31 | 8.3 |
| P | 34 | 25 | 5.5 | 13 |
| F | 33 | 40 | 14 | 17 |
| Mn | 17 | 15 | 9.1 | 20 |
| C | 17 | 1,275 | 32 | 19 |
| S | 8.1 | 75 | 75 | 37 |
| Cl | 3.6 | 5.1 | 0.28 | 4.2 |
| N | 1.4 | 41 | 6.4 | – |
| Zn | 11 | 1.5 | 0.25 | 0.31 |
| B | 0.91 | 9.1 | 3.2 | 1.9 |
| Co | 0.42 | 0.32 | – | – |
| Mo | 0.015 | 0.027 | – | – |

biological origin such as the skeletal remains of microflora and amorphous organic material.

The burial of carbon in sediments is an important part of the geobiological cycle, contributing to the accretion of oxygen in the atmosphere. This will be discussed in more detail later.

Compared with our estimates of the time constant of the hydrologic cycle, the sedimentary cycle is a long one with the mean life of sediments being in the neighborhood of $3.5 \times 10^8$ years. This is still less than $^1/_{10}$ the age of the Earth but it is long in comparison with the rate of biological evolution. For this reason and because biological processes strongly influence the sedimentary cycle, we would expect to find in the sedimentary record some evidence of this biological change. The most striking biological effect should be in the rate of weathering but there should also be recognizable differences in the chemical composition of weathering products. The first great biological invention, of course, was life itself. Presumably this supplied a new source of chelating agents which would hasten mobilization and transport of metallic ions, particularly the less soluble ones such as iron. It also resulted in the increased rate of burial of reduced carbon resulting from photosynthesis. This made possible the accumulation elsewhere of more oxidized products of photosynthesis, first sulfur and then oxygen. These will be discussed in more detail when the cycling of these particular elements is considered.

## 2.3 The Magmatic Cycle

Superimposed upon the sedimentary and hydrologic cycles, influencing both of them and in turn being influenced by both is the magmatic cycle, whereby through the processes of ocean floor spreading and subduction, magmatic material is transported across the ocean floors and buried under the continental plates. This cycle has an approximate time constant of 200 million years, of the same order of magnitude as that of the sedimentary cycle. Along with the burial of the basaltic ocean floor some of the sedimentary burden which it had picked up on its trek across the ocean is also buried. This results in the deep burial of reduced carbon and other materials of the ocean floor mud, further contributing to the possibility of accretion of oxygen in the atmosphere and indirectly altering the composition of sedimentary materials. Interstitial water and dissolved salts included in the sediments are also buried, altering the composition of the ocean and the sediments.

At some later time, and presumably with a shorter time constant, the volatile constituents of this buried material, particularly $H_2O$, $CO_2$, $H_2S$, and assorted volatile nitrogenous compounds, are returned to the surface by continental vulcanism and again are able to participate in the hydrologic, sedimentary, and geobiological cycles. The altered composition of the sediments from the characteristics of the original igneous or sedimentary materials from which they were derived would be expected to be evident in a differing composition of lava from these continental volcanos as contrasted with that of oceanic lava flows. Such differences in lava composition are discernible but confused by other processes of fractionation in the magma and the differences in depth from which various lava flows originate.

The above description is a highly simplified synthesis of a complex process. It is presented here only as backdrop for a discussion of the geobiological processes and cannot be dealt with in detail.

## 2.4 The Geobiological Cycles

The appearance of life on the planet influenced the hydrologic, sedimentary and magmatic cycles, at first only modestly but to an increasing extent as successive biological innovations evolved (Broda 1975b). The ancient sedimentary record is vague but evidence is strong that some simple life forms existed at least $2.7 \times 10^9$ years ago. From virtually the beginning this life had to be powered by solar energy. In other words, photosynthesis of one sort or another has always been a part of life though it need not have been the three basic classifications of photosynthesis we now recognize, the photolytic splitting of organic compounds, hydrogen sulfide, or water. More probably, these three common classifications of photosynthesis are simply manifestations of electron sources for the photosynthetic process.

Photosynthesis made possible the dismutation of available compounds into some which were more oxidized and some which were more reduced than thermodynamic equilibration would dictate. Thus, the combination of carbon dioxide and water could be converted into more reduced carbon (carbohydrate or hydrocarbon) and more oxidized water ($O_2$). Carbon dioxide and hydrogen sulfide could be converted to more reduced carbon (carbohydrate or hydrocarbon) and more oxidized hydrogen sulfide (sulfur). The existence of such a thermodynamic disequilibrium in turn made possible the existence of heterotrophic organisms which gain their energy by carrying on the reverse process, the oxidation of the more reduced carbohydrate or hydrocarbon with the more oxidized sulfur or oxygen or other electron acceptors.

Other elements from the weathering of igneous rocks such as reduced iron provided additional electron sources for reduction of the available electron acceptors. Before any appreciable accretion of oxygen in the atmosphere could take place, ferrous iron in the oceans or in the soil would first have to have been oxidized.

The biological cycle of photosynthesis producing oxygen and the heterotrophic oxidation of the products of photosynthesis is a rapid one compared with the sedimentary cycle. For this reason the accretion of oxygen in the atmosphere was dependent more on the rate at which reduced compounds of biological origin could be buried than it was upon the rate of photosynthesis itself. Initially photosynthesis was largely in the oceans and other bodies of water or on soils which were regularly kept moist. In either case the rate of photosynthesis was limited by the availability of other elements necessary to support life and so the total photosynthetic rate could not have been much greater than the present marine rate which presumably is phosphorus-limited.

The one great biological innovation which changed this, shifted the center of gravity of photosynthesis to the continents and resulted in a much greater overall photosynthetic rate, was the evolution of vascular tissue. Once plants with a vascular system evolved it became possible to mine the soil more deeply for mineral elements necessary to support life and thereby made possible large plants with a higher photosynthetic capability. This in turn greatly slowed physical weathering, made possible the accumulation of a significant soil layer on the continents and slowed the rate at which the more reduced material from the weathering of igneous rocks was delivered to the oceans as a sink for atmospheric oxygen. This remarkable development took place during Silurian and Devonian times and resulted in an Earth similar to the one we see today.

It is at this point that we can undertake to piece together available information on the cycling of elements of biological interest with reasonable confidence that although these cycles have been altered by the course of evolution they have been operating with their essential features as we now know them for nearly 500 million years.

It is not possible to assign any hierarchy of importance to the various elements essential for plant growth. Each element is utilized, transported, precipitated or otherwise transformed in a manner unique to its chemistry and biological function (Deevey 1970). The cycles of some have received more attention from the scientist than others, either because of their complexity, their intrinsic interest, or their susceptibility to modification by human activity. The cycles of hydrogen, oxygen, and carbon have not been subject to much study from an agronomic stand-point aside from the obvious problems of water management and a few studies of carbon dioxide enrichment of atmospheres over plants. Recently the cycle of carbon has become of interest, not as a plant physiological problem but because of the potential implications which the burning of fossil fuels and release of carbon dioxide may have to the atmospheric level of $CO_2$ and the Earth's heat budget.

Many of the elements have comparatively simple cycles in that they are taken up by the plant from the soil, perform their function in the plant and are returned to the soil upon death and decay of the plant or animal material. Nitrogen and sulfur have more complex cycles for several reasons. In addition to their functional role in the plant they are participants in the energy metabolism of microorganisms in the soil and are alternately oxidized or reduced depending upon soil conditions. The nitrogen cycle is made even more complex by the release of nitrogen gas to the atmosphere where it is no longer available to most plants and must be "fixed" before it can again participate in biological processes.

Phosphorus is of interest because in many environments, particularly the ocean, the availability of phosphorus is a primary limitation on productivity.

# 3 The Nitrogen Cycle

## 3.1 Overall Cycle Features

In addition to the characteristics which the nitrogen cycle shares with most other elements of the biosphere, the chemistry of nitrogen is such as to make the cycle considerably more complex than most (SÖDERLUND and SVENSSON 1976, DELWICHE 1970). The nitrogen cycle is outlined in Figs. 2,3.

Nitrogen forms several gaseous compounds including several oxides, ammonia, some volatile amines and nitrogen gas ($N_2$). Consequently, the atmosphere plays an important role in the nitrogen cycle with gaseous nitrogen being by far the major nitrogen pool excluding nitrogen which is buried in sediments. Secondly, the oxidation-reduction range of nitrogen from a valence of $+5$ in nitrate ion to one of $-3$ in ammonia and amino nitrogen, is well within the oxidation range of all plants and most microorganisms (Table 2). As a consequence nitrogen undergoes alternate oxidation and reduction in a number of energy reactions of plants and microorganisms in addition to its structure and catalytic roles in all organisms.

The nitrogen cycle can be looked upon as two superimposed cycles, one involving the soil–plant–soil cycle already mentioned which is a comparatively rapid one with the mean turnover time being a few decades or less depending upon the ecosystem involved, and the other involving the loss of gaseous nitrogen to the atmosphere. This nitrogen is no longer available to most plants, with only a few microorganisms or in some cases microorganisms associated with plants (or some animals) being capable of fixing atmospheric nitrogen and making it available to the biosphere. Because the atmospheric reservoir

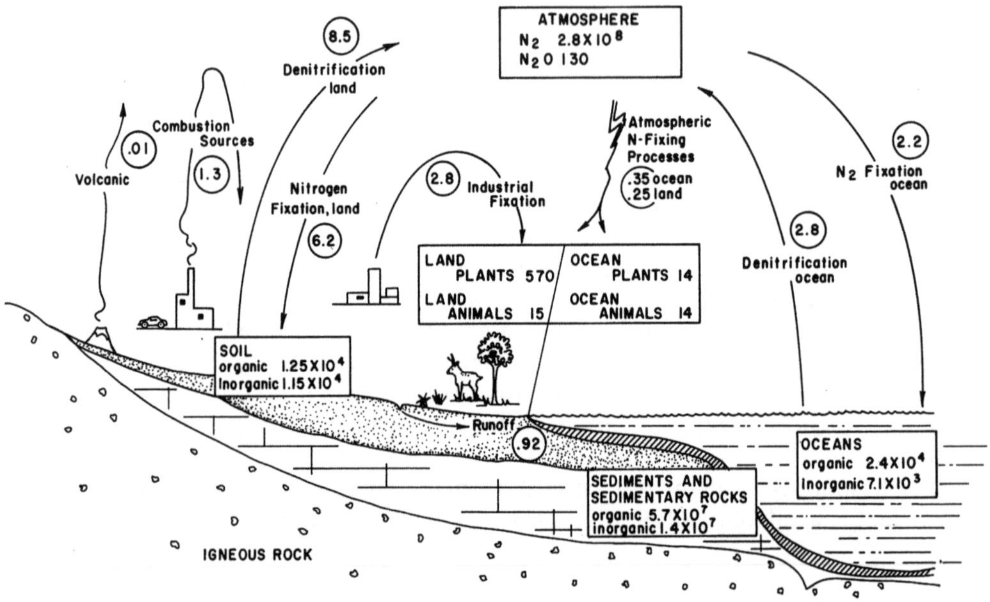


**Fig. 2.** Detailed nitrogen cycle. Reservoirs (pools) are shown in rectangles and transfer rates in circles. Units used are Gigagram atoms of N ($10^{12}$ gram atoms). Many values for the nitrogen cycle are uncertain, as for example, ocean processes, the land and ocean reservoirs of organic N and others. (Values given here are obtained from various sources: SÖDERLUND and SVENSSON 1976, HARDY and HAVELKA 1975, DELWICHE 1970)

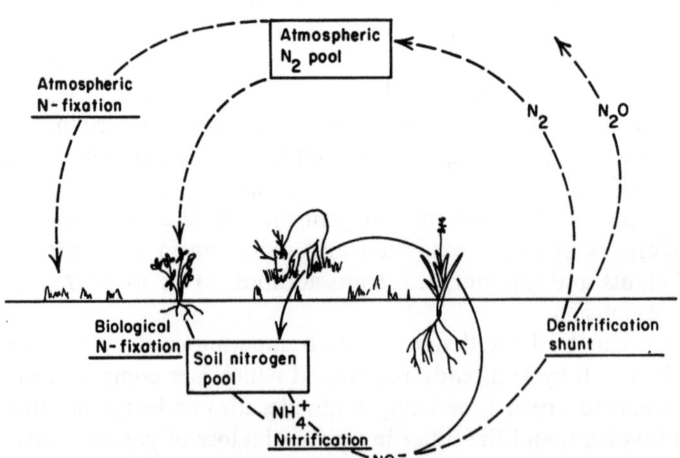


**Fig. 3.** Schematic outline of the nitrogen cycle. Processes shown by *solid lines* are typical of many elements. Those shown by *broken lines* are unique to the nitrogen cycle

**Table 2.** Oxidation states of nitrogen and typical compounds of interest to the biologist

| Valence | Typical compound | |
|---|---|---|
| +5 | $NO_3^-$ | Nitrate ion |
| +3 | $NO_2^-$ | Nitrite ion |
| +1 | [HNO] | Nitroxyl |
| 0 | $N_2$ | Nitrogen Gas |
| −1 | $HONH_2$ | Hydroxylamine |
| −3 | $NH_3$ | Ammonia |

is so large this second portion of the cycle is a ponderous one requiring $2 \times 10^7$ years or more before a given molecule is returned to soil or the sea. Actually, because this cycle is superimposed upon the other sedimentary magmatic cycles mentioned earlier, the residence time of a molecule in the atmosphere is even greater.

Overall elements of the nitrogen cycle are outlined in Fig. 2.

### 3.2 Nitrification

Most of the nitrogen of plants and animals is at the reduced or −3 valence level of ammonium or amino nitrogen. When it is released to the soil by the decomposition of plant and animal tissues, and if oxygen is present, there is energy available in the oxidation of ammonium ion to nitrate ion and a number of "nitrifying" organisms use this as their energy source for growth (ALEEM 1977, ALEXANDER 1965). These organisms are classically looked upon as "autotrophs" because they do not require organic compounds for growth (ALEEM 1970). There are some "heterotrophic" organisms (ones which require an organic substrate) which are also capable of nitrification (ALEXANDER et al. 1960). As a result of this nitrifying process most of the inorganic nitrogen in soil is as nitrate ion ($NO_3^-$) which is the soil nitrogen species most likely to be taken up by plants. Once taken up by the plant this nitrate must again be reduced requiring the expenditure of energy either from the oxidation of organic materials or utilizing directly the energy captured in photosynthesis.

There are other consequences of the nitrification reaction. Ammonium ion as a cation is more likely to be held on the exchange complex of the soil near its surface, particularly if the soil has significant quantities of organic matter or layered clay minerals. Once oxidized to nitrate, however, the nitrogen is more mobile and more readily moved through the soil by rainfall where it is delivered into the root zone. If plants are absent, as under fallow conditions, the nitrate ion is more likely to be leached through the root zone and appear in groundwaters unless some other fate befalls the ion. When nitrogen fertilizers containing ammonium ion are used, efforts are taken to minimize this possibility of loss. These include the timing and placement of nitrogenous fertilizers, the use of "slow release" fertilizers which liberate the ammonium ion to the soil slowly, or the use of agents to inhibit the activity of nitrifying organisms.

Since ammonium ion, whether from the decomposition of organic materials or from applied nitrogen fertilizers, is in partial equilibrium with gaseous ammonia, there is also the possibility of a volatilization of ammonia to the atmosphere. Because ammonia is highly water-soluble, its residence time in the atmosphere is short and it is soon returned to the soil in rainfall or in combination with sulfate ion or other anions as dry fallout.

The nitrification reaction is usually thought of as taking place in two steps because of the nature of the metabolic processes involved. The first of these is the oxidation of ammonium ion to nitrite ion ($NO_2^-$). This reaction is carried on by several species of *Nitrosomonas* and several other genera including some heterotrophs as mentioned above (BARTHOLOMEW and CLARK 1965).

$$NH_4^+ + 1.5\,O_2 \rightarrow NO_2^- + H_2O + 2\,H^+$$

$$\Delta G' \text{ (at pH 7)} \simeq -65 \text{ kcal.}$$

The second step in the reaction, the oxidation of nitrite ion to nitrate is carried on by organisms of the genus *Nitrobacter* or other organisms including some heterotrophs.

$$NO_2^- + 0.5\,O_2 \rightarrow NO_3^-$$

$$\Delta G' \text{ (at pH 7)} \simeq -18 \text{ kcal.}$$

Nitrous oxide ($N_2O$) is also sometimes produced in the nitrification reaction (BREMNER and BLACKMER 1978), under some circumstances in considerable quantities. Since the nitrogen of nitrous oxide is intermediate in valence level between that of nitrate and ammonia, the formation of gaseous nitrous oxide short circuits the nitrification sequence, not only depriving *Nitrobacter* of a substrate but also losing nitrogen from the system. This may escape to the atmosphere or, as will be discussed below, may participate in other microbial reactions. Nitrous oxide production in nitrification may be more prevalent in the ocean than it is on land (HAHN 1974).

### 3.3 Denitrification

In addition to the possibility of being lost to groundwaters by leaching, nitrate ion may undergo other reactions which exclude it from plants. If there is organic material present and the oxygen supply is low, as for example in waterlogged soils or highly eutrophied waters, the nitrate can play a quite different role. Some microorganisms, in the absence of sufficient oxygen, can utilize nitrate ion instead of oxygen as "electron acceptor" for the oxidation of organic materials in order to obtain their energy. In this process the nitrate ion is reduced first to nitrite and eventually to nitrogen gas or nitrous oxide (DELWICHE and BRYAN 1976). Both of these commonly escape to the atmosphere, the $N_2$ entering the large atmospheric pool and the nitrous oxide a much smaller and much shorter-lived atmospheric pool. The atmosphere contains about 0.35 parts per million $N_2O$ on a volume basis. Much of this, perhaps most, is of biological origin although there are other processes in the atmosphere which can result in $N_2O$ production (ROBINSON and ROBBINS 1970). Its mean residence time in the atmosphere is comparatively short, between 10 and 100 years (JUNGE 1972, 1974). Among the atmospheric reactions in which nitrous oxide can participate are those involving the formation and destruction of ozone ($O_3$). The function of ozone as an ultraviolet shield makes of interest the effect which any human influence on the rate of production of $N_2O$ might have. The use of nitrogenous fertilizers and the concentration of organic wastes in urban areas and other management practices presumably could increase the rate of $N_2O$ released to the atmosphere. Some practices such as the draining of wetlands could result in a lessened rate of $N_2O$ production. Some of these will be discussed in more detail below.

There are other manifestations of denitrification including its role in the problem of "bloat" in ruminants and gas production in foods and other stored materials.

Nitric oxide (NO) is also a possible product of denitrification, particularly under acid conditions. When plant materials which are high in nitrate are subjected to fermentation as in the ensiling process, nitric oxide, which is highly toxic, may be formed in concentrations sufficient to be hazardous to humans. Silo fillers' disease, an example of this category, occurs frequently enough to be of clinical interest and is sometimes fatal.

### 3.4 Nitrogen Fixation

Nitrogen fixation (or dinitrogen fixation) is discussed in detail in Chaps. II.1 and II.2, this volume and is dealt with only briefly here. To replace the nitrogen which is lost from the soil or from waters by denitrification, sedimentation or other processes, new nitrogen from the atmosphere must be fixed and brought into the biological system (DELWICHE 1970, BRILL 1977). Ionizing events in the atmosphere such as electrical discharge and cosmic radiation result in the formation of some compounds of nitrogen (largely oxides) which are brought to the earth by rainfall. This replenishes some of the nitrogen supply. Most of the requirement is met by biological reactions. Organisms such as *Azotobacter* free living or in association with grasses such as *Paspalum* are capable of fixing nitrogen under aerobic conditions and these make some contribution to soil nitrogen. With some tropical grasses this contribution may be a major one. In environments low in oxygen such as rice paddies and swamps, anaerobic or microaerophilic microorganisms, also free-living, can make sizable contributions. Examples here are various clostridia and other bacteria. Blue-green algae, growing either independently or in lichen associations also fix nitrogen.

Perhaps the greatest contribution to the biological pool, however, is the nitrogen fixed by microbial association with higher plants. Although there are a number of such associations, the best-known are those of the legumes which bear root nodules supporting *Rhizobium*. There are an estimated 600 genera and 13,000 species of legumes, most of them presumably capable of nitrogen fixation in this sort of symbiotic arrangement. Much of the tropical rain forest is populated with legumes and the widespread use of agricultural legumes has long been a contributor to the human food supply.

A considerable number of other higher plants representing at least 10 plant families also bear root nodules and are capable of nitrogen fixation in association with microorganisms (BOND 1976, TORREY 1978). In most cases the microorganism is an actinomycete, but this group has been studied in much less detail than have the legumes and our knowledge of them is limited (TORREY 1976). They are, however, significant contributors to the nitrogen budget of many plant associations, particularly the chaparral and deciduous forest lands of temperate zones. Attention is now being given this group, and their nitrogen fixation potential for both food and fiber production undoubtedly will be exploited.

The transfer of the nitrogen fixation capability to other plants, particularly cereals, by genetic manipulation is a possibility now being given some attention (see Chap. II.1).

The intensification of agricultural management, particularly in Europe and the Americas, but in other countries as well, has resulted in a great expansion of the use of industrially fixed nitrogen for fertilizer. Approximately 50 million metric tons of nitrogen are now fixed industrially and an additional approximately 20 million metric tons are fixed inadvertently by combustion processes annually. This fixation plus that contributed by leguminous agricultural crops probably is about equal to the nitrogen fixed annually by "natural" processes before human intervention. Thus, at least the fixation portion of the nitrogen cycle has been doubled. Denitrification probably has not kept pace with this new nitrogen input, although other management practices such as the concentration of wastes in urban areas and the delivery of soil nitrogen to coastal waters by erosion probably also have increased denitrification (BROADBENT and CARLTON 1978).

The nitrogen fixation reaction, whether it is carried on industrially or by biological systems, requires considerable energy, particularly under aerobic conditions for the biological system. Biological fixers accomplish this by the use of photosynthetic energy (their net dry matter yield is probably reduced accordingly) and so they have the advantage that they can be used at locations where logistic, economic or other factors prohibit the use of industrially fixed nitrogen. The energy limitation on the fixation of nitrogen apparently is imposed by the kinetic barrier of the high activation energy of the dinitrogen molecule, a barrier which neither the biological system nor the industrial process has overcome. (The high temperature reaction of nitrogen with oxygen presumably could circumvent this problem but because the resultant product would be a nitrate weighing at least four times as much as anhydrous ammonia per unit nitrogen, the energy cost of its transport and other problems in its use would outweigh the energy advantages over ammonia synthesis.)

### 3.5 Human Influences

A number of human activities, including the drainage of wetlands, the use of nitrogenous fertilizers, the fixation of nitrogen in combustion reactions such as automobiles, the concentration of wastes in urban areas and others, have greatly altered the nitrogen cycle. What the significance of these changes might be is difficult to evaluate. Each of these changes can be viewed as alarming or insignificant depending upon the bias of the analyst, but in most cases available information is not sufficient to yield reliable objective analysis. Probably in most cases the negative consequences outweigh the positive ones and therefore these perturbations should be minimized. Some of the consequences of nitrogen mismanagement, such as the eutrophication of streams and estuaries, are sufficiently obvious as to result in corrective action, whereas others are more subtle and some perhaps have not been recognized (DELWICHE 1981).

The use of nitrogenous fertilizers, particularly in heavy applications, appears to result in an increased rate of denitrification. Under most field conditions about 90% of the nitrogen lost in denitrification is released as $N_2$ and about 10% is released as $N_2O$. (The figures range from 0% to 20% in most cases studied.) With heavy applications of nitrogen fertilizers the ratio of $N_2O$ to nitrogen may be higher. This added increment of $N_2O$ to the atmospheric budget can result in some reduction of concentration of ozone and a slight lowering of the ozone "layer" in the stratosphere (CRUTZEN and EHALT 1977). The ozone layer and the atmospheric $N_2O$ content have been under study in this context for a comparatively short time and so we are uncertain of trends at present, but there probably has been some increase in the concentration of atmospheric $N_2O$. Where organic wastes are concentrated, such as in feedlots and sewage disposal systems, there is also opportunity for increased denitrification and increased $N_2O$ production, as well as for the volatilization of ammonia to the atmosphere.

Excessive use of nitrogen can also cause higher nitrate concentration in foods and feeds with potentially deleterious effects. Nitrates have always been a part of the human diet, but this does not mean that higher levels are permissible. Some nitrites are formed by the reduction of nitrate in the intestine and these can lead to the formation of nitrosamines which are potentially carcinogenic. Data to date dealing with the problem are limited. Similarly, domestic water supplies have always contained some nitrate ion. In some locations the nitrate ion concentration of groundwaters has increased significantly as a result of fertilizer use or from sewage disposal systems or other sources. Legumes also can contribute to groundwater nitrate. Acceptable levels of nitrate in groundwaters are likewise difficult to establish, but at least for some individuals high nitrate waters can be hazardous.

The intensive management of agricultural lands probably will continue and increase so the problem appears to be one of gathering the information necessary to minimize the negative consequences of this intensive management and then implementing the policies which the information dictates.

## 4 The Sulfur Cycle

### 4.1 Comparison with the Nitrogen Cycle

An outline of the sulfur cycle is given in Fig. 4. Sulfur experiences the same cycling from soil to plant and back again, as do other elements essential to plants. Like nitrogen, because it has various oxidation levels (Table 3), sulfur

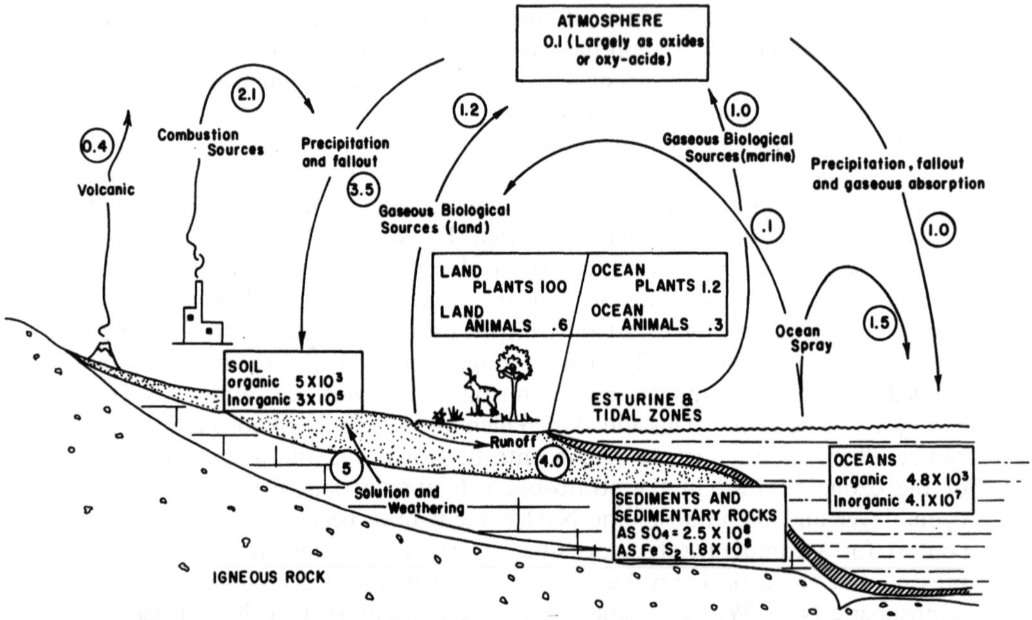


**Fig. 4.** The sulfur cycle. Reservoirs (pools) are shown in rectangles, transfer rates in circles. Units are gigagram atoms ($10^{12}$ g at). (Sources: BOLIN and CHARLESON 1976, GARRELS et al. 1975)

**Table 3.** Oxidation states of sulfur and some typical compounds

| Valence | Typical compound | |
|---|---|---|
| +6 | $H_2SO_4$ | Sulfuric acid |
| +4 | $H_2SO_3$ | Sulfurous acid |
| +2 | $H_2SO_2$ | Sulfoxylic acid |
| 0 | S | Sulfur |
| −2 | $H_2S$ | Hydrogen sulfide |

participates in the energy metabolism of a number of organisms. It also forms volatile compounds such as oxides, [sulfur dioxide ($SO_2$), sulfur trioxide ($SO_3$), and others], hydrogen sulfide ($H_2S$), and various methylated derivatives. The atmosphere, therefore, is an important medium for the cycling of sulfur, but unlike nitrogen, the primary reservoir of sulfur is not the atmosphere. Most compounds of sulfur are metabolized by at least some organisms and so the system is more tightly coupled than is that of nitrogen (BOLIN 1976, BORMANN and LIKENS 1979). Although sulfur is not required in as large quantities by most organisms as is nitrogen, it is still a major plant constituent and there are many examples of sulfur deficiency. The most common form of inorganic sulfur in the soil is sulfate ion ($SO_4^{2-}$) which is highly mobile. It is therefore somewhat surprising that since sulfate is readily transported to the oceans, sulfur deficiency is not more common than it is. The answer must be attributed to the many processes which there are for the volatilization of sulfur and its return through the atmosphere to the land. The oxidation-reduction levels of sulfur range from its most oxidized form as sulfate ion in which sulfur has a valence of $+6$ to that of hydrogen sulfide ($H_2S$) in which sulfur has a valence of $-2$. This range like that of nitrogen also represents eight electrons (Table 3).

### 4.2 Microbial Oxidation

As with nitrogen most of sulfur in plants is reduced, as SH groups at the oxidation-reduction level of hydrogen sulfide. Upon decomposition of plant and animal material, as sulfide ion is returned to the soil, it also can serve as energy substrate for autotrophic organisms. Various organisms such as *Thiobacillus thiooxidans* can oxidize sulfide ion either to sulfur ($S_2$) or all the way to sulfate ion ($SO_4^{2-}$). Other organisms can carry on the full oxidation sequence, or some of them some steps of it. The energy release is considerable as shown by the following reactions:

$$H_2S + 0.5\,O_2 \rightarrow H_2O + S$$

$$\Delta G° = -52.1 \text{ kcal mol}^{-1}$$

$$S + 1.5\,O_2 + H_2O \rightarrow H_2SO_4$$

$$\text{at pH 7 } \Delta G' = -145.5 \text{ kcal mol}^{-1}.$$

As is the case with nitrogen, available sulfur in the soil is usually in its most oxidized form as sulfate ion. The energy released in the overall oxidation of sulfide to sulfate is even greater than that released in the oxidation of ammonium ion to nitrate. Conversely, when sulfate ion is taken up by plants an equivalent amount of energy and more must be expended to again reduce it to the level of sulfide, its form in plant protein.

### 4.3 Sulfate Reduction

These energy arguments apply to the conditions under which most plants live where molecular oxygen is available at a pressure of approximately 0.2 atmospheres. If oxygen is absent, sulfate ion can serve as "electron acceptor" by microorganisms for the oxidation of whatever soluble organic materials may be available. This is a situation comparable with that which results in denitrification of nitrate ion. Sulfate ion and various reduced compounds between it and sulfide ion ($S^{2-}$) are formed, with the organic substrate being converted to carbon dioxide and water. Energy yielded in the reaction goes to do the work of the organisms involved, including the synthesis of new cell material. The sulfides of a number of elements, notably those of iron, are quite insoluble and common constituents of anoxic soils and muds. Deposits of sulfur usually are the consequences of sulfide ion oxidation or are juvenile in origin. Hydrogen sulfide ($H_2S$) commonly is volatilized in the reaction and escapes to the atmosphere as gas. In the atmosphere it is autooxidized to oxides of sulfur or sulfate ion. This is one of the major avenues of return of sulfur from tidal flats or estuarial waters to the atmosphere and eventually to the land.

### 4.4 Patterns of Sulfur Movement

Since most sulfates are comparatively soluble and sulfate ion therefore relatively mobile in the soil, its general pattern of movement is into groundwaters and eventually to the sea. Processes of sulfate reduction with the yield of gaseous hydrogen sulfide result in the return of sulfur to the atmosphere and its redistribution to the soil in precipitation. Although sulfur, like other elements, is rather tightly cycled from the soil back to the plant, this process of sulfur return is particularly important in areas of abundant rainfall where the leaching of mobile elements can be significant. In more arid climates, where leaching takes place less rapidly, sulfates may accumulate in the soil and under conditions of extreme aridity, as in desert lands, sulfates may be present in large quantities along with nitrates and chlorides. In the extreme, precipitates of gypsum ($CaSO_4$) are formed.

The significance of the volatilization and transport of sulfur resulting from sulfate reducing processes frequently is overlooked. The abundance of sulfur and chlorine in the earth's crust is the same order of magnitude, yet the concentration of sulfur in the ocean is only about $^1/_{20}$ that of chlorine. Either the sulfur is precipitated as sulfides or volatilized and returned to the land. On the other hand, the requirement of land plants for sulfur is perhaps a thousand times the requirement for chlorine and were it not for this mechanism of return of sulfur to the land, sulfur deficiency would be much more prevalent than it is.

### 4.5 Human Influences

The effects of human activity on the sulfur cycle have been at least as striking as those on the nitrogen cycle. Sulfur has not been extensively mined and transported to the soil as a fertilizer, nor indeed have deficiencies of sulfur on agricultural soils been recognized extensively. The intensification of agricultural management with demands for exogenous supplies of mineral elements has occurred more or less simultaneously with the industrial revolution. The fossil fuels which powered the industrial revolution (coal, natural gas and petroleum) were formed under reducing conditions and all of them contain significant quantities of sulfur as sulfides. Their combustion releases to the atmosphere oxides of sulfur which eventually are returned to the Earth in precipitation. This human influence can perhaps be looked upon as a positive one, helping to meet what might otherwise have been a more common sulfur deficiency. In many cases this probably is true but there are also discernible negative consequences. In the atmosphere the oxides of sulfur are eventually oxidized all the way to $SO_4^{2-}$ which in combination with water yields sulfuric acid (GRANAT et al. 1976). Sulfuric acid is highly dissociated so that the resultant precipitation is acidic (LIKENS et al. 1979). Downwind from industrial centers acid precipitation has influenced weakly buffered soils and water bodies. Examples are known from the Appalachian Mountains and from Scandinavian lakes where the pH has been lowered to 4 or less resulting in the killing of fish and other life. Weakly buffered soils in these environments have been correspondingly affected. Not all of the acidity is attributable to sulfate ion, some of it resulting from nitrate ion also from combustion sources as well as other processes of nitrogen volatilization.

Recently alarming damage, in particular in coniferous forests, has been confirmed in different regions of Central Europe, popularly described as acid rain causing "Waldsterben" (forest decimation). In this case also $SO_2$ and $H_2SO_4$ arising directly or indirectly from power plants working with coal and from other technical sources are very probably the prevailing pollutants. Considerable work has still to be invested to analyze the whole complex of injurious factors. For more details the reader is referred to Chapter 14, Volume 12D, this Series.

# 5 The Phosphorus Cycle

## 5.1 Oxidation and Reduction

Phosphorus, belonging to the same subgroup as nitrogen in the periodic table and possessing many of the general characteristics of nitrogen, nevertheless has a strikingly different biological role and geochemical or geobiological behavior (PIERROU 1976). Among the less abundant elements of the lithosphere (about 0.1% of igneous rocks) it commonly occurs in the +5 valence state except under most strongly reducing conditions such as the metal and sulfide phases of meteorites and possibly some strongly reducing biological systems. Phosphorus tends to be a geochemical companion of titanium, an element which is chemically quite different. This coherence is largely the consequence of a coincidental similarity of some of their respective minerals.

Phosphorus like nitrogen exists at a number of valence levels from the reduced (−3) level of the gas phosphine to the oxidized (+5) level of phosphate ion (Table 4). Unlike nitrogen, however, phosphorus is normally in its most

**Table 4.** Oxidation states of phosphorus and some typical compounds

| Valence | Typical compound | |
|---|---|---|
| +5 | $H_3PO_4$ | Phosphoric acid |
| +3 | $H_3PO_3$ | Phosphorous acid |
| +1 | $H_3PO_2$ | Hypophosphorous acid |
| 0 | P | Phosphorus |
| −1 | [PH] | Unstable gas |
| −2 | $P_2H_4$ | Hydrogen diphosphide |
| −3 | $PH_3$ | Phosphine |

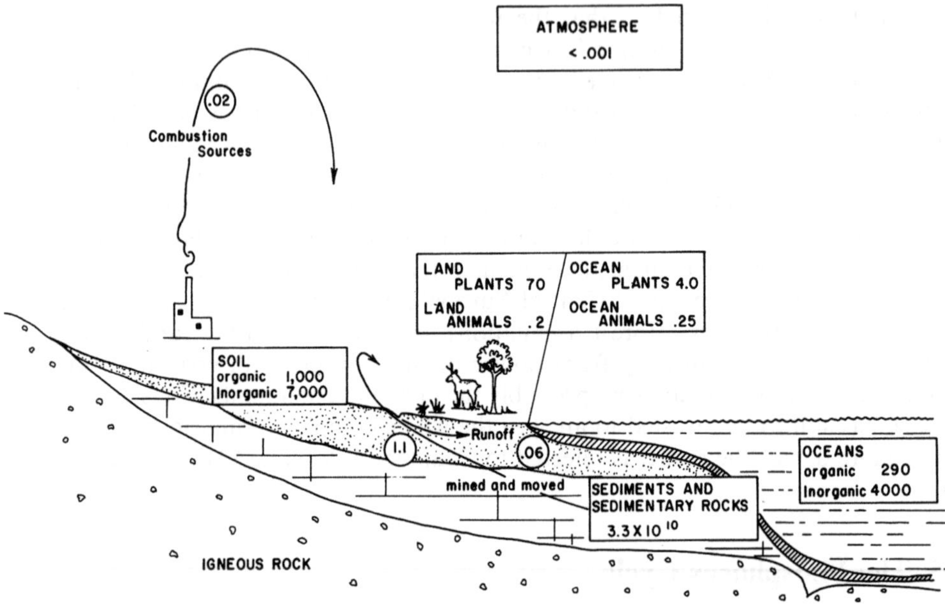


**Fig. 5.** The phosphorus cycle. Reservoirs (pools) are shown in *rectangles*, transfer rates in *circles*. Units are in gigagram atoms ($10^{12}$ g at). (Sources: GARRELS et al. 1975, STUMM 1972, LIKENS and BORMAN 1972, DELWICHE and LIKENS 1977

oxidized state ($PO_4^{3-}$) and is comparatively nonmobile in the soil. Whereas almost all salts of nitrates and many of the sulfates are quite soluble and hence mobile in the soil–water system, most phosphates are sparingly soluble, with some, such as apatite, being particularly resistant to weathering. This chemical stability of the phosphate ion and the relative immobility of phosphate have resulted in the development of some interesting biological characteristics geared to survival in an environment where a critical life support material was frequently in short supply. The phosphorus cycle is shown in Fig. 5.

## 5.2 Movement and Transport in the Biosphere

Because of its triple negative charge as well as the low solubility of most of its salts, phosphorus is tenaciously held in the soil, and its transport through the soil and into waters is slow compared with most of the elements essential to plants. For this reason and because normally it does not form volatile compounds, it is closely cycled in the biosphere and its concentration in seawater is exceedingly low (about 0.1 ppm or $3 \times 10^{-6}$ mol $l^{-1}$). The low seawater concentration is commonly considered to be the principal limiting factor on seawater productivity although other elements, particularly iron, may also be limiting.

In soil, although its concentration is always comparatively low in the soil solution, phosphorus is most available between pH's 6 and 7. Under more alkaline conditions it tends to be bound as insoluble phosphates or held on the exchange complex tightly; and at lower pH values, as the concentrations of iron and aluminum become greater, these elements tend to bind the $H_2PO_4^-$ or $HPO_4^{2-}$ ions as insoluble complexes. Consequently, in both more acid and more alkaline soils, there tends to be some selection for those plant species which are more able to extract phosphorus from this system of limited availability. Frequently, this appears to include symbiotic associations of microorganisms. Mycorrhizae, as their name implies, are fungi closely associated with the roots of some plant species and appear to provide such a function. Such associations are most common under conditions of low nutrient supply, particularly where phosphorus, zinc, or other elements are limiting. The fungus appears to assist in the solution and transport of this limiting ion supply.

Plants require about twice as much phosphorus as sulfur on a molar basis, so that it is fortunate that phosphorus is not transported to the sea more rapidly than it is. There is no mechanism whereby it would be returned to the land from the sea, and the supply in soil must be maintained by the weathering of rocks. Igneous rocks contain about 0.3% phosphorus on a weight basis so that about 330 g of rock would be required to yield 1 g of P. Of that phosphorus which does reach the ocean, most is concentrated in deposits on the continental shelf or continent margins, near zones of upwelling in subtropical climates. The peninsula of Florida and neighboring regions contain large quantities of high grade phosphatic ores. This rock phosphate consists of fluorapatite, $Ca_5(PO_4)_3F$ and the corresponding chlorine derivative, mixed with carbonate precipitates.

Although under most circumstances inorganic phosphorus is in its oxidized form, under highly reducing conditions such as in marshes and anoxic muds, phosphorus can be reduced biologically to the level of phosphine gas ($PH_3$) analogous with ammonia. There is evidence that "marsh gas" (mostly methane) contains some phosphine gas or its methylated analog, the autooxidation of which when it reaches the atmosphere being responsible for the glow frequently associated with marsh gas, at least in legend (the will-o-the-wisp). It is unlikely that this phenomenon is a significant mechanism for the transport of phosphorus in the biosphere.

### 5.3 Human Influences

A small quantity of phosphorus is released to the atmosphere by the burning of fossil fuels (about $5 \times 10^{11}$ g annually) and more by the combustion of plant material (about $2.5 \times 10^{12}$ g annually) (DELWICHE and LIKENS 1977). The quantity from fossil fuel combustion is less than 1% on a molar basis of the nitrogen and sulfur emissions so that its significance in acid precipitation is negligible.

The annual use of approximately $4 \times 10^{13}$ grams of phosphorus as fertilizer does not in itself constitute a major perturbation of the phosphorus cycle, largely because of the comparative insolubility and immobility of phosphorus in the soil. The misuse of such fertilizers can result in the loss of phosphates to streams with resultant eutrophication, but the more likely source of phosphorus contamination is erosional clay particles whether or not phosphorus fertilizers have been used. On the other hand, because phosphorus is less mobile, streams and other water bodies are usually low in phosphorus so that even a slight addition of phosphorus to the system can result in a considerable change from natural conditions.

One possible consequence of the use of phosphatic fertilizers may deserve some attention. The rock phosphate deposits from which they are derived are all of secondary origin and because of the close similarity between phosphorus and uranium, there is a tendency for uranium compounds to coprecipitate with phosphates in these rocks. Depending on the processes used for the manufacture of fertilizers, it is possible to transport uranium to agricultural soils by this means. Some fields have received an increment of uranium by the use of rock phosphate fertilizers, but present methods for the preparation of more highly processed "superphosphate" fertilizers tend to avoid this difficulty.

For information on the phosphorus load in the waste waters of industry and households see Chapter 14, Volume 12D, this series.

# 6 Other Elements

## 6.1 Biological Cycling

Most of the other mineral elements required by plants also have complex and in many cases poorly understood cycles. These have received less attention than have the cycles of nitrogen, phosphorus, sulfur and carbon, largely because examples of their deficiency are less prevalent or because perturbations of their cycles by human activity have been less obvious. Each of these elements has its own unique chemical and biological characteristics, and each of their cycles has its own unique behavior (BOWEN 1966, AHRENS 1979).

## 6.2 The Special Significance of Iron and Aluminum

Although, as indicated above, aluminum is present in comparatively high concentrations in igneous rocks, and therefore an abundant product of weathering, it is not an essential element for any organism as far as is now known. In retrospect, it is possible to rationalize this inessentiality of aluminum. It is commonly found in its +3 valence state and in aqueous media cannot be further reduced. Consequently, it would not be expected to participate in any biological oxidation-reduction reaction involving energy metabolism. Neither would it be expected to be a constituent of any electron transport coenzyme and so its only remaining role would be that of some structural unit. This is a role which the element appears not to have played, even though it is reasonably ubiquitous in its distribution.

Indirectly, however, aluminum has had an influence on the biological world. Both in the aluminosilicates of layered clay minerals and as derivatives of its oxides, aluminum is a common soil constituent. The layered clay minerals have a comparatively high cation exchange capacity. The sesquioxide clays which will be discussed below are made up of oxides of aluminum and iron and do not have a strong cation adsorption capability.

Iron can exist in either the ferrous ($Fe^{2+}$) level, or ferric ($Fe^{3+}$) level within the normal range of biological oxidation-reduction potentials. The oxidized ferric ion forms highly insoluble ferric hydrate precipitates, particularly above pH 7, and so its availability in more alkaline soils frequently may be limiting. Some soils (for example oxisols) may be made up largely of colloidal ferric hydrate particles and these present special problems of management for agricultural purposes. These sesquioxide clays which may also contain large quantities of the oxides or hydrated oxides of aluminum are widespread in their distribution and particularly common in tropical and subtropical climates. Their comparatively low cation exchange capacity and their common association with the soils of higher rainfall regions in the tropics is of some interest (MALAVOLTA et al. 1974). Ecosystems existing on these sesquioxide clays usually carry most of the mineral element pool in the vegetation itself or in the small amount of litter remaining on the forest floor. The soils are not "fertile" in the sense of their function as mineral element reservoirs, but they nevertheless support abundant vegetation in many locations. The conversion of native ecosystems on these soils to agricultural production which is seasonal in nature can have disappointing results because of the absence of exchange capacity and an attendant mineral element pool.

The low solubility of ferric hydroxide at pH 8 provides a special mechanism for the transfer of oxidizing capability to depths in an anoxic medium. Oxidized ferric clays may be colloidally dispersed in river waters flowing to the sea. In higher concentrations of salt as in seawater the colloid is flocculated and the particles resulting can sink. As they settle to anoxic zones the ferric oxide can serve as electron acceptor for biological oxidations and be reduced to ferrous ion which then will combine with sulfide ion and be precipitated as pyrite or other reduced sulfides. This provides a more rapid mechanism for transferring electron acceptors to depth than does the slower diffusion of oxygen. The large amount of reduced iron which must have been present in the ocean in early pre-Cambrian times had to be removed either by precipitation or oxidation before any significant accumulation of oxygen could have taken place in the atmosphere. Consequently, this process may have had an important role in these early phases of atmosphere evolution.

### 6.3 Hydrogen Ion

Many climatic and vegetation patterns influence the pH of a soil. Cool climates with adequate rainfall where the decomposition of soil organic matter is slow or incomplete, result in a somewhat acid leaching of the upper layers of the soil leaving behind the less soluble silica and silicates. These soils frequently have a low cation exchange capacity, and like the highly leached sesquioxide clays of tropical and subtropical climates, maintain much of the mineral budget in the vegetation and litter. Conversion of these soils to seasonal cultivation may have the same consequences of low productivity. As a result, "swidden" agriculture involving the burning of vegetation to prepare an agricultural field which is then abandoned after 2 or 3 years, has been practiced in widely differing areas such as the jungles of New Guinea and the Boreal Forests of Scandinavia. In more arid climates, alkaline soil pH's are common, and these can result in problems of a different sort. Both the accumulation of salts and toxicities or deficiencies resulting from the high pH are common. Over the comparatively narrow range of pH from 4.5 to 8.0, the relative availability of various ions can vary greatly. In general, neutral to slightly acidic pH's are favorable for most plants, although some are more tolerant to alkaline pH's and others to more acid conditions.

If aluminum is present in the soil, as it frequently is, it can cause toxic effects at pH's below 5.5. Aluminum toxicity is rather nonspecific, much like other "heavy metal" toxicities encountered when concentrations of a number of metals such as copper, zinc, mercury, and others are high. Aluminum is also more soluble at very high pH's (in the neighborhood of 10) but many other problems would be encountered in attempting to grow plants under such alkaline conditions.

## 6.4 Characteristics of Sediments

The figures given in Table 1 for the elemental composition of various rocks and other materials must be interpreted with caution. They represent mean values useful for geochemical calculations, but values for a particular rock might vary widely from these. Processes of metamorphism can greatly change the composition of a parent material. Such rocks as serpentine which have been highly altered by water and heat frequently show concentrations of some metals and depletions of others. Soils derived from serpentine rocks commonly are relatively high in magnesium (when compared with calcium) and frequently are deficient in molybdenum. Other elements, some of them toxic to plants, can be concentrated by metamorphic processes (MACKENZIE et al. 1979). Examples of these are selenium, mercury, boron, and others. The figures given here are useful when interpreted as part of the soil-forming process. The weathering of a limestone, for example, yields largely calcium ion, which is mobile, and carbon dioxide, which is lost to the atmosphere. The resulting soil is made up of the "impurities" in the limestone, secondary clay minerals formed at the site, and the organic residues of plants and animals. As a result, soils formed from limestone frequently are shallow even though they resulted from the weathering of many meters of rock.

These are merely examples of the complexities which element cycles might exhibit. Space does not permit dealing with them in detail here but the cycles of these "other" elements, although they have received little attention, cannot be looked upon as simple. BOWEN (1966) provides detailed treatment of some of these.

## 6.5 Passive Cycling

All of the elements which play an essential role in the structure or metabolism of plants are actively cycled in the sense that their biological role is responsible for their transfer from one compartment to another. But the plant is not a completely selective organism and many ions which play no known role in either its structure or metabolism are nevertheless cycled passively by being taken up by the plant and returned by whatever route to the soil. Sometimes the cycles of these elements are of importance because of their toxicity to plants and the consumers of plants or because of their interaction with other elements which are essential to plants. Even essential elements frequently are passively cycled in the sense that they may be taken up in luxury or even toxic quantities because of the inability of the plant to exclude them. Other passively cycled elements are some of the heavy metals such as mercury, cadmium and silver. Sometimes the passive cycling of these elements serves as an indicator of their presence in the soil. Known accumulators of gold, uranium and other metals are used as "indicator" plants, the analysis of their tissues providing information regarding concentration in the soil.

In some cases the similarity of the chemistry of an element to that of an essential element lends it a particular toxicity. Selenium is an example of this. It frequently accompanies sulfur and is substituted for sulfur in some of the amino acid and other sulfur compounds of plants and animals resulting in toxic effects. Some plants are more tolerant of selenium than others and can be accumulator plants for this element.

The process of passive ion cycling (as well as the cycling of required elements) commonly results in the concentration of these materials near the soil surface. The heavy metals in particular are comparatively immobile in soil and older soils formed on sedimentary materials frequently show high surface concentrations of them.

## 6.6 Possibilities of Deficiency

Soil formation is a complex process involving factors of parent material and climate particularly and all soils change in characteristics with time. The more mobile elements – nitrogen, sulfur, potassium, calcium, and magnesium – tend to move downward if the water supplied in rainfall exceeds that lost in runoff or evaporative loss. Other elements such as iron, which are less soluble, can still be moved downward through the root zone as chelates of organic compounds. As these elements are moved downward organic material frequently is slowly oxidized and the elements again abandoned as insoluble inorganic compounds some distance down in the soil. The consequence is formation of a hardpan layer of iron oxide or iron oxide-cemented secondary clay minerals. This layer can be quite impervious to water, changing the drainage characteristics of the system, lessening leaching and causing greater evaporative loss. The hardpan can prevent the penetration of roots depending upon the depth at which it is located, resulting in less plant growth and a further deterioration of the quality of the site. The most important element in this process is time, placing certain limitations on the productive age of a soil.

Some generalizations can be made regarding the likelihood of essential element deficiencies. Old soils are more likely to show mineral deficiencies than are young soils. Highly leached soils, either because of a light texture or heavy precipitation, are more likely to show mineral deficiencies than others. Soils with a high sesquioxide clay content or other soils with a low cation exchange capacity are more likely to show deficiencies than others. Soils in climates of abundant rainfall are more likely to show deficiencies in soluble monovalent ions such as potassium than are those in more arid environments. Soils evolved on parent material that has been subject to high-temperature solution and metamorphosis such as those formed on serpentine rock, are more likely to exhibit either deficiency or toxicity problems because of the tendency of the serpentinization process to concentrate some elements and remove others. Unfortunately, these generalizations do not always hold true and the nutrient status of the soil can be determined only by analysis and study.

## 7 "Open" Versus "Closed" Agricultural Systems

One of the major effects on the cycles of elements required by plants has been that resulting from the shift in patterns of food production and utilization (DELWICHE 1981). Urban centers have always been a part of civilization but the combination of population growth and the industrialization of agriculture have made this population concentration particularly significant during the past half-century. Most population centers are bordering the seas or on large rivers. Products of agriculture which are exported to urban centers eventually appear as organic wastes, either from domestic consumption or from industrial processes. These organic wastes are processed with varying degrees of effectiveness or sometimes not processed. In either case degradation of the organic wastes results in the liberation of the inorganic elements which they contain and these eventually are transferred to groundwaters or to rivers, estuaries, and the ocean.

In addition to the eutrophication resulting from increased levels of inorganic elements in these waters, the net result of the process is a continuous transfer of elements required by plants from the soil on which the plants were grown to waters and eventually the ocean. Soil-forming processes are slow compared with the rate of ion utilization by plants and the concentration of these ions in soils is steadily diminished. The loss of fertility is particularly evident for the elements required in largest amounts such as nitrogen, phosphorus, sulfur, and potassium, but all nutrients are affected. To make up for this loss, fertilizers are used requiring energy for the fixation of nitrogen, the mining and transport of minerals and the delivery of the fertilizer to the soil.

The waste organic materials, in addition to the nutrient elements they contain, are potential sources of energy if they can be gathered economically and provided the technology can be developed for the efficient utilization of this energy. Ideally, energy should be extracted for fuel and the mineral elements recovered so that they can be returned to the agricultural system. Transport of the organic materials back to the soil directly is impractical in most cases because of the energy requirement as well as the labor cost.

This opening of the production-utilization chain is unlikely to be reversed by any process of de-urbanization and so a technological process which is energy efficient appears to be the only solution.

Even the dispersion of populations to smaller communities would only appear to be a solution to the problem by diluting large points of concentration to a larger number of smaller points of concentration (villages) and these would not assure the return of the essential elements to the soil.

## References

Ahrens LH (1979) Origin and distribution of the elements. Pergamon, New York London

Aleem MIH (1970) Oxidation of inorganic nitrogen compounds. Annu Rev Plant Physiol 21:67–90

Aleem MIH (1977) Energy coupling in chemolithotrophic bacteria. Symp Soc Gen Microbiol 27:351–381

Alexander M (1965) Nitrification. In: Bartholomew WV, Clark FE (eds) Soil nitrogen. Am Soc Agron, Madison

Alexander MK, Marshall KC, Hirsch P (1960) Autotrophy and heterotrophy in nitrification. Trans 7th Int Congr Soil Sci, Madison 2:586–591

Bartholomew WV, Clark FE (eds) (1965) Soil nitrogen. Am Soc Agron, Madison

Berner RA (1979) Kinetics of nutrient regeneration in anoxic marine sediments. In: Ahrens LH (ed) Origin and distribution of the elements. Pergamon, New York London

Bolin B (1976) Transfer process and time scales in biogeochemical cycles. In: Svensson BH, Söderland R (eds) Nitrogen, phosphorus and sulfur – global cycles. SCOPE Rep 7. Ecol. Bull (Stockholm) 22:17–22

Bolin B, Charlson RJ (1976) On the role of the tropospheric sulfur cycle in the shortwave radiation climate of the earth. Ambio 5:47–54

Bond G (1976) The results of the IBP survey of root-nodule formation in non-leguminous angiosperms. In: Nutman PS (ed) Symbiotic nitrogen fixation in plants, vol. VII. Int Biol Progr 443–474, Cambridge Univ Press

Bormann FH, Likens GE (1979) Pattern and process in a forested ecosystem. Springer, Berlin Heidelberg New York

Bowen HMJ (1966) Trace elements in biochemistry. Academic Press, London New York

Bremner JM, Blackmer AM (1978) Nitrous oxide: Emission from soils during nitrification of fertilizer nitrogen. Science 199:295–296

Brill WJ (1977) Biological nitrogen fixation. Sci Am 236(3):68–81

Broadbent FE, Carlton AB (1978) Field trials with isotopically labeled nitrogen fertilizer. In: Nielsen DR, MacDonald JG (eds) Nitrogen in the environment, vol I. Academic Press, London New York

Broda E (1975a) The history of inorganic nitrogen in the biosphere. J Mol Evol 7:87–100

Broda E (1975b) The evolution of the bioenergetic process. Pergamon, Oxford

Crutzen PJ, Ehalt DH (1977) Effects of nitrogen fertilizers and combustion on the stratospheric ozone layer. Ambio 6(2–3):112–117

Deevey ES Jr (1970) Mineral cycles. Sci Am. 223:149–158

Delwiche CC (1970) The nitrogen cycle. Sci Am 223(3):137–146

Delwiche CC (1981) Nitrogen fertilizer. In: Bower FA, Ward RB (eds) Stratospheric ozone and man. CRC Press, Boca Raton

Delwiche CC, Bryan BA (1976) Denitrification. Annu Rev Microbiol 30:241–262
Delwiche CC, Likens GE (1977) Biological response to fossil fuel combustion products. In: Stumm W (ed) Global chemical cycles and their alterations by man. Dahlem Konferenzen, Berlin
Garrels RM, Mackenzie FT (1971) Evolution of sedimentary rocks. Norton, New York
Garrels RM, Mackenzie FT, Hunt C (1975) Chemical cycles and the global environment. Kaufman, Los Altos, Cal
Gosz JR, Likens GE, Bormann FH (1973) Nutrient release from decomposing leaf and branch litter in the Hubbard Brook Forest, New Hampshire. Ecol Monogr 43(2):173–191
Granat L, Rodhe H, Hallberg RO (1976) The global sulfur cycle. In: Svensson BH, Söderlund R (eds) Nitrogen, phosphorus and sulfur – global cycles. SCOPE Rep 7. Ecol Bull (Stockholm) 22:89–134
Hahn J (1974) The North Atlantic Ocean as a source of atmospheric $N_2O$. Tellus 26:160–168
Hardy RWF, Havelka UD (1975) Nitrogen fixation research: A key to world food? Science 188:633–643
Horn MK, Adams JAS (1966) Computer-derived geochemical balances and element abundances. Geochim Cosmochim Acta 30:279–297
Junge CE (1972) The cycle of atmospheric gases–natural and man made. 1972. QJR Meteorol Soc 98:711–729
Junge CE (1974) Residence time and variability of tropospheric trace gases. Tellus 26:477–488
Likens GE, Bormann FH (1972) Nutrient cycling in ecosystems. In: Wiens J (ed) Ecosystem structure and function. Oregon State Univ Press, Corvallis
Likens GE, Wright RF, Galloway JN, Butler TJ (1979) Acid rain. Sci Am 241(4):43–51
Mackenzie FT, Lantzy RJ, Paterson V (1979) Global trace metal cycles and predictions. Math Geol 11(2):99–142
Malavolta E, Haag HP, Mells FA, Brasil MOC (1974) Nutricão Minerale Adubacão De Plantas Cultivadas. Quazzelli, São Paulo
Penman HL (1973) The water cycle. In: Hamilton CL (ed) Chemistry in the environment. Freeman, San Francisco
Pierrou V (1976) The global phosphorus cycle. In: Svennson BH, Söderlund R (eds) Nitrogen phosphorus and sulfur – global cycles. SCOPE Rep 7. Ecol Bull (Stockholm) 22:89–134
Poldervaart A (1955) Chemistry of the earth's crust. Geol Soc Am Spec Pap 62:119
Robinson E, Robbins RC (1970) Gaseous nitrogen compound pollutants from urban and natural sources. J Air Pollut Contrib Assoc 20:303–306
Söderlund R, Svensson BH (1976) The global nitrogen cycle. In: Svensson BH, Söderlund R (eds) Nitrogen, phosphorus and sulphur – global cycles. SCOPE Rep 7. Ecol Bull (Stockholm) 22:23–72
Stumm W (1972) The acceleration of the hydrogeochemical cycling of phosphorus. In: Dyrssen D, Jagner D (eds) The changing chemistry of oceans. Wiley and Sons, New York
Taylor SR (1964) Abundance of chemical elements in the continental crust: A new table. Geochim Cosmochim Acta 28:1273–1285
Torrey JG (1976) Initiation and development of root nodules of *Casuarina* (Casuarinaceae). Am J Bot 63:335–344
Torrey JG (1978) Nitrogen fixation by actinomycete-nodulated angiosperms. BioSci 28:586–592
Wazer van F (ed) (1961) Phosphorus and its compounds, vol II. Wiley-Interscience, New York

# II. Inorganic Nitrogen Nutrition

# II.1 Physiology, Biochemistry and Genetics of Dinitrogen Fixation

H. BOTHE, M.G. YATES, and F.C. CANNON

## 1 The Nitrogen-Fixing Organisms and the Nitrogenase Reactions

### 1.1 Introduction

(Di)nitrogen fixation is the terminus technicus for the conversion of the dinitrogen molecule to ammonia according to the following equation: $6\,H^+ + 6\,e^- + N_2 = 2\,NH_3$. This reduction can proceed either chemically or biologically in a reaction catalyzed by the enzyme complex nitrogenase. The direct comparison of the data for chemical and biological nitrogen fixation reveals the importance of the latter reaction. The chemical reaction, known as the Haber-Bosch process, is used to supply nitrogen fertilizers for agriculture. Estimates of the global consumption of fertilizers in 1974 were $30 \times 10^6$ (BURNS and HARDY 1975) or $40 \times 10^6$ t $year^{-1}$ (Rothamsted Subject Day 1975), and this latter compendium stated that the world-wide consumption had increased tenfold since 1950. The Haber-Bosch process is highly energy-consuming since it proceeds at a pressure of some 200 atm and temperatures between 400°–600 °C where molecular hydrogen and nitrogen are converted to ammonia in the presence of a metal catalyst and with a yield of approximately 13%. It is said (SPRENT 1979) that about 1.5 kg of fuel oil is needed to deliver 1 kg of fertilizer N to the farm. In comparison, the rate of biological nitrogen fixation (in $10^6$ t $yr^{-1}$) was calculated to be 175 (BURNS and HARDY 1975), 170–270 (SÖDERLUND and SVENSSON 1976) and 150 (U.S. Council of Agricultural Science and Technology, Cast, Anon, 1976). These values are probably underestimates, since the contributions of the nitrogen-fixing organisms to the nitrogen balance in the sea are virtually unknown. Though the biological reaction requires the expenditure of cell energy as discussed in Section 2, it proceeds at 1 atm of pressure and at ambient temperatures. Nitrogen fixation by living cells is thus a fine example of the potential of an enzyme-catalyzed reaction. It is often said that biological nitrogen fixation, next to photosynthesis, is the most important reaction in nature. This may be the reason that nitrogen fixation has become a popular scientific subject in the past few years, resulting in a vast number of publications annually including a diversity of books (see e.g. POSTGATE 1971, 1978; HARDY 1977–1979, QUISPEL 1974, AYANABA and DART 1977, BURNS and HARDY 1975, SPRENT 1979) and reviews (WINTER and BURRIS 1976; ZUMFT and MORTENSON 1975, MORTENSON and THORNELEY 1979, ZUMFT 1976 etc.). Nitrogen fixation and its relationship to photosynthesis has been described by STEWART and QUEBEDEAUX in Chaps. 6 and 7, respectively, Vol. 6, this Series.

Relatively few groups of microorganisms contribute to the overall global nitrogen fixation rate (Table 1). Since WILSON's (1958) impressive review on nitrogen fixation in Volume 8 of the first Encyclopedia of Plant Physiology, the names of several organisms in the list have changed. Some microorganisms, previously considered to fix nitrogen, proved to be excellent scavengers of combined nitrogen (e.g. from dust) and thrive on this nitrogen source. It was not until the introduction of more sensitive assay methods ($^{15}N$-analysis, $C_2H_2$-reduction) that such organisms could be distinguished from true nitrogen-fixers. Even now the list is probably incomplete. New nitrogen-fixing organisms have been found recently and will presumably be detected in future. Two points immediately emerge from looking at Table 1: Firstly, there is no example of a nitrogen-fixing eukaryotic organism. All such candidates (yeasts) did not withstand critical examination by the more sensitive assay methods (MILLBANK 1969). Secondly, the capability for nitrogen fixation is relatively scarce but spread widely among the different groups of prokaryotes which have little else in common. Since there are no generally accepted systematic classifications for bacteria at present, nitrogen-fixing microorganisms are often listed according to the physiological characteristics of the nitrogen-fixing process (as in Table 1).

**Table 1.** List of selected genera of nitrogen-fixing organisms

| Genus or type | Species |
|---|---|
| **A. Free-living nitrogen-fixers** | |
| 1. Obligate aerobes | |
| *Azotobacter* | *vinelandii, chroococcum, paspali, beijerinckii* |
| *Beijerinckia* | *indica* |
| *Derxia* | *gummosa* |
| *Azotococcus* | *agilis* |
| 2. Obligate aerobes that fix nitrogen only at low oxygen tensions | |
| *Azospirillum* | *brasilense, lipoferum* |
| *Xanthobacter* | *autotrophicus, flavus* |
| *Thiobacillus* | *ferro-oxidans* |
| *Rhizobium* | cowpea group |
| *Methylosinus* | *sporium* |
| *Methylococcus* | *capsulatus* |
| 3. Facultative anaerobic bacteria that fix nitrogen only under anaerobiç conditions | |
| *Klebsiella* | *pneumoniae* |
| *Bacillus* | *polymyxa, macerans* |
| *Propionibacterium* | *shermanii, petersonii* |
| *Escherichia* | *intermedia* |
| *Citrobacter* | *freundii* |
| *Enterobacter* | *cloacae, agglomerans* |
| *Erwinia* | *herbicola* |

**Table 1** (continued)

| | Genus or type | Species |
|---|---|---|
| 4. Obligate anaerobes | | |
| | *Clostridium* | *pasteurianum, butyricum* |
| | *Desulfovibrio* | *gigas, desulfuricans, vulgaris* |
| | *Desulfotomaculum* | *ruminis* |
| 5. Phototrophic bacteria | | |
| | a) Rhodospirillaceae | |
| | *Rhodospirillum* | *rubrum, tenue, fulvum, molischianum, photometricum* |
| | *Rhodopseudomonas* | *palustris, viridis, capsulata, spheroides* |
| | b) Chromatiaceae | |
| | *Chromatium* | *vinosum, gracile, minus, violacea* |
| | *Thiocystis* | *violacea* |
| | *Thiocapsa* | *roseopersicina, pfennigii* |
| | *Amoebobacter* | *roseus* |
| | *Ectothiorhodospira* | *spaposhnikovii* |
| | c) Chlorobiaceae | |
| | *Chlorobium* | *limicola, vibrioforme* |
| | *Pelodiction* | *luteulum* |
| 6. Blue-green algae or cyanobacteria | | |
| | a) Unicellular, aerobically fixing strains | |
| | *Gloeothece* (formerly *Gloeocapsa*) | *alpicola* (5 strains) |
| | *Aphanothece* | |
| | b) Filamentous, heterocystous forms which fix aerobically and anaerobically | |
| | *Nostoc* | *muscorum, commune* |
| | *Anabaena* | *cylindrica, variabilis* |
| | *Aphanizomenon* | *flos – aquae* |
| | *Cylindrospermum* | Various strains |
| | *Calothrix* | Various strains |
| | c) Filamentous, non-heterocystous forms which fix microaerobically | |
| | *Plectonema* | *boryanum* |
| | *Oscillatoria* | Various strains |
| | *Pseudanabaena* | Various strains |
| | *Lyngbya* | Various strains |
| | *Phormidium* | Various strains |

**B. Symbiotic nitrogen-fixing systems and associations**

| Host family | Host genus | $N_2$-fixing microorganism |
|---|---|---|
| a) *Rhizobium* symbioses | | |
| Leguminosae | Most species | *Rhizobium* species |
| Ulmaceae | *Parasponia* | *Rhizobium* |

**Table 1** (continued)

| Host family | Host genus | $N_2$-fixing microorganism |
|---|---|---|
| b) Non-*Rhizobium* symbioses | | |
| Betulaceae | *Alnus* | |
| Myricaceae | *Myrica, Comptonia* | |
| Eleagnaceae | *Eleagnus, Hippophae* | |
| | *Shepherdia* | In all cases |
| Rhamnaceae | *Ceanothus, Trevoa, Discaria* | *Actinomycetes "Frankia"* |
| Rosaceae | *Dryas, Cercocarpus, Purshia* | |
| Coriariaceae | *Coriaria, Colleta* | |
| Casuarinaceae | *Casuarina* | |
| c) Lichens | *Collema, Peltigera* | *Nostoc* |
| | *Dendriscocaulon* | *Scytonema* |
| d) Liverworts | *Anthoceros, Blasia* | *Nostoc sphaericum* |
| | *Cavicularia* | |
| e) Waterfern | *Azolla* | *Anabaena azollae* |
| f) Cycads | *Cycas, Ceratozamia, Encephalartos, Macrozamia, Dioon* etc. | *Nostoc, Anabaena* |
| g) Higher plants | | |
| Haloragaceae | *Gunnera* | *Nostoc punctiforme* |
| h) Associative symbioses and casual associations | | |
| Phyllosphere | Leaves | *Azotobacter* spp. |
| Rhizosphere | Roots of grasses | |
| | *Paspalum notatum* | *Azotobacter paspali* |
| | *Zea mays* | *Azospirillum* |

Compiled mainly from data by BURNS and HARDY 1975, SPRENT 1979, POSTGATE 1978, SIEFERT 1976, RIPPKA et al. 1979

## 1.2 Nitrogen Fixation by Free-Living Organisms

Among the aerobic free-living forms, *Azotobacter* is the most investigated example. *Azotobacter* is present in almost every soil though never abundant. It rarely meets optimal conditions for growth and nitrogen fixation which include a sufficient supply of carbohydrates and oxygen. *Azotobacter* is said to have the highest respiratory rate among bacteria. On the other hand, it must protect its oxygen-sensitive nitrogenase from damage by the gas (see Sect. 4).

The facultative anaerobic bacteria can grow in air when provided with combined nitrogen. Nitrogen fixation, however, requires oxygen exclusion (e.g. *Klebsiella pneumoniae, Bacillus polymyxa*). In *K. pneumoniae,* nitrogenase has been studied extensively, both biochemically and genetically. Many of the techniques developed for *Escherichia coli* can be directly transferred to study the genetics of the nitrogen-fixing system in *Klebsiella* (see Sect. 7).

Among the strict anaerobic bacteria, *Clostridium pasteurianum* is the classical organism in studies on nitrogen fixation. It was used in investigations by the Du Pont group in the early 1960's which resulted in the description of the first nitrogenase preparation with reproducibly high activities (CARNAHAN et al. 1960). Moreover, it was in the course of these investigations that the electron carriers ferredoxin (MORTENSON et al. 1962) and flavodoxin (KNIGHT et al. 1966) were discovered rather than in studies on photosynthesis or nitrite reduction.

The capability for nitrogen fixation is widespread among photosynthetic bacteria. An extensive survey by SIEFERT (1976) showed 24 nitrogen-fixing strains out of 26 Rhodospirillaceae, 16 out of 19 Chromatiaceae and 13 out of 17 Chlorobiaceae.

Many of the blue-green algae, or cyanobacteria (according to bacteriologists; STANIER and COHEN-BAZIRE 1977) fix nitrogen. Under aerobic growth conditions, nitrogenase in the filamentous forms is restricted to heterocysts which are specialized cells. Whereas all heterocystous forms so far investigated are capable of fixing nitrogen, the reverse is not true. STEWART and LEX (1970) showed that the filamentous, non-heterocystous *Plectonema boryanum* fixed nitrogen when grown anaerobically or, more correctly, microaerobically, since blue-green algae evolve oxygen by photosynthesis. Several non-heterocystous strains of *Oscillatoria, Pseudanabaena* and the LPP (= *Lyngbya, Phormidium, Plectonema*) group (RIPPKA and WATERBURY 1977, RIPPKA et al. 1979) are now known to fix nitrogen microaerobically. The remarkable exceptions among the unicellular blue-green algae are *Gloeocapsa* (WYATT and SILVEY 1969, renamed *Gloeothece,* RIPPKA et al. 1979) and *Aphanothece* (SINGH 1973) which fix nitrogen under air. There is no other example of an unicellular nitrogen-fixing blue-green alga. The blue-green algal-like inclusions (termed cyanelles) which are found in the eukaryotic *Cyanophora paradoxa, Glaucocystis* and *Glaucosphaera* do not fix $N_2$ (BOTHE and FLOENER 1978).

Non-photosynthetic microaerophilic bacteria include *Xanthobacter flavus* (formerly *Mycobacterium flavum*), *Xanthobacter autotrophus* (formerly *Corynebacterium autotrophicum*) and the *Azospirillum* species, see also Chapter II.3, this Volume. In addition, some strains of *Rhizobium* which were reported to fix nitrogen in the free-living state need microaerobic conditions.

## 1.3 Symbiotic Nitrogen Fixation

The overwhelming majority of the 12,000 species of the Leguminosae form nodules containing *Rhizobium* which fix nitrogen. Nodules infected by *Rhizobium* have also been found in the non-legume *Parasponia* (TRINICK 1973, AKKERMANS et al. 1978). Nodules containing nitrogen-fixing actinomycetes, called *Frankia,* occur at the roots of several plants which bear no systematic relationship. Blue-green algae are symbiotic partners of a diversity of eukaryotic hosts (see Table 1 and the contribution by QUISPEL Chap. II.2, this Vol.). The associa-

**Table 2.** Estimates of the contributions of nitrogen-fixing microorganisms in the field

| Plant | | Associative organisms | Fixation rate |
|---|---|---|---|
| 1. Rhizobium symbioses | | | |
| Soybean: | *Glycine max* | *R. japonicum* | 40–200 |
| Peanut: | *Arachis hypogaea* | Cowpea *Rhizobium* | 70–240 |
| Field bean: | *Vicia faba* | *R. leguminosarum* | 45–300 |
| Red clovers: | *Trifolium* sp. | *R. trifolii* | 45–350 |
| Lucerne: | *Medicago sativa* | *R. meliloti* | 50–460 |
| 2. Actinomycetes symbioses | | | |
| Alder grove: | *Alnus* sp. | *Frankia* | 50–150 |
| Dunes with: | *Casuarina equisetifolia* | *Frankia* | 58 |
| Bog myrtle: | *Myrica gale* | *Frankia* | 27 |
| Sea buckthorn: | *Hippophae rhamnoides* | *Frankia* | 180 |
| 3. Blue-green algal symbioses | | | |
| Fern: | *Azolla* | *Anabaena azollae* | 50–80 |
| Lichen: | *Lobaria* | *Nostoc* (+*fungus*) | 5–10 |
| Cycad: | *Macrozamia riedlei* | *Nostoc* | 18 |
| Higher plant: | *Gunnera* | *Nostoc* | 58 |
| 4. Associative symbioses and casual associations | | | |
| Grasses: | *Paspalum notatum* | *Azotobacter paspali* | 10 |
| | *Zea mays* (maize) or *Saccharum officinarum* (sugar cane) | *Klebsiella, Bacillus, Azospirillum, and Beijerinckia* | <3 |
| 5. Free living organisms | | | |
| Blue-green algae | | | 80–120 |
| *Azotobacter, Clostridium, Klebsiella* and other free-living $N_2$-fixers | | | <1 |

Rates are given in kg N fixed $ha^{-1}$ $yr^{-1}$. The data are grossly estimated from field studies and are therefore subject to large variations (climate, soil, coverage by the plants). Table 2 is compiled principally from data cited by SPRENT 1979, POSTGATE 1978, VINCENT 1974, MISHUSTIN and SHIL'NIKOVA 1971 and the Rothamsted Experimental Station (Rothamsted Subject Day 1975)

tive symbiosis between free-living nitrogen fixing bacteria and plants are discussed by DÖBEREINER (Chap. II.3, this Vol.).

Table 2 contains gross estimates of the maximum nitrogen fixation rates by different organisms in the field. It is generally accepted that legumes contribute most to global nitrogen fixation. The waterfern *Azolla* containing the blue-green alga *Anabaena azollae* forms a green manure which increases the nitrogen supply substantially in rice fields of southeast Asia (see SPRENT 1979). The contributions of the free-living organisms to global nitrogen fixation are marginal. Probably the only organisms of serious ecological importance are the blue-green algae which populate many different environments and can improve the nitrogen balance significantly.

### 1.4 Substrates of Nitrogenase

Nitrogenase catalyzes the reduction of a variety of substrates (Table 3), most being small molecules which, like $N_2$, contain triple bonds. The apparent $K_m$ values depend on the organisms assayed and whether intact cells or purified nitrogenase is tested. Nevertheless, dinitrogen clearly has the highest affinity for nitrogenase (Table 3). The $K_m$ value for proton reduction may be as low but this cannot be determined. The reduction of $C_2H_2$ to $C_2H_4$, discovered by DILWORTH (1966) and SCHÖLLHORN and BURRIS (1967), has become a standard technique for determining nitrogen fixation activities. These two gases are easily separated and determined quantitatively by gas chromatography where less than 10 pmol of $C_2H_4$ in the assays can be detected accurately. This method is quick and inexpensive compared with other nitrogenase assays (e.g. $^{15}N$ incorporation by mass spectrometry). For detailed descriptions of the techniques the reader should consult articles by BURRIS (1972, 1975) and HARDY and HOLSTEN (1977). All nitrogen-fixing bacteria except the methane oxidizer, *Methylosinus,* when grown on methane, will reduce acetylene to ethylene. In methane-oxidizing organisms acetylene inhibits the oxidation of $CH_4$ to $CH_3OH$ (DEBONT and MULDER 1974), thus preventing the supply of energy and reducing equivalents to nitrogenase.

The acetylene reduction test is therefore a useful qualitative indication of nitrogen-fixing ability. However, it is often unwise to make quantitative assessments of nitrogen-fixing activity in the field on the basis of a 6- and 2-electron requirement for $N_2$ and $C_2H_2$ reduction respectively. Quantitative field tests

**Table 3.** The different substrates of nitrogenase

| Substrate utilized | Product formed | Electrons transferred | Apparent $K_m$ (M) | Relative rate |
|---|---|---|---|---|
| 1. $N_2$ | $2NH_3$ | 6 | $2\times10^{-5}$–$1\times10^{-4}$ | 1 |
| 2. $N_3^-$ | $NH_3$, $N_2$ | 2 | $2\times10^{-4}$–$1\times10^{-3}$ | 3 |
| 3. $N_2O$ | $N_2$, $H_2O$ | 2 | $1\times10^{-3}$ | 3 |
| 4. HCN | $CH_4$, $NH_3$<br>($CH_3NH_2$) | 6<br>4 | $2\times10^{-4}$–$4\times10^{-3}$ | 0.6 |
| 5. $CH_3CN$ | $C_2H_6$, $NH_3$ | 6 | $5\times10^{-1}$ | 0.004 |
| 6. $CH_3NC$ | $CH_3NH_2$, $CH_4$ | 6 | $2\times10^{-4}$–$2\times10^{-3}$ | 0.8 |
| 7. $C_2H_2$ | $C_2H_4$ | 2 | $4\times10^{-4}$–$1\times10^{-3}$ | 3–5 |
| 8. $2H^+$ | $H_2$ | 2 | Not determined | 4 |
| 9. 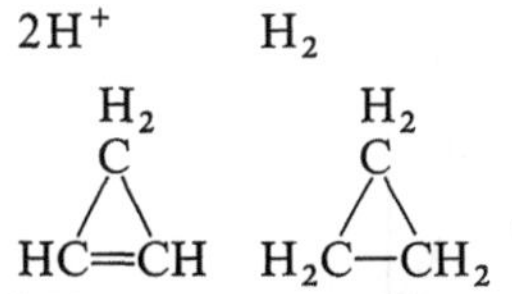 | , $CH_3CH{=}CH_2$ | | | |

These data are adopted from tables given by DALTON and MORTENSON 1972, ZUMFT and MORTENSON 1975, ZUMFT 1976, BURNS and HARDY 1975, MCKENNA and HUANG 1979 where the original references are given.

have shown that the $C_2H_2/N_2$ conversion factor varies from 2 to 25 (HARDY et al. 1973). Possible reasons for such variations include: (1) a greater proportion of electrons are diverted to $H_2$ production when $N_2$ rather than $C_2H_2$ is the substrate. (2) $C_2H_2$ inhibits several metabolic processes, including nitrification, denitrification, methane oxidation, methanogenesis and bacterial hydrogenase activity. It also inhibits the processes which oxidize the ethylene which is formed naturally in soils. In addition, repeated exposures to acetylene may induce conformational changes in nitrogenase which enhance the rate of acetylene reduction (see YATES 1980). Such effects must distort the simple $C_2H_2:N_2$ conversion factor.

It has been suggested that nitrogenase may have evolved to detoxify acetylene or cyanides present during the anaerobic phase of the earth's development. After this period combined nitrogen became limiting and the enzyme developed the ability to reduce $N_2$ (SILVER and POSTGATE 1973). An alternative theory (POSTGATE 1974) is that nitrogenase is a relatively young enzyme evolved on plasmids which were incorporated into the genetic material of various organisms. Either possibility would explain the rather haphazard distribution of nitrogenase amongst prokaryotic organisms, but both theories lack substantial evidence for their support.

## 2 Biochemistry of Nitrogen Fixation

### 2.1 Introduction

Nitrogenase consists of two iron-sulphur proteins (see Fig. 1), which, individually, have no detectable enzymic activity but together catalyze ATP hydrolysis and all substrate reductions. One of these is called the MoFe protein; it contains up to 36 Fe atoms, an almost equivalent number of acid labile sulphide ions, and 2 Mo atoms per 220,000 M.W. The other, the Fe protein, contains one [4 Fe-4 S] cluster per ~60,000 M.W. Both proteins are oxygen-sensitive, the Fe protein extremely so. Thus an anaerobic environment is necessary to assay nitrogenase activity. A quick and simple assay involves sodium dithionite as

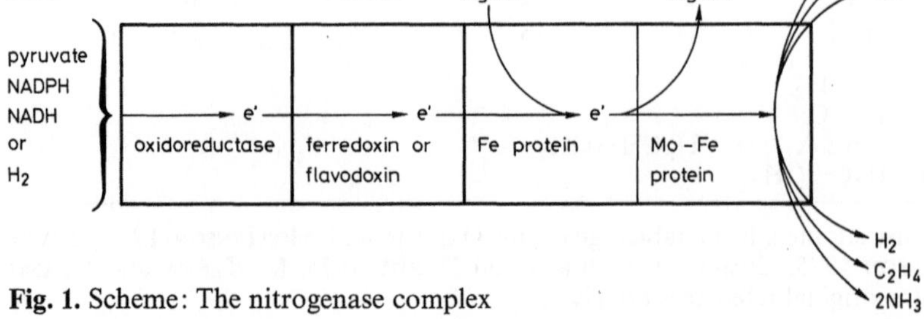


Fig. 1. Scheme: The nitrogenase complex

the reductant, an ATP-generating system consisting of creatine phosphokinase, ADP and phosphocreatine, nitrogenase proteins in an anaerobic environment at neutral pH and acetylene as the reducible substrate. The ATP-generating system is included because ADP inhibits nitrogenase activity and must be kept at a negligible concentration to ensure a linear reduction rate.

Nitrogenase proteins have to be purified under strict anaerobic conditions. These include sparging all buffers with $O_2$-free $N_2$, using low concentrations of sodium dithionite to react with the remaining traces of $O_2$ and dithiothreitol to protect sulfhydryl groups. When these precautions are taken, active nitrogenase proteins can be isolated by conventional techniques of enzyme purification, including ion exchange and molecular sieve chromatography. Nitrogenase proteins are strongly coloured (brown), stable at room temperature and are one of the major protein constituents in $N_2$-fixing bacteria. These three properties ease purification procedures considerably. Review articles covering the early developments of the biochemistry of nitrogen fixation include those by HARDY and BURNS (1968), BURRIS (1971), ZUMFT and MORTENSON (1975) and EADY and SMITH (1979). Purification procedures for nitrogenase have been reviewed by EADY (1980).

## 2.2 Nomenclature of Nitrogenase Proteins

The MoFe protein of nitrogenase is called component 1, molybdoferredoxin, azofermo or nitrogen reductase; the Fe protein is called component 2, azoferredoxin, azofer or nitrogenase reductase. The names nitrogen- and nitrogenase-reductase were given in an attempt to assign biochemical roles (HAGEMAN and BURRIS 1978a). The nomenclature used in this review is mechanistically noncommittal: involving the initials of the parent bacterium and the order of elution from DEAE cellulose. Thus Av1, Kp1 and Cp1 represent the MoFe proteins and Av2, Kp2 and Cp2 the Fe proteins of *Azotobacter vinelandii, Klebsiella pneumoniae* and *Clostridium pasteurianum* respectively (EADY et al. 1972).

## 2.3 Physicochemical Properties of Nitrogenase Proteins

Nitrogenase proteins have been isolated from more than 20 organisms; ten MoFe proteins and seven Fe proteins have been partly characterized (Table 4). Important basic criteria include molecular weights, subunit contents and numbers of metal atoms and acid labile sulphur groups per protein. The molecular weights of the MoFe proteins range from 180,000 to 270,000, they contain 18 to 36 g atoms of Fe and 1 to 2 g atoms of Mo $mol^{-1}$; acid-labile sulphides, when determined, are usually 2 to 3 g ions $mol^{-1}$ below the Fe values. It is not known whether the range of molecular weights and metal contents reported reflect species differences, impure preparations or the inadequacy of analytical methods. The most active preparations, obtained either from *A. vinelandii, K. pneumoniae* or *C. pasteurianum* contain >30 g atoms Fe and 2 g atoms Mo/mol. It is now generally accepted that the MoFe proteins have four subunits of

**Table 4.** Properties of highly purified nitrogenase proteins[c]

| | MoFe proteins | | | | | |
|---|---|---|---|---|---|---|
| | Cp1 | Bp1 | Kp1 | Rr1 | Ac1 | Anc1[b] |
| Molecular weights | | | | | | |
| By gel filtration | 210,000 | 215,000 | 220,000 | 230,000 (215,000) | 216,000 | 230,000 |
| By ultracentrifugation | 201,000 | | 200,400 | | | |
| By disc electrophoresis | | 240,000 | 217,000 | 234,000 | 230,000 | 216,000 |
| From amino acid composition | 221,800 | | 229,000 | | 227,000 | |
| Sum of subunit Mol.wts. | 220,000 | | 221,000 | | 240,000 | |
| Subunit Composition | Two types | Not known | Two types | Not known | Probably 2 types | Two types |
| Mol wts of the subunits | 50,700 59,500 | | 51,300 59,600 | 58,500 (56,000) | 60,000 Approx. | 52,800 55,000 |
| Mo g atom/mol | 2 | 2 | 2.01 | 1.7 (2) | 1.9±0.3 | 2 |
| Fe g atom/mol | Approx. 24 | 33 | 32.5±3 | 20 (25–30) | 22±2 | 20 |
| Acid-labile sulfide g ion/mol | Approx. 24 | 21 | 16.7±1[a] | (19–22) | 20±2 | 20 |
| $O_2$ sensitivity $t_{1/2}$ min | | | 10 | | 10 | |

[a] This value refers to a Kp1 preparation containing 17.5 g atoms Fe/mol
[b] HALLENBECK et al. (1979)
[c] Adapted from Yates 1980

two types ($\alpha_2\beta_2$). Analytical evidence, involving sodiumdodecylsulphate-polyacrylamide gel electrophoresis (SDS-PAGE) followed by peptide and amino acid analysis of the separated subunits, as well as genetical evidence, indicating the presence of separate genes for the two subunit types, support this conclusion.

The Fe proteins of nitrogenase from most sources contain two identical subunits when analyzed by SDS-PAGE. This has been confirmed by amino acid sequence analysis of Cp2 (TANAKA et al. 1978). The one exception is the Fe protein from *R. rubrum* which contains two non-identical subunits of molecular weights 31,500 and 30,000 respectively (LUDDEN and BURRIS 1976). Molecular weights of Fe proteins range from 55,000 to 75,000 (Table 4).

All nitrogenase proteins are acidic, containing a preponderance of glutamic and aspartic acid residues. An arithmetical analysis of their relatedness shows that they are quite similar (EADY and SMITH 1979), although Cp1 is distinguished

| | | | Fe proteins | | | | | |
|---|---|---|---|---|---|---|---|---|
| Xa1 | Rj1 | Rl1 | Cp2 | Bp2 | Kp2 | Rr2 | Ac2 | Xa2 |
| 235,000 | 180,000 | 184,000 | 56,000 | 55,000 | 62,000 | (65,000) | 63,000 | |
| | | | 36,000 | | 68,000 | | | |
| 237,000 | | | | 62,000 | | 61,500 | | |
| 223,500 | | | | | 67,800 | | 57,000 | 82,900 |
| 237,400 | 228,000 | 228,000 | 58,000 | | 69,200 | | 61,600 | 73,000 |
| Two types | ? | ? | One type | One type | One type | Two types | One type | One type |
| 61,200<br>57,500 | 55,000 | 57,000 | 27,000 | | 34,600 | 30,000<br>31,500 | 30,800 | 36.500 |
| 2.2 | 1.3 | | | | | | | |
| 23.1 | 28.2 | | 4 | 3.2 | 4 | 3.5<br>(4.15) | 4 | 3.8 |
| 20.1 | 26.2 | | 4 | 3.6 | 3.85 | 3.9 | 3.9 | 2.4 |
| | | | | | | >0.5 | 0.75 | |

Cp *Clostridium pasteurianum;* Bp *Bacillus polymyxa;* Kp *Klebsiella pneumoniae;* Rr *Rhodospirillum rubrum;* Ac *Azotobacter chroococcum;* Anc. *Anabaena cylindrica;* Xa *Xanthobacter autotrophicus;* Rj *Rhizobium japonicum;* Rl *Rhizobium leguminosarum* bacteroids

by its relatively low tryptophan content. These similarities may account in part for the ability of MoFe protein from one species to combine with the Fe protein from another to produce active nitrogenases. Sequence analysis shows that Cp2 protein has 273 amino acid residues. The basic amino acids occur randomly while the acidic residues occur in clusters. It has no sequence homology with other types of iron-sulphur proteins (e.g. ferredoxins). A partial sequence of the $\alpha$ and $\beta$ subunit of Av1 has also been made (LUNDELL and HOWARD 1978).

## 2.4 Metal Clusters in Nitrogenase Proteins

Nitrogenase proteins can be studied by several spectroscopic techniques. These include electron paramagnetic resonance (EPR), Mössbauer, nuclear magnetic resonance (NMR), extended X-ray absorption edge and fine structure spectroscopies and ultraviolet (UV)-visible spectrophotometry.

Metal clusters in iron-sulphur proteins are, usually, of two types: 2- and 4-iron although some proteins apparently contain 3-iron clusters (see Sect. 3). The Fe atoms are linked to each other by acid-labile sulphide bridges and to the protein by cysteine residues. The [2 Fe-2 S] clusters are planar and the [4 Fe-4 S] type are cubane. The former undergoes oxidation and reduction between the fully oxidized, ferric-ferric state and the partly reduced ferric-ferrous state. [4 Fe-4 S] clusters can, theoretically, be in any one of five oxidation levels, designated 0 to 4 (latest nomenclature, Eur J Biochem 93:427–430 1978). States 1 and 3 are paramagnetic and have epr signals; states 0, 2 and 4 are diamagnetic and EPR-silent. According to the most recent publications there are two major types of iron centres in the MoFe protein of nitrogenase: the iron-molybdenum cofactor (FeMoco), and cubane type [4 Fe-4 S] clusters. The number of clusters of each type within the molecule is uncertain but tentative conclusions are possible on the basis of evidence which will be discussed in the following sections.

## 2.5 EPR and Mössbauer Spectroscopy on the MoFe Protein

The MoFe protein of nitrogenase possesses a unique EPR spectrum with g values close to 4.3, 3.7 and 2.01. It is observed only below 30 K and is probably a single rhombic EPR signal associated with one type of Fe, probably that in the FeMo cofactor. Early Mössbauer studies by SMITH and LANG (1974) on Kp1 protein containing 18 g atoms Fe $mol^{-1}$ showed the presence of 3 Fe species containing 2, 8 and 8 Fe atoms respectively. MUNCK et al. (1975) made similar findings with Av1 containing 18 g atoms Fe $mol^{-1}$.

Recent studies (for references see HUYHN et al. 1979) have reinterpreted the Mössbauer data using Cp1, Kp1 and Av1. The general conclusions are that dithionite-reduced MoFe proteins should contain $30 \pm 2$ g atoms Fe $mol^{-1}$ in four states.

These include (1) 12 Fe atoms associated with the EPR signal (called $M_{EPR}$), (2) a component, prominent at 4.2 K, containing 41% of the total Fe, (3) a $Fe^{2+}$ component (13% total Fe) and (4) an unidentified state (S) containing 2 Fe $mol^{-1}$. Components (2) and (3) together form 3 to 4 [4 Fe-4 S] clusters, each containing 3 Fe from component (2) and 1 Fe from component (3). These P clusters, as they are called, are new structures since they are EPR-silent when oxidized or reduced, but their Mössbauer spectra indicate that they are closely related to conventional [4 Fe-4 S] clusters.

Cluster displacement techniques, using organic solvents to extract metal clusters from proteins and UV-visible spectrophotometry or EPR spectroscopy to identify them, have confirmed the Mössbauer data. ORME-JOHNSON et al. (1977) extracted Cp1 with o-xylyl-$\alpha$, $\alpha'$-dithiol in the presence of DMSO (80%) in tris buffer, pH 8.0, and associated the extruded Fe-S clusters with the apoproteins of *B. polymyxa* ferredoxin [4 Fe-4 S] or adrenodoxin [2 Fe-2 S]. From the resulting EPR spectra they calculated 3 [4 Fe-4 S] clusters in Cp1 containing 24 g atoms Fe $mol^{-1}$. KURTZ et al. (1979) analyzed extruded cores of Cp1 spectrophotometrically and found 3.4 to 4 (interpreted as 4) 4 Fe-4 S centres/$\alpha_2\beta_2$ subunit. Because the clusters were not so readily extruded as [4 Fe-4 S] clusters in ferredoxins, the authors speculated that the P-cluster of MoFe proteins may not be bound to the protein by cysteine residues.

## 2.6 The FeMo Cofactor and the Fe Protein

Many molybdo-enzymes (e.g. xanthine oxidase, nitrate reductase and nitrogenase) contain Mo cofactors; the nitrogenase cofactor also contains Fe. Mutants of *A. vinelandii* and *K. pneumoniae* containing inactive MoFe proteins which are activated by acid-treated Av1 or Kp1 have been used to isolate the FeMo cofactors (called FeMoco) from these organisms (SHAH and BRILL 1977, SMITH 1980). The cofactor contains the Mössbauer spectral features of $M_{EPR}$ and its EPR spectrum is similar to that of the holoprotein (RAWLINGS et al. 1978, SMITH 1980).

FeMoco is unstable and $O_2$-sensitive. It contains Mo, Fe and acid-labile sulphur, but their proportions are disputed: that from Av1 has a Mo:Fe:$S^{2-}$ ratio of 1:8:6 according to SHAH and BRILL (1977) or 1:6:6 to 8 according to HUYHN et al. (1979). FeMoco from Kp1 has a ratio of 1:8:4 (B.E. SMITH personal communication). In some preparations it has an associated peptide.

The Mo atom in FeMoco or nitrogenase cannot be studied by Mössbauer, nor, because it is either diamagnetic or masked by Fe signals, by EPR. Extended X-ray absorption spectroscopy (EXAFS), however, has yielded interesting data about the environment of Mo in both FeMoco and the MoFe protein. The absorption edge and fine structure spectra of Cp1, Av1 and FeMoco are very similar and show the presence of Mo-S and Mo-Fe ligands but not Mo=O unless the protein is damaged by $O_2$. Fourier transform analysis of EXAFS data indicate that Mo is bound to 3 or 4 S atoms at 2.36 Å and 2 or 3 Fe atoms at 2.7 Å spacings. There may also be one or two S atoms bound with longer Mo-S bonds (CRAMER et al. 1978). Chemical complexes containing structures analogous to [4 Fe-4 S] clusters with Mo replacing one Fe have been synthesized (e.g. CHRISTOU et al. 1978). Their EXAFS spectra are very similar to that of nitrogenase but their chemical compositions do not correspond with FeMoco.

To summarize, present knowledge of the structure of the MoFe protein indicates that it is an $\alpha_2\beta_2$ iron-sulphur protein containing 2 Mo atoms and 30±2 Fe atoms/M.W. 220,000. The Mo atoms and 12–16 Fe atoms form the FeMo cofactor(s) which is EPR-active and is thought to contain the active site of nitrogenase. Sixteen Fe atoms occur in [4 Fe-4 S] clusters which are EPR-silent and which may not be bound to the protein by cysteine residues. Two other Fe atoms are spectroscopically inert and may be involved in the quaternary structure (SMITH and LANG 1974). This analysis conforms to a symmetrical structure. It is not known whether FeMoco contains 1 or 2 Mo $mol^{-1}$.

All the Fe in the Fe protein of nitrogenase is present as a single [4 Fe-4 S] cluster. Evidence from four different spectroscopic techniques indicate that this is so: (1) Mössbauer spectroscopy; SMITH and LANG (1974) showed that the

Mössbauer spectrum of reduced Kp2 was similar to that of reduced [4 Fe-4 S] ferredoxin; (2) cluster displacement techniques: ORME-JOHNSON et al. (1977) extracted Cp2 Fe and showed that it was similar to that of *C. pasteurianum* ferredoxin which has [4 Fe-4 S] clusters; (3) Linear electric field effect (LEFE), which measures the shift in the EPR spectrum under the influence of a strong electric field. This shift is large when the charge around the centre is non-symmetrical as in a [4 Fe-4 S] cluster. Such was the case with Cp2 (ORME-JOHNSON et al. 1977). (4) Magnetic circular dichroism (MCD): the MCD of reduced Av2 or Kp2 was similar to that of a [4 Fe-4 S] ferredoxin (STEPHENS et al. 1979).

So much evidence in favour of a single [4 Fe-4 S] cluster raised two interesting questions. Firstly, how is a single cluster distributed between two identical subunits? Secondly, the oxidized Fe protein will take up two electrons $mol^{-1}$ upon reduction, one rapidly and the other slowly. Only the rapid phase correlates with the appearance of the EPR spectrum (rhombic with g values close to 2.05, 1.94 and 1.87) associated with the [4 Fe-4 S] cluster. Neither the nature of this second electron-accepting site nor its relevance to enzyme activity is known. Further evidence indicating its presence is that the EPR spectrum of the Fe protein integrates to less than 1 electron $mol^{-1}$ which suggests the presence of two interacting paramagnetic centres (LOWE 1979). It is difficult to envisage the nature of the second centre since all the Fe atoms appear to be associated with the [4 Fe-4 S] centre.

## 2.7 Nitrogenase Proteins in Photosynthetic Organisms

Some nitrogenase proteins have been purified from symbiotic and photosynthetic sources (Table 4). MoFe proteins from symbiotic systems and from *A. cylindrica* conform with the general pattern. Homogeneous preparations of Fe proteins have not yet been obtained from these organisms.

The Fe protein from the photosynthetic bacterium *R. rubrum* is unusual in that it has non-identical subunits, is associated with equimolar concentrations of adenine, ribose and phosphate ions and requires a protein-activating factor which may function to remove these ligands (LUDDEN and BURRIS 1978).

## 2.8 The Mechanism of Nitrogenase Activity

As mentioned earlier, nitrogenase requires two component proteins, a source of electrons, ATP, $Mg^{2+}$ plus an ATP-generating system, and an anaerobic environment. In the absence of any added reducible substrate, such as $N_2$ or $C_2H_2$, nitrogenase will convert protons to $H_2$. This ATP-dependent $H_2$ evolution also occurs, though to a lesser degree, in the presence of added reducible substrates, thus complicating kinetic studies of their reduction. The questions to be resolved are obvious: they include (a) the roles of the two component proteins; (b) the roles of ATP and ADP; (c) the number and nature of the reducing sites and (d) the mechanism of $N_2$ reduction.

### 2.8.1 The Roles of the Two Proteins

The present opinion of the function of the two proteins is as follows: the Fe protein accepts electrons from the electron carrier (see Sect. 3) and then binds Mg ATP which causes a conformational change to lower the redox potential; the Fe protein·Mg ATP then complexes with the MoFe protein and transfers one electron to the latter with the concomitant hydrolysis of Mg ATP to form Mg ADP. The electrons are then transferred to the reducible substrate by the MoFe protein. The series of steps are depicted by the following equations:

$$Fe_{ox} \overset{e^-}{\rightleftarrows} Fe_{red} \tag{1}$$

$$Fe_{red} + Mg\,ATP \rightarrow Fe_{red} \cdot Mg\,ATP \tag{2}$$

$$Fe_{red} \cdot Mg\,ATP + MoFe \cdot N_2 \rightarrow Fe_{red} \cdot Mg\,ATP \cdot MoFe \cdot N_2 \tag{3}$$

$$Fe_{red} \cdot Mg\,ATP \cdot MoFe \cdot N_2 \rightarrow Fe_{ox} \cdot Mg\,ADP \cdot MoFe_{super\text{-}red} \cdot N_2 + Pi \tag{4}$$

$$Fe_{ox} \cdot Mg\,ADP \cdot MoFe_{super\text{-}red} \cdot N_2 \rightarrow Fe_{ox} \cdot + Mg\,ADP + MoFe + NH_3 \tag{5}$$

There are many unresolved questions in this scheme: why are 2 to 2.5 ATP's hydrolyzed per electron transferred? Does the complex between the two proteins dissociate after each electron transfer or is it longer-lived than the enzyme turnover? What is the rate-limiting step responsible for a relatively slow enzyme turnover? What is the stoichiometry of complex formation? Other questions will emerge later.

There are several lines of evidence for the role of the Fe protein. (1) The oxidized Fe-protein is rapidly reduced by sodium dithionite (YATES et al. 1975), a reduction which is inhibited by ADP ($K_i = 9\ \mu M$; THORNELEY and YATES unpublished data), whereas the oxidized MoFe protein is reduced slowly and no further than the half-reduced state. Using stopped-flow spectrophotometry THORNELEY (1975) obtained direct evidence that Kp2 transferred electrons to Kp1 in an ATP-dependent reaction. This substantiated earlier deductions from EPR measurements (see ZUMFT and MORTENSON 1975). EADY et al. (1979) showed that ATP was hydrolyzed concomitantly with this electron transfer (2.3 Mg ATP hydrolyzed per electron transferred). (2) Evidence that Mg ATP and Mg ADP bind to the Fe protein is summarized as follows: Mg ATP increases the sensitivity of the Fe protein to $O_2$, towards the sulfhydryl group reagent 5,5′-dithiobis-(2-nitrobenzoic acid) (DTNB) and towards Fe-chelating reagents. Mg ADP, on the other hand, has little effect on these sensitivities. Either Mg ATP or Mg ADP change the EPR spectrum of the Fe protein from a rhombic to an axial form and also change the mid-point potential from $-280$ to $\sim -400$ mV (see ZUMFT and MORTENSON 1975). TSO and BURRIS (1973) measured the level of $^{14}$C-ATP or $^{14}$C-ADP in Cp2 after equilibration between an aqueous phase and Sephadex G 50. They showed that $Mg^{2+}$ was necessary to bind both nucleotides and calculated two binding sites for Mg ATP with a dissociation constant ($K_D$) of 16.7 μM and one for Mg ADP ($K_D$ 5.2 μM). They concluded that Mg ATP inhibited Mg ADP binding, whereas Mg ADP inhibited ATP binding at one site but enhanced binding at the other site. WYETH and ORME-JOHNSON (personal communication) have repeated these experiments but obtain three to five binding sites for Mg ATP. BISHOP et al. (1977) studied the binding of $Mn^{2+}$ or $Mg^{2+}$ to Kp2 by enhancement of the proton NMR relaxation rate. They showed four sites of Mn or Mg-binding and evidence for ternary complexes involving metal, protein and ATP or ADP. However, they could not distinguish the number of nucleotide binding sites. To summarize, although the evidence is not conclusive, it is reasonable to assume that the Fe protein possesses at least two binding sites for Mg ATP and one or more for Mg ADP.

The MoFe protein is capable of existing in three oxidation states, two of which are associated with enzyme activity. These are the EPR signal state and a "super-reduced" state obtained during enzyme activity by ATP-mediated electron transfer from the Fe protein. An oxidized state, obtained by treating the EPR-signal-state with thionine, is not considered to be part of the active process because the mid-point potential of this reaction is not conserved: i.e. it varies between MoFe proteins from different sources (O'DONNELL and SMITH 1978).

Addition of acetylene or a change in pH can effect the EPR spectrum of the MoFe protein which indicates that either protons or acetylene bind to the MoFe protein but not to the Fe protein. This is evidence that the MoFe protein may contain the substrate-reducing site. $CN^-$ binds to both nitrogenase proteins, but this may be a non-specific effect (BIGGINS and KELLY 1973). There is no evidence that $N_2$ binds to either of the isolated proteins.

Electron transfer between the nitrogenase proteins cannot occur without complex formation beforehand. The stoichiometry and lifetime of this complex are questions which have received much attention since they bear on the location and nature of the active site. Ultracentrifuge experiments showed that the nitrogenase proteins from *K. pneumoniae* or *A. chroococcum* aggregated in equimolar complexes (EADY 1973). Titration data, measuring nitrogenase activity when the Fe protein concentration was constant and the MoFe protein concentration increased, indicated an equimolar ratio for maximum activity. Studies of the dilution effect, which is a disproportionately low activity at low enzyme concentration, attributed to dissociation of the component proteins, also indicated a 1:1 molar complex. However, titration data also exist supporting the presence of an active 2:1 Fe protein:MoFe protein ratio (BERGERSEN and TURNER 1973, HAGEMAN and BURRIS 1978b). The latter authors implied that the nature of the reducible substrates may influence the stoichiometry of the complex. All these approaches can be criticized on the grounds that these apparent molar ratios may be influenced by the percentages of inactive proteins in preparations; such experiments should be performed with proteins of the highest specific activity.

The occurrence of the dilution effect and the fact that reductant-free Kp1 and Kp2 form a complex, which suggests that the two nitrogenase proteins are complexed at the end of the catalytic cycle, tentatively support the concept of a long-lived complex. However, HAGEMAN and BURRIS (1978a), using a high ratio of Av1:Av2 to slow down electron-flow, found that ATP hydrolysis was linear whereas a time lag occurred before $H^+$ was reduced to $H_2$. They concluded that electrons were transferred singly and that dissociation of Av1 and Av2 occurred after each transfer, otherwise the second electron would transfer rapidly following the first and $H_2$ would be evolved immediately. From these observations they proposed that the MoFe protein should be called nitrogen reductase and the Fe protein nitrogenase reductase.

### 2.8.2 Evidence for Interaction of Mg ATP and Mg ADP with the MoFe Protein

HAGEMAN and BURRIS (1978a) asserted that ATP is associated only with electron transfer from the Fe- to the MoFe-protein. However, nitrogenase possesses a reductant-independent ATPase which requires both proteins (JENG et al. 1970). Secondly, ATP hydrolysis is almost independent of electron transfer at high ratios of MoFe protein to Fe protein or at low temperatures (see MORTENSON and THORNELEY 1979). Thirdly, the ratio of moles of ATP hydrolyzed per electron transferred varied with reducible substrate in the presence of antibody to the MoFe protein (RENNIE et al. 1978). Finally, chemical analogues of ATP alter the electron distribution between $H_2$ production and $C_2H_2$ reduction by nitrogenase (see MORTENSON and THORNELEY 1979). These observations suggest, but do not unequivocally prove, a role for ATP associated with the MoFe protein.

THORNELEY and CORNISH-BOWDEN (1977) found that the kinetics of inhibition by Mg ADP of the electron transfer between the proteins were different from those of $H_2$ production suggesting that Mg ADP inhibited the latter activity at a different site, possibly on the MoFe protein.

### 2.8.3 The Nature of the Active Site(s)

Nitrogenase is capable of reducing several substrates (Table 3) but kinetic studies suggest that the active site for each substrate may not be identical. The results of these studies indicate, according to RIVERA-ORTIZ and BURRIS (1975), that there are four different substrate-reducing sites: (1) $N_2$ and, possibly, $N_2O$; (2) azide, HCN and methylisocyanide; (3) $C_2H_2$; (4) $H^+$. Whether the recently discovered substrate cyclopropene (MCKENNA and HUANG 1979) is reduced at yet a fifth site is not known. Finally, kinetic data are available to suggest that, at high concentrations, acetylene induces a second reduction site in vitro (DAVIS et al. 1979) and in vivo (WALKER and YATES 1978a). Alternatively, substrates may be reduced by different (substrate-induced) forms of the enzyme (SMITH BE et al. 1976).

The chemical nature of the reducing sites for nitrogenase substrates has not been resolved, but there is considerable evidence that metal atoms, probably Fe and Mo, are involved. Mössbauer and EPR spectroscopy and stopped-flow spectrophotometry all indicate that Fe atoms are oxidized and reduced during the catalytic cycle. LOWE et al. (1978), from EPR and Mössbauer data, have devised an elegant scheme explaining the oxidation/reduction cycles of FeS clusters during $H^+$ and $C_2H_2$ reduction. However, it is likely that Mo is the metal involved in the active substrate reducing site, whereas FeS clusters are involved in electron transfer. However, evidence for the location and function of Mo is circumstantial: demolybdo-Cp1 is inactive (see ZUMFT and MORTENSON

1975) and addition of FeMoco (the FeMo cofactor) activates extracts from mutants of *A. vinelandii* or *K. pneumoniae* defective in one of the nitrogen fixation genes (*nif* B) (SHAH and BRILL 1977a, SMITH 1980). Organometallic complexes where $N_2$ is bound to Mo can be reduced to yield $NH_3$.

### 2.8.4 Pathways of $N_2$-Reduction

$NH_4^+$ or $NH_3$ was the only known intermediate of $N_2$-fixation until THORNELEY et al. (1978) devised conditions for accumulating an intermediate which yielded hydrazine upon treatment of the enzyme with acid or alkali. The fact that nitrogenase catalyzed $N_2$-dependent hydrogen-deuterium (HD) exchange led several investigators to suggest that an unstable di-imide-type enzyme-bound intermediate occurred which reacted with $D_2$ according to the equation:

$$\mathrm{E{-}\underset{H}{N}{=}\underset{H}{N}} + D_2 \longrightarrow \mathrm{E{-}N_2} + 2\,\mathrm{HD}$$

(NEWTON et al. 1977). However, THORNELEY et al. (1978) thought it more likely that the intermediate which gave rise to hydrazine was an enzyme-bound complex of type $\mathrm{E{-}Mo{\equiv}N{=}NH_2}$. A chemical model to explain how $N_2$ bound to Mo can be reduced to ammonia has been proposed by CHATT et al. (1978):

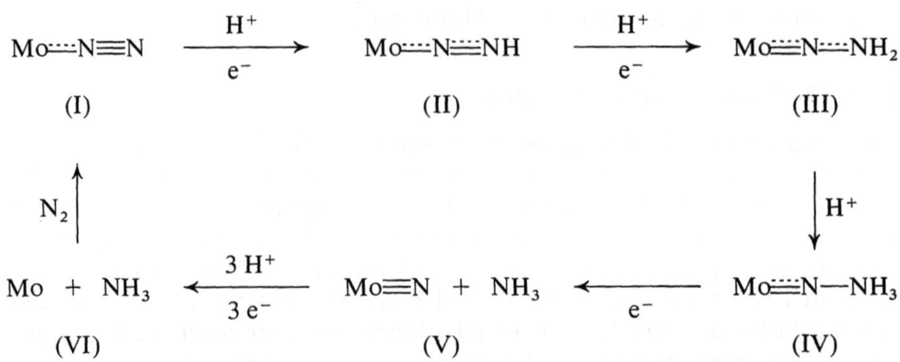


The scheme envisages the reduction of the Mo in six one-electron steps, the electrons being provided via the iron-sulphur clusters in the enzyme. In this way, the Mo need not change its oxidation by as many as six redox states during the catalytic cycle.

The model must explain ATP-dependent $H_2$ evolution by nitrogenase. CHATT (1980) envisaged the active site in a cleft in the molecule which limited proton access. Mo at this site is bound to hydrogen atoms which are replaced by $N_2$.

This means an obligatory evolution of 1 mol $H_2$ per $N_2$ reduced according to the equation:

$$N_2 + 8\,H^+ + 8e^- \rightarrow 2\,NH_3 + H_2$$

In the absence of $N_2$, or other reducible substrates, protons combine with $H^-$ to yield $H_2$. CO, $H_2$ or $D_2$ inhibited $N_2$ reduction by displacing $N_2$ from this site. CHATT suggested that this fully reduced ($3\ H^-$) state of the enzyme is the only state in which $N_2$ can be reduced and in which $D_2$ can bind and form deuteride ions which, in turn, are displaced by $N_2$ to give HD exchange. When CO is bound the protons react with the third $H^-$ to discharge $H_2$.

The scheme does not satisfactorily explain why very little or no HD exchange occurs in the absence of $N_2$. Nevertheless, it is an attractive working model based on the observed properties of the enzyme and on characterized organometallic complexes involving Mo and N in various oxidation states.

# 3 Electron Transport to Nitrogenase

## 3.1 Introduction

This subject has been reviewed in detail by several authors (YATES 1977, EVANS and PHILLIPS 1975, JOCH and VALENTINE 1972). In experiments with isolated nitrogenases, the most commonly used reductant is sodium dithionite which does not require additional electron transport components in the assays. Natural carriers which transfer the electrons to nitrogenase are ferredoxin and flavodoxin. These proteins can be substituted by the low redox potential electron carriers methyl- or benzylviologen. Natural electron donors which reduce ferredoxin or flavodoxin and thus couple cell metabolism to nitrogen fixation are pyruvate, NAD(P)H and, possibly, formate or hydrogen (see Table 5). The reactions between electron donors and carriers are enzyme-catalyzed.

**Table 5.** The electron transport to nitrogenase in some representative microorganisms

| Organism | Electron carrier | Electron donor | Enzyme |
|---|---|---|---|
| 1. *Clostridium pasteurianum* | Ferredoxin (flavodoxin) | Pyruvate<br>NADH | Pyruvate: fd oxidoreductase<br>NADH: fd oxidoreductase |
| 2. *Bacillus polymyxa* | 2 ferredoxins | Pyruvate | Pyruvate: fd oxidoreductase<br>Formate: $CO_2$ oxidoreductase (?) |
| 3. *Azotobacter* sp. | Flavodoxin<br>2 ferredoxins | NADH (?) | Possibly NADH: flavodoxin oxidoreductase |
| 4. *Rhizobium* sp. | Ferredoxin (flavodoxin) | NADH (?) | NADH: ferredoxin oxidoreductase |
| 5. *Mycobacterium flavum* | 2 ferredoxins | ? | ? |
| 6. *Klebsiella pneumoniae* | Flavodoxin? | ? | ? |
| 7. Blue-green algae | Ferredoxin (flavodoxin) | NADPH<br>Pyruvate<br>Photo-system I, $H_2$ | NADPH: ferredoxin oxidoreductase<br>Pyruvate: ferredoxin oxidoreductase<br>Hydrogenase via photosystem I |

### 3.2 Ferredoxins (for a review see YOCH and CARITHERS 1979)

These iron-sulphur proteins are active not only in nitrogen fixation but also in a variety of redox reactions. Examples are the photosynthetic $NADP^+$-reduction and the cyclic photophosphorylation of chloroplasts, the assimilatory thiosulphonate and nitrite reductions in plants, the conversion of nitrate to nitrite in some organisms (blue-green algae, *Azotobacter*) as well as the hydrogenase and the pyruvate phosphoroclastic reactions (see Sect. 3.4) of *Clostridium pasteurianum*. Such reactivity requires an electronegative redox potential of about −400 mV, which is a characteristic feature of ferredoxins. Electron transfer is mediated through [FeS] clusters in the protein (see Sect. 2). Ferredoxins contain either one [2 Fe-2 S], one [4 Fe-4 S] or two [4 Fe-4 S] clusters. In the redox reactions, they usually operate between the +2 and +1 oxidation levels (see Sect. 2). They are paramagnetic when reduced, have a typical EPR-spectrum with a Landé-Factor $g<2$ and are diamagnetic when oxidized. Other types of ferredoxins have been isolated from nitrogen-fixing *A. vinelandii* (SWEENEY et al. 1975), *X. flavus* (BERNDT et al. 1978) and *Rhodospirillum rubrum* (YOCH et al. 1977). These are paramagnetic in the oxidized state and EPR-silent in the reduced form. They operate between the +2 and +3 oxidation states. Some organisms (*B. polymyxa, X. flavus, R. rubrum, Azotobacter*) form constitutively more than one soluble ferredoxin which can act independently as electron carriers to nitrogenase when assayed in vitro. The physiological roles of these different proteins have not been established in many cases. However, in *R. rubrum* ferredoxin I is formed only in the light, whereas ferredoxin II is present both in the light and during heterotrophic growth in the dark. In addition to these soluble ferredoxins, *R. rubrum* contains two iron-sulphur proteins (ferredoxin III and IV) which are integral parts of the chromatophore membranes (SHANMUGAM et al. 1972, YOCH et al. 1977). Other nitrogen-fixing microorganisms also have membrane-bound iron-sulphur centres with unresolved biological and chemical characteristics.

### 3.3 Flavodoxins (for a review see MAYHEW and LUDWIG 1975)

These are small proteins of about 20,000 m.w. having one molecule of FMN in the prosthetic group. They are distinguished from other flavoproteins by the formation of a stable radical or semiquinone form in high yields. Consequently, flavodoxins have the two different redox couples: the oxidized/semiquinone and semiquinone/hydroquinone forms. Only the latter redox couple, having a potential of −350 to −500 mV, is sufficiently electronegative to act in lieu of ferredoxin. In all reactions tested (including nitrogen fixation) flavodoxins substitute for ferredoxins, and, in most cases, with similar specific activities. The physiological role of flavodoxin has not been established in all cases. In some organisms (*Clostridium*, blue-green algae, *Rhodospirillum*) flavodoxin is only formed under iron deficiency when the biosynthesis of ferredoxin is limited. In *Azotobacter* and the non-nitrogen-fixing *Escherichia coli* both ferredoxin and flavodoxin are formed constitutively. Surprisingly, *Azotobacter* appears to be the only organism which may utilize flavodoxin exclusively as the electron carrier to nitrogenase (for the details see YATES 1980). *Xanthobacter flavus,* on the other hand, forms two ferredoxins and no flavodoxin (BOTHE and YATES 1976). Also in *Rhizobium*, ferredoxin is predominantly formed whereas the flavoprotein is synthesized in small amounts and only under some culture conditions (CARTER et al. 1980).

### 3.4 Electron Donors (for a review see YATES 1977, 1980, BOTHE et al. 1980)

The best-known electron donor is pyruvate (see Table 5) which, in the presence of coenzyme A, is split to acetylcoenzyme A and $CO_2$; the remaining two electrons reduce ferredoxin. This pyruvate phosphoroclastic reaction requires the participation of the enzyme pyruvate: ferredoxin oxidoreductase which was purified and characterized in some detail from *Clostridium acidiurici* (UYEDA and RABINOWITZ 1971). Acetylcoenzyme A can be converted to acetylphosphate and further to ATP catalyzed by phosphotransacetylase and acetate kinase. Thus the pyruvate cleavage provides reduced ferredoxin and ATP for nitrogenase and other reactions. Although ideal for nitrogen fixation, the pyruvate phosphoroclastic reaction occurs only in *Clostridium* species, in *Bacillus polymyxa* and in blue-green algae. Pyruvate is also the best electron donor to nitrogenase in *K. pneumoniae,* though the manner of the pyruvate cleavage has not been established. Even in *C. pasteurianum* pyruvate is not the only electron donor. Fermentation balances indicated that approximately 25% of the electrons to reduce ferredoxin came from NADH (THAUER et al. 1969). The enzyme involved, NADH: ferredoxin oxidoreductase, requires acetylcoenzyme A as an allosteric effector. Ferredoxin was also reduced by $H_2$ plus hydrogenase or formate plus formate-dehydrogenase by Clostridial extracts. However, this is believed to be an artefact of cell-free systems, since hydrogenase favours $H_2$-evolution and formate dehydrogenase the formate cleavage in vivo (THAUER et al. 1969).

In the aerobically growing filamentous blue-green algae *Nostoc* and *Anabaena* nitrogen fixation takes place in the heterocysts. Neighbouring vegetative cells supply these specialized cells with carbohydrates which are degraded via the hexosemonophosphate shunt to provide NADPH. Heterocysts were shown to contain high levels of the two key enzymes of this pathway, glucose-6 phosphate dehydrogenase and 6-phosphogluconate dehydrogenase. Since NADPH: ferredoxin oxidoreductase is also present in heterocysts, a major route of electron donation is likely to comprise glucose-6-phosphate, the dehydrogenase, NADPH, NADPH:ferredoxin oxidoreductase, ferredoxin, nitrogenase (see BOTHE 1970 and several publications thereafter). An additional or alternative pathway may involve pyruvate as an electron donor to nitrogenase. Pyruvate: ferredoxin oxidoreductase was observed in blue-green algae (LEACH and CARR 1971) and was found to be particularly active under nitrogen-fixing conditions (BOTHE et al. 1974). The presence of this enzyme in heterocysts, however, remains to be established. Nitrogen fixation is light-stimulated in isolated heterocysts. Light is presumably necessary to meet the ATP requirements for nitrogen fixation by cyclic photophosphorylation, whereas the reducing equivalents may be generated in the dark. However, the involvement of photosystem I in the transfer of electrons to nitrogenase cannot be ruled out. Photosystem I is definitely involved in electron donation to nitrogenase when $H_2$ and hydrogenase are electron donors in heterocysts (see Sect. 6). In these cells, the electrons cannot directly come from the photosynthetic water-splitting reaction, since they lack photosystem II. In the unicellular, non-heterocystous alga, *Gloeocapsa*, nitrogen fixation is more dependent on the operation of the two photosynthetic reactions than in heterocystous forms (Gallon 1980).

The pathways of electron transport to nitrogenase in *Azotobacter, Rhizobium* or *X. flavus* and surprisingly *Klebsellia pneumoniae* are less well understood. There are two conflicting theories of electron transfer to nitrogenase in *Azotobacter*. According to Benemann et al. (1971), the main electron donor is NADPH. These authors observed low, but positive $C_2H_2$-reduction activities with the system NADPH, NADPH:ferredoxin oxidoreductase (from spinach), *Azotobacter* ferredoxin and flavodoxin, unidentified factor(s) and nitrogenase. The rates which they reported were approximately 1% of the nitrogen-fixing activity in vivo, and the spinach reductase could not be substituted by an enzyme native to *Azotobacter*. NADH also supports nitrogenase activity: Yates (1971) observed significant $C_2H_2$-reduction activities using NADH, a homogeneous dehydrogenase from *Azotobacter chroococcum,* benzylviologen and nitrogenase. However, benzylviologen could not be replaced by ferredoxin, flavodoxin or any other component from *Azotobacter*. Since the redox potential of NAD(P)H/NAD(P)$^+$ is $-320$ mV, the reduction of ferredoxin or flavodoxin requires a high level of NAD(P)H in the cells. However, Haaker et al. (1974) did not find a correlation between the NAD(P)H content and the nitrogenase activity in intact cells. Moreover, high NAD(P)H levels inhibit key enzymes of *Azotobacter* (Yates and Jones 1974). To circumvent such difficulties, Haaker et al. (1974) postulated that an energized membrane was the driving force to generate reductants for nitrogenase in aerobes. According to the Mitchell hypothesis, the proton motive force consists of the $\Delta$ pH and $\Delta\,\psi$ terms. Since Laane et al. (1979a, b) did not find a correlation between proton gradients and nitrogenase activity, they assumed that the membrane potential was the driving force for generating reductant to nitrogenase in *Azotobacter* and *Rhizobium leguminosarum.* This latter assumption was substantiated by the observation that the membrane de-energizing agent 4:5:6:7 tetrachloro-2-trifluoromethyl-benzimidazole (TTFB) inhibited nitrogenase activity but not respiratory ATP-formation in *Azotobacter*. This hypothesis, though easily reconcilable with modern thinking, is difficult to verify experimentally. In particular, reduction of flavodoxin beyond the semiquinone state by the membrane-bound NADH: ferredoxin oxidoreductase (Laane et al. 1978) has not been demonstrated in vitro. The situation is even less resolved in *Rhizobium.* Poly-$\beta$-hydroxybutyrate, the main carbon-storage compound of nodules, was thought to be the electron donor for nitrogenase. However, neither poly-$\beta$-hydroxybutyrate nor the activity of the NAD$^+$-linked $\beta$-hydroxybutyrate dehydrogenase correlated with nitrogenase activity during different experimental conditions (Evans and Phillips 1975).

In conclusion, many aspects of the electron transfer to nitrogenase have still to be elucidated. It has become clear, however, that different organisms utilize different pathways and that one organism does not restrict itself to only one electron donor.

# 4 Mechanisms to Protect Nitrogenase Against Damage by Oxygen

## 4.1 In Free-Living Organisms

Aerobic nitrogen-fixing bacteria require oxygen to produce ATP and must, therefore, possess mechanisms to prevent oxygen from inhibiting or damaging nitrogenase. Even in *Azotobacter,* the most oxygen-tolerant of nitrogen-fixers, nitrogenase is inhibited by high concentrations of $O_2$. This can easily be demonstrated by measuring nitrogenase activity at increasing oxygen tensions when a "bell-shaped" curve of activity versus $pO_2$ is produced, increasing to a maximum as the level of available ATP increases and then declining as excess $O_2$ inhibits nitrogenase activity. The range of oxygen tensions at which the nitrogenase can function increases with the population density and the carbon-substrate concentration. Nitrogenase activity can also be "switched off" abruptly by increasing the oxygen solution rate to a level faster than the culture can respire.

The mechanisms which protect nitrogenase from damage by $O_2$ in aerobes have been reviewed exhaustively (see YATES 1977). The simplest concept is that of respiratory protection: the organism respires rapidly enough to maintain an anaerobic environment at the nitrogenase site. *Azotobacter* has the highest known respiratory quotient amongst bacteria and has a branched electron transport chain which is largely uncoupled to ATP production at high $O_2$ tensions, thereby allowing an increased respiration rate.

A second protection mechanism against $O_2$ in *Azotobacter* is known as conformational protection. This is indicated by:

1. nitrogenase in crude extracts is oxygen-tolerant whereas the purified enzyme is oxygen-sensitive;
2. nitrogenase activity is inhibited but apparently not damaged by a sudden increase in the oxygen solution rate in vivo.

This protection is provided by an equimolar complex of the nitrogenase proteins and a third iron-sulphur protein previously known as the Shethna [2 Fe-2 S] protein (SHETHNA 1970). $Mg^{2+}$ ions are necessary for maximum stability of this complex. It has been suggested that this complex also regulates nitrogenase activity as follows: the complex forms between the oxidized proteins and is enzymically inactive; it dissociates and nitrogenase becomes active only when the [2 Fe-2 S] protein is reduced (SCHERINGS et al. 1977; for a recent review see ROBSON and POSTGATE 1980). Other organisms which may also possess some form of conformational protection include *X. flavus* (BIGGINS and POSTGATE 1970) and *K. pneumoniae* (HILL 1976).

Another defence against excess $O_2$ is the formation of slime, which may decrease the oxygen solution rate in liquid culture. Hill (1971) suggested that massive colony formation by *Derxia gummosa* which involved slime production under nitrogen-fixing conditions, also protected against excess $O_2$ during growth on agar. However, Wilcockson (1977) found no correlation between slime production and $O_2$ by nitrogen-fixing *Klebsiella* on solid agar.

The mechanism whereby oxygen causes damage to nitrogenase is not resolved. The most favoured hypothesis at present is that the products of $O_2$ metabolism, hydrogen peroxide, superoxide- and hydroxyl-radicals, are the damaging agents. However, although aerobic organisms possess superoxide dismutase and catalase, enzymes which convert these products to $O_2$ and $H_2O$, there is no convincing evidence that they serve to protect nitrogenase.

### 4.2 The Heterocysts of Blue-Green Algae

In the filaments of blue-green algae, the heterocysts can easily be distinguished from the bluish vegetative cells by their thick envelope and their more greenish colour due to the absence (or reduced content) of phycocyanin. There is little doubt among the different experimenters that heterocysts are the site of nitrogenase under aerobic growth conditions. Heterocysts are often said to provide an anaerobic compartment for nitrogenase. Clearly, since they are devoid of the photosynthetic water-splitting system II, nitrogenase inside these cells need not be protected from oxygen evolved photosynthetically. It has been suggested that nitrogenase is protected against $O_2$ from the air by four unique glycolipids, detected in the laminated layer of the heterocyst envelope, which may bind oxygen, thus providing a diffusion barrier (Lambein and Wolk 1973). Indeed, nitrogen fixation is sensitive to air in *Anabaena* mutants lacking these glycolipids (Haury and Wolk 1978). However, heterocysts must have access to $O_2$, since they can respire and support $N_2$ fixation in the dark by oxidative phosphorylation. These observations clearly indicate that oxygen must reach the inside of heterocysts. The gas does not necessarily penetrate the cell envelope, although this cannot be ruled out, but can come from vegetative cells to which heterocysts are connected by plasma bridges called microplasmodesmata (Fay and Lang 1971). Thus, heterocysts, like other aerobes, must protect nitrogenase from damage by oxygen inside the cell. Little information is available about the nature of such protective mechanisms; respiratory protection probably plays an essential role. Site I of the respiratory chain is NADPH-linked in blue-green algae. This is consistent with the findings that the major route of hexose degradation in heterocysts is the oxidative pentose phosphate cycle where the activity levels of the glucose-6-phosphate- and the 6-phosphogluconate dehydrogenases are particularly high. Thus the oxidation of NADPH by respiration may remove oxygen from the nitrogenase site. The Knallgas reaction (see Sect. 6) may also protect nitrogenase from damage by $O_2$. Recent investigations (Haury and Wolk 1978, Gotto et al. 1979) with $O_2$-sensitive mutants from *Anabaena* indicate that the protective mechanisms for nitrogen fixation in heterocysts involve different phenomena and that there may be several sites sensitive to oxygen.

### 4.3 The Role of Leghaemoglobin in Legume Nodules

Leghaemoglobin is abundant in legume nodules and, although it was recognized many years ago to be necessary for effective nitrogen fixation, it is only recently, through the work of APPLEBY, BERGERSEN, WITTENBERG and their colleagues, that its role is understood (APPLEBY et al. 1976). It has two roles: it facilitates the transport of oxygen from the bacteroid envelope to the bacteroid surface and also acts as an "oxygen buffer" to prevent too great a variation in the concentration of oxygen at the bacteroid surface. In this second function it could act to protect nitrogenase from damage by oxygen, but current opinion is that the oxygen concentration within the nodule is too low to cause damage to nitrogenase.

One unresolved problem is the location of leghaemoglobin: is it confined within the bacteroid envelope or also present in the cell cytoplasm? Some investigators have observed leghaemoglobin in nodule electron micrographs to be present both within the bacteroid envelope and in the cell cytoplasm, but its presence in the latter may have been due to damaged and "leaky" bacteroid envelopes.

## 5 Regulation of Nitrogenase Activity and Biosynthesis

Both biosynthesis and activity of the enzyme are subject to complex regulations which are not fully understood, and which may differ between organisms.

### 5.1 Regulation of Nitrogenase Biosynthesis

Nitrogenase, generally, is only formed in the absence of combined nitrogen ($NO_3^-$, $NH_4^+$ or organic nitrogen) in the culture medium. Earlier physiological studies of *nif* regulation in *K. pneumoniae* which showed that nitrogenase was repressed in the presence of ammonia have been substantiated by results which show that nitrogenase polypeptides are not synthesized in cultures grown on ammonia (EADY et al. 1978). Furthermore, the synthesis of polypeptides recently assigned to other *nif* genes is also similarly repressed (ROBERTS et al. 1978). TUBB and POSTGATE (1973) demonstrated that ammonia represses nitrogenase synthesis at the transcriptional level using inhibitors of initiation of mRNA synthesis (rifampicin) and of protein synthesis (chloramphenicol) during derepression of samples taken from fully repressed, sulphate-limited continuous cultures.

The construction of small amplifiable plasmids carrying *nif* DNA sequences has made it possible to detect *nif*-specific mRNA directly using the Southern hybridisation technique (SOUTHERN 1975). JANSSEN et al. (1980) detected mRNA which hybridized to DNA restriction fragments representing six of the seven known *nif* operons in cultures derepressed for *nif*. The *nifJ* operon was not examined since it had not been cloned at the time of the experiment. In contrast *nif*-specific mRNA was not detected in ammonia-grown cultures which were repressed for *nif*. The conclusion from these results is that the *nif* operons for which mRNA was detectable under the experimental conditions used are not transcribed in vivo in ammonia-grown cultures.

$N_2$ does not induce the synthesis of the enzyme, since this occurs when the organisms are incubated under argon. Nitrogenase, therefore, is derepressed in the absence of fixed N. There is evidence that glutamine synthetase (GS) rather than ammonia regulates nitrogenase biosynthesis. GS plays a key role in ammonia incorporation by nitrogen-fixing organisms where it catalyzes the reaction:

$$\text{glutamate} + NH_4^+ + \text{ATP} \rightarrow \text{glutamine} + \text{ADP} + \text{Pi} \qquad (1)$$

Glutamate is reformed from glutamine by glutamate synthase (=L-glutamine: 2-oxoglutarate amidotransferase, abbreviated GOGAT) (MEERS et al. 1970):

$$\text{L-glutamine} + \alpha\text{-ketoglutarate} \xrightarrow{2e^-} 2\ \text{L-glutamate} \qquad (2)$$

The reductant may either be NADPH, NADH or reduced ferredoxin. Other ammonia-assimilating pathways (e.g. those catalyzed by glutamate or alanine dehydrogenases) are of minor physiological significance in $N_2$-fixing bacteria. The apparent $K_M$-value of these enzymes for $NH_4^+$ lie between $1 \times 10^{-4}$ and $1 \times 10^{-3}$ M, concentrations which are probably higher than those normally present in $N_2$-fixing cells.

That GS regulates nitrogenase biosynthesis is indicated by studies with mutants (e.g. see STREICHER et al. 1974) and inhibitors (see BISHOP et al. 1975). Glutamine-requiring mutants from *K. pneumoniae* that lack active glutamine synthetase do not express nitrogenase activity. When such mutants reacquire glutamine synthetase as a result of conjugation with an episome carrying the genes of this enzyme, they gain the ability to synthesize nitrogenase. In addition *Klebsiella* mutants which were unable to regulate GS by adenylylation were able to synthesize nitrogenase even in the presence of $NH_4^+$. L-methionine-DL-sulfoximine and L-methioninesulfone, both analogues of glutamine, specifically block glutamine synthetase. These compounds also derepress nitrogenase synthesis in *Klebsiella, Rhizobium* or *Anabaena* when grown in the presence of $NH_4^+$.

The mechanism by which glutamine synthetase affects nitrogenase synthesis is not clearly understood. Its activity is controlled by its degree of adenylylation; the deadenylylated enzyme being the active form. At high concentration $NH_4^+$ ions are assimilated via glutamate dehydrogenase. This results in a high ratio of glutamine to α-ketoglutarate which, in turn, inhibits the uridylyl transferase involved in the control of glutamine synthetase activity. This prevents deadenylylation of glutamine synthetase and, consequently, derepression of nitrogenase. Apparently, $NH_4^+$ does not repress nitrogenase synthesis via glutamine synthetase in all systems (BISHOP et al. 1976). Carbamylphosphate was also reported to regulate nitrogenase biosynthesis (see YATES 1977). However, it is not clear at present whether this compound acts by producing $NH_4^+$.

Repression of nitrogenase biosynthesis by oxygen has been studied in *K. pneumoniae* and *A. chroococcum*. The method used was to pulse-label the organism with either $^{14}C$- or $^{35}S$-labelled amino acids during derepression of nitrogenase. In both instances the repression by $O_2$ followed different kinetics from that by $NH_4^+$. Moreover, mutants of *K. pneumoniae* exist in which nitrogenase synthesis is repressed by $O_2$ but not by $NH_4^+$ (EADY et al. 1978, ROBSON 1979). These observations indicate clearly that the mechanism of repression by $O_2$ is different from that by $NH_4^+$.

There is some evidence to suggest that lack of Mo also represses nitrogenase synthesis (see YATES and EADY 1980) but other evidence indicates that lack of Mo merely results in the synthesis of inactive nitrogenase polypeptides. It is also possible that active MoFe protein of nitrogenase, or a Mo storage protein act as regulators (see POSTGATE et al. 1981).

## 5.2 Regulation of Nitrogenase Activity

Ammonium ions do not inhibit nitrogenase activity in vitro but do inhibit in several organisms in vivo, particularly in the photosynthetic nitrogen-fixing bacteria *Rhodopseudomonas capsulata, R. palustris* and *Rhodospirillum rubrum.* Complete inhibition of nitrogenase occurs within minutes of adding $NH_4^+$ to the cells (see MEYER et al. 1978). A possible explanation is that $NH_4^+$ causes an increase in glutamine synthetase activity thereby depleting the intracellular ATP pool. The situation in *Azotobacter* is more complex: addition of $NH_4^+$ causes a partial inhibition of nitrogenase activity which is reversible in some instances (see YATES and EADY 1980). LAANE et al. (1979a) showed that $NH_4^+$ de-energized membranes in *A. vinelandii* and proposed that $NH_4^+$ inhibited the energized-membrane-mediated electron transfer to nitrogenase in this organism. By contrast, $NH_4^+$ did not inhibit membrane energization nor nitrogenase activity of bacteroids, nor does it inhibit nitrogenase in *K. pneumoniae* or *C. pasteurianum* (see YATES and EADY 1980).

Inhibition by oxygen has been considered earlier (Sect. 4), but it is important to distinguish between damage to the enzyme and fine control of activity. Most aerobic bacteria can tolerate only low levels of $O_2$ when fixing $N_2$; high levels of $O_2$ damage nitrogenase within the cell. The exception to this rule is *Azotobacter* where nitrogenase activity is apparently undamaged at high $O_2$ levels because of conformational protection. This implies that $O_2$, which may inhibit nitrogenase activity by auto-oxidation of the electron donor flavodoxin hydroquinone (YATES and JONES 1974) may also regulate activity through the oxidation states of the proteins comprising the conformationally-protected complex (Sect. 4.1).

Nitrogenase activity is energy-intensive, hydrolyzing 12–15 molecules of ATP per molecule of $N_2$ reduced. Growth yield measurements indicate that this value may be even higher in vivo. Because it has a low catalytic turnover rate ($\sim$0.5 s) it is necessary for bacterial cells to synthesize large amounts of the enzyme: it may comprise as much as 15% of protein in crude cell supernatant (ORME-JOHNSON et al. 1977). Perhaps the most direct control of nitrogenase activity is by the ATP/ADP ratio: ADP is a potent inhibitor of nitrogenase activity and its apparent $K_i$ is lower than the $K_M$ for ATP. These properties mean that a small increase in Mg ADP concentration, within the right range, should cause a rapid inhibition of nitrogenase activity (Thorneley and Cornish-Bowden 1977).

Depending on the organism, ATP may be provided by fermentation, oxidative phosphorylation or – in the phototrophs – from cyclic photophosphorylation. In blue-green algae, the reducing equivalents to nitrogenase are not provided directly through the photolysis of water and via the two photosystems. Nitrogen-fixation rates in these organisms are dependent on the size of the reductant pool of carbohydrates (Lex and Stewart 1973). Carbohydrates are probably the major limiting factor in all nitrogen-fixing organisms outside laboratory-controlled conditions. This is particularly true of legume-*Rhizobium* symbiosis (see Hardy et al. 1978). $CO_2$-enriched atmospheres increased the dry weight of plants, nodules and the specific nitrogen-fixing activity of legumes. Thus approaches to improve the availability of photosynthate to the nodule would enhance nitrogen fixation and nitrogen input to the plant. Such studies, aiming to construct more efficient plant–*Rhizobium* symbioses may be more rewarding than attempts to find mutants with higher nitrogenase activities.

## 6 The Hydrogenase-Nitrogenase Relationship

Recent compendia in this field are from Schlegel and Schneider (1978) and a special issue of Biochimie (Vol. 60:3, 1978).

The reaction $2\,H^+ + 2e = H_2$ is catalyzed by enzymes. Cells may contain three different proteins that catalyze either uptake, or evolution or both uptake and evolution of molecular $H_2$ under different physiological conditions. Three clearly distinguishable proteins have been described to occur in $N_2$-fixing microorganisms:

a) The "classical" hydrogenase from *Clostridium,* and other anaerobes or facultative anaerobes, couples to ferredoxin, flavodoxin or a variety of redox dyes. It has a molecular weight of about 60,000, mediates an exchange reaction with $D_2O$, is sensitive to inhibition by carbon monoxide and contains three [4 Fe:4 S] centres of which only one is believed to undergo oxidation/reduction during catalysis. Under physiological conditions, the enzyme catalyzes the evolution of molecular hydrogen to dispose of excess reducing equivalents generated during fermentation.

b) The unidirectional $H_2$-evolution catalyzed by nitrogenase can easily be distinguished from the hydrogenase-dependent formation of the gas by its insensitivity to carbon monoxide and dependence on ATP. The mechanism of this ATP-dependent reduction of protons is not fully understood at present. Nitrogenase is believed to have at least two different $H_2$-evolving sites (Sect. 2). Clearly, all hydrogen formation catalyzed by nitrogenase is a loss of energy and reducing power for the organisms. It should be kept in mind that $H_2$-evolution always occurs during nitrogenase activity in vitro, though with variable rates, but is often marginal or zero in intact organisms.

c) The third type of enzyme activity is by the unidirectional, $H_2$-uptake hydrogenase first observed in *Azotobacter* by Phelps and Wilson (1941). This enzyme is present in all aerobic nitrogen-fixing organisms and is membrane-bound in *Azotobacter*, blue-green algae and bacteroids. High levels of such an enzyme have recently been observed in the anaerobic bacterium *C. pasteurianum*, where it is obviously soluble (CHEN and BLANCHARD 1978). The uptake hydrogenase activity is not restricted to cells that are actually fixing $N_2$, but $N_2$-fixing conditions enhance specific activities. Enzyme levels also generally increase after exposure to molecular hydrogen. At present several groups are attempting to purify such enzymes to homogeneity (see SCHLEGEL and SCHNEIDER 1978). The enzymes from *Rhizobium*, *Azotobacter* and blue-green algae are stable to oxygen when particulate and do not (or only poorly) couple with ferredoxin, flavodoxin or other low potential natural electron carriers.

There are four possible roles for the uptake hydrogenase in $N_2$-fixing aerobes (mainly suggested by DIXON 1972). Some of these have been verified experimentally:

a) Since $H_2$ is produced by nitrogenase, this gas could be reoxidized by the uptake hydrogenase. Such a $H_2$-utilization could be coupled to respiratory oxygen consumption. It could provide the organisms with ATP and thus minimize the loss of energy caused by the ATP-dependent $H_2$-formation by nitrogenase. Indeed, ATP formation by the oxyhydrogen reaction was demonstrated in particles from *Azotobacter* and heterocysts from *Anabaena*. Such a function of hydrogenase, to recycle energy lost by nitrogenase, may be particularly important under carbon-limited conditions as shown for *Azotobacter*, *Rhizobium* and blue-green algae (see WALKER and YATES 1978b, BOTHE et al. 1978).

b) Hydrogen uptake catalyzed by hydrogenase could protect nitrogenase from damage by oxygen by means of the Knallgas reaction and aid respiratory protection of nitrogenase at low carbon substrate concentration. Indeed, $C_2H_2$-reduction was found to be more stable to oxygen when hydrogen was present in assays with intact *Azotobacter*, *Anabaena variabilis* or heterocysts isolated from the latter organism.

c) Since high concentrations of molecular hydrogen were shown to inhibit $N_2$-reduction, hydrogenase could act to remove hydrogen produced by nitrogenase in vivo. Such a function is unlikely, since $H_2$ evolved by nitrogenase probably disperses too rapidly to the environment and does not reach the concentrations required to have any effect on nitrogenase in nature (WALKER and YATES 1978b).

d) $H_2$ could be an electron donor to nitrogenase. Hydrogenase, isolated from *Clostridium pasteurianum*, was used to provide electrons to cell-free nitrogenase from *Clostridium*, *Azotobacter* or *Anabaena*. $H_2$ may act as an electron donor to nitrogenase in *Azotobacter chroococcum* in vivo, particularly when grown under low carbon substrate concentrations (WALKER and YATES 1978a, b). In photosynthetic bacteria, the electron donation from $H_2$ to nitrogenase requires light activation: under such conditions, $H_2$ was shown to be the best electron donor for $C_2H_2$-reduction in heterocysts isolated from *Anabaena* (EISBRENNER and BOTHE 1979). These authors postulated two different pathways of $H_2$-utilization in blue-green algae: an oxygen-linked one proceeding in the dark and a light-dependent one requiring the strict exclusion of oxygen. A common intermediate of both pathways is presumably plastoquinone, which, however, is not necessarily the immediate electron acceptor from hydrogenase. This observation that $H_2$ and hydrogenase feed in electrons close to the plasto-

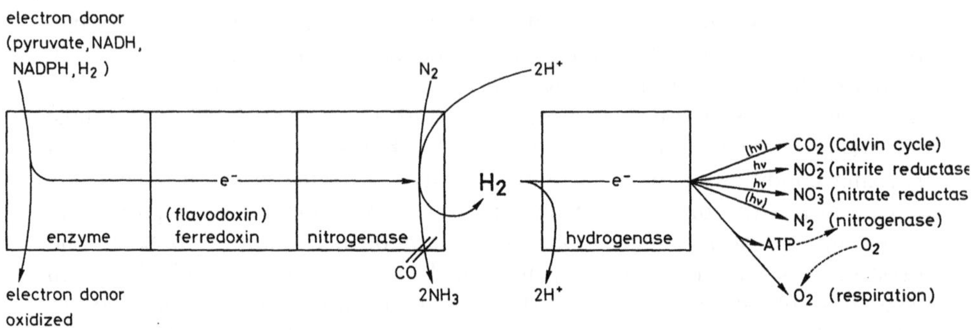


**Fig. 2.** Scheme: The nitrogenase-hydrogenase relationship

quinone site is consistent with the fact that electron donation from $H_2$ to nitrogenase is dependent on photosystem I in blue-green algae. It also explains why hydrogenase cannot catalyze the evolution of $H_2$ in vivo.

In addition to the uptake hydrogenase, a reversible hydrogenase was reported to occur in *Anabaena* (HALLENBECK and BENEMANN 1978). This blue-green alga is the only aerobic nitrogen-fixing organism reported to contain both enzymes. However, BOTHE et al. (1980) were unable to confirm this result and considered such low activity to be an artefact of an isolated enzyme catalyzing the uptake of molecular hydrogen under physiological conditions.

As already mentioned, uptake hydrogenase is not restricted to nitrogen-fixing microorganisms. Besides nitrogen fixation, $H_2$ can support several other reactions (Fig. 2). In green algae, hydrogen acts as an electron donor in photosynthetic $CO_2$-fixation, termed photoreduction (GAFFRON 1940). Such a reaction is not demonstrable in the blue-green algae commonly used in the laboratory but was shown to occur in *Oscillatoria limnetica* (BELKIN and PADAN 1978). This alga is capable of anoxygenic photosynthesis and can fix $CO_2$ at the expense of $H_2S$ or $H_2$. $H_2$-dependent $CO_2$-fixation has recently been demonstrated in *Rhizobium japonicum* and *Derxia gummosa* which synthesize ribulose-1,5-bisphosphate carboxylase under chemoautotrophic growth conditions (SIMPSON et al. 1979, PEDROSA et al. 1980). $H_2$ and hydrogenase can also support nitrate and nitrite reductions in blue-green algae where these reactions are strictly dependent on light activation (BOTHE et al. 1980).

$H_2$-formation by microorganisms has attracted special attention in the past years. In particular, BENEMANN and co-workers (see WEISSMAN and BENEMANN 1977) suggested that photoautotrophically growing blue-green algae might evolve sufficient hydrogen and oxygen to be used as a source for fuel. Among the different investigators, there is a general agreement that all hydrogen evolution is due to nitrogenase in intact cells of aerobic nitrogen-fixing organisms. However, such yields are variable, and depend on the strain used and the culture conditions employed. For example, batch-grown cultures of *Azotobacter chroococcum* did not evolve significant amounts of $H_2$, whereas $O_2$-, $N_2$- as well as C-limited cultures showed considerable $H_2$-production (WALKER and YATES 1978b). *Nostoc muscorum* apparently evolved relatively large quantities of $H_2$

at 22° but not at 32 °C (ERNST et al. 1979). Clearly, the optimal conditions for maximal $H_2$-production have not yet been worked out and the physiology is not fully understood. A major factor influencing $H_2$-formations is the uptake hydrogenase. According to ALBRECHT et al. (1979), *Rhizobium* strains which evolve the gas fail to synthesize an active hydrogenase, whereas the enzyme is effective in those cells which do not produce the gas. The evolution rate can be increased by adding inhibitors of $H_2$-uptake activity. For example, CO and high concentrations of $C_2H_2$ unmask the evolution of hydrogen from the usually faster uptake of the gas by hydrogenase in batch-grown *Azotobacter* (SMITH LA et al. 1976) and *Anabaena* (BOTHE et al. 1977). Similarly, the rate of $H_2$-formation can be increased in *Anabaena* when the utilization of the gas by the Knallgas reaction is prevented by strict exclusion of oxygen from the assays (ERNST et al. 1979).

It is now clearly established that aerobic nitrogen-fixing microorganisms have the potential for hydrogen formation, but it should be kept in mind that the maximal rate is limited by the activity of nitrogenase in the cells. An approximate value for the maximum possible rate of $H_2$ production can be deduced from the rate of $C_2H_4$ production which is an approximate indication of the maximum electron utilisation by nitrogenase. $C_2H_2$ reduction rates in blue-green algae are $^1/_5$ to $^1/_{10}$ of the $CO_2$-fixation activity which is in accord with the requirements of the cell for carbon and nitrogen. Photosynthetic rates are generally 100–250 μmol $CO_2$ fixed $h^{-1}$ $mg^{-1}$ chlorophyll. Thus $H_2$ productions of more than 25 μmol reported in the literature cannot be sustained by the organisms over a longer period. This low, theoretically achievable maximum value is the reason for our belief that blue-green algae or other aerobic nitrogen-fixing organisms are unlikely to produce sufficient $H_2$ to provide power utilizable for mankind.

In contrast, another application of the studies on the hydrogenase-nitrogenase relationship is conceivable. *Rhizobium* strains that possess uptake hydrogenase were shown to fix nitrogen more effectively and to grow faster than strains without an active enzyme (ALBRECHT et al. 1979). Many *Rhizobia* have poor uptake hydrogenase activities. Therefore studies to construct new *Rhizobia* legume symbioses which have a high efficiency for $H_2$-utilization could be useful.

# 7 The Molecular and Genetic Characterization of Nitrogen Fixation Genes

## 7.1 Introduction

The genetics of nitrogen-fixing systems developed rapidly during the 1970's and the majority of genetic studies were carried out with *Klebsiella pneumoniae*. The relatively few studies of nitrogen fixation genes (*nif*) in *Azotobacter* (SHAH et al. 1973, GORDON and BRILL 1972, BISHOP and BRILL 1977), *Rhizobium* (MAIER and BRILL 1976), *Rhodopseudomonas* (WALL et al. 1975, MARRS et al. 1977) and cyanobacteria (STEWART and SINGH 1975) have been mainly concerned with the isolation and preliminary characterization of *nif* mutants and the development of genetic transfer systems. Our review of recent progress in *nif* genetics will therefore be confined mainly to *K. pneumoniae*. Further details and useful background information on the genetics of nitrogen fixation were published in HOLLAENDER et al. (1977).

The initial transductional (STREICHER et al. 1971) and transconjugational (DIXON and POSTGATE 1972) transfer of *nif* showed that the genes for histidine biosynthesis (*his*) and *nif* were linked in *K. pneumoniae*. Large numbers of *K. pneumoniae nif* mutants were subsequently isolated and characterized and the mutations in all of these have been mapped in one region close to the operator end of *his*. Furthermore, the isolation of $nif^+$ hybrids (DIXON and POSTGATE 1972) by the intergeneric transfer of *his* and *nif* from *K. pneumoniae* to a non-nitrogen-fixing *Escherichia coli* indicated that all the genes specific to nitrogenase synthesis were located near *his*.

## 7.2 The *nif* Genes

The close linkage of *his* and *nif* facilitated the construction of self-transmissible plasmids which carry the *his-nif* region of the *K. pneumoniae* genome (CANNON et al. 1976 and DIXON et al. 1976). Plasmid pRD1 carries the transfer and antibiotic resistance genes of the parent plasmid, RP4 and also the chromosomal genes *gnd rfb his nif shiA*. This plasmid and several derivatives have been used for complementation analyses and fine structure mapping of *nif* mutations. pRD1 has also been used to study the expression of *nif* genes in other bacteria (POSTGATE and KRISHNAPILLAI 1977), and was the source of DNA for cloning the *nif* genes.

Fifteen *nif* genes (*nifQ, nifB, nifA, nifL, nifF, nifM, nifV, nifS, nifU, nifN, nifE, nifK, nifD, nifH* and *nifJ*) have now been identified primarily by complementation tests between plasmid and chromosomally located *nif* mutations (DIXON et al. 1977, ELMERICH et al. 1978, MACNEIL et al. 1978, MERRICK et al. 1978, 1980). These mutations included *nif* point mutations on the *K. pneumoniae* chromosome, on pRD1 and its derivatives and insertion mutations obtained by bacteriophage Mu- and transposon-(Tn4, Tn7 and Tn10) directed mutagenesis. *Nif* deletions, derived from insertion mutants, were used for fine structure mapping of *nif* point mutations. The gene order obtained from these studies was, *his ... nifQBALFMVSUNEKDHJ* (Fig. 3). Complementation analysis of insertion mutations which exert transcriptional polarity indicated that the 15 *nif* genes were arranged in 7 operons (Fig. 3).

Genetic studies of polar mutations in the multicistronic *nif* operons have shown that these operons are transcribed in the same direction as the *his* operon of *K. pneumoniae*. RNA-DNA hydridization studies using separated strands of DNA which represented the *his* and six of the seven *nif* operons showed that *his*-specific and *nif*-specific mRNA synthesized in vivo hybridized exclusively to one strand. The conclusions from these results is that those *nif* operons

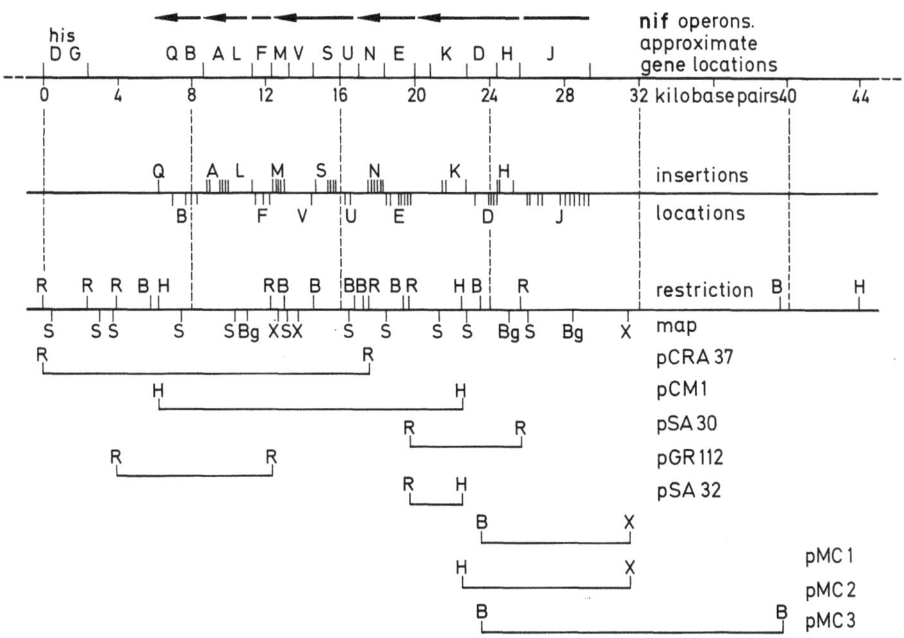


**Fig. 3.** Maps of; (1) *his* and *nif* genes, (2) *nif* insertion mutations (±250 basepairs), and (3) restriction sites in the *his-nif* region of the *K. pneumoniae* chromosome. *Arrows* indicate *nif* operons and their direction of transcription – *nifF* and *nifJ* are monocistronic operons. Symbols for restriction endonucleases are: *R* EcoR1; *H* HindIII; *B* BamH1; *S* SalI; *X* XhoI; *Bg* BglII. *pCRA37, pCM1, pSA30, pGR112, pSA32, pMC1, pMC2* and *pMC3* are small amplifiable plasmids carrying the *nif* DNA restriction fragments indicated

for which detectable levels of mRNA were present under the conditions used are transcribed from the same DNA strand as the *his* operon (Janssen et al. 1980).

### 7.3 *Nif* Gene Products

*K. pneumoniae* nitrogenase is composed of two redox proteins (see Sect. 2). The two identical sub-units of Kp2 are specified by *nifH*. The two non-identical subunits of Kp1 which have molecular weights of 56,000 and 60,000 are the products of *nifD* and *nifK* respectively. The low molecular weight co-factor (FeMoco) which can be extracted from Kp1, partly restores activity to nitrogenase in cell extracts of *nifB, nifE* and *nifN* mutants (Roberts et al. 1978). Polypeptides with molecular weights of 46,000 and 50,000 have been assigned to *nifE* and *nifN* (Roberts et al. 1978). Nitrogenase activity, which is impaired in *nifM* mutants, is partially restored in cell extracts of these mutants by the addition of purified Kp2.

The products of *nifJ* (120,000 m.w.) and *nifF* (17,000 m.w.) are thought to be involved in the electron transport system for nitrogenase. The low levels

**Table 6.** *Nif* gene product

| Gene | Polypeptide Molecular weight | Function |
|---|---|---|
| *nifJ* | 120,000 | Electron input to nitrogenase |
| *nifH* | 35,000 | Kp2 structural gene |
| *nifD* | 56,000 | Kp1 $\alpha$ subunit |
| *nifK* | 60,000 | Kp1 $\beta$ subunit |
| *nifE* | 46,000 | Synthesis or processing of FeMoCo |
| *nifN* | 50,000 | Synthesis or processing of FeMoCo |
| *nifU* | Unknown | Unknown |
| *nifS* | Unknown | Unknown |
| *nifV* | Unknown | Unknown |
| *nifM* | Unknown | Kp2 processing |
| *nifF* | 17,000 | Electron transfer |
| *nifL* | 45,000 | Regulation |
| *nifA* | 57,000 | Activation of *nif* transcription |
| *nifB* | Unknown | Synthesis or processing of FeMoCo |
| *nifQ* | Unknown | Unknown |

of nitrogenase activity in vivo of *nifJ* and *nifF* mutants are significantly enhanced in cell extracts which are assayed in vitro using sodium dithionite as an electron donor for nitrogenase. Pyruvate can also act as an electron donor to nitrogenase in extracts of wild-type but not of *nifF* and *nifJ* mutants. Pyruvate-supported activity can be restored in extracts of *nifF* but not *nifJ* mutants by the addition of *A. chroococcum* flavodoxin which suggests that the *nifF* product is an electron-transport protein (HILL and KAVANAGH 1980).

The biochemical phenotype of *nifA* and *nifL* mutants suggests that the products of these genes have regulatory functions. The *nifA* product is required for the derepression of the *nif* genes since all *nif*-specific proteins so far identified are absent in *nifA* mutants. Complementation studies suggest that the *nifA* product acts as a positive activator of *nif* transcription. A specific regulatory function has not yet been assigned to the *nifL* product mainly because the phenotype of the *nifL* mutations characterized so far is complicated by their polarity on *nifA*. Polypeptides of 57,000 and 45,000 molecular weight have been assigned to *nifA* and *nifL* respectively (F.C. CANNON unpublished).

The products of the remaining genes, *nifQ*, *nifU*, *nifS* and *nifV* have not yet been identified and their functions are unknown. Genetic and biochemical characterization of *nifQ*, *nifU* and *nifV* mutants are complicated by their high levels of nitrogenase activity compared to other mutants. The molecular weights and probable functions of most of the *nif* gene products are listed in Table 6.

### 7.4 Cloning of *K. pneumoniae nif* Genes

The source of DNA for the primary cloning of *nif* genes was pRD1, a plasmid with a molecular weight of $10^8$. This phase of cloning involved the following steps (CANNON et al. 1977, 1979); (1) Digestion of pRD1 and cloning vector

DNAs with restriction endonucleases. (2) Ligation of the resulting fragments with T4 polynucleotide ligase. (3) Transformation of *His⁻Nif⁻* derivatives of *K. pneumoniae* with the resulting mixture of recombinant molecules. (4) Selection of transformants carrying genetic determinants on the cloning vectors followed by screening for a *Nif⁺* phenotype. In some cases, transformants carrying a genetic determinant (*his⁺*) on the "passenger" DNA were selected and subsequently screened for *Nif⁺*. This procedure was used to construct a group of plasmids (pCRA37, pCM1, pSA30 and pMC1) which carry overlapping fragments of *nif* DNA and which collectively carry the complete cluster of *K. pneumoniae nif* genes (Fig. 3). A small amplifiable plasmid (M.W. $18.5 \times 10^6$) carrying the complete cluster of *nif* genes was constructed by combining the *nif* DNA components of pCM1 and pMC2 on the cloning vector, pACYC177 (Fig. 3 and unpublished results of M.C. CANNON and F.C. CANNON). The entire *nif* region has also been cloned by inserting two HindIII restriction fragments (Fig. 3) from pRD1 into a cloning vector (PÜHLER et al. 1979).

### 7.5 A Physical Map of *nif* Genes

The sites at which several restriction endonucleases cleave the *nif* region of the *K. pneumoniae* chromosome were mapped using plasmids carrying cloned DNA fragments (Fig. 3). It was then possible to determine the distribution of *nif* genes on DNA restriction fragments using genetic and molecular techniques (CANNON et al. 1979). The precise locations ($\pm 250$ basepairs) of *nif* insertion mutations were subsequently obtained by determining the molecular distances between the points of insertion and restriction endonuclease cleavage sites which had previously been mapped on the DNA of the *his-nif* region (RIEDEL et al. 1979).

The locations of 43 *nif*::Tn5, 9 *nif*::Tn10 and 17 *nif*::Mu mutations are shown in Fig. 3. The map of *nif* genes derived from these results and the order of the mutations examined are the same as those obtained from genetic fine structure mapping. The minimum size of the *nif* gene cluster derived from these results is 23 kilobases which is in good agreement with that obtained from P1 transduction studies (20 kilobases) by KENNEDY (1977). The maximum separation between insertions in two adjacent genes (*nifK-E*) is 1.6 kilobases – this and smaller gaps between other genes could accommodate small nif genes which have not yet been identified. The boundaries between *nifL* and *F*, *nifN* and *E*, and *nifD* and *H* were accurately determined, because in these cases, insertions in the two adjacent genes mapped within 250 base pairs of each other. From 13 insertions mapped in *nifJ*, an expected size of 124,000 (m.w.) was determined for the gene product which is in good agreement with the size of the *nifJ* product (120,000 m.w.) obtained from SDS polyacrylamide gel electrophoresis (MERRICK et al. 1978, ELMERICH et al. 1978, ROBERTS et al. 1978). However, useful estimates of the expected size of gene products could not be determined for other *nif* genes because the number of insertions mapped were insufficient. KLIPP and PÜHLER (personal communication) have recently developed a procedure by which the size of *nif* polypeptides and their DNA coding regions can be accurately determined.

It has been possible from our knowledge of the physical locations of *nif* genes to clone small fragments of *nif* DNA coding for *nif* promoters, or individual or groups of genes such as the seven *nif* operons (unpublished results of M.C. CANNON and F.C. CANNON). These plasmids are suitable for, (1) transcription studies in vivo and in vitro with purified *nif* promoters, (2) sequencing of *nif* DNA, and (3) identification of *nif* polypeptides and mapping their DNA coding regions.

### 7.6 Interspecies Homology of Nitrogenase Genes

Biochemical studies of nitrogenase from different prokaryotes has suggested a high probable degree of sequence homology between their corresponding DNA's (EADY 1977). If the DNA sequences which encode nitrogenase in these organisms were conserved they should therefore hybridize.

NUTI et al. (1979) reported hybridization between restriction fragments of plasmid DNA from *Rhizobium leguminosarum* and DNA encoding part of the structural genes for *K. pneumoniae* nitrogenase. These results provide evidence that at least some of the *Rhizobium nif* genes are carried on plasmids indigenous to the strains examined. MAZUR et al. (1980) used cloned fragments of *K. pneumoniae nif* genes in similar hybridization studies to identify and clone *Anabaena* 7120 *nif* genes. Their results suggest that the order of *nif* genes is different in these two organisms. RUVKUN and AUSUBEL (1980) found that part of the structural genes for *K. pneumoniae* nitrogenase hybridized to total DNA from 19 out of 21 divergent species of nitrogen-fixing bacteria but not to DNA from 12 different non-nitrogen-fixing species. These results suggest that some nitrogenase genes have either been highly conserved or exchanged between different nitrogen-fixing species in their recent evolutionary history.

These hybridization results demonstrate that a combined molecular and genetical analysis of *nif* genes in $Nif^{+}$ bacteria on which little or no *nif* genetic studies have been done is now possible. Probes of *Klebsiella nif* DNA-encoding homologous parts of the structural genes for nitrogenase could be used to clone *nif* genes from other microbial genera, and all the *nif* mutants of *K. pneumoniae* which have been characterized in the past decade are potential hosts for examining such cloned material.

*Acknowledgements.* The authors are grateful for support from NATO research grant 856 and from the Deutsche Forschungsgemeinschaft (to H.B.).

## References

Akkermans ADL, Abdulkadir S, Trinick MJ (1978) $N_2$-fixing root nodules in Ulmaceae: *Parasponia* or (and) *Trema?* Plant Soil 49:711–715

Albrecht SL, Maier RJ, Hanus FJ, Russell SA, Emerich DW, Evans HJ (1979) Hydrogenase in *Rhizobium japonicum* increases nitrogen fixation by nodulated soybeans. Science 203:1255–1257

Appleby CA, Bergersen FJ, Macnicol PK, Turner GL, Wittenberg BA, Wittenberg JB (1976) Role of leghaemoglobin in symbiotic $N_2$ fixation. In: Newton WE, Nyman

CJ (eds) Proc 1st Int Symp Nitrogen Fixation, Vol I. Washington State Univ Press, Pullman, pp. 274–292

Ayanaba A, Dart PJ (eds) (1977) Biological nitrogen fixation in farming systems of the tropics. Wiley and Sons, New York, pp. 1–377

Belkin S, Padan E (1978) Hydrogen metabolism in the facultative anoxygenic cyanobacteria (blue-green algae) *Oscillatoria limnetica* and *Aphanothece halophytica*. FEBS Lett 94:291–294

Benemann JR, Yoch DC, Valentine RC, Arnon DI (1971) The electron transport in nitrogen fixation by *Azotobacter* III. Requirements for NADPH-supported nitrogenase activity. Biochim Biophys Acta 226:205–212

Bergersen FJ, Turner GL (1973) Kinetic studies of nitrogenase from soyabean root nodules. Biochem J 131:61–75

Berndt H, Lowe DJ, Yates MG (1978) The nitrogen-fixing system of *Corynebacterium autotrophicum*. Eur J Biochem 86:133–142

Biggins DR, Kelly M (1973) Interaction of nitrogenase from *Klebsiella pneumoniae* with ATP or cyanide. Biochim Biophys Acta 205:288–299

Biggins DR, Postgate JR (1970) Nitrogen fixation by extracts of *Mycobacterium flavum* 301. Use of natural electron donors and oxygen sensitivity of cell-free preparations. Eur J Biochem 19:408–415

Bishop PE, Brill WJ (1977) Genetic analysis of *Azotobacter vinelandii* mutant strains unable to fix nitrogen. J Bacteriol 130:954–956

Bishop PE, McParland RH, Evans HJ (1975) Inhibition of the adenylylation of glutamine synthetase by methionine sulfone during nitrogenase repression. Biochem Biophys Res Commun 67:774–781

Bishop PE, Guevara JG, Engelke JA, Evans HJ (1976) Relation between glutamine synthetase and nitrogenase activities in the symbiotic associations between *Rhizobium japonicum* and *Glycine max*. Plant Physiol 57:542–546

Bishop EO, Lambert MD, Orchard D, Smith BE (1977) Nitrogenase of *Klebsiella pneumoniae*: water proton NMR relaxation studies on the binding of divalent metal ions and nucleotides to the iron protein. Biochim Biophys Acta 482:286–300

Bothe H (1970) Photosynthetische Stickstoffixierung mit einem zellfreien Extrakt aus der Blaualge *Anabaena cylindrica*. Ber Dtsch Bot Ges 83:421–432

Bothe H, Floener L (1978) Physiological characterization of *Cyanophora paradoxa*, a flagellate containing cyanelles in endosymbiosis. Z Naturforsch 33c:981–987

Bothe H, Yates MG (1976) The electron transport to nitrogenase in *Mycobacterium flavum*. Arch Mikrobiol 107:25–31

Bothe H, Falkenberg B, Nolteernsting U (1974) Properties and function of the pyruvate: ferredoxin oxidoreductase from the blue-green alga *Anabaena cylindrica*. Arch Microbiol 96:291–304

Bothe H, Tennigkeit J, Eisbrenner G, Yates MG (1977) The hydrogenase-nitrogenase relationship in the blue-green alga *Anabaena cylindrica*. Planta 133:237–242

Bothe H, Distler E, Eisbrenner G (1978) Hydrogen metabolism in blue-green algae. Biochimie 60:277–289

Bothe H, Neuer G, Kalbe I, Eisbrenner G (1980) Electron donors and hydrogenase in nitrogen-fixing microorganisms. In: Stewart WDP, Gallon JR (eds) Nitrogen fixation. Academic Press, London New York, pp. 83–112

Burns RC, Hardy RWF (1975) Nitrogen fixation in bacteria and higher plants. Springer. Berlin Heidelberg New York, pp. 1–189

Burris RH (1971) Fixation by free-living micro-organisms: enzymology. In: Postgate JR (ed) The chemistry and biochemistry of nitrogen fixation. Plenum, London New York, pp. 106–160

Burris RH (1972) Nitrogen fixation, assay methods and techniques. In: Colowick SP, Kaplan NO (eds) Methods in enzymology, Vol 24B. Academic Press, London New York, pp. 415–431

Burris RH (1975) The acetylene reduction technique. In: Stewart WDP (ed) Nitrogen fixation by free-living organisms. Int Biol Programme, Vol VI. Cambridge Univ Press, London, pp. 249–257

Cannon FC, Dixon RA, Postgate JR (1976) Derivation and properties of F-prime factors in *Escherichia coli* carrying nitrogen fixation genes from *Klebsiella pneumoniae*. J Gen Microbiol 93:111–125

Cannon FC, Riedel GE, Ausubel FM (1977) Recombinant plasmid that carries part of the nitrogen fixation (*nif*) gene cluster of *Klebsiella pneumoniae*. Proc Natl Acad Sci USA 74:2963–2967

Cannon FC, Riedel GE, Ausubel FM (1979) Overlapping sequences of *Klebsiella pneumoniae nif* DNA cloned and characterised. Mol Gen Genet 174:59–66

Carnahan JE, Mortenson LE, Mower HF, Castle JE (1960) Nitrogen fixation in cell-free extracts of *Clostridium pasteurianum*. Biochim Biophys Acta 44:520–535

Carter KR, Rawlings J, Orme-Johnson WH, Becker RR, Evans HJ (1980) Purification and characterization of a ferredoxin from *Rhizobium japonicum* bacteroids. J Biol Chem 255:4213–4223

Chatt J (1980) Chemistry relevant to the biological fixation of nitrogen. In: Stewart WDP, Gallon JR (eds) Nitrogen Fixation. Academic Press, London New York, pp. 1–18

Chatt J, Dilworth JR, Richards RL (1978) Recent advances in the chemistry of nitrogen fixation. Chem Rev 78:589–625

Chen JS, Blanchard DK (1978) Isolation and properties of an undirectional $H_2$ oxidizing hydrogenase from the strictly anaerobic $N_2$-fixing bacterium *Clostridium pasteurianum* W 5. Biochem Biophys Res Commun 84:1144–1150

Christou G, Garner CD, Mabbs FE (1978) Crystal structure of tris(tetra-n-butylammonium) tri-μ-benzenethiolato-bis-{tri-μ-sulphido-[$\mu_3$-sulphido-tris(benzenethiolatoiron)]-molybdenum}[$Bu^n_4N$]$_3$[{($PhSFC_3$)$MoS_4$}$_2$($SPh$)$_3$]; an $Fe_3MoS_4$ cubic cluster dimer. J Chem Soc Chem Commun 17:740–741

Cramer SP, Gillum WO, Hodgson KO, Mortenson LE, Steifel EI, Chisnell JR, Brill WJ, Shah VK (1978) The molybdenum site of nitrogenase 2. A comparative study of MoFe proteins and the iron-molybdenum cofactor by X-ray absorption spectroscopy. J Am Chem Soc 100:3814–3819

Dalton H, Mortenson LE (1972) Dinitrogen ($N_2$) fixation (with a biochemical emphasis). Bacteriol Rev 36:231–260

Davis LC, Henzl MT, Burris RH, Orme-Johnson WH (1979) Iron-sulphur clusters in the molybdenum-iron protein component of nitrogenase: Electron paramagnetic resonance of the carbon monoxide-inhibited state. Biochemistry 18:4860–4869

DeBont JAM, Mulder EG (1974) Nitrogen fixation and co-oxidation of ethylene by a methane-utilizing bacterium. J Gen Microbiol 83:113–121

Dilworth MJ (1966) Acetylene reduction by nitrogen-fixing preparations from *Clostridium pasteurianum*. Biochim Biophys Acta 127:285–294

Dixon R, Kennedy C, Kondorosi A, Krishnapillai V, Merrick M (1977) Complementation analysis of *Klebsiella pneumoniae* mutants defective in nitrogen fixation. Mol Gen Genet 157:189–198

Dixon RA, Postgate JR (1972) Genetic transfer of nitrogen fixation from *Klebsiella pneumoniae* to *Escherichia coli*. Nature 237:102–103

Dixon RA, Cannon FC, Kondorosi A (1976) Construction of a P plasmid carrying nitrogen fixation genes from *Klebsiella pneumoniae*. Nature (London) 260:268–271

Dixon ROD (1972) Hydrogenase in legume root nodule bacteroids, occurrence and properties. Arch Mikrobiol 85:193–201

Eady RR (1973) Nitrogenase of *Klebsiella pneumoniae*: Interaction of the component proteins studied by ultracentrifugation. Biochem J 135:531–535

Eady RR (1977) Aspects of the evolution of nitrogenase. In: Leigh GJ (ed) The evolution of metalloenzymes, metalloproteins and related materials. Symposium Press, London, pp 67–84

Eady RR (1980) Methods for studying nitrogenase. In: Bergersen FJ (ed) Methods for evaluation of biological nitrogen fixation. Wiley and Sons, New York, pp. 213–264

Eady RR, Smith BE (1979) Physico-chemical properties of nitrogenase and its components. In: Hardy RWF, Bottemeley F, Burns RC (eds) A treatise on dinitrogen fixation, Sect I, II. Inorganic and physical chemistry and biochemistry. Wiley and Sons, New York, pp. 399–490

Eady RR, Smith BE, Cook KA, Postgate JR (1972) Nitrogenase of *Klebsiella pneumoniae*: Purification and properties of the component proteins. Biochem J 128:655–675

Eady RR, Thorneley RNF, Lowe DJ (1979) Nitrogenase of *Klebsiella pneumoniae*. A pre-steady-state burst of ATP hydrolysis coupled to electron transfer between the component proteins. FEBS Lett 95:211–213

Eady RR, Issack R, Kennedy C, Postgate JR, Ratcliffe H (1978) Nitrogenase synthesis in *Klebsiella pneumoniae*. Comparison of ammonium and oxygen regulation. J Gen Microbiol 104:277–285

Eisbrenner G, Bothe H (1979) Modes of electron transfer from molecular hydrogen in *Anabaena cylindrica*. Arch Microbiol 123:37–45

Elmerich C, Houmard J, Sibold L, Manheimer I, Charpin N (1978) Genetic and biochemical analysis of mutants induced by bacteriophage Mu DNA integration into *Klebsiella pneumoniae* nitrogen fixation genes. Mol Gen Genet 165:181–189

Ernst A, Kerfin W, Spiller H, Böger P (1979) External factors influencing light-induced hydrogen evolution by the blue-green alga, *Nostoc muscorum*. Z Naturforsch 34c:820–825

Evans HJ, Phillips DA (1975) Reductants for nitrogenase and relationship to cellular electron transport. In: Stewart WDP (ed) Nitrogen fixation by free-living microorganisms. Int Biol Programme, vol VI. Cambridge Univ Press, Cambridge, pp 389–420

Fay P, Lang NJ (1971) The heterocysts of blue-green algae. I. Ultrastructural integrity after isolation. Proc R Soc London Ser B 178:185–192

Gaffron H (1940) Carbon dioxide reduction with molecular hydrogen in green algae. Am J Bot 27:273–283

Gallon JR (1980) Nitrogen fixation by photoautotrophs. In: Stewart WDP, Gallon JR (eds) Nitrogen fixation. Academic Press, London New York, pp. 197–238

Gordon JK, Brill WJ (1972) Mutants that produce nitrogenase in the presence of ammonia. Proc Natl Acad Sci USA 69:3501–3503

Gordon JK, Brill WJ (1974) Derepression of nitrogenase synthesis in the presence of excess $NH_4^+$. Biochem Biophys Res Commun 59:967–971

Gotto JW, Tabita FR, van Baalen Ch (1979) Mutants of *Anabaena* strain CA altered in their ability to grow under nitrogen-fixing conditions. J Bacteriol 140:327–332

Green M, Wilson PW (1953) Hydrogenase and nitrogenase in *Azotobacter*. J Bacteriol 65:511–517

Haaker H, de Kok A, Veeger C (1974) Regulation of dinitrogen fixation in intact *Azotobacter vinelandii*. Biochim Biophys Acta 357:344–357

Hageman RV, Burris RH (1978a) Nitrogenase and nitrogenase reductase associate and dissociate with each catalytic cycle. Proc Natl Acad Sci USA 75:2699–2702

Hageman RV, Burris RH (1978b) Kinetic studies on electron transfer and interaction between nitrogenase components from *Azotobacter vinelandii*. Biochemistry 17:4117–4124

Hallenbeck PC, Benemann JR (1978) Characterization and partial purification of the reversible hydrogenase of *Anabaena cylindrica*. FEBS Lett 94:261–264

Hallenbeck PC, Kostel PJ, Benemann JR (1979) Purification and properties of nitrogenase from the cyanobacterium *Anabaena cylindrica*. Eur J Biochem 98:275–284

Hardy RWF (1977–1979) A treatise on nitrogen fixation Vol I–IV. Academic Press, London New York

Hardy RWF, Burns RC (1968) Nitrogen fixation. Annu Rev Biochem 37:331–358

Hardy RWF, Holsten RD (1977) Methods for measurement of dinitrogen fixation. In: Hardy RWF, Gibson AH (eds) A treatise on dinitrogen fixation, vol IV. Wiley and Sons, New York, pp. 451–486

Hardy RWF, Holsten RD, Jackson EK, Burns RC (1968) The acetylene-ethylene assay for $N_2$-fixation: laboratory and field evaluation. Plant Physiol 43:1185–1207

Hardy RWF, Burns RC, Holsten RD (1973) Applications of the acetylene-ethylene assay for measurements of nitrogen fixation. Soil Biol Biochem 5:47–81

Hardy RWF, Havelka UD, Quebedeaux B (1978) Increasing crop productivity: the problem strategies, approach and selected rate-limitations related to photosynthesis. In: Hall DO, Coombs J, Goodwin TW (eds) Proc 4th Int Congr Photosynthesis. Biochem Soc, London, pp. 695–720

Haury JF, Wolk CP (1978) Classes of *Anabaena variabilis* mutants with oxygen-sensitive nitrogenase activity. J Bacteriol 136:688–692

Hill S (1971) Influence of oxygen concentration on the colony type of *Derxia gummosa* grown in nitrogen-free media. J Gen Microbiol 67:77–83

Hill S (1976) Influence of atmospheric oxygen concentration on acetylene reduction and efficiency in intact *Klebsiella pneumoniae*. J Gen Microbiol 93:335–345

Hill S, Kavanagh E (1980) Some biochemical properties of *nifF* and *nifJ* mutants relevant to electron transport to nitrogenase in *Klebsiella pneumoniae*. J Bacteriol 141:470–475

Hollaender A, Burris RH, Day PR, Hardy RWF, Helinski DR, Lamborg MR, Owens L, Valentine RC (eds) (1977) Genetic engineering for nitrogen fixation. Plenum, New York London

Huyhn BH, Munck E, Orme-Johnson WH (1979) Nitrogenase XI: Mössbauer studies on the cofactor centers of the MoFe protein from *Azotobacter vinelandii* OP. Biochim Biophys Acta 527:192–203

Janssen KA, Riedel GE, Ausubel FM, Cannon FC (1980) Transcriptional studies with cloned nitrogen-fixation genes. In: Orme-Johnson WH, Newton WE (eds) Nitrogen fixation, vol I. Univ Park Press, Baltimore, pp. 85–93

Jeng D, Morris JA, Mortenson LE (1970) The effect of reductant in inorganic phosphate release from adenosine-5-triphosphate from purified nitrogenase of *Clostridium pasteurianum*. J Biol Chem 245:2809–2813

Kennedy C (1977) Linkage map of nitrogen fixation (*nif*) genes in *Klebsiella pneumoniae*. Mol Gen Genet 157:199–204

Knight E Jr, D'Eustachio AJ, Hardy RWF (1966) A flavoprotein with ferredoxin activity from *Clostridium pasteurianum*. Biochim Biophys Acta 113:626–628

Kurtz DM, McMillan RS, Burgess BK, Mortenson LE, Holm R (1979) Identification of iron-sulfur centers in the iron-molybdenum proteins of nitrogenase. Proc Natl Acad Sci USA 76:4986–4989

Laane C, Haaker H, Veeger C (1978) Involvement of the cytoplasmatic membrane in nitrogen fixation by *Rhizobium leguminosarum* bacteroids. Eur J Biochem 87:147–153

Laane C, Haaker H, Veeger C (1979a) On the efficiency of oxidative phosphorylation in membrane vesicles of *Azotobacter vinelandii* and of *Rhizobium leguminosarum* bacteroids. Eur J Biochem 97:369–377

Laane C, Krone W, Konings WN, Haaker H, Veeger C (1979b) The involvement of membrane potential in nitrogen fixation by bacteroids of *Rhizobium leguminosarum*. FEBS Lett 103:328–332

Lambein F, Wolk CP (1973) Structural studies on the glycolipids from the envelope of the heterocysts from *Anabaena cylindrica*. Biochemistry 12:791–798

Leach CK, Carr NG (1971) Pyruvate: ferredoxin oxidoreductase and its activation by ADP in the blue-green alga *Anabaena variabilis*. Biochim Biophys Acta 245:165–174

Lex M, Stewart WDP (1973) Algal nitrogenase, reductant pools and photosystem I activity. Biochim Biophys Acta 292:436–443

Lowe DJ (1979) Simulation of the electron-paramagnetic resonance spectrum of the iron-protein of nitrogenase: A prediction of the existence of a second paramagnetic centre. Biochem J 175:955–957

Lowe DJ, Eady RR, Thorneley RNF (1978) Electron paramagnetic resonance on nitrogenase of *Klebsiella pneumoniae*. Evidence for acetylene and ethylene-nitrogenase transient complexes. Biochem J 173:277–290

Ludden PW, Burris RH (1976) Activating factor for the iron protein of nitrogenase from *Rhodospirillum rubrum*. Science 194:424–426

Ludden PW, Burris RH (1978) Purification and properties of nitrogenase from *Rhodospirillum rubrum* and evidence for phosphate, ribose and an adenine-like unit covalently bound to the iron protein. Biochem J 175:251–259

Lundell DJ, Howard JB (1978) Isolation and partial characterisation of two different subunits from the molybdenum iron protein of *Azotobacter vinelandii* nitrogenase. J Biol Chem 253:3422–3426
MacNeil T, MacNeil D, Roberts GP, Supiano MA, Brill WJ (1978) Fine-structure mapping and complementation analysis of *nif* (nitrogen fixation) genes in *Klebsiella pneumoniae*. J. Bacteriol 136:253–266
Maier RJ, Brill WJ (1976) Ineffective and non-nodulating mutant strains of *Rhizobium japonicum*. J. Bacteriol 127:863–869
Marrs B, Wall JD, Gest H (1977) Emergence of the biochemical genetics and molecular biology of photosynthetic bacteria. Trends Biochem Sci 2:105–108
Mayhew SG, Ludwig ML (1975) Flavodoxins. In: Boyer PO (ed) The Enzymes. Academic Press, London New York, pp. 57–118
Mazur BI, Rice D, Haselkorn R (1980) Identification of blue-green algal fixation genes by using heterologous DNA hybridisation probes. Proc Natl Acad Sci USA 77:186–190
McKenna CE, Huang CW (1979) In vivo reduction of cyclopropene by *Azotobacter vinelandii* nitrogenase. Nature 280:609–610
Meers JL, Tempest DW, Brown CM (1970) "Glutamine (amide) 2-oxyglutarate amino transferase oxidoreductase (NADP)", an enzyme involved in the synthesis of glutamate by some bacteria. J Gen Microbiol 64:187–194
Merrick M, Filser M, Kennedy C, Dixon R (1978) Polarity of mutations induced by insertion of transposons Tn5, Tn7 and Tn10 into the *nif* gene cluster of *Klebsiella pneumoniae*. Mol Gen Genet 165:103–111
Merrick M, Filser M, Dixon R, Elmerich C, Sibold L, Houmard J (1980) The use of translocatable genetic elements to construct a fine-structure map of the *Klebsiella pneumoniae* nitrogen fixation (*nif*) gene cluster. J Gen Microbiol 117:509–520
Meyer J, Kelley BC, Vignais PM (1978) Nitrogen fixation and hydrogen metabolism in photosynthetic bacteria. Biochimie 60:245–260
Millbank JW (1969) Nitrogen fixation in moulds and yeasts – a reappraisal. Arch Mikrobiol 68:32–39
Mishustin EN, Shil'nikova VK (1971) Biological fixation of atmospheric nitrogen. MacMillan, London, pp. 1–420
Mortenson LE, Thorneley RNF (1979) Structure and function of nitrogenase. Annu Rev Biochem 48:387–418
Mortenson LE, Valentine RC, Carnahan JE (1962) An electron transport factor from *Clostridium pasteurianum*. Biochem Biophys Res Commun 7:448–452
Munck E, Rhodes H, Orme-Johnson WH, Davies LC, Brill WJ, Shah VK (1975) Nitrogenase VIII. Mössbauer and e.p.r. spectroscopy. The MoFe protein component from *Azotobacter vinelandii* OP. Biochim Biophys Acta 400:32–53
Newton WE, Bulen WA, Hadfield KL, Stiefel EI, Watt GD (1977) HD formation as a probe for intermediate in $N_2$ reduction. In: Newton WE, Postgate JR, Rodriguez-Barrueco C (eds) Recent advances in nitrogen fixation. Academic Press, London New York, pp. 119–130
Nordlund S, Eriksson U, Baltscheffsky H (1977) Necessity of a membrane component for nitrogenase activity in *Rhodospirillum rubrum*. Biochim Biophys Acta 462:187–195
Nuti MP, Lepidi AA, Prakash RK, Schilperoort RA, Cannon FC (1979) Evidence for nitrogen fixation (*nif*) genes on indigenous *Rhizobium* plasmids. Nature 282:533–535
O'Donnell M, Smith BE (1978) Electron-paramagnetic-resonance studies on the redox properties of the molybdenum-iron protein of nitrogenase between +50 and −450 mV. Biochem J 173:831–839
Okon S, Houchins JP, Albrecht SL, Burris RH (1977) Growth of *Spirillum lipoferum* at constant partial pressures of oxygen, and the properties of its nitrogenase in cell-free extracts. J Gen Microbiol 98:87–93
Orme-Johnson WH, Davis LC, Henzl MT, Averill BA, Orme-Johnson NR, Munck E, Zimmerman R (1977) Components and pathways in biological nitrogen fixation. In: Newton WE, Postgate JR, Rodriguez-Barrueco C (eds) Recent advances in nitrogen fixation. Academic Press, London New York, pp. 131–178

Pedrosa FO, Döbereiner J, Yates MG (1980) $H_2$-dependent growth and autotrophic $CO_2$-fixation by *Derxia*. J Gen Microbiol 119:547–551
Peterson RB, Burris RH (1976) Conversion of acetylene reduction rates to nitrogen fixation rates in natural populations of blue-green algae. Anal Biochem 73:404–410
Phelps AS, Wilson PW (1941) Occurrence of hydrogenase in nitrogen-fixing organisms. Proc Soc Exp Biol 47:473–476
Postgate JR (ed) (1971) The chemistry and biochemistry of nitrogen fixation. Plenum Press, New York London, p. 1–326
Postgate JR (1974) Evolution within nitrogen-fixing systems. In: Carlile MJ, Skehel JJ (eds) Evolution within the microbial world. Cambridge Univ Press, Cambridge, pp. 263–292
Postgate JR (1978) Nitrogen fixation. Stud Biol No 92. Arnold, London, pp 1–67
Postgate JR (1982) Biological nitrogen fixation: fundamentals. Phil Trans R Soc Lon B 296:375–385
Postgate JR, Krishnapillai V (1977) Expression of *Klebsiella nif* and *his* genes in *Salmonella typhimurium*. J Gen Microbiol 98:379–385
Postgate JR, Eady RR, Dixon RA, Hill S, Kahn D, Kennedy C, Partridge P, Robson R, Yates MG (1981) Some aspects of the physiology of dinitrogen fixation. In: Bothe H, Trebst A (eds) Biology of inorganic nitrogen and sulfur. Springer, Berlin Heidelberg New York, p. 103–115
Pühler A, Burkardt HJ, Klipp W (1979) Cloning in *Escherichia coli* the genomic region of *Klebsiella pneumoniae* which encodes genes responsible for nitrogen fixation. In: Timmis KN, Pühler A (eds) Plasmids of medical, environmental and commercial importance. Developments in genetics, vol I. Elsevier/North-Holland, Amsterdam, pp. 435–447
Quispel A (1974) The biology of nitrogen fixation. North-Holland Res Monogr Frontiers Biol, Amsterdam Oxford, pp. 1–769
Rawlings J, Shah VK, Chisnell JR, Brill WJ, Zimmerman R, Munck E, Orme-Johnson WH (1978) Novel metal clusters in the iron-molybdenum cofactor of nitrogenase. J Biol Chem 253:1001–1004
Rennie RJ, Funnell A, Smith BE (1978) Immunochemistry of nitrogenase as a probe for the enzyme mechanism: Evidence for multiple enzyme forms and an MgATP binding site on the MoFe protein. FEBS Lett 91:158–161
Riedel GE, Ausubel FM, Cannon FC (1979) Physical map of chromosomal nitrogen fixation (*nif*) genes of *Klebsiella pneumoniae*. Proc Natl Acad Sci USA 76:2866–2879
Rippka R, Waterbury JB (1977) The synthesis of nitrogenase by non-heterocystous cyanobacteria. FEMS Lett 2:83–86
Rippka R, Deruelles J, Waterbury JB, Herdman M, Stanier RY (1979) Generic assignments, strain history and properties of pure cultures of cyanobacteria. J Gen Microbiol 111:1–61
Rivera-Ortiz JM, Burris RH (1975) Interactions among substrates and inhibitors of nitrogenase. J Bacteriol 123:537–545
Roberts GP, MacNeil T, MacNeil D, Brill WJ (1978) Regulation and characterisation of protein products coded by the *nif* (nitrogen fixation) genes of *Klebsiella pneumoniae*. J Bacteriol 136:267–279
Robson RL (1979) $O_2$-repression of nitrogenase synthesis in *Azotobacter chroococcum*. FEMS Microbiol Lett 5:259–262
Robson RL, Postgate JR (1980) Oxygen and hydrogen in biological nitrogen fixation. Annu Rev Microbiol 34:183–207
Ruvkun GB, Ausubel FM (1980) Interspecies homology of nitrogenase genes. Proc Natl Acad Sci USA 77:191–195
Scherings G, Haaker H, Veeger C (1977) Regulation of nitrogen fixation by Fe-S protein II in *A. vinelandii*. Eur J Biochem 77:621–630
Schlegel HG, Schneider K (1978) Hydrogenases: Their catalytic activity, structure and function. Goltze, Göttingen, pp. 1–453

Schöllhorn R, Burris RH (1967) Acetylene as a competitive inhibitor of $N_2$-fixation. Proc Natl Acad Sci USA 58:213–216
Shah VK, Brill WH (1977) Isolation of an iron-molybdenum cofactor from nitrogenase. Proc Natl Acad Sci USA 74:3249–3253
Shah VK, Davis LC, Gordon JK, Orme-Johnson WH, Brill WJ (1973) Nitrogenaseless mutants of *Azotobacter vinelandii*: Activities, cross-reactions and EPR spectra. Biochim Biophys Acta 292:246–255
Shanmugan KT, Buchanan BB, Arnon DI (1972) Ferredoxins in light- and dark-grown photosynthetic cells with special reference to *Rhodospirillum rubrum*. Biochim Biophys Acta 256:477–486
Shethna YI (1970) Non-heme (iron-sulfur) proteins of *Azotobacter vinelandii*. Biochim Biophys Acta 205:58–62
Siefert E (1976) Die Fixierung von molekularem Stickstoff bei phototrophen Bakterien am Beispiel von *Rhodopseudomonas acidophila*. Thesis, Univ Göttingen, pp. 1–149
Silver WS, Postgate JR (1973) Evolution of asymbiotic nitrogen fixation. J Theor Biol 40:1–10
Simpson FB, Maier RJ, Evans HJ (1979) Hydrogen-stimulated $CO_2$-fixation and coordinate induction of hydrogenase and ribulosebisphosphate carboxylase in a $H_2$-uptake positive strain of *Rhizobium japonicum*. Arch Microbiol 123:1–8
Singh PK (1973) Nitrogen fixation by the cellular blue-green alga *Aphanothece*. Arch Microbiol 92:59–62
Smith BE (1980) Studies on the iron molybdenum cofactor from the nitrogenase MoFe protein of *Klebsiella pneumoniae*. In: Newton WE, Otsuka S (eds) Molybdenum chemistry of biological significance. Plenum, New York London, pp. 179–190
Smith BE, Lang G (1974) Mössbauer spectroscopy of nitrogenase components from *Klebsiella pneumoniae*. Structural assignments and mechanistic conclusions. Biochem J 137:169–180
Smith BE, Eady RR, Thorneley RNF, Mortenson LE (1976) Nitrogenases from *Klebsiella pneumoniae* and *Clostridium pasteurianum*. Kinetic investigations of cross reactions as a probe of the enzyme mechanism. Biochem J 157:439–447
Smith LA, Hill S, Yates MG (1976) Inhibition by acetylene of conventional hydrogenase in nitrogen-fixing bacteria. Nature (London) 262:209–210
Söderlund R, Svensson BH (1976) The global nitrogen cycle. In: Svensson BH, Söderlund R (eds) Nitrogen, phosphorus and sulfur-Global cycles. Scope Rep 7. Ecol Bull (Stockholm) 22:23–74
Southern EM (1975) Detection of specific sequences among DNA fragments separated by gel electrophoresis. J Mol Biol 98:503–517
Sprent JI (1979) The biology of nitrogen-fixing organisms. McGraw-Hill, Maidenhead, pp. 1–196
Stanier RY, Cohen-Bazire G (1977) Phototrophic procaryotes: the cyanobacteria. Annu Rev Microbiol 31:225–274
Stephens PJ, McKenna CE, Smith BE, Nguyen HT, McKenna MC, Thomson AJ, Devlin F, Jones JB (1979) Circular dichroism and magnetic circular dichroism of nitrogenase proteins. Proc Natl Acad Sci USA 76:2585–2589
Stewart WDP, Lex M (1970) Nitrogenase activity in the blue-green alga *Plectonema boryanum* strain 594. Arch Microbiol 73:250–260
Stewart WDP, Singh HN (1975) Transfer of nitrogen-fixing *nif* genes in the blue-green alga *Nostoc muscorum*. Biochem Biophys Res Commun 62:62–69
Streicher SL, Gurney EG, Valentine RC (1971) Transduction of nitrogen fixation genes in *Klebsiella pneumoniae*. Proc Natl Acad Sci USA 50:1174–1177
Streicher SL, Shanmugam KT, Ausubel FM, Morandi C, Goldberg RB (1974) Regulation of nitrogen fixation in *Klebsiella pneumoniae*: evidence for a role of glutamine synthetase as a regulator of nitrogenase synthesis. J Bacteriol 120:815–821
Sweeney WV, Rabinowitz JC, Joch DC (1975) High and low reduction potential 4Fe–4S* clusters in *Azotobacter vinelandii*. J Biol Chem 250:7842–7847
Tanaka M, Mitsura H, Yasunobu KT, Mortenson LE (1978) The amino acid sequence

of *Clostridium pasteurianum* iron protein, a component of nitrogenase. J Biol Chem 252:7081–7100
Tel-Or E, Luijk LW, Packer L (1977) An inducible hydrogenase in cyanobacteria enhances $N_2$-fixation. FEBS Lett 78:49–51
Thauer RK, Rupprecht E, Jungermann K, Decker K (1969) Hydrogen formation from NADH in cell-free extracts of *Clostridium kluyveri*. FEBS Lett 4:108–112
Thorneley RNF (1975) Nitrogenase of *Klebsiella pneumoniae*. A stopped-flow study of magnesium-adenosine triphosphate-induced electron transfer between the component proteins. Biochem J 145:391–396
Thorneley RNF, Cornish-Bowden A (1977) Kinetics of nitrogenase from *Klebsiella pneumoniae*. Heterotrophic interactions between magnesium adenosine-5-diphosphate and magnesium-adenosine-5′-triphosphate. Biochem J 165:255–262
Thorneley RNF, Eady RR, Lowe DJ (1978) Biological nitrogen fixation by way of an enzyme-bound dinitrogen-hydride intermediate. Nature 272:557–558
Trinick MJ (1973) Symbiosis between *Rhizobium* and the non-legume, *Trema aspera*. Nature 244:459–460
Tso M-Y, Burris RH (1973) The binding of ATP and ADP by nitrogenase components from *Clostridium pasteurianum*. Biochim Biophys Acta 309:263–270
Tubb RS (1974) Glutamine synthetase and ammonia regulation of nitrogenase synthesis in *Klebsiella*. Nature 251:481–485
Tubb RS, Postgate JR (1973) Control of nitrogenase synthesis in *Klebiella pneumoniae*. J Gen Microbiol 79:103–117
Tyler B (1978) Regulation of the assimilation of nitrogen compounds. Annu Rev Biochem 47:1127–1162
Uyeda K, Rabinowitz JC (1971) Pyruvate-ferredoxin oxidoreductase: studies on the reaction mechanism. J Biol Chem 246:3120–3125
Vincent JM (1974) Root-nodule symbiosis with *Rhizobium*. In: Quispel A (ed) The biology of nitrogen fixation. North-Holland Res Monogr Frontiers Biol Vol 33. Amsterdam Oxford
Walker CC, Yates MG (1978a) $H_2$ evolution and acetylene effects in $N_2$-fixing *Azotobacter chroococcum* under different nutrient limitations. Proc Steenbock-Kettering Int Symp Nitrogen Fixation, Madison, Wisconsin, Abstract A-2
Walker CC, Yates MG (1978b) The hydrogen cycle in nitrogen-fixing *Azotobacter chroococcum*. Biochimie 60:225–231
Wall JD, Weaver PF, Gest H (1975) Genetic transfer of nitrogenase-hydrogenase activity in *Rhodopseudomonas capsulata*. Nature 258:630–631
Weissman JC, Benemann JR (1977) Hydrogen production by nitrogen-starved cultures of *Anabaena cylindrica*. Appl Environ Microbiol 33:123–131
Wilcockson J (1977) The apparent influence of atmospheric oxygen concentration on nitrogenase activity and slime production in *Klebsiella pneumoniae* grown in solid medium. J Gen Microbiol 101:311–317
Wilson PW (1958) Asymbiotic nitrogen fixation. In: Ruhland W (ed) Encyclopedia of plant physiology, Vol VIII. Springer, Berlin Göttingen Heidelberg, pp. 9–47
Winter RH, Burris HC (1976) Nitrogenase. Annu Rev Biochem 45:409–426
Wyatt JT, Silvey JKG (1969) Nitrogen fixation by *Gloeocapsa*. Science 165:908–909
Yates MG (1971) Electron transport to nitrogenase in *Azotobacter chroococcum*. Eur J Biochem 24:347–357
Yates MG (1977) Physiological aspects of nitrogen fixation. In: Newton WE, Postgate JR, Rodriguez-Barrueco C (eds) Recent advances in nitrogen fixation. Academic Press, London New York, pp. 219–270
Yates MG (1980) Biochemistry of nitrogen fixation. In: Miflin BJ (eds) The biochemistry of plants, Vol V. Academic Press, London New York, pp. 1–64
Yates MG, Eady RR (1980) The physiology and regulation of nitrogen fixation. In: Subba-Rao NS (eds) Recent advances in nitrogen fixation. Oxford and IBH Publ, New Delhi, pp. 88–120
Yates MG, Jones C (1974) Respiration and nitrogen fixation in *Azotobacter*. Adv Microbiol Physiol 11:97–136

Yates MG, Lowe DJ, Thorneley RNF (1975) Nitrogenase of *Azotobacter chroococcum*: Inhibition by ADP of the reduction of oxidised Fe protein by sodium dithionite. FEBS Lett 60:89–93
Yoch DC, Carithers RT (1979) Bacterial iron-sulfur proteins. Microbiol Rev 43:384–421
Yoch DC, Valentine RC (1972) Ferredoxin and flavodoxins of bacteria. Annu Rev Microbiol 26:139–162
Yoch DC, Carithers RP, Arnon DI (1977) Isolation and characterization of bound iron-sulfur proteins from bacterial photosynthetic membranes. J Biol Chem 252:7453–7460
Zumft WG (1976) The molecular basis of biological dinitrogen fixation. Struct Bonding 29:1–66
Zumft WG, Mortenson LE (1975) The nitrogen-fixing complex of bacteria. Biochim Biophys Acta 416:1–52

# II.2 Dinitrogen-Fixing Symbioses with Legumes, Non-Legume Angiosperms and Associative Symbioses

A. QUISPEL

## 1 Introduction

As fixation of dinitrogen, $N_2$, is limited to certain prokaryotic organisms, other organisms like higher plants can only profit from $N_2$-fixation if they live in close proximity to such prokaryotes. On the other hand, $N_2$-fixing organisms with a heterotrophic type of metabolism will need a good provision of organic substrates in order to obtain the energy for the highly energy-demanding process of $N_2$-fixation. Both types of organisms can therefore benefit from mutual associations and the more intimate these associations are, the better results of their interactions may be expected. The examples of symbiotic associations which will be discussed in this chapter range from rather loose associations through systems with more intimate cellular and intercellular contacts to intracellular endosymbioses in which, in the highly developed types of root nodule symbioses, the prokaryotic microsymbionts function more or less as $N_2$-fixing organelles in a well-adapted functional organ as part of the whole plant.

In many natural ecosystems symbiotic associations with $N_2$-fixing organisms play an important role in the N economy. In agriculture the leguminous nodulated plants were used as food producers and green manurial systems already in antiquity. In our century the demands for N are partially covered by fertilizers produced by chemical-technological processes of $N_2$ fixation. However, the increasing demands for N in food production and the increased costs of fossil energy in the production of fertilizers recently have stimulated research on biological, and especially, on symbiotic $N_2$ fixation. This is reflected in many recent symposia (NUTMAN 1976, NEWTON and NYMAN 1976, NEWTON et al. 1977, DÖBEREINER et al. 1977, NEWTON and ORME-JOHNSON 1980). The most recent symposium (GIBSON and NEWTON 1981) could only partially be included in this review. As the former edition of this encyclopedia contained an excellent and very comprehensive survey of $N_2$-fixing symbioses (ALLEN and ALLEN 1958) and recent extensive summaries are available (QUISPEL 1974, HARDY and SILVER 1977) we will restrict ourselves especially to those results of recent physiological research which are representative for present interest and progress in this field. The topics to be discussed will be chosen with regard to their relevance to

*Abbreviations.* ALA $\delta$-amino levulinic acid; ALAD $\delta$-amino levulinic acid dehydrogenase; ALAS $\delta$-amino levulinic acid synthetase; ATP adenosine triphosphate; DNA deoxyribonucleic acid; EPS exopolysaccharide; FITC fluoresceine isothiocyanate; GDH glutamate dehydrogenase; GOGAT glutamate synthase; GS glutamine synthetase; IAA indole acetic acid; ICA indole carbonic acid; LPS lipopolysaccharide; 2-i-P isopentenyl adenine; PEP phosphoenolpyruvic acid; PHBA polyhydroxybutyric acid; RNA ribonucleic acid.

our understanding of the symbiotic interactions which lead to an efficient $N_2$-fixing system. The influence of environmental factors will only be discussed in relation to these interactions, as the ecological aspects as such will be the main topic of another chapter in this encyclopedia (GIBSON and JORDAN Chap. 8, Vol. 12C). For fundamental aspects of the biochemistry of the $N_2$-fixing process we refer to BOTHE et al. (Chap II.1, this Vol.).

# 2 Description of the Main Symbiotic Dinitrogen-Fixing Systems

## 2.1 Associative Symbioses

There is no need to discuss the rather loose symbiotic associations in the rhizosphere and the phyllosphere of higher plants as the next chapter of this volume by DÖBEREINER (Chap. II.3) will be devoted to this subject. There are some reasons to mention them in this chapter. The first step in the establishment of more intimate symbioses such as root nodules of leguminous plants consists of an accumulation of the bacteria in the rhizosphere. It is tempting to speculate that these more complicated symbioses developed during evolution from more associative types of symbioses. This may be indicated by the fact that even in rhizosphere associations bacteria tend to enter the outer root cell layers and occasionally even inner cells of the cortex and the endodermis. Some data on $N_2$ fixation values were already given in Table 2 in BOTHE et al. (Chap. II.1, this Vol.).

## 2.2 Symbioses with Cyanobacteria

### 2.2.1 Distribution

Symbioses with $N_2$-fixing Cyanobacteria (blue-green algae) are found with quite different types of organisms: with fungi in different lichens, in liverworts like *Anthoceros, Blasia* and *Cavicularia,* in *Sphagnum*, in the water fern *Azolla* and in the Cycadeae and as only representative of the Angiospermae, *Gunnera* (Haloragaceae). All symbiotic Cyanobacteria belong to the genera *Nostoc* or *Anabaena.* For a more extensive discussion we refer to MILLBANK (in QUISPEL 1974) and reviews by STEWART in NEWTON et al. (1977) and by PETERS et al. in NEWTON and ORME-JOHNSON (1980).

### 2.2.2 Description and Development

In lichens the Cyanobacteria are either the only "phycobiont", or they are localized in special organs, called cephalodia, in those lichens where the main phycobiont is a green alga such as *Trebouxia.* Though there is a close juxtaposition of the fungal and cyanobacterial cells, no haustoria are formed. Some authors emphasize the thick extracellular sheath which surrounds the Cyanobacteria. In most symbioses with green lower or higher plants the Cyanobacteria occupy cavities which already were formed before invasion with the algae. In many cases growth of these cavities is stimulated by the presence of the cyanobacteria. In *Azolla* the correlated growth of the cavities and the *Anabaena* cells indicates the regulatory influence of some common factor. Contact between the Cyanobacteria and the host cells may be stimulated by the induced development of hairs (*Azolla*) or papillae (liverworts). Such structures act in the transfer of substances between the symbiotic partners. Though in most cases the Cyanobacteria remain intercellular, in *Sphagnum* and especially in *Gunnera* they live intracellularly after dissolving host cell walls.

In Cyanobacteria heterocysts are regarded as special sites for $N_2$ fixation. We refer to the chapter by STEWART in the second photosynthesis volume (Vol. 6) of this Encyclopedia and reviews by FOGG et al. (1973) and STEWART (in QUISPEL 1974).

In most symbiotic Cyanobacteria the percentage of heterocysts is unusually high. In lichens a high percentage of heterocysts is only observed in cephalodia, which is explained by the good C-nutrition and consequently high C/N ratio due to the photosynthesis of the green algae.

#### 2.2.3 $N_2$ Fixation ($C_2H_2$ Reduction)

In all cases $N_2$ fixation has been demonstrated by the $^{15}N_2$ and/or the $C_2H_2$ reduction method.

### 2.3 Root Nodules with Actinomycetes: Actinorhizas

#### 2.3.1 Distribution

Root nodules with actinomycete-like organisms as endophytes, nowadays termed actinorhizas, were first described for *Alnus glutinosa* by WORONIN in 1866 and HILTNER in 1896. The importance of the nodules as sites of $N_2$ fixation was established by KREBBERS in 1932 and ROBERG in 1934. The search after plants with actinorhizas was highly stimulated by BOND (see e.g. in QUISPEL 1974, NUTMAN 1976, NEWTON et al. 1977). The lists of plants in Table 1 will certainly prove to be incomplete as regularly new additions are discovered. It is evident from this table that actinorhizas are found among plants belonging to taxonomically unrelated families. Even within one genus, species with and without actinorhizas are known, sometimes depending on the geographical situation (e.g. the European *Dryas octopetala* which is the only *Dryas* species without root nodules). With the only exception of *Datisca* all actinorhizas are found on woody plants. Their ecological importance is considerable as many of the plants are pioneers in natural ecosystems. Their practical importance for forestry was emphasized by GORDON et al. (1979).

**Table 1.** Incidence of $N_2$-fixing root nodules in Angiospermae. (Data based on ALLEN and ALLEN 1958 and AKKERMANS in GORDON et al. 1979)

| Micro-symbiont | (Sub)family | Approximate number of | | Reported number of nodulated | | Nodulated genera |
|---|---|---|---|---|---|---|
| | | genera | species | genera | species | |
| *Frankia* | Casuarinaceae | 1 | – | 1 | – | *Casuarina* |
| | Coriariaceae | 1 | – | 1 | – | *Coriaria* |
| | Betulaceae | 2 | – | 1 | – | *Alnus* |
| | Datiscaceae | 1 | 2 | 1 | 2 | *Datisca* |
| | Myricaceae | 2 | – | 2 | – | *Myrica* |
| | Rosaceae | 124 | – | 4 | – | *Dryas*<br>*Purshia*<br>*Cercocarpus*<br>*Rubus* |
| | Elaeagnaceae | 3 | – | 3 | – | *Elaeagnus*<br>*Hippophae*<br>*Shepherdia* |
| | Rhamnaceae | 58 | – | 3 | – | *Ceanothus*<br>*Discaria*<br>*Colletia* |
| *Rhizobium* | Ulmaceae | 16 | 230 | 1 | 2 | *Parasponia* |
| | Leguminosae: | | | | | |
| | Mimosoidae | – | 1,500 | – | 127 | |
| | Caesalpinoidae | – | 1,300 | – | 27 | |
| | Papilionatae | – | 10,000 | – | 952 | |

### 2.3.2 Description

Actinorhizas are typical metamorphosized roots with an apical meristem and a centre vascular bundle, surrounded by an endodermis. The enlarged cortex contains the cells with the endophyte. Immediately behind the meristem the newly formed cortex cells are penetrated by hyphae growing from the already infected cells. More to the base senescence and disintegration occurs. The meristems branch in a dichotomous way, giving rise to a system of lobes with a clustered or coralloid form called rhizothamnia. In some plants like *Myrica, Comptonia* and *Casuarina* the apical meristem may give rise to normal, though negatively geotropic, roots. In the latter plant the repeated initiation of lateral roots leads to the formation of a complex truncate system. We refer to SCHAEDE (1962) for a survey of the older literature on nodule cytology. SCHAEDE's light microscopical studies of the nodules of *Alnus, Hippophae* and *Myrica* from 1933 to 1938 still form the base for all further work on actinorhizae. Studies on ultrastructural aspects are available for *Alnus glutinosa, A. crispa, Myrica cerifera, M. pennsylvanica, Hippophae rhamnoides, Datisca cannabina, Purshia tridentata, Comptonia peregrina, Ceanothus integerrimus* (see e.g. BECKING et al.

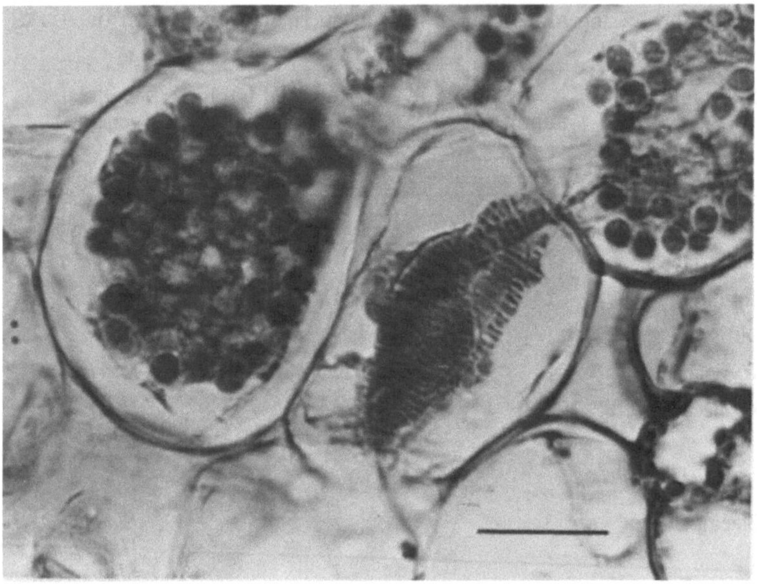

**Fig. 1.** Cross-section through cells of a spore (+) rhizothamnion of *Alnus glutinosa* showing cells with hyphae and vesicles and the typical sporangia formation. (Photo C. VAN DIJK) Bar = 10 μm

1964, VAN DIJK and MERKUS 1976, STRAND and LAETSCH 1977, and articles in TORREY and TJEPKEMA 1979, and GORDON et al. 1979). Notwithstanding the taxonomic diversity of the plants studied the pictures of the endophyte-containing cells show a surprising uniformity (Fig. 1). The cells are filled with dense clusters of hyphae which at the cell surface swell into vesicular (most species) or club-shaped (Myricaceae and *Coriaria*) structures. Only in *Casuarina* are such swellings absent or inconspicuous, while in *Datisca* they are situated in the central part of the cells. The vesicles are characterized by incomplete septae and inclusions with an unknown character called striated bodies. All hyphae and vesicles are surrounded by a capsule which is continuous with the host cell walls and can be dissolved by pectic enzymes. Most probably these capsules are of host cell origin, as is the plasmalemma, which is present at the side of the host-cell cytoplasm (LALONDE and DEVOE 1976).

In some cases a third type of cell is observed, originally called bacteroid, bacteria-like cells or granulae. They are typically formed in thickenings of hyphae which are characterized by transversal cell divisions (VAN DIJK and MERKUS 1976). The detailed descriptions leave no doubt that these swellings are sporangia and the small cells formed are the spores of the endophytes. This spore formation inside the nodules appears to be more the exception than the rule. It has been observed up till now in only part of the population of *Alnus glutinosa* in the Netherlands, in Scottish *Myrica gale,* but not in *Myrica gale* in the Netherlands, in *Dryas drummondii* from Alaska and American *Purshia tridentata*. Most nodules are practically devoid of spores, though they may be never entirely absent (VAN DIJK in GORDON et al. 1979).

### 2.3.3 Infection and Development

Initiation and development of actinorhizal root nodules was first studied in *Alnus glutinosa* by POMMER and TAUBERT in 1959 (QUISPEL 1974) and more recently by ANGULO-CARMONA (1974). There are no observations on a development of the Actinomycetes in the rhizosphere. Pictures of polar adhesion and encapsulation on the root hairs (LALONDE in NEWTON et al. 1977) appear to be not specific and are not correlated with subsequent infections. Infection is presumed to take place in deformed root hairs. Nodule formation is a two-step process. After penetration, cell divisions are initiated in the cortex after which the newly formed cells are occupied by the endophytes. These swellings we call the primary nodules. As they are already $N_2$-fixing small nodules the term prenodules is less preferable. Under the influence of these primary nodules the nearby formation of lateral roots is initiated. These root initials are infected from the primary nodules and are transformed into the definite nodule system, the rhizothamnia. Studies with *Casuarina, Ceanothus, Comptonia* and *Myrica gale* confirm this general pattern of development (see TORREY and TJEPKEMA 1979).

### 2.3.4 $N_2$ Fixation ($C_2H_2$ Reduction)

With only a few exceptions, $N_2$ fixation has been confirmed in the plants mentioned in Table 1 by $^{15}N_2$ uptake or $C_2H_2$ reduction. Typical examples of $C_2H_2$ reduction range from 9–92.0 $\mu m\ h^{-1}\ g^{-1}$ fresh weight for seedlings of *Alnus glutinosa*.

No differences were observed between $C_2H_2$ reduction of nodules with or without spores. This indicates that spores are not $N_2$-fixing cells (AKKERMANS and VAN DIJK in NUTMAN 1976).

## 2.4 Leguminous Root Nodules with *Rhizobium*

### 2.4.1 Distribution

Root nodules produced by an infection with bacteria belonging to the genus *Rhizobium* are widespread among plants belonging to the families of the Leguminosae (Table 1). Their agronomical and ecological importance is considerable (HARDY and GIBSON 1977).

The bacterial origin of the root nodules and their role in the fixation of atmospheric dinitrogen was established by the classical work of HELLRIEGEL and WILLFAHRT in 1888. In the same year Beyerinck isolated the responsible bacteria. The older literature was extensively summarized in the excellent books of FRED et al. (1932) and WILSON (1940) and the review of ALLEN and ALLEN (1958).

### 2.4.2 Description

Leguminous root nodules are more or less spherical to cylindrical structures, in the latter case sometimes branched into finger-like or coralloid forms. In most cases they are exogenous to the root pericycle, either one-sided laterally adhered to the roots or encircling them. Exceptional nodule types are formed, e.g. in *Arachis hypogaea,* (CHANDLER 1978). All nodules are formed by some meristematic activity. In many plants such as, e.g. peas, the meristematic activity continues some time at the apex of the nodule. In such nodules the youngest, recently infected cells are found immediately behind the meristem, while old, more or less senescent parts are present at the base. In other plants such as, e.g. soybeans, the meristematic activity is not persistent after the nodules are formed. The development of the nodules up till the senescent state then takes place more or less synchronously in all cells of the nodule. All nodules contain some small vascular bundles which form the connection with the vascular system of the root.

The main part of the healthy, actively $N_2$-fixing, nodules is occupied by the bacteroid-containing cells. These are large, parenchymatous cells which are densely packed with the actively $N_2$-fixing cells of *Rhizobium* called bacteroids. Compared to normal bacteria, as they are observed in *Rhizobium* cultures, these bacteroids are swollen, and in most cases more or less pleomorphic. They will be discussed more extensively in Section 5.2 (Fig. 3). In all cases the bacteroids are surrounded by an envelope membrane or peribacteroid membrane, which is derived from the host cell plasmalemma. Another characteristic of these cells is the presence of the red pigment leghemoglobin (see Sect. 5.5). The characteristics of the bacteroid-containing cells can be best understood by studying their development.

Microscopical and ultrastructural descriptions of leguminous root nodules and their development are available for species of *Pisum, Trifolium, Medicago, Glycine, Arachis, Lupinus, Phaseolus,* and members of the cowpea group. We refer for recent discussions and detailed references to KIJNE (1975a, b), KIJNE and PLANQUÉ (1979), NEWCOMB (1976), ROBERTSON et al. (1978a, b), TU (1979) and WERNER and MÖRSCHEL (1978).

### 2.4.3 Infection and Nodule Development

If rhizobia from an infective strain are inoculated on the roots of a susceptible host plant they start developing in the rhizosphere and get adsorbed to the root hairs. Then the normally straight-growing root hairs are deformed, during which a phase called "marked curling" is especially important (YAO and VINCENT 1976). The metabolized cell wall, which is correlated with this deformation, is the place where the bacteria enter the root hair by way of an invagination of the cell wall producing an infection thread. Such threads contain a mucilaginous mass in which the bacteria multiply and which is surrounded by a cell wall deposited by the host cell (Fig. 2). In some plants, like lupin or *Arachis,* infection threads could not be observed. Inward growth of the infection thread is preceded by the host cell (root-hair) nucleus. For a more extensive description of this infection process we refer to DART in QUISPEL (1974) and HARDY and SILVER (1977). If the infection thread reaches the opposite cell walls the bacteria multiply in the intercellular space from which new infection threads are formed in the next cell. In this way the infection threads grow towards the inner part

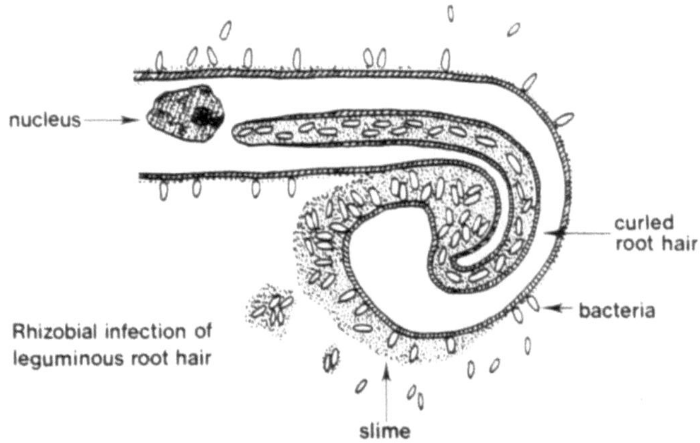


**Fig. 2.** Rhizobial infection of leguminous root hair: schematic representation of the binding of the bacteria in the slime at the root hair tip and the invagination of the infection thread. (Courtesy of J.W. KIJNE)

of the cortex. In most cases the infection threads grow towards the nuclei of the host cells which are swollen, synthesize DNA but do not yet divide (LIBBENGA and BOGERS, in QUISPEL 1974). In peas, cell divisions are only initiated in the inner cortex of the infected roots (LIBBENGA and BOGERS, in QUISPEL 1974); in other plants the central or outer cells of the cortex are involved. These cell divisions are the beginning of the nodule meristem.

Up till now the bacteria are surrounded by a cell-wall deposit and thus are still intercellular. The change to the intracellular situation is limited to the cells which are formed by the meristems, either the meristematic centres formed in the cortex, or, during further growth of the nodules, the apical meristems. In these cells the cell wall deposit around the infection thread becomes very thin and finally is absent. Now the bacteria are taken up by the host-cell cytoplasm in a process of endocytosis of the plasmalemma. The endocytotic membrane encloses the bacteria and grows with the multiplying bacteria till they are transformed into the bacteroids. Bacteroids are still surrounded by the membrane, now called peribacteroid membrane. Either every bacteroid is surrounded by its own peribacteroid membrane or, as in soybeans, the peribacteroid membrane encloses a small group of bacteroids. This phase of development more or less coincides with the synthesis of leghaemoglobin (Sect. 5.5).

When the bacteroids leave the infection threads by endocytosis, more or less dramatic events take place in the host cell cytoplasm. The vacuoles may become filled with granulous material resembling cytoplasm with ribosomes, which in peas coincides with a disrupted vacuole membrane (KIJNE 1975a). At a later phase the cell recovers and new vacuoles are formed. There is a marked enlargement of the cells with an increase of ribosomes, endoplasmatic reticulum, mitochondria and plasmids. In the fully developed bacteroid-containing cell (Fig. 3) the mitochondria occupy the outward surface of the cytoplasmic cell contents. The nuclei swell, while the nuclear membranes are unaltered and are surrounded by the expanding endoplasmatic reticulum. The role and func-

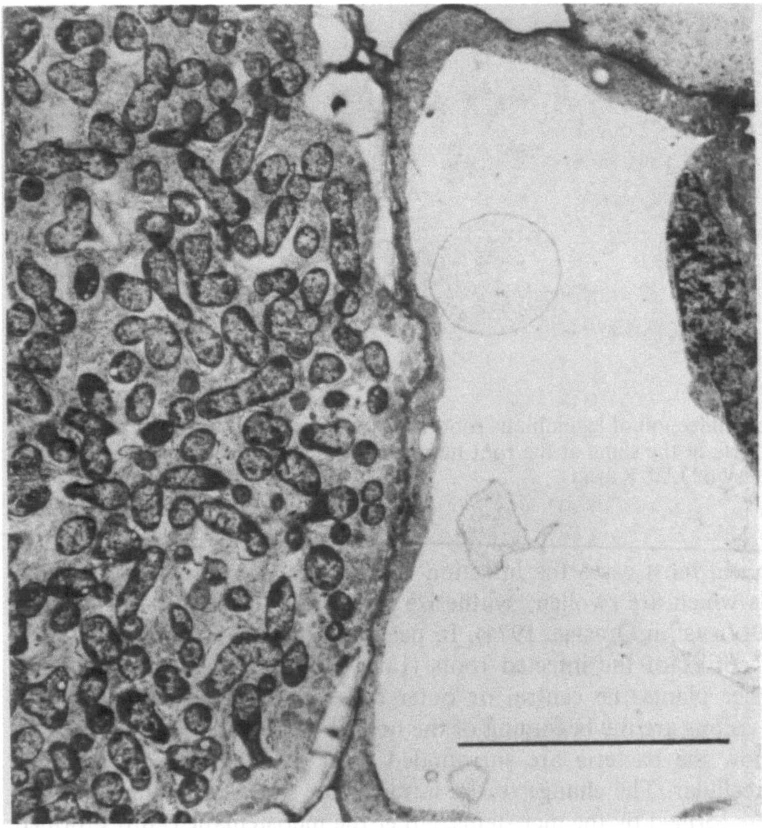

**Fig. 3.** Electron microscopic photograph of a root nodule cell from a nodule of *Pisum sativum* filled with bacteroids. The adjoining cell remained uninfected. (KIJNE 1975a) Bar = 5 μm

tion of small vesicles is uncertain. They may be formed from the endoplasmatic reticulum and the dictyosomes, and might play a role in the synthesis and the removal of the infection thread cell wall material. This latter activity might be indicated by the presence of vesicles with fibrillous material. Another function might be the transport to or from the peribacteroid membranes (ROBERTSON et al. 1978a).

In older nodules or at the base of nodules with an apical meristem senescence takes place. Depending on the host-bacterial strain combination studied, degeneration starts with the bacteroid, with the host cell protoplasm or there is a simultaneous breakdown. Senescent nodules are characterized by the green biliverdin-like substances derived from the breakdown of leghaemoglobin.

The normal sequence of events of infection and nodulation only occurs in compatible combination of host plants in *Rhizobium* strains. In incompatible or less compatible combinations either no infection is possible (non-infective combinations) or the infection and nodulation process is somewhere interrupted or develops in an abnormal way. Growth of infection threads may be interrupted, the bacteria may remain in the infection threads or, if endocytosis is possible, they do not change into typical bacteroids, no synthesis of leghaemoglobin may be possible or there are disfunctions in the $N_2$-fixing or assimilatory system. Very often a too early disintegration and degeneration of bacteroids takes place. Typical examples of ultrastructural events leading to abnormal nodulation were described by NEWCOMB et al. (1977). In all cases where no or insufficiently $N_2$-fixing bacteroids are formed the nodules are ineffective.

Effectivity may be further influenced by environmental factors. Especially the presence of bound N, e.g. in the form of higher concentrations of $NH_4^+$ or $NO_3^-$ inhibits infection and nodulation as well as the $N_2$ fixing activity of the nodules formed. Some possible explanations for this inhibiting effect will be discussed later (e.g. in Sect. 6).

#### 2.4.4 $N_2$ Fixation ($C_2H_2$ Reduction)

The presence of $N_2$ fixation, already extensively proven by the practice of agricultural N yields and older experiments with growth on N-poor soils or nutrient solutions, has been extensively confirmed by numerous $^{15}N_2$ experiments and determinations of $C_2H_2$ reduction under laboratory and field conditions. $N_2$-fixing activity of course depends on the plant – *Rhizobium* combination studied and the environmental conditions. Some extrapolations to annual N increases under field conditions are based on the $N_2$ ($C_2H_2$) reduction as measured under field conditions, the environmental situation and the life span of the active root nodules (see Table 2, Chap. II.1, this Vol.).

### 2.5 Non-Leguminous Root Nodules with *Rhizobium*

The only non-leguminous plants in which the presence of effective root nodules with *Rhizobium* is well-documented belong to the genus *Parasponia* of the family Ulmaceae (TRINICK 1979, TRINICK and GALBRAITH 1976). They were initially identified as *Trema* species, which however do not possess root nodules. The root nodules look more like those of the non-legumes, as they have a central vascular bundle. The bacteria remain in the infection threads though their cell wall deposits may become very thin and finally even disappear. Yet no real endocytosis was observed nor was formation of bacteroids obvious. The leaf nodules which are found in the leaves of many non-leguminous tropical plants will not be discussed in this chapter, as previous claims of $N_2$ fixation could not be confirmed by modern studies (BECKING in QUISPEL 1974).

## 3 The Dinitrogen-Fixing Micro-Symbionts: Isolates and Cultures

### 3.1 Introduction

Studies of symbiotic associations must be based on experiments with the symbiotic system and on experiments with the isolated symbiotic partners. Usually there are few problems in obtaining the plants without their micro-organisms.

It may be far more difficult to obtain cultures of the micro-symbionts. Moreover, it soon appears that isolated micro-symbionts may differ in many aspects from these organisms in the symbiotic situation. It is necessary to make a clear distinction between isolations of the micro-symbionts out of the symbiotic situation and further cultivation of these isolates. Isolates may be obtained after micro-dissection or crude homogenization after which the cells of the micro-symbiont can be freed from debris of the partner by the usual methods for the preparation of cell organelles like filtration, centrifugation, buoyant density centrifugation or chromatography on suitable Sephadex columns. After isolation we can try to make cultures in suitable nutrient media. This is not yet possible for all micro-symbionts. When cultures are obtained it is always necessary to prove the identity of the cultivated organisms with the micro-symbionts by application of Koch's criteria. The cultivated organisms must be able to infect axenic plants belonging to the specific symbiotic partner after which they can again be isolated from this reconstructed symbiosis.

### 3.2 Cyanobacteria

Though the Cyanobacteria from most symbiotic systems described above (Sect. 2.1) can be readily isolated, cultivation and reinfection was seldom successful. Only from some liverworts were Cyanobacteria isolated, cultivated and used for successful re-inoculations. Reinfections of *Azolla* plants with *"Anabaena azollae"* have never been successful up till now, so that their identity with the endophytic *Anabaena* from *Azolla* must be doubted (for references see PETERS et al. in NEWTON and ORME-JOHNSON 1980).

Freshly isolated *Nostoc* from the liverworts *Anthoceros punctatus* and *Blasia pusilla* did not show any photosynthetic activity. This suggests that in the symbiosis they obtain organic substrates from their partners, the photosynthetically active liverworts. They were able to reduce $N_2$ with an active nitrogenase, but the assimilatory enzymes glutamine synthetase (GS) and glutamate dehydrogenase (GDH) were hardly or not active. All $NH_3$, which was formed by $N_2$ fixation, was excreted into the medium (STEWART and RODGERS 1977). Similar observations were made with isolated Cyanobacteria from lichen-cephalodia and *Azolla* (PETERS et al. l.c.). The high activity of nitrogenase coincided with the high percentage of heterocysts.

After cultivation, as it was possible with the liverworts', *Nostoc* strains, photosynthetic activity recovers, the percentage of heterocysts and nitrogenase activity decreases, while the activity of GS and the use of the $NH_3$ for protein synthesis recovers.

There are only few data about the specificity of the Cyanobacteria. The liverworts studied by STEWART and RODGERS could be infected by an isolate from *Gunnera,* while infections with isolates from Cycadeae were unsuccessful. The cultivated cyanobacterial microsymbionts are in no way different from other heterocyst-forming cyanobacteria. We refer to Chapter 6, Volume 6, this Series.

## 3.3 Frankia, the Endophyte from the Actinorhizas

### 3.3.1 Isolation and Cultivation

In most cases crushed nodule suspensions can be used for infection and nodulation experiments. Such suspensions thus contain the endophytes in a vital, infective form. Clusters of hyphae with vesicles can be isolated by filtration through Nylon filters, by Sephadex G50 chromatography or sucrose density gradient centrifugation (for references AKKERMANS et al. in NEWTON et al. 1977, AKKERMANS et al., BAKER et al. in GORDON et al. 1979). Though many disputable claims have been made about successful, but mostly non-infective cultures of the endophytes, it is only since 1978 that infective cultures forming effective root nodules were obtained by QUISPEL and TAK for *Alnus glutinosa,* by TORREY and coworkers for *A. rubra, A. viridis* and *Comptonia peregrina* and by LECHEVALIER for *Myrica pennsylvanica* (further references in TORREY and TJEPKEMA 1979, GORDON et al. 1979) while an infective, but not effective, strain was isolated from *Elaeagnus umbellata,* BAKER et al. 1979). All cultivated strains grow as thin hyphae with a diameter less than 1 µm, consisting of typical prokaryotic cells, sometimes forming small vesicle-like cells but most characterized by sporangia of the same type as described inside root nodules by VAN DIJK and MERKUS (1976) (Fig. 4). A quite similar organism was already described by POMMER in 1959 as an infective culture from *A. glutinosa* nodules, which alas could not be maintained. We may honour POMMER as haven given the first

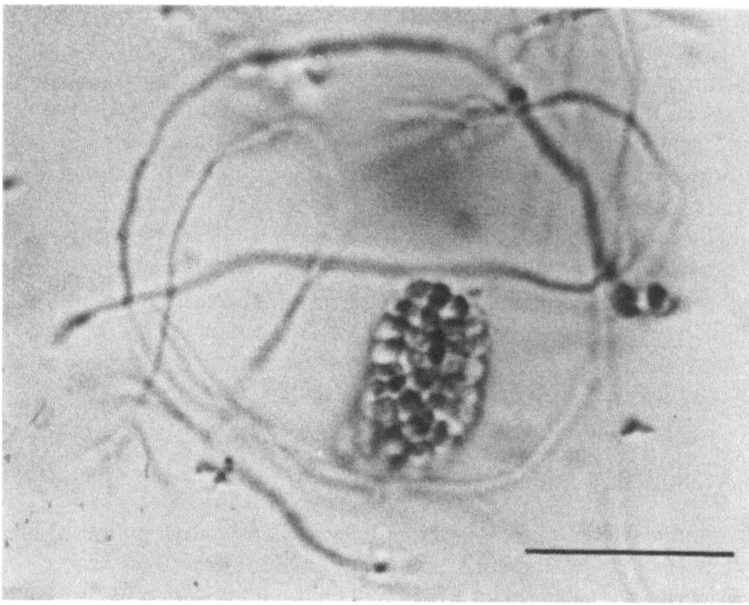

**Fig. 4.** Cultivated form of a *Frankia,* isolated from spore (−) nodules of *Alnus glutinosa.* Note the formation of a sporangium by actinomycete-like hyphae. (Photo J. VALSTAR) BAR = 10 µm

description of the real endophytes of Actinorhizas. The generic name *Frankia* is now generally accepted (BECKING in BERGEY 1974). Its taxonomic position will be near the Dermatophyllaceae (VAN DIJK and MERKUS 1976, LECHEVALIER and LECHEVALIER in GORDON et al. 1979).

### 3.3.2 Specificity

Most studies on specificity of the endophytes were done before cultivated *Frankia* strains were available, and were based on cross-inoculation experiments with crushed nodule suspension. Infection and nodulation appeared to be possible if plants were inoculated with crushed nodules from the same or related species. Positive results were obtained in cross-inoculations between the Elaeagnaceae: *Elaeagnus, Hippophae* and *Shepherdia,* but no results were obtained between *Alnus, Myrica, Casuarina, Ceanothus, Coriaria* and other unrelated species (BOND in QUISPEL 1974). Positive results were claimed for cross-inoculation between *Alnus glutinosa* with *Myrica gale* (RODRIGUEZ-BARRUECO and BOND in NUTMAN 1976) and *Elaeagnus umbellifera* (MIGUEL et al. in DÖBEREINER et al. 1977). Not all related species yield good results. Sometimes cross-inoculation between different species of one genus like *Alnus* or *Myrica* or between geographically distinct plants yields non-effective nodules (BOND). VAN DIJK (1978) showed that the presence or absence of spore formation in the nodules [Spore (+) or Spore (−) type] is a characteristic of the endophyte strains used for inoculation. This is the more remarkable since all *Frankia* strains which are isolated from spore (−) types of nodules abundantly produce sporangia and spores in cultures.

Cross-inoculation with the isolated and cultivated *Frankia* strains has shown that cross-inoculations have positive results when the strain from *Comptonia peregrina* is inoculated on roots of different *Alnus* species (LALONDE in TORREY and TJEPKEMA 1979, LALONDE and CALVERT in GORDON et al. 1979). *Frankia* cultures from *A. rubra* infect *Myrica* spp. but no *Ceanothus* plants (BERRY and TORREY in GORDON et al. 1979).

The *Frankia* cultures from *A. glutinosa* infect *Myrica gale* but are not infective to *Hippophae rhamnoides*. The structures, which are formed inside the nodules: vesicles or club-shaped swellings, are specific for the host plant and are not determined by the *Frankia* strains (LALONDE and CALVERT, BERRY and TORREY in GORDON et al. 1979). They thus cannot be used for taxonomic purposes as was done by BECKING (in BERGEY 1974).

### 3.3.3 Nutrient Requirements

The strains which were isolated up till now can be divided into three classes:

1. The strains isolated by TORREY and collaborators from, e.g. *Comptonia* and American *Alnus* species grow well, though slowly, on simple nutrient solutions with salts, yeast extract and casamino acids. According to LALONDE and CALVERT (in GORDON et al. 1979), they show a much better growth after addition of commercial lecithin to the nutrient solutions.

2. The strains isolated from *Alnus glutinosa* spore (−) type are absolutely dependent on the presence of root lipids.
3. Some other strains, like an isolate from the spore (+) type of *Alnus glutinosa*, needed these root lipids only during isolation (QUISPEL and BURGGRAAF in GIBSON and NEWTON 1981).

   All strains grow well on Tween 80 or fatty acids as C and energy source (BLOM 1980).

#### 3.3.4 Metabolic Activities

Cultivated *Frankia* strains reduce $C_2H_2$ and fix $^{15}N_2$ so that they can form an active nitrogenase. This formation of nitrogenase is repressed by ammonium, but its activity is remarkably insensitive to oxygen. The correlation between nitrogenase induction and the formation of vesicles in these cultures is a strong indication that these vesicles are the sites of nitrogenase activity and might to some extent be comparable to the heterocysts of Cyanobacteria (TJEPKEMA et al. 1980). Isolated clusters of hyphae with vesicles from root nodules of *A. glutinosa* contain nitrogenase, an active hydrogenase, glutamate dehydrogenase and glutamate synthase but no glutamine synthetase, which is present in the plant cell filtrate together with enzymes of citrulline biosynthesis (AKKERMANS et al. in NEWTON et al. 1977, GORDON et al. 1979).

### 3.4 *Rhizobium*

#### 3.4.1 Isolation and Description

After the first isolation of *"Bacterium radicicola"* by Beyerinck, strains of *Rhizobium*, the generic name given by FRANK in 1890, were isolated from root nodules of many leguminous plants. No special difficulties were encountered in isolation, cultivation and reinfection experiments. In most cases rather simple nutrient solutions may be used, such as yeast-extract/mannitol. Completely defined media must contain a suitable sugar, the usual salts and some vitamins like biotin or thiamin. For some strains, and especially for single cell cultures, more complicated media are recommended. There are no special characteristics to identify *Rhizobium* strains besides their possibility to infect roots of susceptible plants. Non-infective strains therefore can only be identified as rhizobia if it is absolutely certain that they originated, e.g. by mutation, from infective strains. For a more extensive discussion of all aspects of *Rhizobium* we refer to VINCENT (1970, in QUISPEL 1974, in HARDY and SILVER 1977).

*Rhizobium* species are aerobic non-spore-forming bacteria, consisting of Gram-negative short rods with one or more flagellae. On agar media they form white or occasionally pink colonies with a more or less slimy surface, especially on carbohydrate-containing media. The cells contain a nucleoid and many ribosomes, spherical electron-dense bodies which most probably consist of glycogen, and older cells contain polyhydroxybutyric acid (PHB) as a reserve material.

The cell envelope has the composition and structure which is typical for Gram-negative bacteria. At the outside of the cytoplasmic membrane a small layer of mucopeptide is covered by an outer membrane with protein, lipids, polysaccharides and lipopolysaccharide (LPS). Several studies were made of the LPS to study the possible importance for taxonomic classifications, to explain the morphological changes during bacteroid formation (see Sect. 5.2) and more recently to find antigenic binding sites for lectins (see Sect. 4.1). For further references see PLANQUÉ et al. 1979, CARLSON et al. 1978). The sugar analysis indicates a composition which is not essentially different from the best-studied LPS of coliform bacteria. Differences between strains may be considerable, between strains isolated from one cross-inoculation group no less than between strains from different groups. The outer membrane may be covered by a polysaccharide capsule and extra cellular slime (EPS). In liquid nutrient solutions this EPS can be precipitated from the medium. Both PHB and EPS may function as reserve substances.

Periplasmic enzymes, such as alkaline phosphatase, have been found (VAN BRUSSEL 1978, GLENN and DILWORTH 1979) in *R. leguminosarum*. They can be liberated from the cells by osmotic shock methods. Binding activities, suggesting the presence of binding proteins, have been found for glucose, glutamine and other substrates (VAN BRUSSEL in preparation).

### 3.4.2 Taxonomy

*Rhizobium* belongs to the family Rhizobiaceae (BECKING in BERGEY 1974). The only other genus of this family is *Agrobacterium* of which *A. tumefaciens,* the responsible agent of crown galls, is the best-known representative. Both bacteria are so much related that some bacteriologists prefer to classify all of them in the genus *Rhizobium.*

There is a marked difference in growth velocity between different *Rhizobium* species on the usual nutrient solutions. This leads to a sharp distinction between fast-growing and slow-growing species. This difference is reflected in other characteristics, like the insertion of flagellae, acid-production, sugar-substrate specificity, metabolic pathways, DNA-composition, serological differences and phage specificity. All further distinctions between *Rhizobium* "species" are based on host specificity. It is usual to classify the leguminous plant genera as well as their specific bacteria in cross-inoculation groups (Table 2). *Rhizobium* strains isolated from plants of a certain cross-inoculation group are infective for other plants. However, there are many exceptions. Nevertheless, these cross-inoculation groups and the *Rhizobium* "species" based on these groups have proven their practical usefulness.

### 3.4.3 Metabolism

*Rhizobium* strains are all aerobic bacteria with a respiratory type of metabolism. They cannot live anaerobically by fermentation, but some strains can use $NO_3$ for anaerobic dissimilatory $NO_3$ reduction. For most strains the presence of the Embden-Meyerhof pathway is doubtful. Certainly many strains can use the Pentose Phosphate pathway and the Entner-Doudoroff pathway, the latter being the predominant one for most slow growers (for further references see

**Table 2.** Cross-inoculation groups in the *Rhizobium* symbioses, in which mutual infections by the corresponding *Rhizobium* are possible

| Group | *Rhizobium* | Plants which are regularly infected | Occasional infections |
|---|---|---|---|
| *Fast growing* | | | |
| Pea group | *R. leguminosarum* | *Pisum, Vicia, Lens Lathyrus, Cicer* | *Trifolium* |
| Clover group | *R. trifolii* | *Trifolium* | |
| Bean group | *R. phaseoli* | *Phaseolus* | |
| Medicago group | *R. meliloti* | *Medicago, Melilotus, Trigonella* | |
| *Slow growing* | | | |
| Soybean group | *R. japonicum* | *Glycine* | |
| *Lupinus–Lotus* group | *R. lupini* | *Lupinus, Ornithopus, Lotus, Anthyllis, Astragalus, Caragana, Ononis, Dorycnium* | |
| Cow-pea miscellany specialized | *R.* strains | *Stylosanthes, Centrosema, Desmodium, Lotononis, Leucaena* | |
| promiscuous | *R.* strains | *Vigna sinensis, Phaseolus lathyroides Ph. atropurpurea, Arachis hypogaea Parasponia* | Other *Phaseolus* species *Medicago, Glycine* |

VINCENT and BERGERSEN in QUISPEL 1974, RONSON and PRIMROSE 1979). In *R. japonicum* a direct oxidation of gluconate has been postulated while the presence of the tricarbonic acid cycle has been demonstrated.

At normal $O_2$ concentrations the respiratory pathway consists of cytochromes b, c and a-a3. Under reduced $O_2$ concentration cytochrome a-a3 is no longer formed, while there is a production of protoporphyrin IX, the haem of leghaemoglobin, due to an increase of the activities of the enzymes $\delta$-aminolevulinic synthetase (ALAS) and -dehydrogenase (ALAD), (AVISSAR and NADLER 1978).

Different substrates can be used as respiratory substrates, hexoses, pentoses and organic acids from the tricarbonic acid cycle (SKOTNICKI and ROLFE 1978). Many catabolic enzymes, as well as some transport systems, are induced by their specific substrates (DE VRIES 1980). Because a typical catabolite repression by glucose is absent, simultaneous use of different substrates is possible (DE HOLLANDER and STOUTHAMER 1979), while growth, and as we shall see in Section 3.4.4, $N_2$ fixation, may be stimulated by the combination of different substrates like pentoses, glycerol and succinic acid (WILCOCKSON and WERNER 1979). A quite unexpected substrate for some *Rhizobium* strains was recently discovered by HANUS et al. (1979). In these strains $H_2$ can induce hydrogenase and ribulose bisphosphate carboxylase. They can use the energy obtained by the oxidation of $H_2$ for $CO_2$ assimilation in the Calvin pathway like typical hydrogen bacteria.

### 3.4.4 $N_2$ Fixation ($C_2H_2$ Reduction)

Up till 1975 nobody succeeded in demonstrating $N_2$ fixation ($C_2H_2$ reduction) in cultivated rhizobia. After successful efforts to inoculate rhizobia into plant callus cultures, it was observed that mere contact with these plant cells was sufficient to induce nitrogenase activities in some *Rhizobium* strains (see CHILD and KURZ 1978). This led in 1975 to the independent discovery in five laboratories that these strains reduced $C_2H_2$ in certain nutrient solutions either on agar media (see GIBSON et al. 1976) or in stationary liquid of chemostat cultures (EVANS and KEISTER 1976). Up till now $C_2H_2$ reduction has been demonstrated in a number of strains of slow-growing rhizobia, mainly belonging to the cowpea group and *R. japonicum*. Requirements are a nutrient solution with a suitable sugar, e.g. arabinose instead of glucose, the simultaneous presence of succinic acid or a related organic acid and the presence of a nitrogenous substance such as glutamate, glutamine or $NH_4^+$ salts in concentrations below 0.1 mM. The most important prerequisite for nitrogenase activity is the establishment of a very low $O_2$ concentration. In agar and stationary liquid cultures this low $O_2$ concentration is obtained by the respiration of the cells.

The fact that $C_2H_2$ reduction is first detected at the end of the exponential growth period might be explained by the decrease of $O_2$ in dense bacterial populations. It has been suggested that the combination of sugars and organic acids enables the optimal respiratory activity to remove most of the $O_2$. Another explanation is that, due to certain regulatory mechanisms in this situation, the energy of respiration is chanelled to $N_2$ fixation instead of the biosynthesis of cell material (WILCOCKSON and WERNER 1979). Indeed $N_2$ fixation is only detected in phases of the growth curve where growth is severely impeded, so that most of the $NH_3$ formed is excreted into the medium (O'GARA and SHANMUGAM 1976). Several authors observed that in $N_2$-fixing cultures the bacterial cells may be swollen and pleomorphic, more or less resembling the bacteroids in the nodules (PANKHURST and CRAIG 1978, VAN BRUSSEL et al. 1979, WILCOCKSON and WERNER 1979). This again is an indication of the absence of normal cell growth. The nitrogenase activity in cultures of *Rhizobium* appears to be rather insensitive to the presence of $NH_4^+$ in the medium as compared to other $N_2$-fixing bacteria. This is related to the low $O_2$ concentrations; at higher $O_2$ content nitrogenase activity is far more sensitive to $NH_4^+$ (BERGERSEN and TURNER 1978).

The original observations on the induction of nitrogenase activity in *Rhizobium* by plant callus cultures is not only explained by the provision of the right substrates or the establishment of the optimal low $O_2$ concentration, as suggested by CHILD and KURZ (1978). REPORTER developed a system in which *Rhizobium* cultures were separated from callus cell suspensions. In this system *Rhizobium* strains were activated to reduce $C_2H_2$, which otherwise did not show any nitrogenase activity. Two active principles were produced by these callus cultures which stimulated nitrogenase activity in the bacteria. It was claimed that the production of such active factors was induced by the *Rhizobium* cells or nutrient solutions in which *Rhizobium* had been cultivated (BEDNARSKI and REPORTER 1978).

The development of nitrogenase activity in *Rhizobium* cultures is a real *de novo* synthesis. This is shown by inhibition with chloramphenicol or rifampicin (WERNER 1978) and incorporation of L-$^{35}$S-methionine in nitrogenase component 1 (LIM et al. 1979). The latter experiments showed that nitrogenase synthesis is repressed by high $O_2$ concentration and derepressed under microaerophilic conditions. Derepression is moreover inhibited by $NO_3$, most probably through $NO_2$, and by cyclic GMP.

The $NH_3$ formed is further assimilated by the enzymes glutamine synthetase (GS). *Rhizobium* and the closely related *Agrobacterium* species are unique in possessing two quite distinct forms of GS: GS I which is inactivated by adenylylation in the presence of non-limiting concentrations of $NH_3$, and GS II, which is not adenylylated but is very sensitive to higher oxygen concentrations, higher temperatures and certain chemical agents. There are no indications about a regulatory role of the GS operon for the induction of nitrogenase synthesis as has been postulated for *Klebsiella* (references LUDWIG 1978, RANGA RAO et al. 1978).

$N_2$ fixation ($C_2H_2$ reduction) and N assimilation in fresh isolates from root nodule bacteroids will be extensively discussed in Sections 5.4 and 5.5

### 3.4.5 Genetics

The use of mutants which are defective in infectivity or effectivity opens up very promising possibilities for analysis. Especially when non-infective mutants are obtained, it is essential to start with strains characterized by genetic markers to be sure that the mutants which are obtained really are rhizobia and not contaminants.

Mutants are usually obtained with nitrosoguanidine as mutagenic agent. Most isolated mutants are auxotrophs without further defects in their symbiotic behaviour (JOHNSTON and BERINGER in HOLLÆNDER 1977). More interesting are mutants which are symbiotically defective (e.g. DENARIÉ et al. in NUTMAN 1976), or even show better symbiotic performance (MAIER and BRILL 1978). A very promising method to obtain mutants using transposon mutagenesis was described by JOHNSTON et al. (1978). Upon transfer of a RP4 plasmid, containing phage Mu with an inserted Tn5 transposon, selection is possible for Kanamycin resistance coded by Tn5. As a result of the suicide character of this plasmid upon transfer to hosts without Mu repressor, the transposon Tn5 is found inserted at other sites (e.g. in the chromosome). A high frequency of auxotrophs was reported. As these mutations can be mapped by mapping the position of the drug-resistant marker, they must be useful for isolation and mapping of symbiotically defective mutants, with a reduction in time-consuming plant tests. Genetic exchange among mutants only recently led to convincing results. Successful transformation depends on a most probably periplasmatic competence factor (DANDEKAR et al. 1978). Only a few successful transduction experiments were published (e.g. by BUCHANAN-WOLLASTON 1979). Most experiments on genetic recombination were performed by conjugation with plasmids e.g. the P-plasmid R 68.45 (JOHNSTON et al. 1978, BERINGER et al. 1978). Many chromosomal genes were transferred, but a gene for host specificity co-transferred with the chromosomal alleles was never tested. High frequencies of transfer of "nodulation genes" higher than those of the transfer of chromosomal markers are strong evidence that such genes are present on a plasmid. Plasmids are indeed found in all *Rhizobium* species tested up till now. Many strains even contain more than one big plasmid. Many, if not all, genes which code for factors essential for infection and subsequent nodulation appear to be located on one of such plasmids in *R. leguminosarum* and other strains. Such plasmids are therefore called sym-plasmids (see articles by BERINGER and PRAKASH et al. in GIBSON and NEWTON 1981).

The presence of *nif* genes on plasmids was made probable by the demonstration of homologous sequences between plasmid DNA from *R. leguminosarum* and *Klebsiella* nif gene DNA (NUTI et al. 1979). For further discussion of *nif* genes we refer to Chapter II.1, this Volume).

# 4 Symbiotic Relations

Most research on the physiological and biochemical aspects of the symbiotic relations, which lead to the establishment of a $N_2$-fixing symbiotic system, was and is based on the root nodule symbiosis of legumes. Therefore this chapter will be mainly based on research with the legume–*Rhizobium* symbiosis with, where possible, incidental comparisons with other symbioses.

## 4.1 Chemotaxis and Rhizosphere Accumulation

There are some indications that *Rhizobium* is chemotactically attracted by roots and root extracts. As non-mobile mutants of *Rhizobium* strains are usually as infective as mobile strains, we must conclude that chemotaxis will play only a minor role (NAPOLI et al. in NEWTON and ORME-JOHNSON 1980). After contact of *Rhizobium* cells with the root surface, they start to multiply in the rhizosphere. Some analyses have been made of the substances which are exudated by the roots and may stimulate bacterial growth (e.g. EGERAAT 1972). All analysis indicated the presence of many amino acids with some smaller peptides. Homoserine may play a dominating role. Though rhizosphere accumulation certainly is an important phase for the infection process, there are no indications that specific differences in root exudate composition and rhizobial growth stimulation play a role in the specific recognition process between leguminous roots and *Rhizobium* strains. Nor is there any evidence that the rhizobia in the rhizosphere already develop nitrogenase activity.

### 4.2 Binding of *Rhizobium* to Root Hairs

LI and HUBBELL in 1969 suggested that the specific recognition between plant roots and bacteria, which determines whether infection will follow, already takes place at a very early phase. It has been repeatedly observed that *Rhizobium* cells may adhere to root hairs or plant callus cells. This adhesion, mostly as a polar binding of one of the tips of the bacterial cell, is not a specific process but may take place at different types of surfaces. On the other hand DAZZO et al. 1976) observed a four to five times as abundant binding to clover root hairs by infective strains of *R. trifolii* than by non-infective strains.

In peas CHEN and PHILLIPS (1976) could not find any difference in binding of infective and non-infective *Rhizobium* strains. In non-leguminous plants the remarkable polar exo-encapsulation threads, described by LALONDE et al. in NEWTON et al. (1977) turned out to be non-specific and no indication for a successful infection (QUISPEL in GORDON et al. 1979). Yet it is very well possible that a microscopically visible adhesion of *Rhizobium* cells to root hairs, though generally non-specific, may include specific recognition aspects which are determinative for the initiation of the infection.

HAMBLIN and KENT (1973) were the first to consider the possibility that phyto-haemagglutinins, presently called lectins, play a role in a specific recognition process between root hairs and *Rhizobium* cells. Lectins are sugar-binding proteins, which are usually extracted from seeds, and which are important tools in medical-biological research as agglutinants for e.g. red blood cells and lymphocytes (see also Chap. 5 in Vol. 13B, Plant Carbohydrates II, of this Encyclopedia). They bind to the surface of such cells and may initiate cell divisions in lymphocytes. Little is known about their possible functions in the plant. BOWLES and KAUSS (1976) suggested that they are involved in cell-wall synthesis as sugar-transporting agents. Indeed their most characteristic aspect is the binding to specific sugar configurations. Binding to polysaccharides, e.g. cell wall constituents, can be prevented or removed by sugars which as haptens compete with the specific groups of the polysaccharides. A very attractive theory about the possible role of lectins in the very early phases of infection was developed by DAZZO and HUBBELL (1975). By serological methods they demonstrated the presence of common cross-reactive antigens on roots of white clover and the cell surface of *R. trifolii*. These antigenic sites were found both on infective and non-infective strains of *R. trifolii*, but infective strains displayed a distinctively greater degree of antigenic reactivity. On the other hand, FITC-labeled antisera, prepared against infective *R. trifolii* strains, were bound to the tips of growing root hairs. These tips were the sites where during infection the most abundant binding of bacteria was observed. A soluble, non-dialyzable clover lectin was capable of binding to the antigens of both root hairs and bacteria. Lectins are typically bivalent, which explains their agglutinating activity. On these data DAZZO and HUBBELL proposed the hypothesis that lectins connect the common antigenic sites and in this way bind the bacteria to the root hair tips (Fig. 5). The lectin was isolated from clover seeds and purified as a protein called trifolin. A similar protein was eluted from clover roots with solutions of the hapten for clover lectin, 2-deoxy-D-glucose. This same

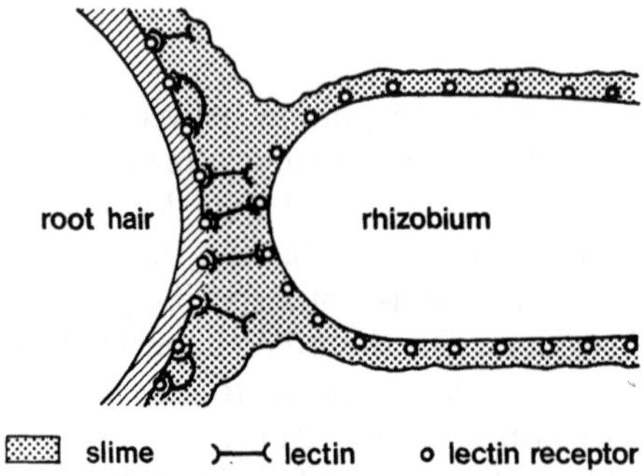


**Fig. 5.** Schematic representation of the lectin-binding theory. (Courtesy of J.W. KIJNE)

hapten prevents binding of clover lectin to *R. trifolii,* as well as the binding of infective bacteria to the root hairs (DAZZO et al. 1976). Antiserum to trifolin binds at the root hair region of the roots of 24-h-old clover seedlings, but does not bind to root hairs of *Medicago sativa* and roots of *Aeschynomene americana* and *Lotus corniculatus* (DAZZO in NEWTON and ORME JOHNSON 1980). The bacterial antigenic determinants are found at the surface in a morphological distinct capsule. This capsular material could be extracted by 1% phenol, which certainly will dissolve part of the LPS as well. The extracted material could be separated into three fractions. Two of these fractions precipitated with anti-clover root serum (DAZZO and BRILL 1979). WOLPERT and ALBERSHEIM observed that LPS from different *Rhizobium* species specifically was bound to lectin from plants belonging to the same cross-inoculation group. No binding occurred between these lectins and the extra-cellular polysaccharides (EPS). More recent work from the same laboratory demonstrated, however, that most mutants which lacked EPS were non-infective but regained infectivity after genetic reversion (NAPOLI et al. in NEWTON and ORME-JOHNSON 1980). Mutants without a capsule had lost their infectivity (BAUER in NEWTON and ORME-JOHNSON 1980). This confusing situation will only be elucidated after a still better insight into the chemistry of the polysaccharides of the outer membrane, capsule and extracellular slime, their situations in the structure of the cell wall and their ontogenetical relations.

Most work on lectins was done with lectins which had been isolated from seeds. It must be emphasized that we are only interested in lectins which are eventually present at the root hair surface. In eluates of pea roots carbohydrate containing lectin was found (KIJNE et al. 1980). The carbohydrate composition resembled in some essential aspects the carbohydrates in a hot water extract of young pea root cell walls. The hypothesis was put forward that lectins, which normally bind sugars for cell wall synthesis, recognize bacterial polysaccharides with the same configuration as part of their "self".

It must be admitted that the theory of DAZZO and HUBBELL is still disputed, as different authors failed to find a correlation between the binding of lectins and specificity of infection. We may again refer to the work of CHEN and PHILLIPS (1976) who observed that less than 10% of infective strains of *R. leguminosarum* bound pea lectin. Though FITC conjugates of antiserum against *R. leguminosarum* indeed adhered to pea root hairs, these conjugates were bound in the same way to roots of many other plants. There are some explanations for such conflicting results. The antigenic sites at the bacterial surfaces may be covered by non-specific slime. BAUER and coworkers (BAUER in HOLLAENDER 1977, in NEWTON and ORME-JOHNSON 1980) observed that correlations between cross-inoculation specificity and lectin binding increased if better-purified lectin preparations were used. The presence of lectin-binding sites on the bacterial surface strongly depends on the phase of development. Best lectin-binding was mostly observed during mid-log phase. In some cases binding of lectin depends on the presence of an additional factor from the plant root culture solution. It must be stressed that correlations between lectin binding and infection specificity must be based on direct observations of the root hair infection. A correlation with nodule formation is not necessary, as many other factors may interfere in later phases of the nodulation process.

### 4.3 Root Hair Deformation and Infection-Thread Formation

Root hair deformation is an essential prerequisite for infection, both in leguminous (see DART in QUISPEL 1974, in HARDY and SILVER 1977) as in non-leguminous infections, (CALLAHAN et al. in TORREY and TJEPKEMA 1979). Infection threads were correctly interpreted by NUTMAN already in 1956, as an invagination process, though we must realize that this invagination can only proceed by means of a continuous deposit of cell-wall substances around the slimy mass of bacteria. Root hair deformation and infection thread formation must be regarded as two aspects of a process of disregulation in the normal process of root hair growth.

Much confusion has arisen around the question of what types of substances are responsible for root hair deformation. Part of this confusion was caused by the fact that no distinction was made between different types of deformations such as curled, helical, branched etc. According to YAO and VINCENT (1976) the type of deformation described as "marked curling" is the only one which is directly related with the subsequent infection thread formation. Soluble factors never produce real "marked curling", which is only obtained after direct cell-to-cell contact with the infective bacteria. It is tempting to relate this effect to the specific binding of *Rhizobium* to root hairs according to DAZZO and HUBBELL's lectin bridge theory. The importance of this binding is not the binding as such but the reactions it evokes in the cells (BAUER in HOLLAENDER 1977). If KIJNE et al. (1980) are right in assuming that plant root lectins erroneously recognize bacterial surface sites as carbohydrates for cell wall synthesis, we may understand why normal cell wall growth is disrupted and channelled into infection thread formation.

### 4.4 Cell Wall Degrading Enzymes

For most phytopathogenic organisms penetration of, for example, the leaf surface demands the activity of cell wall degrading enzymes. Formation and activity of such enzymes is a complex process in which cell-wall substances from the host induce the synthesis of the enzymes in the pathogens, often in a sequential way. On the other hand, the activity of the enzymes is susceptible to inhibition by substances produced in the host cells (ALBERSHEIM and ANDERSON-PROUTY 1975, COOPER in SOLHEIM and RAA 1977).

It is not immediately obvious to expect similar processes in *Rhizobium* infection, as the infection proceeds not by penetration but by invagination. Yet we must realize that growth of the infection thread must be accompanied by cell wall metabolism, probably more of a synthetic than a degradative nature. In the deformation of root hairs, degradative aspects may be dominating.

In much-discussed experiments FAHRAEUS and LJUNGGREN observed an increase of "polygalacturonase" activity in the medium or in superficial phosphate-buffer washings of roots which were infected with *Rhizobium*. This activity was only observed after addition of infective *Rhizobium* strains. The method which they used does not distinguish between enzymes produced by the bacteria or enzymes which are exuded by the roots. As nobody had ever succeeded in demonstrating pectolytic enzymes in *Rhizobium* cultures, they preferred the plant root origin of the enzymes. The specific effect of infective *Rhizobium* strains was explained by the assumption that polysaccharides from the bacteria induced "polygalacturonase" formation at the root surface (LJUNGGREN 1969). Most other authors did not succeed in reproducing these results. The influence of plant growth conditions, as used by the different authors, may explain some of the discrepancies. The original supposition, that *Rhizobium* does not produce pectinolytic enzymes may be erroneous. HUBBELL et al. (1978), when applying a very sensitive agar plate assay, succeeded in demonstrating low levels of pectinolytic enzymes in *Rhizobium* strains. This was only demonstrated in *Rhizobium* strains from temperate regions. In the *Rhizobium* strains which produced pectinolytic enzymes, no correlation with their infectivity was observed. In any case these results open up new possibilities for research and a re-interpretation of older experiments. The results of LJUNGGREN and FAHREAUS may be equally well explained by the usual induction by plant substances of microbial enzyme production. Nothing is known about cell wall metabolism during growth of the infection thread nor about the reason that cell wall deposits stop as soon as the infection reaches the cells which just were formed by the meristem. The role of plant factors is indicated by genetic mutations in the plant genome which lead to continued growth of the infection thread. In other cases growth of the infection threads stops already in the root hair. Nothing is known about the immediate expression of the genes involved (see Sect. 4.6). For a survey of all factors controlling the nodule formation we refer to VINCENT (in NEWTON and ORME-JOHNSON 1980).

## 4.5 The Role of Plant Hormones in Nodule Formation

The role of plant hormones in the initial phases of infection is doubtful. Though indole acetic acid may be formed in the rhizosphere its influence may be at best of a secondary nature. Cytokinins may also be produced in the rhizosphere and may lead to root thickenings and local malformations (PUPPO and RIGAUD 1978). There are no indications that such swellings on roots of leguminous and non-leguminous plants (BERMUDEZ DE CASTRO in NEWTON et al. 1977) bear any resemblance to real root nodules. Certainly plant hormones are of fundamental importance in later phases of the nodulation process (LIBBENGA and BOGERS in QUISPEL 1974). While in peas the cortex cells which are passed by the infection thread react by a swelling of the nuclei due to DNA synthesis, they do not yet divide. The first pattern of cell division is observed ahead of the penetrating infection thread in the inner cortex cells opposite the xylem strands. This pattern could be reproduced in excised root cortex cylinders by application of plant hormones. Addition of auxins only stimulated divisions in the pericycle, combination of auxins with kinetin produced the same pattern of cell divisions in the inner cortex as is observed after *Rhizobium* infection. The remarkable localization of cell divisions to the region opposite the xylem strands indicated that factors from the vascular bundles are limiting the de-differentiation process. Indeed it appeared that a mixture of auxins, kinetin and an extract from the stele initiated cell divisions all over the cortex. The initiation of cell divisions which lead to the nodule meristem therefore could be explained by a gradient hypothesis. According to this hypothesis de-differentiation of cortex cells depends on the optimal interaction of a gradient of auxins and cytokinins from the infected cells with a gradient of as yet unidentified factors from the xylem. The much-discussed polyploid character of most root nodule cells could be explained by the reaction of tetraploid cells, which are already most abundant in the outer cortex, on combinations of auxins and cytokinins with or without root stele extracts (BOGERS et al. 1976). In non-leguminous plants like *Alnus*, the initial cell divisions are observed in the outer and central cortex cells, while in the neighbouring pericycle lateral roots are initiated. ANGULO-CARMONA (1974) concluded that they are formed in excess of normal lateral roots and at sites where normally no roots should have been initiated. As the formation of root initials is stimulated by auxins, it is tempting to suggest that the production of auxins in the infected cells is responsible for the initiation of these lateral roots, which after infection from the primary nodule develop into the rhizothamnia.

It is difficult to demonstrate whether the cells with an infection thread really are centres of auxin and cytokinin production. However, it is a well-known fact that young root nodules are hyperauxinic. This high auxin content may be the result of synthesis by the bacteria or by an increased synthesis of the invaded root cells. Cultivated rhizobia synthesize indole acetic acid (IAA), but only if a precursor such as tryptophan is present in the culture solution. The biochemical pathways for IAA synthesis are basically similar in plants and bacteria, but there is a difference in the production of the side-product indole carbonic acid (ICA). A comparison of the IAA/ICA ratio led to the conclusion that IAA synthesis in root nodules is mainly a plant-cell process (DULLAART 1970). This is in accordance with observations on phytopathogenic organisms which induce a higher IAA synthesis in the host cells. Even if bacteria synthesize part of the IAA, the production of the precursor tryptophan will be the main limiting factor.

A higher production of IAA may be caused by a higher synthesis rate or by a decreased rate of destruction. MENNES (1973) indeed observed that in peas the specific activity of IAA-oxidase is lower in the nodule than in the roots. Differences in IAA-oxidase activity can be explained by differences in phenolic contents. Cytochemical studies indicated that peroxidase activity (which can be used as an indication of IAA-oxidase) is absent in the cytoplasm of the bacteroid-containing cells (OOSTROM et al. 1975). More difficult to understand was the high peroxidase activity in the cells of the infected root cortex area. This poses the problem whether indeed enough IAA may reach the target cells in the inner cortex as is expected on the base of the gradient hypothesis. A comparison with other symbiotic systems shows that hyperauxiny is found in *Alnus* nodules (DULLAART 1970) but not in nodules of Cycadeae. This may reflect the difference between nodules which are formed by the infection and preformed nodules which are invaded by the microsymbiont. Several authors observed an increased content of cytokinins in root nodules as compared with uninfected root tissue. The cytokinins were identified as zeatin and its riboside, which are normal products of root metabolism, and iso-pentenyladenine (2iP) which is produced in *Rhizobium* cultures. Bacterial cytokinin production may be stimulated by roots. In pea nodules a positive correlation was observed between cytokinin activity and the mitotic index in the nodule meristems. It was suggested that the bacteria in the infection thread form 2iP and the meristems are responsible for the production of zeatin and its derivative. As both substances are closely related it might be possible that the meristems convert the 2iP from the bacteria into zeatin. For detailed references see e.g. SYONO et al. (1976), PUPPO and RIGAUD (1978), NEWCOMB in NEWTON and ORME-JOHNSON (1980) and HENSON and WHEELER (1977b) for non-legumes. In animal cells mitotic activity can be stimulated by LPS from bacterial cell walls. KIJNE et al. in SOLHEIM and RAA (1977) reported that *Rhizobium*- and *E. coli* LPS, if added together with auxins, can stimulate cell divisions in root cortex cells and in this respect can replace cytokinins. The active part of LPS is the lipid A moiety. It is not yet known whether LPS stimulates cytokinin production or directly stimulates cell divisions. High contents of gibberellins in root nodules were found in *Lupinus luteus* (DULLAART and DUBA 1970) and *Alnus glutinosa* (HENSON and WHEELER 1977a). DULLAART and DUBA (1970) observed a stimulation by gibberellic acid of the synthesis of IAA from tryptophan by cell-free nodule extracts. This and the well-known auxin-sparing effects of gibberellins makes it worthwhile to pay more attention to gibberellins in the nodulation process.

### 4.6 Miscellaneous Problems

The physiological studies described in the preceding paragraphs were only a small part of the wealth of problems which arise from the description of the infection and nodulation process (see Sect. 2.4.3). The majority of these problems can only be stated with few or no possibilities for experimental approach. Much inspiration might be obtained from comparisons with related phenomena in phytopathology. Certain ultrastructural events bear some resemblance with events during, for example, hypersensitive reactions. The fundamental difference is that hypersensitive reactions lead to disruption of cell structures, while in nodule formation the reactions are regulated and channelled into the formation of new functional structures. Phytopathogens may be inhibited by the production of phyto-alexins which are formed by the host and elicited by glucan-like substances from the cell walls of the infecting micro-organisms (ALBERSHEIM and ANDERSON-PROUTY 1975). Nothing is known of such events during *Rhizobium* infections, but the attention paid to cell-surface binding might lead to some insight in these aspects as well. Here we must not limit ourselves to the events at the root hair surface. A highly crucial moment certainly is the endocytosis of bacteria, as at that moment the cell surface of the bacteria comes in direct contact with the plasmalemma of the host cells. The discovery of a mitogenic activity of rhizobial LPS (KIJNE et al. in SOLHEIM and RAA 1977) may be a first indication of the interactions involved. In the formation of crown gall by the related *Agrobacterium tumefaciens* it is now well established that plasmid DNA of the bacteria is taken up and brought to expression by the plant cells. The first indication that *Rhizobium*-DNA may be taken up by legume roots was obtained by LEPIDI et al. (1976). Our insight into complicated biological systems is always highly increased by comparisons with systems where deviations in the normal process take place. In root nodule symbioses such comparisons are possible with plant–bacteria combinations where the normal infection process is somewhere disturbed. Such deviations may be based on mutations of the genome of either the plants, or the bacteria or to unfavourable environmental conditions. VINCENT (in NEWTON and ORME-JOHNSON 1980) distinguished the following genetically determined moments in infection and nodulation: pre-infection (root surface colonisation, root adhesion, root hair branching and marked curling), infection (formation of the infection thread, development of meristematic activity, release of bacteria, bacteroid development), nodule function (nitrogen fixation, complementary biological and biochemical functions and nodule persistence). Certainly these moments can be further divided into different gene functions. Many of such deviations were studied with *Rhizobium* strains or plant species whose genetical relations were unclear. The methods developed for isolation and genetical analysis of *Rhizobium* mutants (see Sect. 3.4.5) open up possibilities for better genetic-physiological comparisons. The same may be expected from plant molecular-genetic studies. A better study of the genes involved and their biochemical expression will certainly be one of the most important tools for a better understanding of the physiological backgrounds of root nodule formation (VERMA 1980). Such studies already led to the discovery of the de novo synthesis of specific nodule proteins.

# 5 The $N_2$-Fixing System

## 5.1 Introduction

After our discussion of the symbiotic interactions which lead to the development of the active $N_2$-fixing root nodule we will now consider the biochemical and physiological aspects of the structures which are functional for the $N_2$-fixing process in leguminous root nodules. These are the bacteroids, surrounded by the peribacteroid membranes in the bacteroid-containing host cells. In this system the exchange of substances between host and bacteria, which is essential for $N_2$-fixing activity, must take place through a complicated structure of membranes (Fig. 6). Our aim must be to understand the biochemical and transport processes which take place in the bacterial and host cytoplasm and the membrane systems between.

## 5.2 Bacteroids

For practical reasons some authors prefer to call all bacteria inside the root nodules bacteroids. This is an unfortunate definition, as it neglects the great difference between the still-multiplying bacteria and the final forms which are the actively $N_2$-fixing sites in the nodules. The gradual change from bacteria to bacteroids in the transition process may be emphasized by distinguishing bacteria, transforming bacteria and bacteroids, the more so as these phases can be approximately separated by sucrose gradient centrifugation (TE MAY CHING et al. 1977). We must admit that in some plants the morphological differences between bacteria and bacteroids are far more pronounced than in others.

The old dogma that the bacteroids are no longer viable in the sense that they can no longer develop in nutrient solutions has been questioned by GRESHOFF et al. (1978). If the bacteroids of clover and soybean nodules are very carefully prepared from bacteroid-containing protoplasts, isolated by maceration, they are all able to multiply on suitable nutrient media. Bacteroids are only very susceptible, especially to osmotic damage. This coincides with the observation of LAANE et al. (1978) that during preparation of bacteroids, membranes may be damaged by the effect of fatty acids and phospholipase from the plant cells. This can be prevented with bovine serum albumin.

The DNA content of bacteroid cells was determined in bacteroids of many species and compared with the DNA content of bacteria in mid log and early stationary phase. Determinations were made in isolated bacteroids or by microfluorometry in situ (for detailed references see PAAU et al. 1979, SUTTON et al. 1978). The DNA content in bacteroids either equals the content of bacteria or exceeds it by factors varying from 1.5–3 or occasionally even 6 times, depending on the leguminous association studied. It is generally observed that the RNA content, especially the ribosomal RNA in bacteroids, is low (BERGERSEN, in QUISPEL 1974, BISSELING et al. 1979). Protein synthesis has been demonstrated by the incorporation of labeled amino acids in proteins in situ and after isolation. SHAW and SUTTON (1979) observed a change in the labeling pattern of different polypeptides during bacteroid development. After 12 days, two polypeptides were especially labeled which have the molecular weight of the nitrogenase components.

Bacteroids have a normal respiratory activity with a similar glycolytic system as in the cultivated bacteria (e.g. KURZ and LA RUE 1977), a TCA cycle with a partial glyoxylate cycle (STOVALL and COLE 1978) and a terminal electron flow to $O_2$ or $NO_3^-$. There are some changes in the cytochrome content, which, at least partially, coincide with changes in cultivated bacteria due to anaerobic conditions (see Sect. 3.4.3). The pigments P-420 and P-450 are supposed to be coupled to the nitrogenase system (BERGERSEN in QUISPEL 1974). Flavins may be present in high concentration, while a non-haem protein has been found (PHILLIPS et al. 1973). Especially important is the presence of an intact, integrated cytoplasmic membrane, which can be energized. This membrane energization is essential to deliver the energy to the cytoplasmic nitrogenase (LAANE et al. 1979). The mediation between the energized membrane and the nitrogenase may be provided by the flavodoxin which is the most powerful reductant for nitrogenase or by the non-heme iron protein. Though reserve substances like PHBA are broken down during active respiration and $N_2$-fixation, the normal substrate for respiration must be provided by the carbohydrates (sucrose) of the phloem sap. However, all studies on respiratory substrates for bacteroids agree that sugars are far less suitable substrates than TCA-cycle acids such as succinate. As the catabolic enzymes for glucose degradation are present (KURZ and LA RUE 1977) a plausible explanation may be that glucose transport is inhibited. Indeed DE VRIES (1980) observed that bacteroids of *R. leguminosarum* no longer were able to transport the glucose analogue 2-deoxy-D-glucose, which is well transported by bacteria. On the other hand these bacteroids possessed a highly active transport system for succinate. Either the glucose transport system is repressed in bacteroids by the presence of organic acids like malic acid in the plant cells, or this transport is impeded by the absence of binding proteins.

That carbohydrates are not essential for bacteroids is, moreover, indicated by the fact that mutants which are defective in sugar metabolism still form effective root nodules (RONSON and PRIMROSE 1979). If organic acids are the favourite substrates for bacteroids, the surrounding root cell has to metabolize the phloem sucrose to these acids (see Sect. 5.3).

Other transport systems must be present for the transport of amino acids. In *R. leguminosarum* bacteroids, glutamate and aspartate are rapidly used for $CO_2$ production and thus may be considered as respiratory substances. Other amino acids like leucine are, after transport, used for protein synthesis (STAAL in preparation). Since it is doubtful whether the $NH_3$ formed from $N_2$-reduction is used in bacteroids, the question arises whether bacteroids depend on plant amino acids for their protein synthesis. Bacteroids possess a weakened, more permeable, outer membrane. Direct isolation of LPS with the aqueous phenol method showed a far smaller amount of LPS than was observed in bacteria cultivated on different nutrient solutions (VAN BRUSSEL et al. 1977). Further studies by PLANQUÉ et al. (1979) did not indicate essential differences in the sugar and fatty acid composition of LPS of bacteria and bacteroids. There were differences in physical characteristics, for example the behaviour during ultracentrifugation and the solubility. Part of the LPS is already lost during isolation of bacteroids, which certainly does not happen after similar treatments of bacteria. This altered, more soluble LPS and the more permeable structure of the bacteroid outer membrane may have important consequences. It may explain the swollen pleomorphic forms. The solubilized part of the LPS may have effects on the host cells (KIJNE et al. in SOLHEIM and RAA 1977). The increased permeability to macromolecules must have consequences for the localization of periplasmatic enzymes and binding proteins which now may be dissolved in the whole space between bacteroid cytoplasmic membrane and host cell peribacteroid membrane. For the time being we can only speculate about the consequences for the symbiotic relations.

### 5.3 The Bacteroid-Containing Plant Cells

The cytoplasm of the bacteroid-containing plant cells has two important functions: it must provide the bacteroids with the respiratory substrates and most probably the amino acids for protein (nitrogenase) synthesis. On the other hand they must absorb the products of $N_2$fixation, most probably $NH_3$, and assimilate this $NH_3$ into amino acids. Both transport processes must take place in the peribacteroid membrane. Cytochemical evidence, e.g. the presence of adenylcyclase and staining with phosphotungstic acid (TU 1979, ROBERTSON et al. 1978a) is in accordance with the plasmalemma origin. This, of course, does not mean that the peribacteroid membrane is identical with the plasmalemma. ROBERTSON et al. (1978b) prepared peribacteroid membrane fractions after osmotic shock of membrane-surrounded bacteroid preparations.

Their identity was indicated by staining with phosphotungstic acid and silver salts. They consisted of lipoids and protein in a 6.1:1 ratio. As succinic acid and related acids are the obvious substrates for bacteroids, the host cells must produce these acids from the phloem sucrose. Moreover organic acids are necessary to provide the C-skeletons for amino acid synthesis. Already in 1960 MULDER and VAN VEEN observed an increase of $N_2$ fixation in root nodules of peas by increased concentrations of $CO_2$, which promotes organic acid synthesis. More recently a direct correlation was observed between $C_2H_2$ reduction and $CO_2$ binding. Both $CO_2$ binding in vivo and PEP-carboxylase activity in vitro were far higher in nodules than in roots. Studies with differently $^{14}C$ labeled glucose confirmed the role of glycolysis with PEP-carboxylation, with only a minor role of the pentose phosphate pathway (LAING et al. 1979). The low oxygen concentration due to the high respiration of the bacteroids may lead to a shift in enzyme activities towards more fermentative organic acid-producing processes. The organic acids produced, like malic acid, might have inducing or depressing effects on metabolic and transport processes in the bacteroids (DE VRIES 1980).

The study of the plant cell part in the metabolic activities of bacteroid-containing cells is extremely difficult because root nodules consist of so many different cell types. Better progress may be expected from studies with isolated bacteroid-containing protoplasts. Contrary to isolated bacteroids, $C_2H_2$ reduction by these protoplast suspensions was markedly stimulated by sugars. Moreover $C_2H_2$ reduction was stimulated by nucleotide triphosphates and different amino acids, but strongly inhibited by asparagine and glutamine (WOOI and BROUGHTON 1979).

The function of nodule-specific plant proteins, called nodulines, is not yet understood (LEGOCKI and VERMA 1979).

### 5.4 Nitrogenase

The first definite proof that bacteroids are the sites of $N_2$ fixation was given by the work of KENNEDY et al. in 1966 and BERGERSEN and TURNER in 1967. The first cell-free system of bacteroid nitrogenase was obtained by KOCH et al. in 1967. Highly purified preparations can be obtained by purification with DEAE-cellulose and Sephadex G 150 (WHITING and DILWORTH 1974). The nitrogenase resembles the nitrogenase from other $N_2$-fixing organisms in all aspects, so that we can refer to the discussion in Chapter II.1, this Volume).

The older discussions about the question whether bacteroid formation is essential for $N_2$ fixation by *Rhizobium* (BERGERSEN, QUISPEL, in QUISPEL, 1974) have lost part of their meaning since some strains can fix $N_2$ in cultures. Moreover, in *Parasponia* $N_2$ fixation takes place without a conspicuous bacteroid formation. On the other hand in legumes the formation of nitrogenase is only derepressed at a certain phase in the development which more or less coincides with the development of bacteroids. This derepression was studied by many authors by determining nitrogenase activity in bacteria and bacteroids in different phases of development (e.g. BISSELING et al. 1979, PAAU et al. 1979, WERNER and STRIPF 1978). In all cases a clear picture resulted of the development of nitrogenase during bacteroid development. In peas it appeared that the two protein components of nitrogenase were not coordinatedly synthesized, young bacteroids containing relatively more protein II, older bacteroids relatively more protein I (BISSELING et al. 1979).

No direct experimental evidence is available about the factors which lead to derepression of nitrogenase but much may be learned from the regulatory factors which influence (de)-repression in cultivated strains. Microaerophilic conditions may be among the most important factors to derepress nitrogenase (see Sect. 5.6). The role of GS in transcription is uncertain (see Sect. 5.5). Very suggestive are the observations on plant factors which might derepress nitrogenase (BEDNARSKI and REPORTER 1978). Though $NH_3$ is a repressor in most $N_2$-fixing organisms, either directly or by promoting GS-adenylylation, *Rhizobium* appears to be rather insensitive. Though BISHOP et al. (1976) demonstrated a 65% reduction of nitrogenase activity in soybean nodules after a 4-day period in 5 mM $(NH_4)_2SO_4$, this reduction was more probably caused by an effect on nodule formation than by a direct effect on nitrogenase. Other inhibiting effects of $NH_4^+$ are explained by a deviation of ATP supply (HOUWAARD 1978). This aspect will be further discussed in Section 6. For detailed references we refer to DILWORTH (in NEWTON and ORME-JOHNSON 1980).

### 5.5 $NH_3$ Assimilation

In bacteria $NH_3$ is assimilated by the GS-GOGAT pathway with GDH only playing a role at high $NH_3$ concentration. The role of GS in cultivated *Rhizobium* bacteria was already discussed in Section 3.4.4. Especially important aspects are the inactivation of GS by adenylylation which is stimulated at higher $NH_3$ concentrations, and the suggested role of deadenylylated GS in transcription of the *nif* genes.

For plants as well it is now generally accepted that GDH mainly may have an anabolic function, while GS-GOGAT is the most active $NH_3$ assimilatory system (MIFLIN and LEA 1977). Affinities for $NH_3$ are far greater for GS than for GDH. $K_m$ values ranging from 0.02–0.5 mM and from 0.03–0.1 have been reported. There is no evidence in plants for $NH_3$ repression, nor for an adenylylation cascade system of GS, as in bacteria.

In bacteroids GDH is present, but usually in too low concentrations to play a role in $NH_3$ assimilation. Data on GS were rather conflicting (BROWN and DILWORTH 1975), which may result from the use of different assay methods. The often used transferase reaction is not fully specific; it measures activities both of the adenylylated and the deadenylylated form, depending on the addition of $Mg^{2+}$ or $Mn^{2+}$ and has a higher turnover rate than the biosynthetic assay. From the data of BROWN and DILWORTH (1975), we may conclude that the activity of the biosynthetically active de-adenylylation form is small or negligible. When bacteroid samples of different phases of development were compared, it appeared that activity of GS, even when measured with the transferase assay, was negatively correlated with the activity of nitrogenase (PLANQUÉ et al. 1978).

Microaerophilic conditions, as they are supposed to be present in the nodules and essential for nitrogenase derepression, decrease GS activity. As RANGA RAO et al. (1978) observed a complete loss of activity of GS II at $pO_2$ beneath 0.4% it is not probable that this type of GS can play a role in bacteroids. However, further studies on these two forms during bacteroid development are required. Studies on the effect of excess $NH_3$ or N depletion in vivo with intact nodules, did not show any effect on the level of bacteroid GS nor on its level of adenylylation (BISHOP et al. 1976). Glutamate synthase (GOGAT) has been observed in bacteroids of lupin, soybeans and peas, but in most cases the activities were very low. The general conclusion must be that the activities of $NH_3$-assimilating enzymes in bacteroids are too low to match those of the $N_2$-fixing rates and even get lost during nitrogenase derepression. This explains the observation that isolated bacteroids excrete the fixed N as $NH_3$ (for further references see BOLAND et al. in NEWTON and ORME-JOHNSON 1980). All investigations agree that there is a high activity of GS in the supernatant which is left after sedimentation of bacteroids from nodule homogenates and which is considered to represent the plant cell cytosol (e.g. BROWN and DILWORTH 1975, ROBERTSON et al. 1975, SCOTT et al. 1976). Though of plant origin, the activity of GS depends on the development of effective bacteroid tissue. Increase of plant GS correlates with the increase of bacteroid nitrogenase (ROBERTSON et al. 1975, STRIPF and WERNER 1978). No increase or even decrease of GS is observed in ineffective root nodules (WERNER et al. 1979).

Initially controversial data on plant GOGAT may be explained by methodological difficulties. The role of plant GDH is generally considered to be of minor importance. Sometimes high activities were obtained which were even correlated with bacteroid nitrogenase activity (STRIPF and WERNER 1978, KENNEDY 1979). The latter author suggests that the occurrence of both complementary pathways of $NH_3$ assimilation may be important as a regulated response to gradients of $NH_3$ concentrations around bacteroids in nodule tissues.

### 5.6 Oxygen Regulation and Leghaemoglobin

One of the most central problems for all aerobic $N_2$-fixing organisms is the regulation of $O_2$. This $O_2$ is necessary for respiration which produces the energy for $N_2$ fixation though the nitrogenase must be protected against inactivation by $O_2$. In the root nodules we have seen that there is strong evidence that micro-aerophilic conditions are among the most important factors to derepress nitrogenase formation.

TJEPKEMA and YOCUM (1974) concluded from a deflection in the oxygen concentration curve of root nodule respiration that there must be a diffusion barrier against $O_2$ somewhere in the cortex. This was directly confirmed by measurements of the $O_2$ concentration by micro-electrodes in the cortex. BERGERSEN and TURNER (1975) concluded from the $O_2$ concentration effect on respiration of bacteroid suspensions that besides a high-affinity system which is linked with $N_2$ fixation through P-450, and consequently inhibited by N-phenyl imidazole, there is a low affinity respiratory system which might act as a respiratory protector of nitrogenase. There is no evidence for a "switch on–switch off" type of regulation (CRISWELL et al. 1976) (see Chap. II.1, this Vol.).

There is now general consensus that besides these types of $O_2$ protection the most important regulation of $O_2$ provision and $O_2$ protection is provided by leghaemoglobin. Since its discovery by KUBO in 1939 and identification by KEILIN and WANG in 1945 and VIRTANEN in 1947, it was soon recognized that leghaemoglobin content was directly correlated with effectivity. The correlation between development of nitrogenase activity and leghaemoglobin synthesis during nodule development is not as absolute (e.g. NASH and SCHULMAN 1976), but we must be careful, as our methods for measuring the first nitrogenase activity are far more sensitive than the methods for leghaemoglobin determination.

As several extensive reviews summarized our knowledge about the chemistry and physico-chemistry of leghaemoglobin (e.g. APPLEBY in QUISPEL 1974, HARDY and SILVER 1977) we can restrict ourselves here to the remark that leghaemoglobin consists of an apoprotein which has a striking similarity with animal haemoglobins and especially myoglobins, while the haem group is a protohaem IX.

All evidence shows that the apoprotein is coded by plant DNA and synthesized on plant ribosomes. We refer to the cross-inoculation experiments of CUTTING and SCHULMAN, which showed that leghemoglobin is determined by the host plant, and the inhibition of $^{14}C$ incorporation in leghemoglobin by cycloheximide, but not by chloramphenicol (references: DILWORTH in NEWTON and ORME-JOHNSON 1980). On the other hand NADLER and AVISSAR (1977) obtained strong evidence that the bacteroids are the sites of haem synthesis. The enzyme $\delta$-aminolevulinic synthetase (ALAS) could only be demonstrated in the bacteroids while the activity of this enzyme, which plays a role in haem synthesis, increased during leghaemoglobin synthesis. Another enzyme of the haem-synthesis pathway, $\delta$-aminolevulinic dehydrogenase (ALAD), increased in the bacteroids during effective nodule development but decreased in the plant fraction. Plant extracts do stimulate ALA incorporation in haem in bacteroids, which indicates a cooperative effort of plant and bacteroids in haem synthesis. We already saw in Section 3.4.3 that cultivated bacteria are stimulated to synthesize protoporphyrin IX under micro-aerophilic conditions (AVISSAR and NADLER 1978). This confirms an earlier suggestion by NASH and SCHULMAN (1976) that the synthesis of leghaemoglobin might be a reaction to a low oxygen concentration. Under such conditions both ALAS and ALAD activities increase.

The localization of leghaemoglobin is still controversial. Micro-autoradiograms after labelling with $^{59}Fe$ showed the main activity between the bacteroids and the peri-bacteroid membrane. Peroxidase activity, one aspect of leghaemoglobin, was demonstrated at the same site. X-Ray probe analysis, however, indicated the presence of leghaemoglobin in the cytoplasm of the bacteroid-containing cells (references in APPLEBY in QUISPEL 1974). Localization with ferritin-labelled antiserum showed activity in the host cell cytoplasm and at the outside of the peribacteroid membrane (VERMA and BAL 1976). The exact localization must be important for our interpretations of leghaemoglobin function. These interpretations are based on the work of BERGERSEN et al. (1973), WITTENBERG, APPLEBY and co-workers (for recent discussions and references see DILWORTH and WITTENBERG in NEWTON and ORME-JOHNSON 1980). Oxygenated leghaemoglobin stimulates respiration of succinate and nitrogenase activity of bacteroids of *R. leguminosarum* at low $O_2$ concentration in such a way that respiration far better supports $N_2$ fixation with leghaemoglobin than without. Oxygenated leghaemoglobin increases ATP content and the ATP/ADP ratio, which is counteracted by the P-450 inhibitor N-phenylimidazole. Ferric leghaemoglobin and other oxidized ferric haemoproteins are without any effect. Several explanations have been put forward. Leghaemoglobin might act as an oxygen carrier enabling a facilitated diffusion of oxygen to the bacteroids. This function might be especially important as the diffusion barrier in the cortex and respiration in the cortex already decreased the oxygen concentration around the bacteroids. In this way it is difficult to understand why ferrioxy-leghaemoglobin should be inactive. Another explanation claims that leghaemoglobin acts as an $O_2$ buffer and establishes the optimal $O_2$ concentration for nitrogenase derepression and activity while still sustaining enough respiratory activity. It cannot yet be excluded that leghaemoglobin still has a more specific function in direct relations with certain components of the high affinity respiratory pathway which leads to energization of the bacteroid membrane. Finally it is interesting to make a comparison with the non-leguminous *Frankia* nodules. With the exception of one positive claim for *Casuarina* by DAVENPORT in 1960, no haemoglobin has ever been demonstrated, though the haem content may be elevated (references in QUISPEL 1974). Moreover no diffusion barrier to oxygen is present in the cortex (TJEPKEMA in GORDON et al. 1979). Regulation of $O_2$ therefore is far less developed, though the activity of nitrogenase shows that some unknown protective mechanism must be active.

## 5.7 Hydrogen Production and Hydrogen Uptake

Production of hydrogen ($H_2$) is a natural consequence of nitrogenase activity. This ATP-dependent $H_2$ production is a waste of energy which is used for reduction of $H^+$ ions, while it could have been used for $N_2$ reduction. DIXON in 1972 was the first who drew the attention to the possible role of a conventional hydrogenase in re-utilizing this $H_2$ in an oxidative process which again yields ATP and thus makes the nitrogenase reaction more efficient. SCHUBERT and EVANS (1976) therefore introduced the concept of relative efficiency as the factor:

$$1-\frac{H_2 \text{ evolved}}{C_2H_2 \text{ reduced}} \text{ or } 1-\frac{\text{rate of } H_2 \text{ evolution in air}}{\text{rate of } H_2 \text{ evolution in Ar-}CO_2\text{-}O_2}$$

It appeared that in different leguminous symbiotic combinations this factor varied from 0.20 (in *Trifolium subterraneum*) to 0.99 (in cowpea). Very high values were obtained for most actinorhizas, with the exception of *Ceanothus*. These differences were caused by the bacterial partner and were correlated with similar results in the cultivated $N_2$-fixing strains of *R. japonicum*. Oxidation of $H_2$ by hydrogenase could indeed sustain nitrogenase activity and thus partially could regain the lost energy. Hydrogenase is an inducible enzyme which thus can be induced by the $H_2$ produced by nitrogenase. This induction is repressed by succinate, and this repression is in its turn counteracted by cAMP (LIM and SHANMUGAM 1979).

Up till now it remains to be decided whether hydrogenase plays a really important role in determining the efficiency of $N_2$-fixation in the field as compared with other factors. For more detailed discussions and references we refer to EVANS et al. (in NEWTON and ORME-JOHNSON 1980).

# 6 Root Nodules as Part of the Whole Plant

In all symbiotic $N_2$ fixing plant systems the micro-symbiont fixes $N_2$ under circumstances where the $NH_3$ formed is hardly or not at all used for the micro-symbiont itself, but is for the most part transported to the host. On the other hand, the micro-symbionts depend on the host for organic substrates for production of energy and reducing power. These organic substrates must, moreover, produce the C skeletons for amino acid formation. During nodule formation organic substrates are essential for growth of the nodules and the development of the $N_2$-fixing systems. These organic substrates provide energy, C and N, and it is well possible that even in the fully developed bacteroids organic N substances from the host are still necessary for nitrogenase synthesis if the bacteroids are unable to assimilate the $NH_3$ formed. Therefore root nodules must be considered as integrated parts of the whole plant in which growth and $N_2$ fixation are regulated and coordinated by the plant (Fig. 6).

This integration is achieved by the presence of nodular vascular bundles. Markedly developed transport cells around these vascular bundles indicate the important role of transport processes either towards or from the nodules system.

The functioning of this transport system has been extensively studied by PATE and co-workers, who calculated flow-sheets for C and N during different phases of nodule development in peas, cowpea and lupin (e.g. PATE et al. 1979)

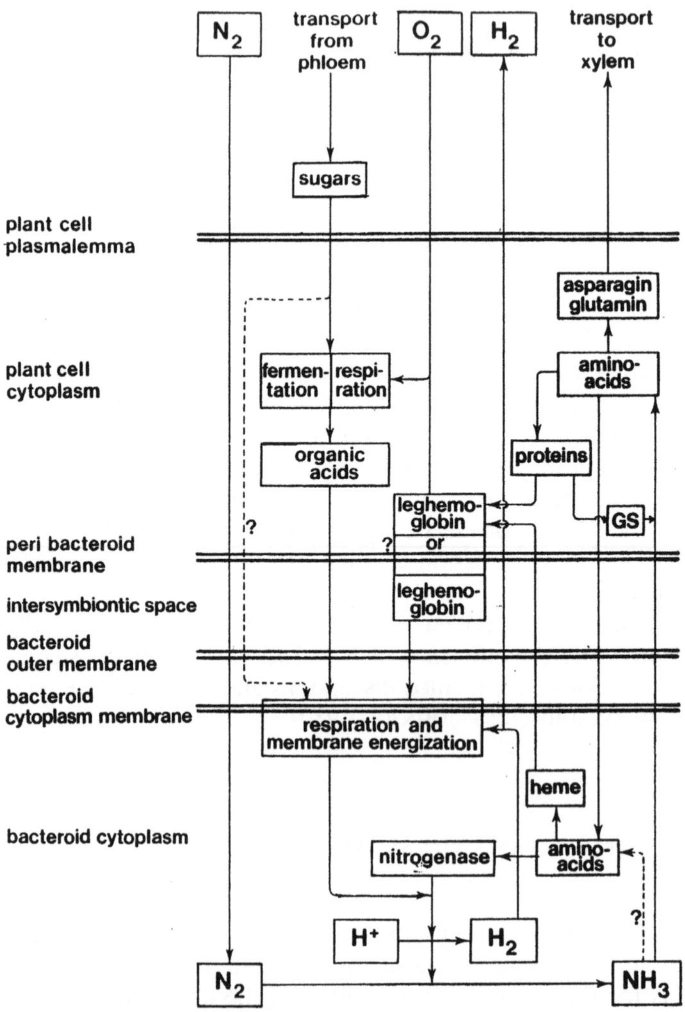


**Fig. 6.** Highly simplified scheme of the main interactions between root nodule cells and bacteroids at the interface between bacteroid and plant cell cytoplasm in its relation to the long-distance transport. Disputed aspects like the use of sugar by bacteroids, the formation of amino acids inside the bacteroids and the localization of leghemoglobin are indicated by *dotted line*

(Fig. 7). We refer to the contribution by PATE in Volume 1, this Series (1975). Dependence on the flow of photosynthetic carbohydrates through the phloem may be direct or indirect after a previous formation of root reserve carbohydrates. There are several indications that this flow of phloem carbohydrates forms the main limiting factor for nodule growth and $N_2$ fixation. Studies on nodulation and $N_2$ fixation during growth and development of leguminous and non-leguminous plants have been made by many investigators (e.g. PATE and co-workers, as above, LAWRIE and WHEELER 1975, BETHLENFALVAY and PHILLIPS 1978, RYLE et al. 1979).

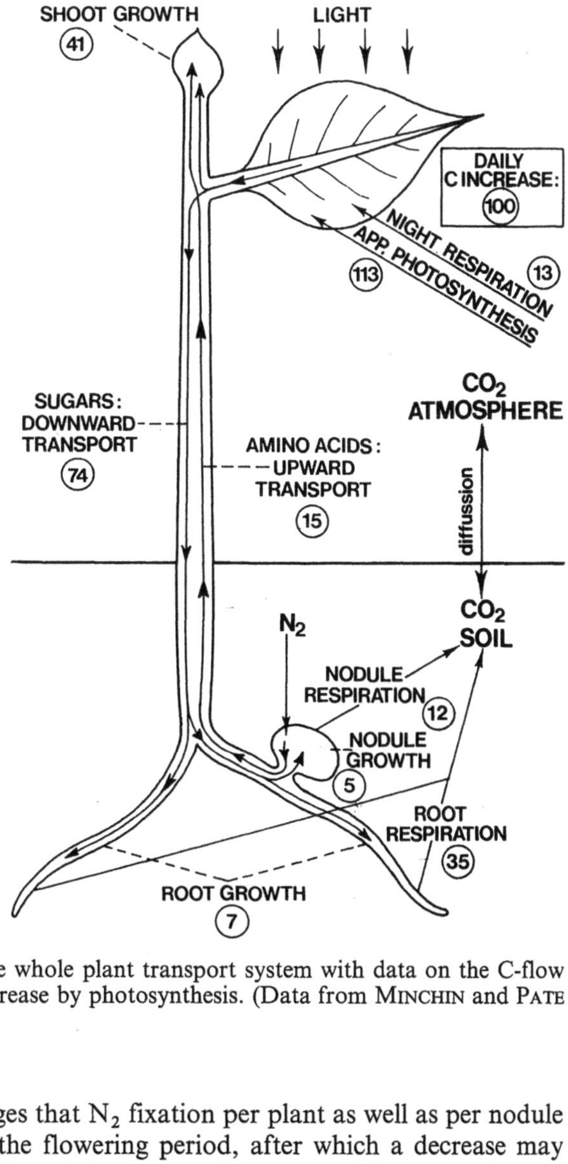


**Fig. 7.** The nodule as part of the whole plant transport system with data on the C-flow in percentages of the daily C increase by photosynthesis. (Data from MINCHIN and PATE 1973)

The general pattern emerges that $N_2$ fixation per plant as well as per nodule increases during growth till the flowering period, after which a decrease may be observed during fruit and seed formation. Removal of flower buds may prevent this decrease. Though there are marked differences between different species and environmental circumstances, this general pattern is easily understood by considering root nodules as sinks in the phloem transport system which competes with other sinks such as growing fruits. Interruption of the phloem transport by excision of nodulated roots or stem-ringing leads to a rapid decrease and even a complete annihilation of $N_2$ fixation (see QUISPEL in QUISPEL 1974).

Rapid reductions in $N_2$ fixation are obtained by darkness, shading or removal of leaves.

Increase of photosynthesis has been obtained by increasing the $CO_2$ supply to the leaves or by decreasing the $O_2$ concentration to reduce photorespiration (HARDY and HAVELKA in NEWTON and NYMAN 1976). Both methods lead to increased $N_2$ fixation which indicates that under normal conditions $N_2$ fixation is limited by photosynthesis.

The conclusion that $N_2$ fixation in root nodules is primarily limited by the direct flow of assimilates appears to be in contradiction to the conclusion based on C-flow sheets that in lupin root, respiration relied heavily on earlier received carbon in the reserve carbohydrates. This might be explained by assuming that the bacteroid tissue is more directly linked to the flow of assimilates in the phloem than the remaining parts of the root or nodule tissue which is responsible for most of the total nodule and root respiration. As the latter respiration is not affected in the same degree by darkening, excision or stem-ringing as $N_2$ fixation or N assimilation, it has been suggested that either a special part of respiration is linked to $N_2$ fixation or N assimilation, or that these processes have a lower affinity for ATP and thus are more susceptible to reduction in ATP production.

The well-know inhibiting effect of $NH_4^+$ has been explained as well by a competition for ATP, in a diversion of energy from the $N_2$-fixing system to the $NH_4^+$ assimilating reactions (HOUWAARD 1978). This is in accordance with the general observation that the inhibiting effect of $NH_4^+$ may be counteracted by increased photosynthesis (WILSON 1940).

All considerations of the relation between root and nodule respiration with nodulation and $N_2$ fixation must take due account of the existence of different pathways of respiration in plants. LAMBERS recently observed that in many different plant tissues a CN-insensitive respiration is found which is not connected with energy production. He considers this part of respiration as an overflow system. If more energy is needed, respiration shifts from this overflow to the normal, energy-producing system. Such shifts occur during development and when $N_2$ is the main N source (DE VISSER and LAMBERS 1979).

Such considerations may be helpful in elucidating the controversies about the differences in energy consumption between $N_2$ and $NO_3^-$-utilizing organisms. While some investigators did not observe a difference in energy demands on theoretical grounds or experimental observations, other authors arrived at the conclusion that $N_2$ fixation is a less efficient, more energy-demanding process (SILSBURY 1977). RYLE et al. (1979), observed a higher respiration of nodulated plants of about 10%–15%, as compared to equivalent non-nodulated plants on $NO_3^-$. As this difference is independent of efficiency of $N_2$ fixation, it must be more related to the respiration of root nodules. A better insight will be obtained if such observations are analyzed with regard to eventual non-energy-producing overflow types of respiration. There are no indications that the removal of $NH_3$ from the $N_2$ fixing nodule parts plays an important role in the regulation of $N_2$ fixation.

The general picture emerges of a plant system in which the flow of assimilates to the root nodules is the main regulating system. The efficient use of these assimilates depends on the efficiency of the respiratory system and on the relative efficiency of the bacteroids which may be related to the $H_2$-recycling possibilities.

## 7 Concluding Remarks

The different symbiotic systems described in this chapter offer a range of evolutionary adaptations. Starting with loose associations in which $N_2$-fixing bacteria may enter cortex cells, the symbioses with Cyanobacteria give us examples of inter- and intracellular organs which for the main part were formed before the invasion with the $N_2$-fixing bacteria. In the real root nodule symbioses the specialized organs are formed under the influence of the infecting microsymbiont. There are several reasons like the less-developed $O_2$ protection system to consider the actinorhizas as a less-adapted system than the legume–*Rhizobium* symbioses. The complicated interactions in these symbioses are schematically summarized in Fig. 6. In all symbiotic systems $N_2$ fixation is directed to the benefit of the host plant. In Cyanobacteria this direction is based on phenotypic reactions by the host plant, as the bacteria themselves are able to use $N_2$ under free-living conditions. In *Rhizobium* the adaptation must moreover have a genetic base as the free-living bacteria profit hardly or not at all from their $N_2$-fixing capacities.

The highly complicated interactions involved in nodule formation and symbiotic $N_2$ fixation must be still far better resolved to understand why they are limited to specialized bacteria and specialized plants and before we can dream of possibilities to extend symbiotic $N_2$ fixation to other plants (HOLLAENDER 1977). Without any doubt the further studies on symbiotic interactions leading to nodule formation and $N_2$ fixation, together with genetic research and selection of plant and bacterial germ plasm will lead to the most efficient systems under different environmental situations and therewith a better use of these extremely important systems for N enrichment in the biosphere.

## References

Albersheim P, Anderson-Prouty A (1975) Carbohydrates, proteins, cell surfaces and the biochemistry of pathogenesis. Annu Rev Plant Physiol 26:31–52

Allen EK, Allen ON (1958) Biological aspects of symbiotic nitrogen fixation. In: Ruhland W (ed) Encyclopedia of plant physiology, vol X. Springer, Berlin Göttingen Heidelberg

Angulo-Carmona A (1974) La formation des nodules fixateurs d'azote chez *Alnus glutinosa* (L.). Vill. Acta Bot Neerl 23:257–303

Avissar YT, Nadler KD (1978) Stimulation of tetrapyrrole formation in *Rhizobium japonicum* by restricted aeration. J Bacteriol 135:782–789

Baker D, Torrey JG, Kidd GH (1979) Isolation by sucrose-density fractionation and cultivation in vitro of Actinomycetes from nitrogen-fixing root nodules. Nature (London) 281:76–78

Becking JH, Boer WE de, Houwink AL (1964) Electron microscopy of the endophyte of *Alnus glutinosa*. Antonie van Leeuwenhoek J Microbiol Serol 30:343–376

Bednarski MA, Reporter M (1978) Expression of rhizobial nitrogenase: influence of plant cell-conditioned medium. Appl Environ Microbiol 36:115–120

Bergersen FJ, Turner GL (1975) Leghemoglobin and the supply of $O_2$ to nitrogen-fixing root nodule bacteroids: presence of two oxidase systems and ATP production at low free $O_2$ concentration. J Gen Microbiol 91:345–354

Bergersen FJ, Turner GL (1978) Activity of nitrogenase and glutamine synthetase in relation to availability of oxygen in continuous cultures of a strain of cowpea *Rhizobium* sp. supplied with excess ammonium. Biochim Biophys Acta 538:406–416
Bergey's Manual of determinative bacteriology (1974) 8th edn. Williams and Wilkins, Baltimore
Beringer JE, Beynon JL, Buchanan-Wollaston AV, Johnston AWB (1978) Transfer of the drug-resistance transposon Tn5 to *Rhizobium*. Nature 276:633–634
Bethlenfalvay GJ, Phillips DA (1978) Interactions between symbiotic nitrogen fixation, combined-N application and photosynthesis in *Pisum sativum*. Physiol Plant 42:119–123
Bishop PE, Guevara JG, Engelke JA, Evans HJ (1976) Relation between glutamine synthetase and nitrogenase activities in the symbiotic association between *Rhizobium japonicum* and *Glycine max*. Plant Physiol 57:542–546
Bisseling T, Bos RC v d, Kammen A v (1978) The effect of ammonium nitrate on the synthesis of nitrogenase and the concentration of leghemoglobin in pea root nodules induced by *Rhizobium leguminosarum*. Biochim Biophys Acta 539:1–11
Bisseling T, Bos RC v d, Weststrate MW, Hakkaart MJJ, Kammen A v (1979) Development of the nitrogen-fixing and protein-synthesizing apparatus of bacteroids in pea root nodules. Biochim Biophys Acta 562:515–526
Blom J (1980) Utilization of fatty acids and $NH_4^+$ by Frankia Avcl 1. FEMS Microbiol Lett 10:143–145
Bogers RJ, Hooijmans-Klappe HTM, Libbenga KR (1976) The first cell cycle in explants from the mature root cortex of 7-day-old *Pisum sativum*. Plant Sci Lett 6:43–48
Bowles DJ, Kauss H (1976) Characterization, enzymatic and lectin properties of isolated membranes of *Phaseolus aureus*. Biochim Biophys Acta 443:360–376
Brown CM, Dilworth MJ (1975) Ammonia assimilation by *Rhizobium* cultures and bacteroids. J Gen Microbiol 86:39–48
Brussel van AAN, Planqué K, Quispel A (1977) The wall of *Rhizobium leguminosarum* bacteroid and free-living form. J Gen Microbiol 101:51–56
Brussel van AAN, Costerton JW, Irvin RF, Korevaar CGN (1978) Localization studies in *Rhizobium leguminosarum* Abstract 115. In: Proc Steenbock-Kettering Int Symp Nitrogen Fixation, Madison
Brussel van AAN, Costeron JW, Child JJ (1979) Nitrogen fixation by *Rhizobium* sp. 32 H1 a morphological and ultrastructural comparison of asymbiotic and symbiotic nitrogen-fixing forms. Can J Microbiol 25:352–361
Buchanan-Wollaston V (1979) Generalized transduction in *Rhizobium leguminosarum*. J Gen Microbiol 112:135–142
Carlson RW, Sanders RE, Napoli C, Albersheim P (1978) Host-symbiont interactions. III Purification and partial characterization of *Rhizobium* lipopolysaccharides. Plant Physiol 62:912–917
Chandler MR (1978) Some observations on infection of *Arachis hypogaea* L. by *Rhizobium*. J Exp Bot 29:749–755
Chen AP, Phillips D (1976) Attachment of *Rhizobium* to legume roots as the basis for specific interactions. Physiol Plant 38:83–88
Child JJ, Kurz WGW (1978) Inducing effect of plant cells on nitrogenase activity by *Spirillum* and *Rhizobium* in vitro. Can J Microbiol 24:143–148
Criswell JG, Havelka VD, Quebedoux B, Hardy RWF (1976) Adaptation of nitrogen fixation by intact soybean nodules to altered rhizosphere $pO_2$. Plant Physiol 58:622–625
Dandekar AM, Vasavadi HA, Modi VV (1978) Partial purification of a competence factor from *Rhizobium japonicum*. Arch Microbiol 118:253–256
Dazzo FB, Brill WJ (1979) Bacterial polysaccharide which binds *Rhizobium trifolii* to clover root hairs. J Bacteriol 127:1362–1373
Dazzo B, Hubbell DH (1975) Cross-reactive antigens and lectin as determinants of symbiotic specificity in the *Rhizobium* clover association. Appl Microbiol 30:1017–1033
Dazzo FB, Napoli CA, Hubbell DH (1976) Adsorption of bacteria to roots as related to host specificity in the *Rhizobium*-clover symbiosis. Appl Environm Microbiol 32:166–171

Dijk C van (1978) Spore formation and endophyte diversity in root nodules of *Alnus glutinosa* (L.) Vill. New Phytol 81:601–615

Dijk C van, Merkus E (1976) A microscopical study of the development of a spore-like stage in the life cycle of the root-nodule endophyte of *Alnus glutinosa* (L.) Gaertn. New Phytol 77:73–91

Döbereiner J, Burris RH, Hollaender A (1977) Limitations and potentials for biological nitrogen fixation in the Tropics. Plenum, New York London

Dullaart J (1970) The bioproduction of indole-3-acetic acid and related compounds in root nodules and roots of *Lupinus luteus* L. and by its rhizobial symbiont. Acta Bot Neerl 19:573–615

Dullaart J, Duba LI (1970) Presence of gibberellin-like substances and their possible role in auxin bioproduction in root nodules and roots of *Lupinus luteus* L. Acta Bot Neerl 19:877–883

Egeraat AWSM v (1972) Pea root exudates and their effects on root nodule bacteria. Meded Landbouwhogesch Wageningen 72–127

Evans WR, Keister DL (1976) Reduction of acetylene by stationary cultures of free-living *Rhizobium* sp. under atmospheric oxygen levels. Can J Microbiol 22:949–952

Fogg GE, Stewart WDP, Fay P, Walsby AE (1973) The blue-green algae. Academic Press London New York

Fred EB, Baldwin IL, McCoy E (1932) Root nodule bacteria and leguminous plants. Univ Wisconsin Press, Madison

Gibson AH, Newton WE (1981) Current perspectives in nitrogen fixation. Aust Acad, Sci Canberra

Gibson AH, Scowcroft WR, Child JJ, Pagan JD (1976) Nitrogenase activity in cultured *Rhizobium* sp. strain 32 H. Nutritional and physical considerations. Arch Microbiol 108:45–54

Glenn AR, Dilworth MJ (1979) An examination of *Rhizobium leguminosarum* for the production of extracellular and periplasmic proteins. J Gen Microbiol 112:405–409

Gordon J, Wheeler CT, Perry DA (eds) (1979) Symbiotic nitrogen fixation in the management of temperate forests. For Res Lab, Oregon State Univ

Greshoff PM, Skotnicki ML, Eady JF, Rolfe BG (1978) Viability of *Rhizobium trifolii* bacteroids from clover root nodules. Plant Sci Lett 10:299–304

Hanus FJ, Maier RJ, Evans H (1979) Autotrophic growth of $H_2$-uptake positive strains of *Rhizobium japonicum* in an atmosphere supplied with hydrogen gas. Proc Natl Acad Sci USA 76:1788–1792

Hardy RWF, Gibson AH (1977) A treatise on dinitrogen fixation, vol IV. Wiley and Sons, New York

Hardy RWF, Silver WS (1977) A treatise on dinitrogen fixation, vol III. Wiley and Sons, New York

Henson LE, Wheeler CT (1977a) Hormones in plants bearing nitrogen-fixing root nodules, gibberellin-like substances in *Alnus glutinosa* (L) Gaertn. New Phytol 78:373–382

Henson LE, Wheeler CT (1977b) Hormones in plants bearing nitrogen-fixing root nodules, distribution and seasonal changes in levels of cytokinins in *Alnus glutinosa* (L) Gaertn. J Exp Bot 28:205–214

Hollaender A (ed) (1977) Genetic engineering for nitrogen fixation. Plenum, New York London

Hollander JA de, Stouthamer AH (1979) Multicarbon-substrate growth of *Rhizobium trifolii*. FEMS Microbiol Lett 6:57–59

Houwaard F (1978) Influence of ammonium chloride on the nitrogenase activity of nodulated pea plants (*Pisum sativum*). Appl Environm Microbiol 35:1061–1065

Hubbell DH, Morales VM, Umali-Garcia M (1978) Pectolytic enzymes in *Rhizobium*. Appl Environm Microbiol 35:210–213

Johnston AWB, Beynon JL, Buchanan-Wollaston AV, Setchell SM, Hirsch PR, Beringer JE (1978) High frequency transfer of nodulating ability between strains and species of *Rhizobium*. Nature 276:634–636

Keister DL, Evans WR (1976) Oxygen requirements for acetylene reduction by pure cultures of rhizobia. J Bacteriol 129:149–153

Kennedy IR (1979) Assimilating fixed nitrogen. Proc Austr Biochem Sci 12:Q 15

Kijne JW (1975a) The fine structure of pea root nodules 1. Vacuolar changes after endocytotic host cell infection by *Rhizobium leguminosarum*. Physiol Plant Pathol 5:75–79

Kijne JW (1975b) The fine structure of pea root nodules 2. Senescence and disintegration of the bacteroid tissue. Physiol Plant Pathol 7:17–21

Kijne JW, Planqué K (1979) Ultrastructural study of the endomembrane system in infected cells of pea and soybean root nodules. Physiol Plant Pathol 14:339–345

Kijne JW, Schaal CAM van der, Vries GE de (1980) Pea lectins and the recognition of *Rhizobium leguminosarum*. Plant Sci Lett 18:65–74

Kurz WGW, La Rue TA (1977) Citric acid cycle enzymes and nitrogenase in nodules of *Pisum sativum*. Can J Microbiol 23:1197–1200

Laane C, Haaker H, Veeger C (1978) Involvement of the cytoplasmic membrane in nitrogen fixation by *Rhizobium leguminosarum* bacteroids. Eur J Biochem 87:147–153

Laane C, Krone W, Konings WN, Haaker H, Veeger C (1979) The involvement of the membrane potential in nitrogen fixation by bacteroids. FEBS Lett 103:328–332

Laing WA, Christeller JT, Sutton WD (1979) Carbon dioxide fixation by lupin root nodules. II. Studies with $^{14}C$ labeled glucose, the pathway of glucose catabolism and the effects of some treatments that inhibit nitrogen fixation. Plant Physiol 63:450–454

Lalonde M, Devoe IW (1976) Origin of the membrane envelope enclosing the *Alnus crispa* var. *mollis* Fern. root nodule endophyte as revealed by freeze-etching microscopy. Physiol Plant Pathol 8:123–129

Lawrie AC, Wheeler CT (1975) Nitrogen fixation in the root nodules of *Vicia faba* L. in relation to the assimilation of carbon. New Phytol 74:429–445

Legocki RP, Verma DP (1979) A nodule-specific plant protein (Nodulin-35) from soybean. Science 205:190–193

Lepidi AA, Nuti MP, Bernacchi G, Neglia R (1976) Physiology of the *Rhizobium*-legume association. I The presence of bacterial DNA within the plant root. Plant Soil 45:555–569

Lim ST, Shanmugam KT (1979) Regulation of hydrogen utilization in *Rhizobium japonicum* by cyclic AMP. Biochim Biophys Acta 584:479–492

Lim ST, Hennecke H, Scott DB (1979) Effect of guanosine-3′-5′-cyclic monophosphate on nitrogen fixation in *Rhizobium japonicum*. J Bacteriol 139:256–263

Ljunggren H (1969) Mechanism and pattern of *Rhizobium* invasion into leguminous root hairs. Physiol Plant Suppl 5

Ludwig RA (1978) Control of ammonium assimilation in *Rhizobium* 32 H1. J Bacteriol 135:114–123

Maier RJ, Brill WJ (1978) Mutant strains of *Rhizobium japonicum* with increased ability to fix nitrogen for soybean. Science 201:448–449

Mennes AM (1973) The indole-3-acetic acid oxidase of *Lupinus luteus* L. Acta Bot Neerl 22:694–729

Miflin BJ, Lea PJ (1977) Amino acid metabolism. Annu Rev Plant Physiol 28:299–329

Nadler KD, Avissar YJ (1977) Heme synthesis in soybean root nodules. I. On the role of bacteroid $\delta$ aminolevulinic acid synthase and aminolevulinic dehydrase in the synthesis of the heme of leghemoglobin. Plant Physiol 60:433–436

Nash DT, Schulman HM (1976) Leghemoglobin and nitrogenase activity during soybean root nodule development. Can J Bot 54:2790–2797

Newcomb W (1976) A correlated light and electron microscopic study of symbiotic growth and differentiation in *Pisum sativum* root nodules. Can J Bot 54:2163–2186

Newcomb W, Syono K, Torrey JG (1977) Development of ineffective pea root nodule: morphogenesis, fine structure and cytokinin biosynthesis. Can J Bot 55:1891–1907

Newton WE, Nyman CJ (eds) (1976) Proc 1st Int Symp Nitrogen Fixation. Wash State Univ Press

Newton WE, Orme-Johnson NR (eds) (1980) 3rd Int Conf Nitrogen Fixation. Univ Park Press, Baltimore

Newton WE, Postgate JR, Rodriguez-Barrueco C (1977) Recent developments in nitrogen fixation. Academic Press, London New York

Nuti MP, Lepidi AA, Prakash RK, Schilperoort PA, Cannon FC (1979) Evidence for

nitrogen fixation (*nif*) genes on naturally occurring plasmids in *Rhizobium*. Nature 282:533–535
Nutman PS (ed) (1976) Symbiotic nitrogen fixation. Int Biol Programm, vol VII. Cambridge Univ Press, Cambridge
O'Gara F, Shanmugam KT (1976) Regulation of nitrogen fixation by rhizobia, export of fixed $N_2$ as $NH_4^+$. Biochim Biophys Acta 437:313–332
Oostrom H, Treurniet FE, Mennes AM (1975) Cytochemical localization of peroxidase during the development of root nodules of *Pisum sativum* L. Z Pflanzenphysiol 74:451–463
Paau AS, Oro J, Cowles JR (1979) DNA content of free-living rhizobia and bacteroids of various *Rhizobium*-legume associations. Plant Physiol 69:402–405
Pankhurst CE, Craig AS (1978) Effect of oxygen concentration, temperature, and combined nitrogen on the morphology and nitrogenase activity of *Rhizobium* spp. strain 32H1, in agar culture. J Gen Microbiol 106:207–219
Pate JS, Layzell DB, MacNeil DL (1979) Modelling the transport and utilization of carbon and nitrogen in a nodulated legume. Plant Physiol 63:730–739
Phillips DA, Daniel RM, Appleby CA, Evans HJ (1973) Isolation from *Rhizobium* of factors which transfer electrons to soybean nitrogenase. Plant Physiol 51:136–138
Planqué K, Vries GE de, Kijne JW (1978) The relation between nitrogenase and glutamine synthetase levels in bacteroids of *Rhizobium leguminosarum* of various age. J Gen Microbiol 106:173–178
Planqué K, Nierop JJ van, Burgers A, Wilkinson SG (1979) The lipopolysaccharide of free-living and bacteroid forms of *Rhizobium leguminosarum*. J Gen Microbiol 110:151–159
Puppo A, Rigaud J (1978) Cytokinins and morphological aspects of French-bean roots in the presence of *Rhizobium*. Physiol Plant 42:202–206
Quispel A (ed) (1974) The biology of nitrogen fixation. Elsevier, North-Holland Amsterdam Oxford New York
Quispel A (1976) Symbiotic nitrogen fixation in non-leguminous plants V. The growth requirements of the endophyte of *Alnus glutinosa*. Acta Bot Neerl 9:380–396
Quispel A, Tak T (1978) Studies on the growth of the endophyte of *Alnus glutinosa* (L.). Vill in nutrient solutions. New Phytol 81:587–600
Ranga Rao V, Darrow RA, Keister DL (1978) Effect of oxygen tension on nitrogenase and glutamine synthetases I and II in *Rhizobium japonicum* 61 A 76. Biochem Biophys Res Commun 81:224–231
Robertson JG, Warburton MP, Farnden KJF, Banks JM (1975) Induction of glutamine synthetase during nodule development in lupin. Austr J Plant Physiol 2:265–272
Robertson JG, Lyttleton P, Bullivant S, Grayston GF (1978a) Membranes in lupin root nodules. I. The role of Golgi bodies in the biogenesis of infection threads and peribacteroid membranes. J Cell Sci 30:129–149
Robertson JG, Warburton MP, Lyttleton P, Fordyce AM, Bullivant S (1978b) Membranes in lupin root nodules. II Preparation and properties of peribacteroid membranes and bacteroid envelope inner membranes from developing lupin nodules. J Cell Sci 30:151–174
Ronson CW, Primrose SB (1979) Carbohydrate metabolism in *Rhizobium trifolii*: identification and symbiotic properties of mutants. J Gen Microbiol 112:77–88
Ryle GJA, Powell CE, Gordon AJ (1979) The respiratory costs of nitrogen fixation in soybean, cowpea and white clover. J Exp Bot 30:135–153
Schaede R (1976) Die pflanzlichen Symbiosen. Fischer, Stuttgart
Scott DB, Farnden KJF, Robertson JGF (1976) Ammonia assimilation in lupin nodules. Nature 263:703–706
Shaw BD, Sutton WD (1977) Polypeptide synthesis by *Rhizobium* bacteroids and bacteria. Biochim Biophys Acta 563:216–226
Silsbury JH (1977) Energy requirements of symbiotic nitrogen fixation. Nature 267:149–150
Skotnicki ML, Rolfe BG (1978) Differential stimulation and inhibitional growth of *Rhizobium trifolii* strain T1 and other *Rhizobium* species by various carbon sources. Microbiology 20:15–28

Solheim B, Raa J (1977) Cell wall biochemistry related to specificity in host-plant pathogen interactions. Univ Forlaget, Oslo 1977
Stewart WDP, Rodgers GA (1977) The cyanophyte-hepatic symbiosis. 2. Nitrogen fixation and the interchange of nitrogen and carbon. New Phytol 78:459–471
Stovall I, Cole M (1978) Organic acid metabolism by isolated *Rhizobium japonicum* bacteroids. Plant Physiol 61:787–790
Strand R, Laetsch WM (1977) Cell and endophyte structure of the nitrogen fixing root nodules of *Ceanothus integerrimus* H and A. Protoplasma 93:165–190
Stripf R, Werner D (1978) Differentiation of *Rhizobium japonicum* II. Enzymatic activities in bacteroids and plant cytoplasm during the development of nodules of *Glycine max*. Z Naturforsch 33c:373–381
Sutton WD, Bos van den RC, Bisseling T (1978) The DNA content of *Rhizobium* (Strain NZP 2257) bacteroids and bacteria. Plant Sci Lett 12:145–149
Syono K, Newcomb W, Torrey JG (1976) Cytokinin production in relation to the development of pea root nodules. Can J Bot 54:2155–2162
Te May Ching, Hedtke S, Newcomb W (1977) Isolation of bacteria, transforming bacteria and bacteroids from soybean nodules. Plant Physiol 60:771–777
Tjepkema JD, Evans HJ (1975) Nitrogen fixation by free-living *Rhizobium* in a defined liquid medium. Biochem Biophys Res Commun 65:625–628
Tjepkema JD, Yocum CS (1974) Measurement of oxygen partial pressure within soybean nodules by oxygen micro electrodes. Planta 119:351–360
Tjepkema JD, Ormerod W, Torrey JG (1980) On vesicle formation and in vitro acetylene reduction by *Frankia*. Nature 287:633–635
Torrey JG, Tjepkema JD (eds) (1979) Symbiotic nitrogen fixation in actinomycete-nodulated plants. Bot Gaz Suppl 140
Trinick MJ (1973) Symbiosis between *Rhizobium* and the non-legume *Trema aspera*. Nature 244:459–460
Trinick MJ (1979) Structure of nitrogen-fixing nodules formed by *Rhizobium* on roots of *Parasponia andersonii* Planck. Can J Microbiol 25:565–578
Trinick MJ, Galbraith J (1976) Structure of root nodules formed by *Rhizobium* on the non-legume *Trema cannabina* var. *scabra*. Arch Microbiol 108:159–166
Tu TC (1979) Alterations in the membranes of bacteroidal cells in soybean root nodules as revealed by freeze-fracturing. Physiol Plant Pathol 15:35–41
Veeger C, Laane C, Scherings C, Zeeland-Wolbers L van (1978) Membrane energization in relation with nitrogen fixation in *Azotobacter vinelandii* and *Rhizobium leguminosarum*. Biochimie 60:237–243
Verma DPS (1980) Expression of host genes during symbiotic nitrogen fixation. In: Leaver CJ (ed) Genome organization and expression in plants. Plenum, New York London
Verma DPS, Bal AK (1976) Intercellular site of synthesis and localization of leghemoglobin in root nodules. Proc Natl Acad Sci USA 73:3843–3847
Vincent JM (1970) A manual for the practical study of root nodule bacteria. IBP Handbook vol 15. Blackwell, Oxford Edinburgh
Virtanen AI (1947) The biology and chemistry of nitrogen fixation by legume bacteria. Biol Rev 22:239–269
Visser R de, Lambers H (1979) Root respiration of *Pisum sativum* L. Activity of the alternate pathway and relation to $N_2$-fixation. Plant Physiol 63 (Suppl):482
Vries GE de (1980) Transport of sugars and organic acids in *Rhizobium leguminosarum* and their role in symbiosis. Thesis, Univ Leiden
Werner D (1978) Differentiation of *Rhizobium japonicum* III. Inhibition of nitrogenase derepression by chloramphenicol and rifampicin concentrations not inhibiting growth. Z Naturforsch 33c:859–862
Werner D, Mörschel E (1978) Differentiation of nodules of *Glycine max*. Ultrastructural studies of plant cells and bacteroids. Planta 141:169–177
Werner D, Stripf R (1978) Differentiation of *Rhizobium japonicum*. I. Enzymatic comparison of nitrogenase repressed and derepressed living cells and of bacteroids. Z Naturforsch 33c:245–252

Werner D, Mörschel E, Stripf R, Winchenbach B (1980) Development of nodules of *Glycine max* infected with an ineffective strain of *Rhizobium japonicum*. Planta 147:320–329
Whiting MJ, Dilworth MJ (1974) Legume root nodule nitrogenase purification, properties and studies on its genetic control. Biochim Biophys Acta 371:337–351
Wilcockson J, Werner D (1979) Organic acids and prolonged nitrogenase activity by non-growing free-living *Rhizobium japonicum*. Arch Microbiol 122:153–160
Wilson PW (1940) The biochemistry of symbiotic nitrogen fixation. Univ Wisconsin Press, Madison
Wooi KC, Broughton WJ (1979) Isolation and metabolism of *Vigna unguiculata* root nodule protoplasts. Planta 145:487–495
Yao PY, Vincent JM (1976) Factors responsible for the curling and branching of clover root hairs by *Rhizobium*. Plant Soil 45:1–16

# II.3 Dinitrogen Fixation in Rhizosphere and Phyllosphere Associations

J. DÖBEREINER

## 1 Introduction

Among the agricultural systems which contribute biologically fixed nitrogen, the symbiotic legume associations described in Chapter II.2 are the most important. With the steadily increasing prices of nitrogen fertilizer however, other less perfect $N_2$-fixing systems are becoming more and more important. Members of the family Gramineae are here of major interest because cereals are the most important food basis all over the world, especially in developing countries. Fortunately it is in the tropics and subtropics, where most poor people live, that chances are best for supplying large parts of the nitrogen needs for productive agricultural systems through biological fixation. Not only are soil temperatures during most of the year suitable for microbial activities and plant growth, but also many tropical plants have developed a specific photosynthetic pathway ($C_4$-dicarboxylic acid pathway) which enables them to convert sun energy twice as effectively as temperate species. Such plants can more easily devote part of their photosynthates to nitrogen fixation. The major tropical forage grasses and cereals, except rice, are in this group.

Nitrogen fixation has been shown in many forage grasses and grains all over the world, but substantial amounts were found under tropical conditions only. Table 1 summarizes a few relevant data obtained from nitrogen balance studies. Many other results reported in the literature (e.g. DÖBEREINER 1978, NEYRA and DÖBEREINER 1977) were obtained by the indirect acetylene reduction assay and are in the same range. Except for sugar cane (see Sect. 2.1) very little or no nitrogen fixation has been found in soil without roots (DÖBEREINER et al. 1972b; LEE et al. 1977a).

## 2 Characterization of Rhizocoenoses

The identification of an association between $N_2$-fixing bacteria and plant roots, termed rhizocoenosis (Intern. Workshop Associative $N_2$ Fixation, Piracicaba, 1979) requires demonstration of plant–bacteria interactions. The simple isolation of one or several species of $N_2$-fixing bacteria from the rhizosphere is insufficient even if numbers of bacteria are high. The description of such systems in this chapter has therefore been limited to associations where plant–bacteria interactions have been shown.

**Table 1.** Nitrogen gains in grass systems as estimated by nitrogen balance studies

| Country or regions | Plant species | Period | N gain ($kg\ ha^{-1}$) | Observation | Reference |
|---|---|---|---|---|---|
| Temperate regions | | | | | |
| U.S.A. (Pennsylvania) | Grassland | 82 years | 3,240 | Field exp. | WHITE et al. 1945 |
| England | Continuous wheat | 72 years | 2,788 | Broadbalk field exp. | JENKINSON 1973 |
| Subtropical regions | | | | | |
| U.S.A. (Texas) | Grassland | 1 year | 22–34 | Field exp. | SMITH et al. 1954 |
| U.S.A. (Georgia) | Zoysia grass (*Z. matrella*) | 2 years | 147 | Pots with 9 kg soil | GIDDENS 1977 |
| U.S.A. (Georgia) | Bahia grass (*Paspalum notatum*) | 2 years | 183 | Pots with 9 kg soil | GIDDENS 1977 |
| Australia | Rye grass (*Lolium rigidum*) | 3 years | 185 | Field exp. | PARKER 1957 |
| Tropical regions | | | | | |
| Nigeria | Stargrass (*Cynodon plecto-stachtris*) | 3 years | 269 | Field exp. | JAIYEBO and MOORE 1963 |
| Nigeria | Taurbagrass (*Eleusine caracana*) | 4 months | 112–146 | Pots with 2.5 kg soil | MOORE 1963 |
| Rhodesia | Varsey grass (*Paspalum urvillei*) | 4 months | 202 | Drums with 100 kg soil | PURCHASE 1978 |
| Philippines | Rice (*Oryza sativa*) | 6 crops[a] | 32 | Pots with 10 kg soil | APP et al. 1980 |
| Philippines | Rice (*Oryza sativa*) | 4 crops[a] | 36 | Pots with 10 kg soil | APP et al. 1980 |

[a] Bacterial $N_2$ fixation only, algae were eliminated with black cloth

## 2.1 Sugar Cane – *Beijerinckia*

Many years ago, rhizosphere and vegetation effects on *Beijerinckia* spp. were shown for sugar cane (DÖBEREINER 1961). Almost all (95%) of the soils collected in continuously planted sugar cane fields contained an abundant *Beijerinckia* population, while only 62% of soils collected under a number of other types of vegetations contained this organism and in much lower numbers. Fields planted with sugar cane for the first time showed steady increases in *Beijerinckia* numbers and the soil on the root surface contained 2 times more of these bacteria than the rhizosphere soil, and 56 times more than the soil between rows. Assays of nitrogenase activity in sugar cane fields confirmed that the major contribution was in the rhizosphere and root surface soil and not in the roots themselves (DÖBEREINER et al. 1972b, RUSCHEL et al. 1978). In addition rain-water can carry leaf exudates into the soil which enhance *Beijerinckia* growth and nitrogenase activities (DÖBEREINER 1961, DÖBEREINER et al. 1972b). Sugar cane seedlings exposed to $^{15}N_2$ showed fixation, incorporation and translocation of nitrogen into the leaves (RUSCHEL et al. 1978); stem cuttings are infected by nitrogen-fixing bacteria which extend into xylem vessels and intercellular spaces of roots of field-grown cane plants (PATRIQUIN et al. 1980). These organisms have not been identified and their role in sugar cane nitrogen fixation is not yet known.

### 2.2 *Paspalum notatum – Azotobacter paspali*

The association of *Paspalum notatum* with *Azotobacter paspali* (DÖBEREINER 1966) seems unique among grass rhizocoenoses because of its extreme specificity, comparable with the most specific *Rhizobium* symbioses. Among 31 cultivars or ecotypes of this species only the cv. 'batatais' associates with *A. paspali* and 98% of a large number (252) of root surface soil samples from this cultivar collected all over Brazil and in several other countries contained the bacteria (DÖBEREINER 1970). Only 3% of the samples from the remaining cultivars and none of 200 samples of other grasses and legumes contained the bacteria. Due to the nature of *Paspalum* as a useless invasion grass, this most interesting rhizocoenosis has unfortunately not been sufficiently studied. From the information available on host plant specificity and also some physiological data (DÖBEREINER and CAMPELO 1971, DÖBEREINER and DAY 1975), it must be concluded that the plant–bacteria interaction is a very close one. Strong washing of the roots did not reduce nitrogenase activity, but surface sterilization eliminated the bacteria. The site of the bacteria was therefore suggested to be the mucigel layer (DÖBEREINER et al. 1972a). *A. paspali* is an aerobic fast-growing bacterium which can be easily isolated from batatais roots (DÖBEREINER 1966), where it occurs in very high numbers. Although the organism has a very narrow pH requirement in culture (MACHADO and DÖBEREINER 1969), it does occur on *P. notatum* roots grown in soil with pH 4.9 (DÖBEREINER 1966). Root exudates from *P. notatum* stimulate the growth of *A. paspali* in culture, and all *A. paspali* strains tested so far are unable to reduce nitrate (PEÑA and DÖBEREINER 1974). Nitrogenase activity in the organism was not inhibited by concentrations of 20 mM $NO_3^-$ and the derepression of nitrogenase in $NH_4^+$-grown cells was the same in the presence and absence of 10 mM $NO_3^-$ (DÖBEREINER and DAY 1975).

### 2.3 *Azospirillum* Rhizocoenoses

*Azospirillum* associations seem very common in tropical regions. High numbers of these bacteria were found in rhizosphere soil and roots of many forage grasses (e.g. *Panicum maximum, Brachiaria* spp., *Pennisetum purpureum, Digitaria* spp.), and in maize, millets, sorghum, rice, wheat and rye in the areas of Central and South Brazil (DÖBEREINER et al. 1976; DA SILVA and DÖBEREINER 1978, NERY et al. 1977). A survey in India revealed *Azospirillum* spp. in roots of rice, sorghum, maize, *Panicum* and a number of other non-Gramineae (LAKSHMI-KURAMI et al. 1976). The organism was also isolated from various grass roots in Mexico (QUINTERO and GARZA 1978), Australia (BERGERSEN personal communication), Belgium (REYNDERS and VLASSAK 1976) and the U.S.A. (SLOGER and OWENS 1976).

The role of *Azospirillum* in $N_2$ fixation of *Digitaria* (DÖBEREINER and DAY 1976) and maize (VON BÜLLOW and DÖBEREINER 1975) was confirmed by highly significant correlations of root segment nitrogenase activities with *Azospirillum* enrichment culture activities obtained with the same root segments. It is also supported by the correlation of numbers of *Azospirillum* within roots with the nitrogenase activity of maize (MAGALHÃES et al. 1979) and rice (WATANABE and CHOLITKUL 1979) roots during the growth cycle.

Koch's postulates have been established in that sterile maize plants grown in agar cultures when inoculated with *Azospirillum brasilense* showed nitrogenase activity and the organism could be reisolated and reinoculated with the same results (BURRIS 1977). Similarly, *Brachiaria*, sorghum and wheat seedlings grown in sterile vermiculite inoculated with *A. brasilense* or *A. lipoferum* reduced $C_2H_2$ in intact systems, but only when the partial oxygen pressure on top of the vermiculite cultures was reduced to less than 0.02 atm. (FREITAS and DÖBEREINER 1978). In none of these experiments, however, was the incorporation of fixed nitrogen into the plant observed.

### 2.3.1 Taxonomy of *Azospirillum* spp.

BEIJERINCK (1925) described as *Spirillum lipoferum*, $N_2$-fixing organisms which he had isolated from muds and soils in Holland, but he was unable to demonstrate $N_2$ fixation in pure culture. BECKING (1963) showed $^{15}N_2$ incorporation with a "*Vibrio* or *Spirillum*" which was most probably the same organism. A large number of strains have now been isolated from all over the world, two (OKON et al. 1976a) or possibly three groups being identified (SAMPAIO et al. 1978) on morphological and physiological characteristics. Based on DNA homology studies with 60 strains, TARRAND et al. (1978) described a new genus, *Azospirillum*, having two species, *A. lipoferum*, (probably identical with Beijerinck's *S. lipoferum*), and *A. brasilense*. The major differential characteristics are absolute biotin requirement and use of glucose and $\alpha$-ketoglutarate as sole carbon sources, and typical large involution forms in older, especially alkaline cultures in *A. lipoferum*.

Both species contain two groups, one which denitrifies ($nir^+$) and another which does not ($nir^-$). The major characteristics of the two *Azospirillum* spp. are summarized in Table 2.

Differentiation between the two species and between the nir groups was also possible by fluorescent antibody techniques (DE-POLLI et al. 1980).

**Table 2.** Characteristics of *Azospirillum* spp. (TARRAND et al. 1978)

| | *A. lipoferum* | *A. brasilense* |
|---|---|---|
| DNA base composition (mol % G+C) | 70 | 70 |
| DNA homology with Sp7 (%) | 35.3±5 | 83±9 |
| DNA homology with Sp59 (%) | 74.8±7 | 42±6 |
| Use of glucose mannitol and α-ketoglutaric acid as C-source | Yes | No |
| Acid from glucose, fructose, ribose, mannitol | Yes | No |
| Pleomorph cells in older especially alkaline cultures | Yes | No |
| Biotin required for growth | Yes | No |
| $NO_3^-$ to $NO_2^-$ (nr) | Yes | Yes |
| $NO_2^-$ to $N_2O$ (nir) | Yes/no | Yes/no |
| $N_2O$ to $N_2$ | Yes | Yes |
| Plants found to be infected | $C_4$ plants | $C_3$ plants |

### 2.3.2 Root Infection

In a symbiotic association like that of legumes the first step is selective multiplication and attachment of the bacteria followed by root hair infection and multiplication of the infecting bacteria within the roots (see Chap. II.2). Adsorption of *A. brasilense* to root hairs and undifferentiated epidermal cells was observed with sterile *Panicum maximum* seedlings (UMALI-GARCIA et al. 1978). Tryptophan is converted by *Azospirillum* to indole acetic acid (VLASSAK and REYNDERS 1978b), a growth substance involved in root infection of legumes. Modification of the bacterial form to large cyst-like structures was observed in monoxenic maize and millet cultures inoculated with some *Azospirillum* strains and not with others (RUSCOE et al. 1978, UMALI-GARCIA et al. 1980). Similar "C-forms" were reported to be formed as a response to excessive oxygen concentrations in sugar cane tissue cultures supporting *Azospirillum* growth (BERG et al. 1980) (Fig. 1).

Infection of sterile seedlings of maize (BURRIS 1977), *Panicum maximum* (UMALI-GARCIA et al. 1978), rice and alfalfa (SUBBA RAO 1981) and *Pennisetum typhoideae* callus tissue (VASIL et al. 1977) in intercellular and intracellular spaces by *Azospirillum* spp. has been demonstrated. Field-grown plants of *Digitaria decumbens* (DÖBEREINER and DAY 1976), maize and *Panicum maximum* (PATRIQUIN and DÖBEREINER 1978) show very similar tetrazolium-reducing bacteria localized in intracellular and intercellular spaces of the cortex and within the stele. Numbers of *Azospirillum* in surface-sterilized roots increased from $10^5$ in young plants to $10^7$ in flowering plants, while the numbers in non-sterilized roots were high during the whole growth cycle (MAGALHÃES et al. 1979). Pronounced variations of nitrogenase activity during the growth cycle, with peak activities at flowering in maize, rice, sorghum and wheat have been shown by many authors (VON BÜLLOW and DÖBEREINER 1975, LEE et al. 1977b, NERY

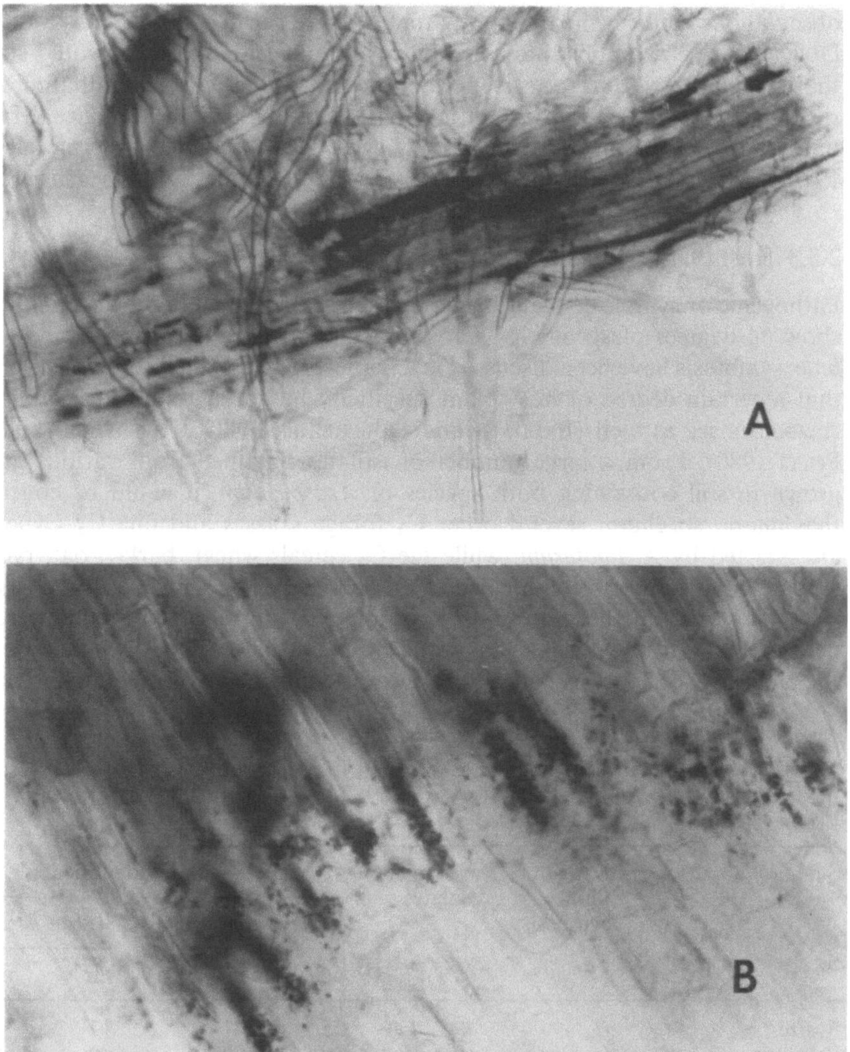


**Fig. 1 A, B.** Infection of wheat seedlings with *Azospirillum brasilense* nir$^-$. Plants were grown in 50-ml test tubes with sterilized vermiculite and basic fertilization with low nitrogen (20 ppm N as $NO_3^-$). Inoculant strain SpBr14 was isolated originally from wheat soil. Nitrogenase activity of 1-month-old plants was 60 nmol $C_2H_4$ $h^{-1}$ per tube. Hand-cut sections were mounted in glycerol after 15 h incubation of the intact plant-vermiculite systems under 0.15% tetrazolium chloride in 0.05 M phosphate buffer. Magnification **A** ×250; **B** ×500. *Dark stained areas* in the intercellular spaces represent *A. brasilense* cells with formazan crystals due to reduction of the tetrazolium salt

et al. 1977). Surface-sterilized roots of maize (VON BÜLLOW and DÖBEREINER 1975), wheat and rice (BALDANI and DÖBEREINER 1980) yielded almost pure *Azospirillum* enrichment cultures while non-sterilized roots contained more mixed populations. Crushing of surface-sterilized maize roots yielded $10^3$–$10^4$ times as many *Azospirillum* cells as intact roots (OKON et al. 1977a). All these observations confirm the internal nature of these rhizocoenoses (PATRIQUIN and DÖBEREINER 1978). Substrate availability and oxygen conditions would suggest that $N_2$ fixation within roots is more important and efficient than that of bacteria situated on the root surface or in soil (BURRIS et al. 1978). In maize (MAGALHÃES et al. 1979), wheat (KAVIMANDAN et al. 1978) and rice (WATANABE et al. 1979), the stems as well as the roots were found to be infected with *Azospirillum*.

### 2.3.3 Host Plant Specificity

Pathogenic or symbiotic plant bacteria associations which involve root infection show all degrees of specificity. Cross-inoculation groups in the legume/*Rhizobium* symbiosis have been discussed in Chapter. II.2. It is therefore not surprising that a certain degree of host plant specificity has been found in *Azospirillum* rhizocoenoses as well (BALDANI and DÖBEREINER 1980, DÖBEREINER and DE-POLLI 1980). From a large number of isolates obtained from various grasses grown in soil containing both species of *Azospirillum*, it could be concluded that maize, sorghum, several other $C_4$ forage grasses and one $C_4$ Cyperacea are infected by *A. lipoferum*, while the $C_3$ cereals wheat, barley, oat, rye and rice are almost exclusively infected by *A. brasilense*. The only exception is sugar cane (Table 3). Furthermore most root-infecting *Azospirillum* strains were not denitrifying ($nir^-$) (DÖBEREINER and DE-POLLI 1980). This discrimination by the plant between species and nir groups was confirmed by differential immuno-

**Table 3.** Host plant specificity in the infection of grass roots by *Azospirillum* spp. in field- and pot-grown plants. (BALDANI and DÖBEREINER 1980, BALDANI et al. 1981, BODDEY 1980)

| Sample | Type of photo-synthesis | No. of isolates | % of root[b] isolates identified as | | | |
|---|---|---|---|---|---|---|
| | | | *A. lipoferum* | | *A. brasilense* | |
| | | | $nir^+$ | $nir^-$ | $nir^+$ | $nir^-$ |
| Maize | $C_4$ | 49 | 55 | 41 | 17 | 3 |
| Sorghum | $C_4$ | 19 | 20 | 80 | 0 | 0 |
| Forage grasses[c] | $C_4$ | 45 | 33 | 70 | 0 | 7 |
| Sugar cane | $C_4$ | 24 | 0 | 0 | 0 | 100 |
| *Cyperus rotundus* | $C_4$ | 7 | 0 | 100 | 0 | 0 |
| Wheat | $C_3$ | 50 | 0 | 0 | 7 | 93 |
| Paddy rice[a] | $C_3$ | 56 | 0 | 0 | 7 | 93 |
| Dry rice | $C_3$ | 22 | 7 | | 13 | 80 |
| Oat | $C_3$ | 5 | 20 | | 0 | 80 |
| Rye | $C_3$ | 22 | 18 | | 0 | 82 |
| Barley | $C_3$ | 11 | 18 | | 0 | 82 |

[a] Includes 34 samples from Trinidad
[b] Roots sterilized in chloramine T for 0.5 to 60 min, depending on the root structure of each species. Soil samples in almost all harvests contained all three *Azospirillum* groups
[c] *Panicum maximum, Cynodon dactylon, Pennisetum purpureum, Digitaria decumbens, Hemarthria altissima*

fluorescence reactions (DE-POLLI et al. 1980). *Azospirillum* strains isolated from rice roots in Trinidad (Table 3) and Senegal (RINAUDO personal communication) were also almost exclusively *A. braisilense* $nir^-$, while the strains isolated from maize roots in Belgium seem from the description to be *A. lipoferum* (VLASSAK and REYNDERS 1978a). These authors also observed a certain specificity of function in that some *Azospirillum* strains (species not mentioned) induced higher nitrogenase activity in maize, and others in *Chicorium intybus*. In contrast, infection and nitrogenase activity in monoxenic grass seedlings seems non-specific (UMALI-GARCIA et al. 1978, BURRIS 1977 and unpublished data from our laboratory). Monoxenic wheat roots adsorbed $10^9$ *Rhizobium* cells per cm of root (SHIMSHICK and HERBERT 1978) confirming lack of specificity in the absence of the soil microflora, also in *Rhizobium* attachment.

When specific *Azospirillum* strains were genetically marked with streptomycin resistance (20 μg $ml^{-1}$) and then inoculated into field-grown maize and wheat, more than 80% of the rhizosphere and root surface population was from the inoculated strains (BALDANI and DÖBEREINER 1980). The identification of the inoculated strains within surface-sterilized roots was complicated by the occurrence of high proportions of low-level streptomycin-tolerant *Azospirillum* within maize and wheat roots. Non-homologous strains also established in the rhizosphere but not within roots.

Proportions of streptomycin-resistant bacteria higher than in soil were observed in legumes and vegetables and in wheat, maize and sorghum (BROWN 1961, DÖBEREINER and BALDANI 1980). Increases of numbers of actinomycetes in the rhizosphere (ROVIRA 1965) and selective assimilation of streptomycin by plant cells (PRAMER, 1955) have been reported. It seems possible, therefore, that cereals can regulate bacterial infection through requirements for antibiotic resistance.

### 2.3.4 Physiology of *Azospirillum*

Better understanding of the physiology of *Azospirillum* spp. in culture media will help in understanding the more complicated nature of the root–bacteria ensemble.

*Azospirillum* is an obligate aerobe and is unable to grow in the absence of air except in the presence of $NO_3^-$ and an organic N source (NEYRA and VAN BERKUM 1977). Rapid aerobic growth occurs with $NH_4^+$ as nitrogen source but growth on $N_2$ is slower (generation time 1 and 5.3–7 h respectively) (OKON et al. 1977a, VOLPON et al. 1981) and does not proceed under air. Oxygen requirements are very narrow, and a dissolved $pO_2$ in the order of 0.003 atm is necessary for a functional nitrogenase (VARGAS 1977, NELSON and KNOWLES 1978). On the other hand, rapidly growing continuous or batch cultures consume so much $O_2$, that once an optical density of about 0.3 is reached (560 nm) *Azospirillum* can be grown under air with rapid stirring and sparging (DÖBEREINER 1977, VOLPON et al. 1981) as long as the input of $O_2$ corresponds to the $O_2$ uptake and the dissolved $pO_2$ remains below the above-mentioned limit. Cultivation of *Azospirillum* in semi-solid media avoids this problem due to the $O_2$ diffusion gradient. The very motile organisms move to the zone with the required $pO_2$, where they multiply. Once $O_2$ demands increase, the bacterial pellicle moves closer to the surface.

The oxygen tolerance of *Azospirillum* changes with the age of the culture, and late log phase organisms are more tolerant to $O_2$ than early log phase cells (STEPHAN et al. 1981, VOLPON et al. 1981). The increased $O_2$ tolerance was more pronounced in *A. lipoferum* and occurred when the glucose or lactate concentration in the medium had dropped to less than 0.05% (STEPHAN et al. 1981, VOLPON et al. 1981). This is in contrast to observations with *Azotobacter* (HILL et al. 1972) and stresses the weak oxygen protection mechanism for nitrogenase through carbon respiration, in *Azospirillum*. The increased $O_2$ tolerance was attributed to a threefold increase in uptake hydrogenase (HUP) activity which reached much higher levels than those observed in *Rhizobium* bacteroids (MAIER et al. 1979) (see Chap. II.1 for explanation of the role of HUP). Hydrogen recycling seems complete in *Azospirillum* spp. and no traces of $H_2$ evolution were observed in the above-mentioned experiments. How far $H_2$ oxidation is responsible for ATP generation and oxygen consumption, as suggested for *Rhizobium* bacteroids (RUIZ-ARGÜESO et al. 1979), is not yet clear.

Efficiency of nitrogen fixation also changes during the growth cycle. Increases from 20 mg N $g^{-1}$ glucose during early log phase to 48 mg during late log phase and beginning stationary phase in *A. lipoferum* and from 15 mg N $g^{-1}$ lactate to 90 mg in *A. brasilense* were observed (VOLPON et al. 1980, STEPHAN et al. 1981). These changes in efficiency were accompanied by changes in specific activity and coincided with the pattern of hydrogenase activity and oxygen tolerance. The high efficiencies confirm earlier reports from our laboratory (DAY and DÖBEREINER 1976) and the changes during the growth cycle explain the discrepancy in the data from various laboratories (OKON et al. 1976b, VARGAS 1977). There was an increase of total PHB (from 24 to 68 µg $ml^{-1}$) during growth, although the percentage decreased (VOLPON et al. 1981) and therefore the given efficiencies cannot be attributed to the use of PHB accumulated earlier in the growth cycle. Whether the high efficiencies operate in bacteria within plant roots needs to be shown. In searching for rhizocoenoses which make substantial contributions to the N economy of the plant, bacteria which fix $N_2$ actively during restricted growth are the kind to look for.

In batch cultures with *A. brasilense* (semi-solid medium) $NH_4^+$ is excreted during early growth, accumulating to concentrations of 0.3 mM (ELMERICH et al. 1978). Later this $NH_4^+$ seems to be assimilated. *A. lipoferum* in $O_2$-limited batch cultures with liquid medium also excretes up to 25% of the fixed nitrogen.

Cell-free extracts containing active nitrogenase were isolated from *A. brasilense* (OKON et al. 1977b), and showed requirements similar to other nitrogenases. The activation factor necessary for a functional *Rhodospirillum rubrum* nitrogenase was necessary for *A. brasilense* (LUDDEN et al. 1978, OKON et al. 1977b). $NH_4^+$ incorporation and regulation of nitrogenase seems similar to that in other $N_2$-fixing organisms and $N_2$-grown cells showed higher glutamine synthetase (GS) and glutamate-oxo-glutarate amino-transferase (GOGAT) activities and lower glutamate dehydrogenase (GDH) activities when compared with $NH_4^+$-grown cells (OKON et al. 1976b).

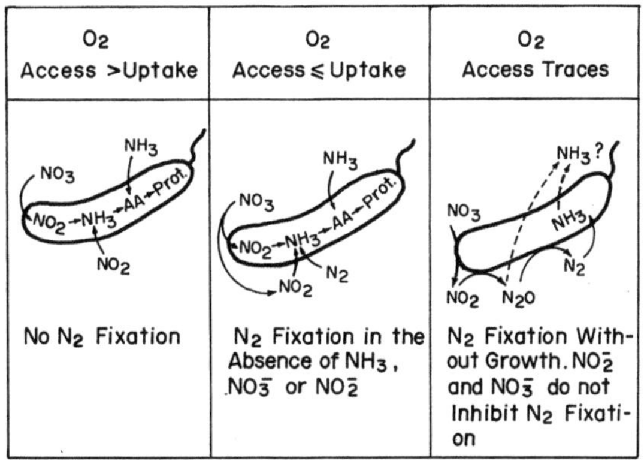


**Fig. 2.** Effect of oxygen on nitrogen metabolism by *Azospirillum* spp. When $O_2$ access is faster than $O_2$ uptake by the cells (*left square*) $NO_3^-$ or $NH_3$ are assimilated and there is no $N_2$ fixation. When $O_2$ access is in equilibrium with respiration (*middle square*) $N_2$ fixation occurs in the absence of mineral nitrogen. When traces or no $O_2$ has access to the cells (*right square*) there is nitrate-dependent nitrogenase activity and denitrification. Notice that $nir^+$ *Azospirillum* strains are able to participate in all steps of the nitrogen cycle except nitrification (for details see text)

Both *Azospirillum* spp. were shown to be capable of autotrophic growth on $H_2$ and $CH_4$ with $N_2$ as sole nitrogen source (SAMPAIO et al. 1981). Figure 2 summarizes the various steps in nitrate metabolism of *Azospirillum* spp. and their interactions with $O_2$. All strains from both species isolated so far are able to assimilate $NO_3^-$ and $NH_4^+$ and there is no $N_2$ fixation with more than 1 mM $NO_3^-$ or $NH_4^+$ (DAY and DÖBEREINER 1976), under aerobic conditions. Under $O_2$-limiting conditions $NO_3^-$ can be dissimilated to $NO_2^-$ which accumulates in the medium in some strains ($nir^-$) (MAGALHÃES et al. 1978, NEYRA et al. 1977). About two thirds of the strains of *A. brasilense* (TARRAND et al. 1978) and most strains of *A. lipoferum* (except root isolates) (DÖBEREINER and BALDANI 1980) can dissimilate $NO_2^-$ to gas (NEYRA et al. 1977). Mutants ($nr^-$, $nir^-$ or both) were isolated by selection for $ClO_3^-$ resistance from both species which in the presence of 10 mM $NO_3^-$ fix as much $N_2$ as without (MAGALHÃES et al. 1979). Monoxenic plants inoculated with $nr^-$ mutants, in the presence of 1 mM $KNO_3$, had twice as much nitrogenase activity than the parent strains (DÖBEREINER and DE-POLLI 1980). Mutants with $nr^+$ but $nir^-$ do not fix $N_2$ in the presence of $NO_3^-$, indicating that the accumulation of $NO_2^-$ and not $NO_3^-$ assimilation is responsible for nitrogenase inhibition. The inhibition by $NO_2^-$ of nitrogenase activity in cell-free extracts (P.B.W. VAN BERKUM personal communication) and that of $nr^-$ as well as $nir^-$ mutants supports this (MAGALHÃES et al. 1978). Sulphate, iron or fumarate cannot replace $NO_3^-$ as terminal respiratory electron acceptor (BOTHE et al. 1981).

Under anaerobic conditions $NO_3^-$ can supply energy for $N_2$ fixation but without supporting growth. This process however was suggested to be transitory only until the assimilatory $NO_3$ reductase is synthesized (BOTHE et al. 1981) when anaerobic growth proceeds with $NO_3^-$ as N source. Anaerobic $NO_3^-$-dependent nitrogenase activity (10–20 mM $NO_3^-$) in rates similar to those observed with $O_2$ was reported by NEYRA and BERKUM (1977) and SCOTT et al. (1979), in both *Azospirillum* species. The activities were higher than those observed of $NO_3^-$-dependent nitrogenase activity in *Rhizobium* bacteroids (ZABLOTOWICZ and FOCHT 1979). Mutants ($nr^-$) showed no anaerobic nitrogenase activity in the presence of $NO_3^-$. The $NO_3^-$-dependent nitrogenase activity was also observed in $nr^+$ $nir^-$ strains of *A. brasilense* (SCOTT et al. 1979). This shows that there is no need for $NO_2^-$ dissimilation to gaseous forms in order to replace $O_2$ for respiration and nitrogenase activity. The same was observed with *Rhizobium* bacteroids (ZABLOTOWICZ et al. 1978). In fact it has not been shown yet that this second step in nitrate dissimilation can be used by *Azospirillum* to drive nitrogenase activity. This apparent contradiction of a nitrogen-fixing organism denitrifying has now also been shown with *Rhizobium* bacteroids (ZABLOTOWICZ and FOCHT 1979). These authors suggest a possible role in avoiding $NO_2^-$ inhibition of $N_2$-ase or as alternative $e^-$ acceptors under limited $O_2$ access. This would, however, not improve the efficiency of these systems because $NO_3^-$ reduction compares unfavorably with $O_2$ in terms of voltage couples (421 and 820 mV respectively) (ZABLOTOWICZ et al. 1978), and 100 mol $NO_3^-$ were reduced for 1 mol $N_2$ fixed (SCOTT et al. 1979). The importance of these characteristics as a basis for studies of complementation of $N_2$ fixation with mineral fertilizer N is obvious.

## 2.4 Associations with Other $N_2$-Fixing Bacteria

A number of less well-defined associations have been described. Under temperate conditions a *Bacillus* sp., probably *B. polymixa,* has been shown to multiply selectively in the rhizosphere of certain wheat lines (NEAL and LARSON 1976) and inoculation of wheat in small pots resulted in increases of plant nitrogen (RENNIE and LARSON 1981). The wheat lines which stimulate *Bacillus* growth are non-leaky lines, that is they accumulate rather than excrete photosynthates and therefore support a localization of the bacteria within the roots. NEAL and LARSON (1976) showed infection of intercellular spaces and middle lamellae of roots.

Many reports are available on paddy rice associations with diazotrophs and there are certainly several organisms involved. *Azospirillum brasilense* has been isolated from surface sterilized roots in large numbers, in many places (BALDANI and DÖBEREINER 1980, BODDEY 1980, RAJARAMAMOHAN et al. 1978, WATANABE et al. 1979) and highest numbers were observed during the peak of nitrogenase activity at flowering; in contrast total numbers of bacteria were high all through the growth cycle of the plant (WATANABE and CHOLITKUL 1979). Generally in the rhizosphere of paddy rice, aerobic $N_2$-fixing bacteria were more numerous than anaerobes, and *Beijerinckia* spp. and *Enterobacter cloaceae* were found in high numbers (YOSHIDA and ANCAJAS 1971, BALANDREAU 1975). Methane-oxidizing diazotrophic bacteria were also found in high numbers, and the large amount of $CH_4$ which can accumulate in rice paddies should not be overlooked as potential energy source for $N_2$ fixation (DE-BONT et al. 1978). Recently WATANABE and BARRAQUIO (1979) reported the occurrence of glucose-utilizing aerobic semi-diazotroph bacteria within rice roots which comprised 81% of the total bacteria flora and which were tentatively classified as *Achromobacter*. Proportions of these bacteria in the stem and rhizosphere soil were 37.5% and 2.4% respectively. Roots of several other water plants contained similar organisms but none of dryland plants did so. Diazotrophic *Spirillum*-like organisms were isolated from roots of *Potamogeton filiformis* (SYLVESTER-BRADLEY 1976) and *Spartina alterniflora*, a $C_4$ marsh grass and were identified as *Compylobacter* sp. (PATRIQUIN et al. 1981). In *Spartina*, root nitrogenase activity was correlated with root sugar concentrations and was $CO_2$-dependent (PATRIQUIN et al. 1981). Additions of sugar or malate did not substantially increase the internal nitrogenase activity, indicating a large carbon pool (BOYLE 1978). Diazotrophic *Enterobacteriaceae* and *Bacillus* sp. have been found in high numbers in several other plants, but plant–bacteria interactions have not been shown.

## 3 Agronomic Aspects

Striking implications of manipulating diazotrophic biocoenoses in increased crop yields or even in fertilizer economy, if ever, can only be expected after the nature of the various systems is better understood, and much has to be done in order to achieve this. There is, however, the possibility of more empirical identification of limiting factors and the search for economically feasible measures to minimize them.

### 3.1 Plant Genotype Effects

The first priority for improved crop production is usually plant breeding. Significant plant genotype differences in nitrogenase activity in intact soil plant systems have been shown in rice (BALANDREAU et al. 1978, LEE et al. 1977b), wheat (NERY et al. 1977, SUBBA RAO 1981), sorghum (NERY 1978, SUBBA RAO 1981) and *Brachiaria* (PEREIRA and DÖBEREINER 1981). There are cereal breeding programs in progress in Brazil (I.E. MARRIEL 1977, personal communication, RUSCHEL and RUSCHEL 1981) and in India (SUBBA RAO 1981).

So far these breeding efforts envisage higher and more stable nitrogenase activity, and there has been some preliminary information on inheritance mechanisms (VON BÜLLOW 1978, RUSCHEL and DA SILVA 1981). These studies however are made difficult by the lack of information about the relevant plant characteristics to be selected for. Although initial screenings were aimed at selecting "leaky" plants which could feed the bacteria in the rhizosphere (BRILL 1978) there is now increasing evidence that the opposite feature is what is wanted. The wheat lines which stimulate *Bacillus* sp. are all root-rot resistant, that is, non-leaky types (RENNIE and LARSON 1981) and PATRIQUIN et al. (1981) have found that plants which accumulate large sugar pools are better $N_2$-fixers than leaky plants.

Another important question is the possibility of breeding for positive or at least non-negative interactions between nitrogen fixation and nitrate assimilation, because even if part of the nitrogen requirements are provided by biological fixation, the amount is unlikely to be sufficient for high yields. Maize genotypes showed negative correlations of nitrogenase activity in excised roots with the nitrate reductase in leaves when high nitrogen fertilizer (80 kg N $ha^{-1}$) levels were applied, but no correlation with 20 kg N $ha^{-1}$. An increase of this limit would permit simultaneously efficient nitrogenase activity and nitrate assimilation. Maize lines selected for high nitrogenase activity at reproductive growth stage reduced as much $C_2H_2$ with 80 as with 20 kg N $ha^{-1}$ (BALDANI et al. 1979). Similar observations were made by TROLLDENIER (1977) with rice. Furthermore application of 20 kg N $ha^{-1}$ increased nitrogenase activity of one out of five *Brachiaria* genotypes (intact soil plant systems) (PEREIRA and DÖBEREINER 1981), while the effect on the others was negative.

### 3.2 Environmental Effects

The identification of environmental factors which limit nitrogenase activity under field conditions is essential for any attempt to find agriculturally viable practices which may increase biological $N_2$ fixation in grasses. A general approach has been presented by BALANDREAU (1975), who monitored weekly during one growing season soil and air temperature, light intensity, photosynthesis, air humidity, and soil nitrogen content with in situ assay of nitrogenase activity in soil plant cores of maize and forage grasses. Weekly variations in nitrogenase activity were mostly dependent on soil N (mineral) and humidity, while oscillations during the day could be attributed to soil temperature and light energy variations. Computer analysis of the interference of minimal and maximal soil and air temperatures, nitrogen content, and soil humidity in *Digitaria* nitrogenase activity confirmed minimum temperatures and $NH_4^+$ content in soil as the major limiting factors (ABRANTES et al. 1976).

### 3.3 Inoculation

The urgency to obtain cereals and grasses which can cover part of their nitrogen needs by biological fixation has precipitated attempts of solving all problems by inoculating with the first strains of *Azotobacter* (RUBENCHIK 1963) or *Azospirillum* (BOUTON et al. 1979 and SMITH et al. 1977) at hand, irrespective of their origin or efficiency. Even in the well-established legume symbiosis, inoculation with specifically selected strains seldom brings about spectacular yield increases. The recent advances discussed in this chapter show clearly the need for better understanding of host strain interactions before it is known which characteristics of *Azospirillum, Bacillus* or other diazotrophic bacteria are of advantage in the various biocoenoses. Plant roots seem to select for nitrite reductase negative (nir$^-$) strains (DÖBEREINER and BALDANI 1980; DÖBEREINER and DE-POLLI 1980) but it is not yet known whether the nitrate reductase (nr) is of advantage to the bacteria, the plant or both. *Azospirillum* strains seem to interact in nitrate assimilation by the plant (DE FREITAS et al. 1981 and VILLAS BOAS and DÖBEREINER 1981). Active denitrification occurs in nir$^+$ *Azospirillum* cultures, but no information as yet is available on the role of such strains in soil, or in the rhizosphere. As much as 7% of the applied $NO_3^-$ fertilizer was lost in 3 days as $N_2O$ or $N_2$ in *Brachiaria* cores (PEREIRA and DÖBEREINER 1981).

Field inoculation of wheat with low level streptomycin-resistant *Azospirillum brasilense* nir$^-$, isolated from sterilized wheat roots, brought about 30% plant nitrogen increases (DÖBEREINER and DE-POLLI 1980). Inoculation with *Azospirillum* strains isolated from maize gave total plant nitrogen increases in the order of 40 kg N ha$^{-1}$, equivalent to those obtained with 60 kg ha$^{-1}$ of fertilizer N (DÖBEREINER and BALDANI 1982). Inoculation effects with empirically selected *Azospirillum* strains in several field experiments in India and Israel corresponded to grain or forage yield increases obtained with 20 to 80 kg of fertilizer N (SUBBA RAO 1981, KAPULNIK et al. 1981) although many other attempts failed. In all these experiments, the inoculation effects increased with low nitrogen fertilizer levels, and a role of *Azospirillum* spp. in nitrate assimilation cannot be ruled out.

# 4 Phyllosphere Associations

According to the newly proposed nomenclature (International Workshop on Associative $N_2$ Fixation, Piracicaba 1979), associations of $N_2$-fixing bacteria with plant leaves were termed phyllocoenoses. RUINEN (1956) reported the very common occurrence of nitrogen-fixing bacteria on leaves of tropical plants and proposed the name phyllosphere for this habitat. To what extent the occurrence of $N_2$-fixing microorganisms in the phyllosphere can be considered an association or a phyllocoenosis is not yet clear. So far no specificity in plant–bacteria interactions has been demonstrated, although the high frequency seems to indicate selective environments on tropical forest leaves. RUINEN (1956) was able to isolate *Beijerinckia* spp. from 192 out of 198 randomly collected leaves of shrubs and trees.

The phyllosphere as characterized by RUINEN includes the leaf surfaces, and leaf sheaths which especially in grasses represent a suitable habitat for microorganisms (RUINEN 1970). Substantial amounts of minerals and photosynthates accumulate on the leaf surface and in leaf sheaths, and dew and light rains which are intercepted by the canopy can transform these substances into a rich culture medium for microorganisms. Sugar concentrations in rainwater collected from broad forest leaves in West Africa were in the order of 10 to 200 ppm (RUINEN 1965). In the sheath water, even higher quantities of sugar can accumulate and 4,000–7,000 ppm were found in young leaves of $C_4$ forage grasses in Indonesia and West Africa (BESSEMS 1973; RUINEN 1970).

### 4.1 Microorganisms in the Phyllosphere

A large variety of $N_2$ fixing bacteria have been isolated from leaves of tropical shrubs, trees and grasses. *Beijerinckia* spp. and *Azotobacter* spp. seem most common, but this could rather reflect the easiness of their isolation and identification (RUINEN 1956, 1961; VASANTARIAN and BHAT 1968, HARRELSON 1969, BHAT et al. 1971). In the damp warm evening air near tropical forests a very characteristic smell of *Beijerinckia* spp. was recognised repeatedly (DÖBEREINER unpublished). Rice sheaths in Java contained predominantly *Azospirillum* and *Beijerinckia* (RUINEN 1974), and the leaf sheaths of mature *Tripsacum laxum*, a tall tropical grass, contained $10^8$ cells of *Azospirillum* and as many of *Klebsiella* sp. when the sugar concentration reached maximal values of 4,000 ppm (RUINEN 1975). This author suggested that $N_2$-fixing bacteria are washed into the soil with sugar-containing leaf and stem washings. We refer to Section 2.3.2 where almost equally high numbers of *Azospirillum* were reported within roots and stems of maize.

Various *Cyanobacteria* were also found on forest leaves in Puerto Rico (HARRELSON 1969) and Java (RUINEN 1974).

A few reports also indicate occurrence of $N_2$-fixing bacteria in the phyllosphere of temperate plants. BESSEMS (1973) found *Klebsiella* spp. in the maize leaf sheaths in Holland, and JONES (1970) found considerable nitrogenase activity (estimated by the acetylene reduction assay) on Douglas fir leaves (*Pseudotsuga douglasii*) but no attempts were reported to identify the responsible organisms.

### 4.2 Nitrogen Fixation in the Phyllosphere

Estimates of nitrogen fixation in the phyllosphere, as in roots, are very variable and range from traces to 70 g N $ha^{-1}$ per day (RUINEN 1974) or 10% of the total nitrogen incorporated into the leaves (HARRELSON 1969). These values may seem small in relation to the large amounts of N which are recycled in a tropical forest where 2 t N $ha^{-1}$ are locked up in the plant system (GREENLAND and NYE 1959). Losses, however, are small in equilibrium systems where nitrogen rarely will remain the limiting factor. In regenerating forests on the other hand 40 to 100 kg N $ha^{-1}$ have been found to be incorporated in ecosystems practically devoid of legume species in Ghana and West Africa (GREENLAND and KOWAL 1960). There it is possible that $N_2$ fixation in rhizocoenoses and in the humus-rich soil is complemented by contributions from the phyllosphere.

## 5 General Conclusion

Nitrogen fixation associated with Gramineae roots (rhizocoenoses) and leaves (phyllocoenoses) is very variable but can reach economically important proportions especially in the tropics. In low nitrogen input agricultural systems this contribution will be most important. Also in regenerating forests and savannas biological nitrogen fixation is the only nitrogen input to replace constant losses through leaching and denitrification. In such environments legumes and nitrogen fixing grasses have competitive advantages and it is there that highest rates of nitrogen fixation should be expected until another element, usually phosphorus, becomes limiting. In contrast, equilibrium forests or savannas will never be nitrogen-limited because there is no other way for nature to add plant nutrients to the ecosystem except by nitrogen. Legumes are seldom nodulated once the climax is reached and these systems will not give us a good measure of the potential of nitrogen fixation by either system.

## References


Abrantes GTV, Day JM, Cruz VC, Döbereiner J (1976) Fatores limitantes da fixação de nitrogênio em campo de *Digitaria decumbens* cv. Transvala. An XV Congr Bras Cie Solo

App AA, Watanabe I, Alexander M, Ventura W, Daez C, Santiago T, De Datta SK (1980) Nonsymbiotic nitrogen fixation associated with the rice plant in flooded soils. Soil Sci 130 5:283–289

Balandreau J (1975) Activité nitrogénasique dans la rhizosphère de quelques Graminées. PhD thesis, Univ Nancy, France

Balandreau J, Ducerf P, Hamad-Fares I, Weinhard P, Rinaudo G, Millier C, Dommergues Y (1978) Limiting factors in grass nitrogen fixation. In: Döbereiner J, Burris RH, Hollaender A (eds) Limitations and potentials for biological nitrogen fixation in the tropics. Basic Life Sciences vol X. Plenum, New York London

Baldani JI, Blaña RAG, Döbereiner J (1979) Efeito do genótipo do milho na atividade da nitrogenase e da nitrato-reductase. Pesqu Agropecu Bras 14:165–173

Baldani JI, Pereira PAA, da Rocha REM, Döbereiner J (1981) Especificidade na infecção de raízes por *Azospirillum* spp. em plantas com via fotossintética $C_3$ e $C_4$. Pesqu Agropecu Bras 16:325–330

Baldani VLD, Döbereiner J (1980) Host plant specificity in the infection of cereals with *Azospirillum* spp. Soil Biol Biochem 12:433–439

Becking JH (1963) Fixation of molecular nitrogen by an aerobic *Vibrio* or *Spirillum* sp. Antonie van Leeuwenhoek J Microbiol Serol 29:326

Beijerinck MW (1925) Über ein *Spirillum,* welches freien Stickstoff binden kann? Zentralbl Bakteriol Parasitkde Abt II 63:353–359

Berg RH, Tyler ME, Novick NJ, Vasil V, Vasil IK (1980) Biology of *Azospirillum*-sugarcane association: enhancement of nitrogenase activity. Appl Environ Microbiol 29:642–649

Bessems EPM (1973) Nitrogen fixation in the phyllosphere of Gramineae. Agric Ses Rep 786:1–68

Bhat JV, Limaye KS, Vasantharajan VN (1971) The effect of the leaf surface microflora on the growth and root exudation of plants. In: Preece TF, Dickinson CH (eds) Ecology of leaf surface microorganisms. Academic Press London New York

Boddey RM (1980) Biological nitrogen fixation in the rhizosphere of lowland rice. PhD thesis, Univ West Indies, Trinidad

Bothe H, Klein B, Stephan MP, Döbereiner J (1981) Transformations of inorganic nitrogen by *Azospirillum* spp. Arch Microbiol 130:96–100

Bouton JH, Smith RL, Schank SC, Burton GW, Tyler ME, Littell RC, Gallaher RN, Quesenberry KH (1979) Response of pearl millet inbreds and hybrids to inoculation with *Azospirillum brasilense.* Crop Sci 19:12–16

Boyle CD (1978) Some characteristics of nitrogenase activity associated with diazotrophs in and around *Spartina alterniflora* roots. MS thesis, Dalhousie Univ, Can
Brill WJ (1978) Genetics and regulation of nitrogen fixation. In: Döbereiner J, Burris RH, Hollaender A (eds) Limitations and potentials for biological nitrogen fixation in the tropics. Basic Life Sciences, vol X. Plenum, New York London
Brown ME (1961) Stimulation of streptomycin-resistant bacteria in the rhizosphere of leguminous plants. J Gen Microbiol 24:369–377
von Büllow JFW (1978) Plant influence in symbiotic nitrogen fixation. In: Döbereiner J, Burris RH, Hollaender A (eds) Limitations and potentials for biological nitrogen fixation in the tropics. Basic life sciences, vol X. Plenum, New York London
von Büllow JFW, Döbereiner J (1975) Potential for nitrogen fixation in maize genotypes in Brazil. Proc Natl Acad Sci USA 72:2389–2393
Burris RH (1977) A synthesis paper on nonleguminous $N_2$-fixing systems. In: Newton W, Postgate JR, Rodriguez-Barrueco C (eds) Recent developments in nitrogen fixation. Academic Press, London New York
Burris RH, Albrecht SL, Okon Y (1978) Physiology and biochemistry of *Spirillum lipoferum*. In: Döbereiner J, Burris RH, Hollaender A (eds) Limitations and potentials for biological nitrogen fixation in the tropics. Basic Life Sciences, vol X. Plenum, New York London
Day JM, Döbereiner J (1976) Physiological aspects of $N_2$-fixation by a *Spirillum* from *Digitaria* roots. Soil Biol Biochem 8:45–50
De-Bont JAM, Lee KK, Bouldin DF (1978) Bacterial oxidation of methane in a rice paddy. In: Environmental role of nitrogen-fixing blue-green algae and asymbiotic bacteria. Ecol Bull (Stockholm) 26:91–96
De-Polli H, Bohlool BB, Döbereiner J (1980) Serological differentiation of *Azospirillum* species belonging to different host-plant specificity groups. Arch Microbiol 126:217–222
Döbereiner J (1961) Nitrogen-fixing bacteria of the genus *Beijerinckia* Derx in the rhizosphere of sugar cane. Plant Soil 15:211–216
Döbereiner J (1966) *Azotobacter paspali* sp. n., uma bactéria fixadora de nitrogênio na rizosfera de *Paspalum*. Pesqu Agropecu Bras 1:357–365
Döbereiner J (1970) Further research on *Azotobacter paspali* and its variety specificity occurrence in the rhizosphere of *Paspalum notatum* Flügge. Zentralbl Bakterid Parasitenkde II 124:224–230
Döbereiner J (1977) Physiological aspects of $N_2$ fixation in grass-bacteria associations. In: Newton W, Postgate JR, Rodriguez-Barrueco (eds) Recent developments in nitrogen fixation. Academic Press, London New York
Döbereiner J (1978) Nitrogen fixation in grass-bacteria associations in the tropics. Proc Isot Biol Dinitrogen Fix. IAEA, Vienna
Döbereiner J, Baldani VLD (1980) Selective infection of maize roots by streptomycin-resistant *Azospirillum lipoferum* and other bacteria. Can J Microbiol 25:1264–1269
Döbereiner J, Baldani JI (1982) Bases cientpificas para uma agricultura biológica. Cienc Cult (Sao Paulo) 37:869–881
Döbereiner J, Campelo AB (1971) Non-symbiotic nitrogen-fixing bacteria in tropical soils. Plant Soil Spec Vol 457–470
Döbereiner J, Day JM (1975) Nitrogen fixation in the rhizosphere of tropical grasses. In: Steward WDP (ed) Nitrogen fixation by free-living micro-organisms. Cambridge Univ Press, London
Döbereiner J, Day JM (1976) Associative symbioses in tropical grasses: Characterization of microorganisms and dinitrogen-fixing sites. In: Newton WE, Nyman CJ (eds) Int Symp $N_2$ Fix. Washington State Univ Press, Pullman
Döbereiner J, De-Polli H (1980) Diazotrophic rhizocoenoses. In: Stewart WDP, Gallon JR (eds) Nitrogen fixation. Academic Press, London New York
Döbereiner J, Day JM, Dart PJ (1972a) Nitrogenase activity of the *Paspalum notatum*-*Azotobacter paspali* association and oxygen sensitivity. J Gen Microbiol 71:103–106
Döbereiner J, Day JM, Dart PJ (1972b) Nitrogenase activity in the rhizosphere of sugar cane and some other tropical grasses. Plant Soil 37:191–196

Döbereiner J, Marriel IE, Nery M (1976) Ecological distribution of *Spirillum lipoferum* Beijerinck. Can J Microbiol 22:1464–1473
Elmerich C, Gauthier D, Houward J (1978) Regulation of nitrogenase biosynthesis in *Azospirillum brasilense*. Proc Steenbock-Kettering Int Symp $N_2$ Fix, Madison
Freitas de JLM, Döbereiner J (1978) Effects of oxygen on nitrogenase activity in a sterilized intact grass system inoculated with *Azospirillum* spp. In: Döbereiner J, Burris, RH, Hollaender A (eds) Limitations and potentials for biological nitrogen fixation in the tropics. Basic Life Sciences, vol X. Plenum, New York London
Freitas de JLM, Pereira PAA, Döbereiner J (1981) Effect of organic matter and *Azospirillum* spp. strains in the metabolism of nitrogen in *Sorghum vulgare*. In: Vose PB, Ruschel AP (eds) Associative $N_2$-fixation, vol I. CRC Press, Boca Raton
Giddens J (1977) Nitrogen accumulation by grasses. Ga Agr Res 19:12–13
Greenland DJ, Kowal JLM (1960) Nutrient content of the moist tropical forest of Ghana. Plant Soil 12:154–174
Greenland DJ, Nye PH (1959) Increases in the carbon and nitrogen contents of tropical soils under natural fallows. J Soil Sci 9:284–299
Harrelson MA (1969) Tropical epiphyllae and nitrogen fixation. PhD thesis, Univ Georgia
Hill S, Drozd JW, Postgate JR (1972) Environmental effects on the growth of nitrogen-fixing bacteria. J Appl Chem Biotechnol 22:541–558
Jaiyebo EO, Moore AW (1963) Soil nitrogen accretion under different covers in a tropical rain-forest environment. Nature 197:317–318
Jenkinson DS (1973) Organic matter and nitrogen in soils of the Rothamsted classical experiments. J Sci Food Agric 24:1149–1150
Jones K (1970) Nitrogen in the Douglas fir, *Pseudotsuga douglasii*. Ann Bot 34:239–244
Kapulnik Y, Sarig S, Nur I, Okon Y, Kigel J, Henis Y (1981) Yield increases in summer cereal crops in Israel fields inoculated with *Azospirillum*. Exp Agric 17:179–187
Kavimandan SK, Subba Rao NS, Mohrir AV (1978) Isolation of *Spirillum lipoferum* from the stems of wheat and nitrogen fixation in enrichment cultures. Curr Sci 47:96–98
Lakshmi-Kurami M, Kavimandan SK, Subba Rao NS (1976) Occurrence of nitrogen fixing *Spirillum* in roots of rice, sorghum, maize and other plants. Indian J Exp Biol 14:638–639
Lee KK, Alimagno B, Yoshida T (1977a) Field technique using the acetylene reduction method to assay nitrogenase activity and its association with the rice rhizosphere. Plant Soil 47:519–526
Lee KK, Castro T, Yoshida T (1977b) Nitrogen fixation throughout growth and varietal differences in nitrogen fixation by the rhizosphere of rice planted in pots. Plant Soil 48:613–619
Ludden PW, Okon Y, Burris RH (1978) The nitrogenase system of *Spirillum lipoferum*. Biochem J 173:1001–1003
Machado WC, Döbereiner J (1969) Estudos complementares sobre a fisiologia de *Azotobacter paspali* e sua dependência da planta (*Paspalum notatum*). Pesqu Agropecu Bras 4:53–58
Magalhães FMM, Patriquin D, Döbereiner J (1979) Infection of field-grown maize with *Azospirillum* spp. Rev Bras Biol 39:587–596
Magalhães LMS, Neyra CA, Döbereiner J (1978) Nitrate and nitrite reductase negative mutants of $N_2$-fixing *Azospirillum* spp. Arch Microbiol 117:247–252
Maier RJ, Hanus FJ, Evans HJ (1979) Regulation of hydrogenase in *Rhizobium japonicum*. J Bacteriol 137:824–829
Moore AW (1963) Nitrogen fixation in latosolic soil under grass. Plant Soil 19:127–138
Neal JL Jr, Larson RI (1976) Acetylene reduction by bacteria isolated from the rhizosphere of wheat. Soil Biol Biochem 8:151–155
Nelson LM, Knowles R (1978) Effect of oxygen and nitrate on nitrogen fixation and denitrification by *Azospirillum brasilense* grown in continuous culture. Can J Microbiol 24:1395–1403
Nery M (1978) Interaction of nitrogen fertilizer with nitrogen fixation ($C_2H_2$) in sorghum. In: Döbereiner J, Burris RH, Hollaender A (eds) Limitations and potentials for bio-

logical nitrogen fixation in the tropics. Basic life sciences, vol X. Plenum, New York London
Nery M, Abrantes GTV, Santos dos O, Döbereiner J (1977) Fixação de nitrogênio em trigo. Rev Bras Ciê Solo 1:15–20
Neyra CA, Berkum Van P (1977) Nitrate reduction and nitrogenase activity in *Spirillum lipoferum*. Can J Microbiol 23:306–310
Neyra CA, Döbereiner J (1977) Nitrogen fixation in grasses. Adv Agron 29:1–38
Neyra CA, Döbereiner J, Lalande R, Knowles R (1977) Denitrification by $N_2$-fixing *Spirillum lipoferum*. Can J Microbiol 23:300–305
Okon Y, Albrecht SL, Burris RH (1976a) Factors affecting growth and nitrogen fixation of *Spirillum lipoferum*. J Bacteriol 125:1248–1254
Okon Y, Albrecht SL, Burris RH (1976b) Carbon and ammonia metabolism of *Spirillum lipoferum*. J Bacteriol 128:592–597
Okon Y, Albrecht SL, Burris RH (1977a) Methods for growing *Spirillum lipoferum* and for counting it in pure culture and in association with plants. Appl Environ Microbiol 33:85–88
Okon Y, Houchins JP, Albrecht SL, Burris RH (1977b) Growth of *Spirillum lipoferum* at constant partial pressure of oxygen, and the properties of its nitrogenase in cell-free extracts. J Gen Microbiol 98:87–93
Parker CA (1957) Non-symbiotic nitrogen-fixing bacteria in soil. III. Total nitrogen changes in a field soil. J Soil Sci 8:48–59
Patriquin DG, Döbereiner J (1978) Light microscopy observations of tetrazolium-reducing bacteria in the endorhizosphere of maize and other grasses in Brazil. Can J Microbiol 24:734–742
Patriquin DG, Gracioli LA, Ruschel AP (1980) Nitrogenase activity of sugar cane propagated from stem cuttings in sterile vermiculite. Soil Biol Biochem 12:413–417
Patriquin DG, Boyle CD, Livingstone DC, McClung CR (1981) Physiology of the diazotrophic rhizocoenosis in salt march cord grass, *Spartina alterniflora* Loisel. In: Vose PB, Ruschel AP (eds) Associative $N_2$-fixation, vol II. CRC Press, Boca Raton
Peña JJ, Döbereiner J (1974) Efecto del nitrato y amonio en la actividad de la nitrogenase de bacterias tropicales fijadoras de nitrógeno atmosférico. Rev Lat Am Microbiol 16:33–44
Pereira PAA, Döbereiner J (1981) A note on nitrogenase and nitrate reductase activities, and denitrification in five *Brachiaria* spp. genotypes. In: Vose PB, Ruschel AP (eds) Associative $N_2$-fixation, vol II. CRC Press, Boca Raton
Pramer D (1955) Absorption of antibiotics by plant cells. Science 121:507–508
Purchase BS (1978) A potential source of nitrogen for Rhodesian agriculture. Rhod Agric 75:99–104
Quintero MJ, Garza T (1978) Isolation of *Spirillum lipoferum* Beijerinck from tropical grasses in Mexico. In: Döbereiner J, Burris RH, Hollaender A (eds) Limitations and potentials for biological nitrogen fixation in the tropics. Basic life sciences, vol X. Plenum, New York London
Rajaramamohan-Rao V, Charyulu PBBN, Nayak DN, Ramakrishna C (1978) Nitrogen fixation by free-living organisms in tropical rice soils. In: Döbereiner J, Burris RH, Hollaender A (eds) Limitations and potentials for biological nitrogen fixation in the tropics. Basic life sciences, vol X. Plenum, New York London
Rennie RJ, Larson RI (1981) Dinitrogen fixation associated with disomic chromosome substitution lines of spring wheat in the phytotron and in the field. In: Vose PB, Ruschel AP (eds) Associative $N_2$-fixation, vol I. CRC Press, Boca Raton
Reynders L, Vlassak K (1976) Nitrogen-fixing *Spirillum* species in Belgian soils. Agriculture 24:329–336
Rovira AD (1965) Interactions between plant roots and soil microorganisms. Annu Rev Microbiol 19:241–266
Rubenchik LI (1963) *Azotobacter* and its use in agriculture. Acad Sci Ukraiman SSR. Artman A (Translater) Natl Sci Found, Washington
Ruinen J (1956) Occurrence of *Beijerinckia* species in the "phyllosphere". Nature 177:220–221

Ruinen J (1961) The phyllosphere. I. An ecologically neglected milieu. Plant Soil 15:81–109

Ruinen J (1965) The phyllosphere. III. Nitrogen fixation in the "phyllosphere". Plant Soil 22:375–394

Ruinen J (1970) The phyllosphere. V. The grass sheath, a habitat for nitrogen-fixing microorganisms. Plant Soil 33:661–671

Ruinen J (1974) Nitrogen fixation in the phyllosphere. In: Quispel A (ed) The biology of nitrogen fixation. Elsevier, North-Holland. Biomedical Press, Amsterdam New York

Ruinen J (1975) Nitrogen fixation in the phyllosphere. In: Stewart WDP (ed) Nitrogen fixation in free-living microorganisms. Cambridge Univ Press, London

Ruiz-Argüeso T, Emerich DW, Evans HJ (1979) Hydrogenase system in legume nodules: A mechanism of providing nitrogenase with energy and protection from oxygen damage. J Bacteriol 137:824–829

Ruschel AP, Silva da WJ (1981) $N_2$-fixing populations in seeds and plant parts of Teosoide and Mazoide plants from *Zea mays* × *Zea mexicana* (Teosinte) crosses. In: Vose PB, Ruschel AP (eds) Associative $N_2$-fixation, vol I. CRC Press, Boca Raton

Ruschel AP, Victoria RL, Salati E, Henis Y (1978) Nitrogen fixation in sugarcane (*Saccharum officinarum* L.). Ecol Bull (Stockholm) 26:297–303

Ruschel R, Ruschel AP (1981) Inheritance of $N_2$-fixing ability in sugar cane. In: Vose PB, Ruschel AP (eds) Associative $N_2$-fixation, vol II. CRC Press, Boca Raton

Ruscoe AW, Newcomb EH, Burris RH (1978) Pleomorphic forms in strains of *Spirillum lipoferum* by light and electron microscopy. Pg. 30th Proc Steenbock-Kettering Int Symp Nitrogen Fixation. Univ Madison, Wisc

Sampaio MJAM, Vasconcelos de L, Döbereiner J (1978) Characterization of three groups within *Spirillum lipoferum* Beijerinck. Ecol Bull (Stockholm) 26:364–365

Sampaio MJAM, Silva da EMR, Döbereiner J, Yates MG, Pedrosa de FO (1981) Autotrophy and methylotrophy in *Derxia gummosa, Azospirillum brasilense* and *A. lipoferum*. In: Gibson AH, Newton WE (eds) Current perspectives in nitrogen-fixation. Proc 4th Int Symp Nitrogen Fixation. Canberra Aust Acad Sci, Canberra, Aust

Scott DB, Scott CA, Döbereiner J (1979) Nitrogenase activity and nitrate respiration in *Azospirillum* spp. Arch Microbiol 121:141–145

Shimshick EJ, Herbert RR (1978) Adsorption of rhizobia to cereal roots. Bioch Biophys Res Commun 84:736–742

Silva da MFS, Döbereiner J (1978) Occurrence of *Azospirillum* spp. in soils and roots. In: Döbereiner J, Burris RH, Hollaender A (eds) Limitations and potentials for biological nitrogen-fixation in the tropics. Basic life sciences, vol X. Plenum, New York London

Sloger C, Owens LD (1976) $N_2$ fixation by a temperate corn-*Spirillum* association. 2nd Int Symp $N_2$-fixation, Salamanca (Poster)

Smith RM, Schank SC, Quesenberry KH, Milam JM, Hubbell DH (1977) Research in nitrogen fixation by grasses after inoculation with *Spirillum*. Univ Fla Dep Agron Annu Rep

Smith RM, Thompson DO, Collier JW, Hervey RJ (1954) Soil organic matter, crop yields, and land use in the Texas Blackland. Soil Sci 77:377–388

Stephan MP, Pedrosa de O F, Döbereiner J (1981) Physiological studies with *Azospirillum* spp. In: Vose PB, Ruschel AP (eds) Associative $N_2$-fixation, vol I. CRC Press, Boca Raton

Subba Rao NS (1981) Response of crops to *Azospirillum* inoculation in India. In: Vose PB, Ruschel A (eds) Associative $N_2$-fixation, vol I. CRC Press, Boca Raton

Sylvester-Bradley R (1976) Isolation of acetylene-reducing spirilla from the roots of *Potamogeton filiformis* from Loch Leven (Kinross). J Gen Microbiol 97:129–132

Tarrand JJ, Krieg NR, Döbereiner J (1978) A taxonomic study of the *Spirillum lipoferum* group, with descriptions of a new genus, *Azospirillum* gen. nov. and two species, *Azospirillum lipoferum* (Beijerinck) comb. nov. and *Azospirillum brasilense* sp. nov. Can J Microbiol 24:967–980

Trolldenier G (1977) Influence of some environmental factors on nitrogen fixation in the rhizosphere of rice. Plant Soil 47:203–217

Umali-Garcia M, Hubbell DH, Gaskins MH (1978) Process of infection of *Panicum maximum* and *Spirillum lipoferum*. Ecol Bull (Stockholm) 26:373–379

Umali-Garcia M, Hubbell DH, Gaskins MH, Dazzo FB (1980) Association of *Azospirillum* with grass roots. Appl Environ Microbiol 39:219–226

Vargas MAT (1977) Effect of oxygen on $N_2$ fixation by *Spirillum lipoferum*. MS thesis, Univ Madison, Wisconsin

Vasantarian VN, Bhat JV (1968) Interrelations of microorganisms and mulberry. II. Phyllosphere microflora and nitrogen fixation in leaf and root surfaces. Plant Soil 28:258–267

Vasil IK, Vasil V, Hubbell DB (1977) Engineering plant cell or fungal association with bacteria that fix nitrogen. In: Hollaender A (ed) Genectic engineering for nitrogen-fixation. Plenum, New York London

Villas Boas FCS, Döbereiner J (1981) Nitrogenase and nitrate reductase activities in rice plants inoculated with various *Azospirillum* strains. In: Vose PB, Ruschel AP (eds) Associative $N_2$-fixation, vol II. CRC Press, Boca Raton

Vlassak K, Reynders L (1978a) Associative dinitrogen fixation in temperate regions. In: Proc Isotopes Biol Dinitrogen fixation. IAEA Vienna

Vlassak K, Reynders L (1978b) Conversion of tryptophan to auxins by *Azospirillum* sp. Proc Steenbock-Kettering Int Symp $N_2$-Fixation, Abstr 44. Madison

Volpon AGT, De-Polli H, Döbereiner J (1980) Growth physiology of *Azospirillum*. Acad Bras Cie 52:651

Volpon AGT, De-Polli H, Döbereiner J (1981) Changes in efficiency of nitrogen fixation in various growth stages of batch cultures of *Azospirillum lipoferum*. In: Vose PB, Ruschel AP (eds) Associative $N_2$-fixation, vol I. CRC Press, Boca Raton

Watanabe I, Barraquio WL (1979) Low levels of fixed nitrogen required for isolation of free-living $N_2$-fixing organisms from rice roots. Nature 277:565–566

Watanabe I, Cholitkul W (1979) Field studies on nitrogen fixation in paddy soils. In: Nitrogen and rice. Int Rice Res Inst, Los Baños, Manila, Philippines

Watanabe I, Barraquio WL, Guzman de MR, Cabrera DA (1979) Nitrogen-fixing (acetylene reduction) activity and population of aerobic heterotrophic nitrogen-fixing associated with wetland rice. Appl Environ Microbiol 37:813–819

White JW, Holben FJ, Richer AC (1945) Maintenance level of nitrogen and organic matter in grassland and cultivated soils over periods of 54 and 72 years. J Am Soc Agron 37:21–33

Yoshida T, Ancajas RR (1971) Nitrogen fixation in the root zones of rice. Proc Soil Soc Am 35:156–157

Zablotowicz RM, Focht DD (1979) Denitrification and anaerobic, nitrate-dependent acetylene reduction in cowpea *Rhizobium*. J Gen Microbiol 111:445–448

Zablotowicz RM, Eskew DL, Focht DD (1978) Denitrification in *Rhizobium*. Can J Microbiol 24:757–760

# II.4 Uptake and Reduction of Nitrate: Bacteria and Higher Plants

L. BEEVERS and R.H. HAGEMAN

## 1 Introduction

Although the primary role of plants in converting $CO_2$ into organic forms of carbon for use by heterotrophic forms of life is recognized, the equally important capacity of plants to convert inorganic nitrogen into organic forms is frequently overlooked. While all animals normally utilize organic nitrogen to satisfy their nutritional needs the non-ruminant animals must be supplied with the pre-fabricated essential amino acids, thus the capability of plants to utilize inorganic nitrate nitrogen or ammonium nitrogen and convert it to organic forms is of great importance. The currently accepted pathway for the assimilation of nitrate and ammonia into organic form by green leaves is as follows.

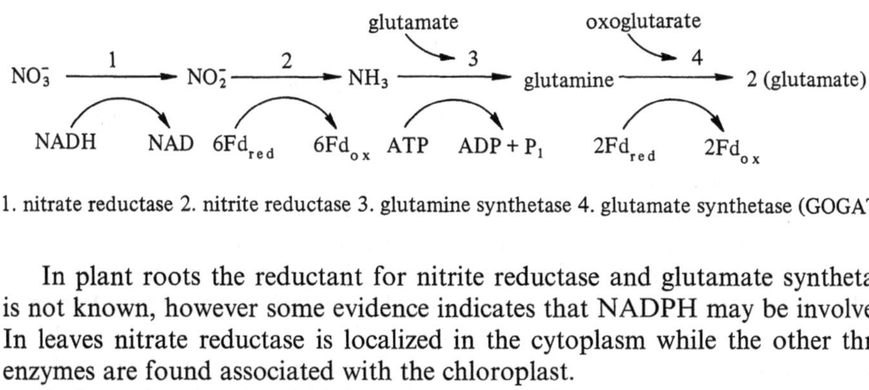


1. nitrate reductase 2. nitrite reductase 3. glutamine synthetase 4. glutamate synthetase (GOGAT).

In plant roots the reductant for nitrite reductase and glutamate synthetase is not known, however some evidence indicates that NADPH may be involved. In leaves nitrate reductase is localized in the cytoplasm while the other three enzymes are found associated with the chloroplast.

In bacteria, algae, and higher plants ammonium nitrogen is converted into organic form primarily by the activities of the enzyme glutamine synthetase and glutamate synthetase (MIFLIN and LEA 1976, TEMPEST et al. 1973). In fungi, incorporation of ammonia into organic form appears to occur via glutamic dehydrogenase (TEMPEST et al. 1973, MIFLIN and LEA 1976, SIMS and FOLKES 1964). Before nitrate nitrogen can be incorporated into organic form it must first be reduced to ammonium. The capacity for nitrate reduction is present in bacteria, fungi, algae and higher plants. Various recent reviews describe the enzymatic reduction of nitrate in higher plants (KESSLER 1976, BEEVERS and HAGEMAN 1980, HEWITT and CUTTING 1979, ZUMFT 1976, GUERRERO et al. 1981, LOSADA et al. 1981); VENNESLAND and GUERRERO (1980) and LOSADA and GUERRERO (1979) review the relationship of nitrate reduction and photosynthesis. PAYNE (1973) and STOUTHAMER (1976) discuss nitrate metabolism in microorganisms.

Because of space limitations, and to minimize overlap with recent reviews, we have restricted coverage in this article to recent findings on the enzymology of nitrate metabolism, but have extended the discussion on nitrate nutrition which, apart from the reviews of HAYNES and GOH (1978), HAGEMAN (1980) and the recent series of studies of the North Carolina group (RUFTY et al. 1982), has received less extensive treatment.

## 2 Available Nitrogen Sources

Although higher plants can utilize gaseous ammonium (HUTCHINSON et al. 1972) and oxides of nitrogen (ROGERS et al. 1979), they obtain the bulk of their nitrogen primarily as nitrate and ammonium ions from the soil. Apart from fertilizer nitrogen and rainfall (lightning and pollution) these ions are derived primarily by mineralization of soil organic matter. The supply of ammonium and nitrate ions varies with the environment, soil type, fertilizer practices and previous cropping history. While plants can absorb soluble organic compounds (e.g. amino acids) from solution, the availability of such compounds to soil-grown plants is considered minimal due to rapid mineralization.

Availability of inorganic nitrogen to the plant can be decreased by leaching of nitrate from the soil or by microbial activity. Inorganic nitrogen (ammonia is the preferred form, JANSSON 1971) utilized by microorganisms for growth is rendered unavailable to the plant. Soil organisms possessing dissimilatory nitrate and nitrite reductive activities release gaseous nitrogen products to the atmosphere (a process termed denitrification).

In contrast to the mobile nitrate anion the ammonium cation may be bound to the negatively charged soil (usually clay) particles. These bound ions are exchangeable and thus available to the plant. The binding of ammonium ions to the clay particles minimizes leaching loss. The retention advantage of ammonium over nitrate ions in most well aerated soils over 5 °C is, however, short-lived because nitrification is rapid. Nitrapyrin (pyridine analog) and related inhibitors of *Nitrosomonas* have been used to prevent nitrification on a commercial scale (PRASAD et al. 1971; HUBER et al. 1977). While the development of a stable highly effective inhibitor of nitrification would appear beneficial with respect to reduction in loss of nitrogen by leaching, the effect on crop production by elimination of nitrate from the soil is debatable (HAGEMAN 1980).

### 2.1 Species Differences in Ammonium and Nitrate Utilization

The differential availability of nitrate and ammonium in the rooting media raises the question of which form of nitrogen permits maximum growth by a particular plant species. Work describing comparative studies aimed at establishing the relative merits of ammonium or nitrate as nitrogen source for plant growth has been reviewed (HAYNES and GOH 1978, HAGEMAN 1980, STREET and SHEAT 1958). Since soils can provide both sources of inorganic nitrogen, reliable assessments of the relative efficiencies of nitrate or ammonium can only be made by using sterile solution or sand culture. However, results from such nutrient culture experiments are frequently difficult to interpret as the removal of nitrate from the medium by the plants creates an alkaline medium, whereas in ammonium medium the external pH declines. The situation is further complicated by the fact that the pH of the media may exert a differential effect on the uptake of ammonium or nitrate and other ions; (VAN DEN DRIESCHE 1978, KIRKBY and HUGHES 1979).

When appropriate precautions are taken to prevent pH fluctuation, it is found that plants vary in their ability to grow on ammonium or nitrate. In general, calcifuge (acid-loving) plants growing naturally in acid soils when little nitrification takes place utilize ammonium in preference to nitrate. In contrast, calcicoles or plants with a wide pH tolerance utilize nitrate preferably (HAYNES and GOH 1978). The differential capability of utilizing nitrate or ammonium in some instances can be related to variations in the potential to produce the enzyme nitrate reductase (STEWART et al. 1974) with plants growing in acidic soil showing limited potential for nitrate reduction.

Although there are demonstrated differences between species in their ability to grow on ammonium or nitrate sources, comparable growth can be frequently obtained on either nitrate or ammonium. However, the growing of plants on ammonium is frequently dependent upon close monitoring of culture conditions. ARNON (1937) demonstrated that ammonium-grown barley plants required higher light intensity, longer photoperiod and additional $Cu^{2+}$ or $Mn^{2+}$ to achieve productivity equivalent to their nitrate-grown counterparts. In view of the less demanding culture conditions for nitrate-grown plants it is generally considered that nitrate is a "safer" nitrogen source for crop growth and production.

With respect to the "safer" aspect, nitrate can be absorbed and stored in various tissues without detriment to the plant (HAGEMAN 1980). Plants also have regulatory mechanisms that minimize nitrate assimilation when they are under stress or are depleted of carbohydrate reserves. Plants grown on nitrate characteristically have high levels of organic acids, predominantly malate, or malate and aconitate. In contrast, plants cannot accumulate ammonium ions as the ammonium ion or its dissociation product ammonia is toxic. Ammonium ions are "detoxified" by union with amino or keto acids to form amides or amino acids. Plants supplied with adequate ammonium salts have low levels of organic acids and reserve carbohydrates. The effects of such differential organic acid contents on plant metabolism are not well studied, however some workers have reported a positive correlation between organic acid level and productivity.

## 2.2 Influence of Ammonium or Nitrate on Cation Uptake

The uptake of ammonium or nitrate is accompanied by changes in accumulation of other ions and this ionic imbalance has been used to account for the differential preference of some plants for ammonium or nitrate. Under conditions of ammonium nutrition there is a decreased intake of cations $Ca^{2+}$ and $Mg^{2+}$ and sometimes $K^+$ (COX and REISENAUER 1973), and plants growing on ammonium frequently develop symptoms of calcium and or magnesium deficiency. Significantly, those species normally adapted for ammonium uptake do not show calcium or magnesium deficiency symptoms when cultivated on ammonium. COX and REISENAUER (1973) attributed the decreased calcium and magnesium uptake during ammonium nutrition to ionic competition at the uptake site. EPSTEIN (1972) suggested that ammonium and potassium may utilize the same transport mechanism. However, the uptake system had only one tenth the affinity for ammonia as for alkali cations.

Nitrate uptake is generally associated with an increased uptake and transport of cations, particularly potassium (JACKSON and WILLIAMS 1968, MINOTTI et al. 1968, BLEVINS et al. 1974, BEN ZIONI et al. 1971). In a nutrient solution where nitrate is the only source of nitrogen, anion (largely $NO_3^-$) exceeds cation uptake. To maintain internal electrical neutrality DIJKSHOORN (1962), suggested that $H^+$ is absorbed or $OH^-$ excreted to balance the excess of $NO_3^-$ absorbed. This transport of $H^+$ or $OH^-$ would create an alkaline external medium and neutralize the charge of the internal nitrate.

KIRKBY and KNIGHT (1977), RAVEN and SMITH (1976), and others indicate that nitrate assimilation produces internal alkalinity according to the following equation.

$$NO_3^- + 8H^+ + 8e^- \rightarrow NH_3 + 2H_2O + OH^-$$

However, DIJKSHOORN (1962) showed that this was valid only for assimilated nitrate that was absorbed as the companion ion of one of the metallic cations (e.g. $K^+$, $Ca^{2+}$ or $Mg^{2+}$). To regulate internal pH, the plant synthesizes organic acids to neutralize the resultant alkalinity (DIJKSHOORN 1962, BEN ZIONI et al. 1971, BLEVINS et al. 1974, ISRAEL and JACKSON 1982). The organic acid synthesized may be predominantly malate, the synthesis of which is regulated by the pH of the cell (DAVIES 1973). In some species, aconitate in addition to malate may be synthesized (NEYRA and HAGEMAN 1976).

## 2.3 Nitrate Uptake

The kinetics of nitrate uptake have not been as extensively studied as those of many other ions. This deficiency probably reflects the lack of a convenient available radioisotope of nitrogen and the fact that the nitrate taken up is metabolized. Measurements of nitrate uptake which have been made have usually involved determinations of nitrate depletion from the surrounding medium. VAN DEN HONERT and HOOYMANS (1955) established that in maize the plot of uptake versus solution nitrate concentration was hyperbolic with a plateau between 0.12 and 0.24 mM; half maximal rate, $K_m$, was at 0.023 mM. Similar hyperbolic relationships between absorption and solution concentration were reported by LYCKLAMA (1963) with a $K_m$ of 0.033 mM. Investigations by RAO and RAINS (1976a) with barley (*Hordeum vulgare*) and by DODDEMA and TELKAMP (1979) with *Arabidopsis thaliana* suggest that nitrate uptake may follow a dual phase relationship with concentration. This is a feature demonstrated for many anions and cations (EPSTEIN 1972). The $K_m$ for nitrate uptake for *Arabidopsis* in Phase I was 0.040 mM and $V_{max}$ at 4.0 μmol $g^{-1}$ $h^{-1}$. For barley at the phase I concentrations below 0.5 mM the $K_m$ was 0.11 mM and $V_{max}$ 2.35 μmol $g^{-1}$ $h^{-1}$. In barley, although nitrate uptake increased at concentrations above 0.5 mM, absorption appeared to be linearly and not curvilinearly related to nitrate concentration. A similar relationship appears in a mutant of *Arabidopsis*. However in the wild-type *Arabidopsis* the Phase II uptake (up to 48 mM $NO_3^-$) had a $K_m$ for nitrate of 25 mM and $V_{max}$ of 700 μmol $g^{-1}$ $h^{-1}$. Although there are reports of an interaction between nitrate uptake and absorption of other anions (EPSTEIN 1953) and a similarity of uptake kinetics, it appears that nitrate uptake is not mediated by the same carrier as other anions. This conclusion is based on the observations that (1) other anions do not always interfere with nitrate uptake (RAO and RAINS 1976a), (2) in nitrogen-deficient plants which show typical chloride uptake nitrate absorption is limited (BLEVINS et al. 1974). A capacity for nitrate uptake develops in the deficient plant during a 3–4 h exposure to nitrate (RAO and RAINS 1976b, BLEVINS et al. 1974, JACKSON et al. 1973, MINOTTI et al. 1968, NEYRA and HAGEMAN 1975). These time-course kinetics for the onset of capacity for nitrate uptake are similar to those for the development of nitrate reductase. The onset of nitrate absorption capacity and buildup of nitrate reductase can be prevented by inhibitors of RNA and protein synthesis (JACKSON et al. 1973). Nitrate uptake is restricted in molybdenum-deficient plants (LYCKLAMA 1963). RAO and RAINS (1976b) also propose the concept of an absorption system positively correlated with nitrate reductase. Collectively these observations have led to the suggestion that nitrate uptake and reduction may be mediated by a common membrane-located protein (BUTZ and JACKSON 1977).

However, the observations that tungstate, which prevents the development of an active nitrate reductase, does not interfere with nitrate uptake (HEIMER and FILNER 1970, 1971, RAO and RAINS 1976b) and the findings that nitrate reductase mutants of *Arabidopsis* can accumulate nitrate (DODDEMA et al. 1978) are inconsistent with the concept of one protein with both carrier and reductase activity.

## 2.4 Influence of Ammonium on Nitrate Uptake and Utilization

Although nitrate uptake is frequently stimulated by and enhances the uptake of readily absorbed cations such as calcium and potassium, it is observed that ammonium may impede the absorption of nitrate in some instances (WEISSMAN 1964, PATE 1973, LYCKLAMA 1963, MINOTTI et al. 1969a, SHEN 1969, DODDEMA et al. 1978, FRITH 1972, FERGUSON and BOLLARD 1969) but not in others (SMITH and THOMPSON 1971, SCHRADER et al. 1972, OREBAMJO and STEWART 1975a). It was originally suggested that the decreased absorption of nitrate in the presence of ammonia was due to an inhibition of nitrate reduction (LYCKLAMA 1963). A lowering of the level of extractable nitrate reductase by ammonium treatment has been reported by SHEN (1969), OREBAMJO and STEWART (1975b),

RADIN (1975) and JOY (1969). A reduced assimilation of nitrate into organic N in the presence of ammonium was reported by FERGUSON (1969) and SCHRADER et al. (1972). Others report that ammonium or products of ammonium assimilation do not interfere with nitrate reductase level (FERGUSON 1969, BEEVERS et al. 1965, MINOTTI et al. 1969a, OAKS et al. 1977). In view of the reduced utilization of nitrate in the presence of ammonium without an alteration in nitrate reductase level it has been suggested (MINOTTI et al. 1969a) that ammonium or to some extent the high acidity adjacent to the cellular boundary membrane resulting from ammonium uptake in excess of nitrate uptake causes an alteration in membrane permeability, thereby restricting the capacity for nitrate absorption.

Despite these conflicting reports on the influence of ammonium on nitrate uptake it is frequently observed that plants grow better on a mixture of inorganic nitrogen sources (SCHRADER et al. 1972, COX and REISENAUER 1973, WEISSMAN 1964, WARNCKE and BARBER 1973). Low levels of ammonium are required for the optimal induction of nitrate reductase and growth of cell tissue cultures of Paul's Scarlet rose (MOHANTY and FLETCHER 1976) and soybeans (BAYLEY et al. 1972).

# 3 Nitrate Reduction

## 3.1 Bacteria

Certain bacteria can utilize nitrate as the sole nitrogen source for the synthesis of all nitrogen-containing components of the cell. The assimilation may occur under both aerobic and anaerobic conditions. In other instances nitrate serves as a terminal electron acceptor under anaerobic conditions and this process has been called nitrate respiration (dissimilation) by PAYNE (1973) and STOUTHAMER (1976). In both cases the end product of nitrate reduction is nitrite. Nitrate reductases from bacteria have been divided into two types – nitrate reductase A, which is membrane-bound and which can use chlorate in addition to nitrate as a substrate, and nitrate reductase B, which is cytoplasmic and is inhibited by chlorate (PICHINOTY and PIECHAUD 1968). Nitrate reductase A plays a role in anaerobic respiration while nitrate reductase B is usually considered assimilatory. However, nitrate reductase B may also serve a respiratory function in facultative anaerobes and it is debatable whether it is always soluble (PICHINOTY 1970).

## 3.2 Dissimilatory Nitrate Reductase

Most of the detailed investigations of bacterial nitrate reductase have been on the A type. The enzyme can be solubilized from the membranes by detergent, organic solvents or by incubating the membranes at alkaline pH at 60 °C

(STOUTHAMER 1976). The solubilized enzymes are characterized as molybdoproteins with iron-sulfur groups. In general the membrane nitrate reductases appear to be composed of two subunits of molecular weight 150,000 and 60,000 (STOUTHAMER 1976, VINCENT and BRAY 1978, VANT RIET et al. 1979). In addition to a membrane-bound nitrate reductase A other bacteria, depending on culture conditions may contain varying amounts of a soluble nitrate reductase (GUERRERO et al. 1973, VILA et al. 1977, DANIEL and GRAY 1976, BURKE and LASCELLES 1979). The soluble enzyme in *Staphylococcus aureus* had a molecular weight of approximately 225,000 and a large subunit of molecular weight 140,000 which has been shown to possess catalytic activity (BURKE and LASCELLES 1979).

Dissimilatory nitrate reductase (Pichinoty Type A) in membrane fractions from bacteria utilizes a variety of respiratory intermediates or reduced pyridine nucleotides for nitrate reduction (STOUTHAMER 1976). In some instances, e.g. *E. coli* and *Klebsiella aerogenes,* nitrate reduction to nitrite is associated with the generation of ATP. In other types, e.g. *Paracoccus denitrificans* and *Bacillus licheniformis,* the reduction of nitrate to nitrite is not accompanied by phosphorylation. In these latter organisms it is considered that phosphorylation is associated with the reduction of nitrite during dissimilatory nitrite reduction (STOUTHAMER 1977, WOOD 1978, VANT RIET et al. 1979).

When respiratory substrates or pyridine nucleotides serve as reductants for dissimilatory nitrate reductase, reduction is inhibited by azide, cyanide and p-chloromercuribenzoate (PCMB). Nitrate reduction mediated by respiratory substrates in some instances is inhibited by n-heptyl hydroxyquinoline-N-oxide (HONO) or dicoumarol. Dissimilatory nitrate reduction by particulate preparations can utilize reduced viologen dyes as electron donors; with the reductants nitrate reduction is insensitive to inhibition by PCMB and (HONO) but is prevented by cyanide and azide. Nitrate reductase assays on solubilized enzymes must be performed with reduced viologen dyes.

When particulate preparations from bacteria grown anaerobically in the presence of nitrate are incubated with respiratory substrates or reduced pyridine nucleotides in the absence of nitrate, a reduced cytochrome b spectrum becomes evident. Cytochrome b is reoxidized when nitrate is added. Thus it appears that the electrons are transferred from these reductants along the respiratory chain components to cytochrome b which in turn transfers electrons to the Mo protein (bacterial nitrate reductase) as shown.

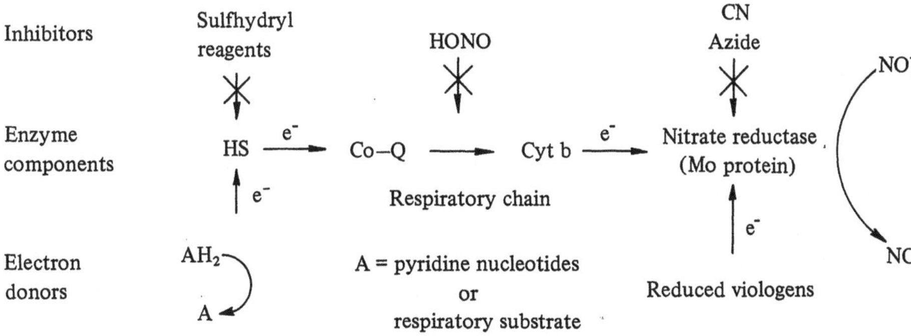

By using reduced viologen dyes as electron donor it has been established that the $K_m$ for nitrate in *K. aerogenes* and *Bacillus licheniformis* is 0.11 mM (Vant Riet et al. 1979).

### 3.3 Assimilatory Nitrate Reduction in Bacteria

Nitrate reductase has been demonstrated by various workers (see Beevers and Hageman 1980) in preparations from photosynthetic bacteria and from the phototrophic bacterium. In the Cyanobacteria, nitrate reductase has been demonstrated in *Anabaena cylindrica, Nostoc muscorum* and *Anacystis nidulans.* Apart from the report of a pyridine nucleotide-dependent nitrate reductase in extracts from *Rhodospirillum rubrum,* most other prokaryotes utilize reduced flavins or reduced ferredoxin. Although a soluble enzyme has been isolated from *Rhodopseudomonas capsulata* with a molecular weight of 180,000, nitrate reductase in these photosynthetic bacteria and cyanobacteria is frequently associated with particles. The enzymes from *Anabaena* and *Anacystis* have been solubilized. The enzyme from *Anacystis nidulans* was ferredoxin-dependent and composed of one polypeptide with molecular weight of 75,000 (Vennesland and Guerrero 1980).

### 3.4 Characterization of Nitrate Reductase from Higher Plants

Although there has been an extensive amount of research on the enzyme since nitrate reductase was first characterized and purified from higher plant extracts by Evans and Nason (1953), there is very little information on its physical properties. This deficiency is partially accounted for by the difficulty experienced in purifying the enzyme (Hageman and Hucklesby 1971). More recent purification techniques have involved affinity chromatography on blue dextran Sepharose or blue Sepharose (Campbell and Smarelli 1978, Notton et al. 1977, Notton and Hewitt 1979, Kuo et al. 1980). These highly purified preparations have a considerably lower specific activity than the enzyme purified from *Chlorella* (Solomonson et al. 1975). The extensively purified enzyme from higher plants had a molecular weight of about 200,000, a sedimentation coefficient of about 8.0, a Stokes radius of about 6.0 nm and a frictional coefficient of 1.55, indicating an assymetric molecule which behaves anomalously on gel filtration (Notton and Hewitt 1979, Kuo et al. 1980). High molecular weights reported on the basis of molecular sieving should be regarded as inaccurate. Sedimentation coefficients of 8.15 have been reported for enzymes prepared from tobacco cell tissue cultures (Wray and Filner 1970) and maize roots (Aslam and Oaks 1976, Wallace and Johnson 1978). A major subunit of molecular weight of about 100,000 was produced during SDS gel electrophoresis of SDS and mercaptoethanol-dissociated nitrate reductase from barley (Kuo et al. 1980). However, a much greater range of polypeptides was produced from spinach (Notton and Hewitt 1979).

Nitrate reduction is mediated by NADH in the majority of higher plant systems (HAGEMAN and HUCKLESBY 1971). The purified enzyme from spinach shows a specific preference for the A type stereisomer of NADH (GUERRERO et al. 1977). Some species are able to utilize NADPH as well as NADH for the reduction of nitrate (EVANS and NASON 1953, HAGEMAN and HUCKLESBY 1971, SHEN et al. 1976, REDINBAUGH and CAMPBELL 1981). This lack of cofactor specificity is due to the presence of a second nitrate reductase which is functional with either NADH or NADPH (CAMPBELL 1976, 1978, JOLLY et al. 1976, SHEN et al. 1976). With soybeans, the NADH nitrate reductase had a molecular weight of 330,000, while the second nitrate reductase which can use either NADH or NADPH had a molecular weight of 220,000 (CAMPBELL 1976, JOLLY et al. 1976). STEWART and OREBAMJO (1979) have reported that the tropical legume *Erythrina senegalensis* has only the non-specific NAD(P)H nitrate reductase. Purified nitrate reductase shows diaphorase activity in which NADH can serve to reduce cytochrome c, ferrocyanide, or other oxidants. Nitrate reduction can be measured independently of the diaphorase by using reduced viologen dyes or reduced flavins. Enzymic nitrate reduction mediated by either reduced pyridine nucleotides or viologen dyes or flavins is inhibited by cyanide or azide, however diaphorase activity is insensitive to these inhibitors (WALLACE and JOHNSON 1978, WRAY and FILNER 1970). Nitrate or cytochrome c reduction mediated by reduced pyridine nucleotides is prevented by p-chloromercuribenzoate and is relatively heat-sensitive; in contrast nitrate reduction mediated by reduced viologens or $FMNH_2$ is insensitive to sulfhydryl inhibitors and is more heat-tolerant (SCHRADER et al. 1968).

The enzyme from higher plants contains flavin (EVANS and NASON 1953). There appears to be variability in the tightness of the association of the flavin with the enzyme. Preparations from some species show strong stimulation of enzymic activity from addition of flavins, whereas preparations from other species demonstrate limited response (HAGEMAN and HUCKLESBY 1971). Kinetic properties of the NADH-dependent enzyme indicate a $K_m$ for nitrate and NADH of 200 μM and 10 μM respectively with optimal activity at pH 7.4. The NAD(P)H-dependent enzyme from soybean (pH 6.5 optimum) and *Erythrina* have characteristically low affinity for nitrate with $K_m$'s for nitrate of 4.5 mM and 8–10 mM having been reported, respectively (CAMPBELL 1976, JOLLY et al. 1976 and STEWART and OREBAMJO 1980). $K_m$'s for NADPH are 15 μM and for NADH 1.4 μM and 3.9 μM.

Purified plant nitrate reductase (CAMPBELL and SMARRELLI 1978 and EAGLESHAM and HEWITT 1975) shows biphasic responses to nitrate, implying negative cooperativity towards nitrate. NADH-dependent nitrate reduction occurs by an ordered ping-pong (by-by) mechanism. This is consistent with reduced pyridine nucleotide being attached and its oxidized form released from a site separated physically from a nitrate-binding site.

Purified nitrate reductase from spinach contains molybdenum (NOTTON and HEWITT 1979) and heme (NOTTON et al. 1977). The heme apparently functions in the enzyme component which shows diaphorase activity (NOTTON et al. 1979). In purified preparations nitrate reduction is stimulated by phosphate (HAGEMAN and HUCKLESBY 1971). The characteristics of nitrate reductase from plants can be represented schematically as follows.

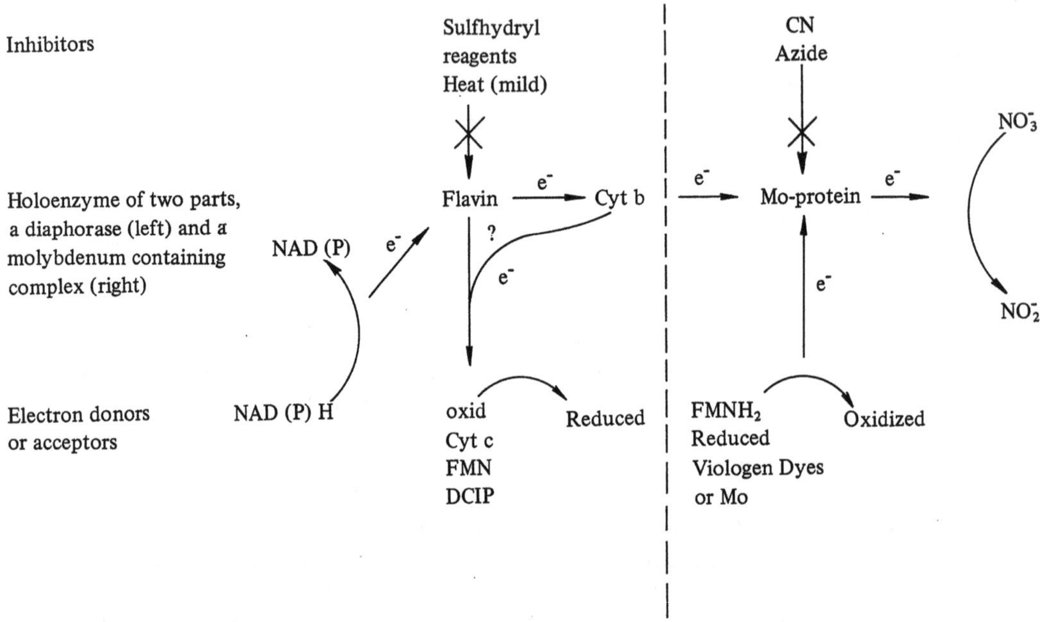


The functional components of the enzyme have not been physically separated in higher plants. However, the observations that tungstate, an analog of molybdenum, inhibits nitrate reductase without impairing diaphorase activity (WRAY and FILNER 1970, RUCKLIDGE et al. 1976) and that an inactive apoprotein with diaphorase activity from molybdenum-deficient plants can be reactivated by mixing with an acid-dissociated product of spinach nitrate reductase are consistent with a two component model. Additionally, the demonstration of tobacco mutants deficient in partial activities of nitrate reductase such as cytochrome c reductase which can be activated by simultaneous extraction with appropriate mutants suggests the in vitro reconstitution of the enzyme from component subunits with partial activities (MENDEL and MULLER 1978).

## 4 Molybdenum in Nitrate Reduction

Although the presence of molybdenum in nitrate reductase from plant extracts has been well documented (HEWITT 1974) and the essentiality of the metal demonstrated from deficiency experiments and tungstate antagonism studies, the mode of operation of molybdenum in reduction is not understood (HEWITT 1974, BEEVERS and HAGEMAN 1980).

Similar uncertainty exists about the role of action of molybdenum in nitrate reduction in bacterial systems (STOUTHAMER 1976).

# 5 Nitrite Reduction

## 5.1 Assimilatory Bacteria

Three different assimilatory nitrite-reducing enzymes have been detected in bacterial extracts (see STOUTHAMER 1976). The bulk of nitrite assimilation appears to be mediated by an NADH-dependent nitrite reductase which has been extensively characterized from *Escherichia coli.* The purified enzyme has a molecular weight of 190,000 and a sedimentation coefficient of 8.5–9.5. On SDS polyacrylamide electrophoresis a subunit of molecular weight of 88,000 is detected. The purified enzyme contains flavin but no detectable heme. Catalylic activity is inhibited by p-chloromercuribenzoate and cyanide. Three molecules of NADH are oxidized for each $NO_2^-$ ion reduced and the product of the reaction is assumed to be ammonium. The enzyme also possesses hydroxylamine reductase activity (COLEMAN et al. 1978).

## 5.2 Dissimilatory Bacteria

When denitrifying bacteria are grown anaerobically on nitrite, nitrogen gas is usually released. However, if cell-free extracts are incubated with nitrite under anaerobic conditions, nitrous oxide and nitric oxide are produced in addition to nitrogen (PAYNE 1973, THAUER et al. 1977). The reduction of nitrite to nitrogen requires both soluble and particulate components and NADH or respiratory intermediates such as succinate or lactate (COX and PAYNE 1973, PAYNE 1973, THAUER et al. 1977). Dissimilation occurs in two steps; nitrite is first reduced to nitric oxide by an apparently soluble dissimilatory nitrite reductase. A second enzmye, nitric oxide reductase, converts nitric oxide to nitrous oxide and a membrane-associated nitrous oxide reductase converts nitrous oxide to nitrogen. The enzymes have not been extensively characterized. Nitrite reductase and nitric oxide reductase usually contain cytochromes which appear to form heme complexes with the nitrogen oxides (PAYNE et al. 1971, COX and PAYNE 1973). The enzymatic conversion of nitrous oxide to nitrogen is inhibited by CO, CN and azide, indicating the involvement of metals; however these have not been characterized (THAUER et al. 1977).

## 5.3 Nitrite Reductase in Plants

Nitrite reductase converts nitrite to ammonium, the six-electron reduction of nitrite occuring without the release of free intermediates. The nitrite reductase from leaves has been extensively purified and shown to be dependent upon reduced ferredoxin as reductant (BEEVERS and HAGEMAN 1980, HEWITT et al. 1976, VENNESLAND and GUERRERO 1980). An enzyme with similar properties has been isolated from non-green tissue (DALLING et al. 1973, HUCKLESBY et al.

1972) although the enzyme from non-green tissue can utilize reduced dyes or ferredoxin as electron donor, the in vivo reductant is not known.

The enzyme, purified from green leaves by a variety of techniques, but more recently by affinity chromatography on ferredoxin-Sepharose (IDA 1977), has a molecular weight of 62,000 and is not dissociated by SDS (IDA 1977, VEGA and KAMIN 1977, HUCKLESBY et al. 1976). Three atoms of iron are present per mol of enzyme, one iron atom is associated with a heme component identified as siroheme (VEGA and KAMIN 1977, HUCKLESBY et al. 1976, MURPHY et al. 1974) and the other two iron atoms are associated with two labile sulfides (i.e. binuclear iron-sulfur center $Fe_2S_2$) (HUCKLESBY et al. 1979, VEGA and KAMIN 1977, APARICIO et al. 1975). In contrast, the work of LANCASTER et al. (1979) indicates that the spinach enzyme has 6 mol of iron and 4 mol of acid-labile sulfur $mol^{-1}$ of siroheme (i.e. the enzyme has a tetranuclear iron-sulfur center).

Electron paramagnetic resonance studies and spectral data indicate the oxidation reduction of iron-sulfur centers and the formation of a NO-heme complex, however other intermediates in the reduction sequence have not been demonstrated (APARICIO et al. 1975, HUCKLESBY et al. 1979, VEGA and KAMIN 1977). The suggested sequence for reduction of nitrite is an initial reduction of siroheme by one electron from ferredoxin which permits the formation of a nitrite–heme complex. Formation of this complex facilitates reduction of the iron-sulfur center. The further transformation of the NO–siroheme to an enzyme $NH_3$ complex involves the addition of five electrons. The following sequence shows that these electrons may be transferred directly from ferredoxin or via the iron center or pairs of electrons may be added simultaneously, one from each source.

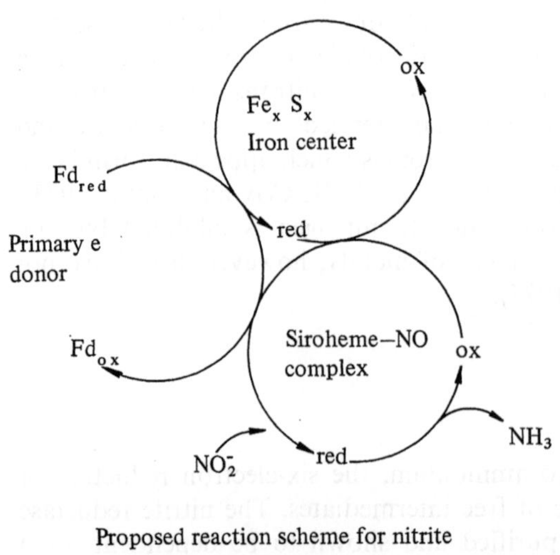


Proposed reaction scheme for nitrite reduction by chlorophyllus tissue.

The sequence of additions and the intermediates formed have not been determined (HUCKLESBY et al. 1979, LANCASTER et al. 1979).

# 6 Location of Enzymes of Nitrate Assimilation in Higher Plants

The observation that living plant cells and organs can be cultured and grown on nitrate as the primary nitrogen source constitutes convincing evidence for the ubiquitous occurrence of nitrate and nitrite reductases in plant tissues. However, in certain organs and tissues the genetic potential for the synthesis of these enzymes may not be expressed as, for example, in the absence of nitrate reductase in roots of *Xanthium pennsylvanicum* (PATE 1973) or the location of nitrate and nitrite reductase in the mesophyll cells but not in the bundle-sheath cells of plants showing $C_4$ photosynthesis (HAREL et al. 1977).

The intracellular localization of nitrate and nitrite reductase has not yet been clearly resolved. The bulk of the studies indicate that nitrite reductase in green leaves is associated with chloroplasts (BEEVERS and HAGEMAN 1980, VENNESLAND and GUERRERO 1980). This chloroplast location is confirmed by the observations that chloroplasts can reduce nitrite in the light (NEYRA and HAGEMAN 1974, MAGALHAES et al. 1974, MIFLIN 1974). A particulate association of nitrite reductase in roots was originally implied from the observation by MIFLIN (1967) and more recently nitrite reductase has been localized from proplastids from roots (DALLING et al. 1972b, EMES and FOWLER 1979).

Nitrate reductase appears to be soluble (HEWITT et al. 1976). Suggestions that the nitrate reductase may be associated with chloroplasts or microbodies in leaf tissue have more recently been interpreted as representing indiscriminate binding of the enzyme to organelle surfaces (DALLING et al. 1972a). However, CAMPBELL and SMARELLI (1979) point out the similarities of certain properties of nitrate reductase and membrane-associated enzymes and suggest that the enzyme is complexed with a membrane from which it is easily solubilized during isolation.

# 7 Provision of Reductant for Nitrate Assimilation in Higher Plants

The reduction of nitrate in photosynthetic tissue has been closely linked to light since the early observations of WARBURG and NEGELEIN (1920) that light stimulated nitrate utilization in *Chlorella*. However, the pyridine nucleotide specificity and the intracellular location of nitrate reductase appears inconsistent with a direct involvement of photosynthetic products (BEEVERS and HAGEMAN 1980, HEWITT et al. 1976).

Current views of the source of NADH for nitrate reduction in leaves have evolved from the use of the in vivo assay for nitrate reductase. Because the assay is conducted under non-physiological conditions (tissue submerged under water, dark, anaerobic) and based upon the accumulation of nitrite, extrapolation of these data to in situ nitrate reduction must be treated with reservations.

The failure of the leaves to assimilate nitrite under dark anaerobic conditions is attributed to the unavailability of reduced ferredoxin. Under in vivo assay conditions in certain species, 3-P-glyceraldehyde and glucose 6-P are more effective than 6-P gluconate or Krebs cycle acids in enhancing nitrite accumulation (KLEPPER et al. 1971, MANN et al. 1978). These results suggest that the oxidation of 3-P-glyceraldehyde rather than the pentose pathway or the mitochondria is the source of NADH for nitrate reduction. ASLAM et al. (1979) demonstrated that barley seedlings can utilize recently fixed photosynthate or stored reserves to reduce nitrate; however they did not indicate the enzyme or enzyme system that generates the reductant. In contrast, MULDER et al. (1959), SAWHNEY et al. (1978), HAGEMAN et al. (1980) and WOO et al. (1980) have shown that addition of various Krebs cycle acids to the in vivo assay enhanced nitrite accumulation, thus implicating mitochondrial involvement in reductant generation. Use of the in vivo assay with roots also results in the accumulation of nitrite (LEE 1979). Because roots lack ferredoxin, such results indicate that the anaerobic conditions impair the generation of reductant for nitrite assimilation.

The reductant for nitrite assimilation in green leaves is ferredoxin. In the light the reduced form is generated via the photosynthetic apparatus. In leaves in the dark, reduced ferredoxin could be produced via the ferredoxin-NADP-oxidoreductase given an ample supply of NADPH. The in situ reductant for root nitrite reductase is not known. BUTT and BEEVERS (1961) found that root supplied with nitrite demonstrated enhanced pentose pathway activity, however in vitro, the root enzyme resembles the leaf enzyme in that it can use reduced ferredoxin or viologen dyes but not NAD(P)H as reductant. Enzymes of the pentose pathway and nitrite reductase are associated with the same particulate fraction from root preparations (EMES and FOWLER 1979) and exposure of roots to nitrate induces a particulate 6-P-gluconate dehydrogenase (EMES et al. 1979). A current hypothesis is that roots contain an unknown compound that serves as an electron carrier between NADPH and nitrite.

# 8 Regulation of Nitrate Reductase in Higher Plants

## 8.1 Substrate

Nitrate reductase is present only at very low levels in most plants (soybean is one exception) not receiving nitrate. The supplying of nitrate results in a great increase in enzyme level. This nitrate-dependent increase in extractable nitrate reductase is prevented by inhibitors of protein and nucleic acid synthesis, indicating that the influence of nitrate reductase is one of induction rather than activation of some inert precursor (HEWITT et al. 1976). This concept was confirmed by ZIELKE and FILNER (1971), who have demonstrated de novo synthesis of the enzyme in response to nitrate.

The requirement for sustained RNA synthesis for induction implies that nitrate may be acting as a coinducer of nitrate reductase in the classical sense proposed by JACOB and MONOD (1961). However, confirmation of this idea has not been provided, and arguments have been proposed that in fungal and bacterial systems the enzyme may be constitutively synthesized at low levels and is synthesized at high levels in the presence of nitrate. The constitutively formed enzyme may be partly responsible for its own repression (STOUTHAMER 1976, COVE and PATEMAN 1969, DANTZIG et al. 1978).

In some instances nitrite may activate a constitutive nitrate reductase component (INGLE et al. 1966, KAPLAN et al. 1978) and organic nitro compounds may stimulate the production of nitrate reductase (SHEN 1972). During rapid vegetative development of cereal crops grown under normal conditions, availability of nitrate in the cells of the leaf blade is a major factor affecting in situ nitrate reduction rate and the diurnal fluctuations of enzyme level (HAGEMAN 1979).

Nitrate reductase was induced in cucumber cotyledons incubated in various organic acids (especially citrate and to a lesser extent by ascorbate) or $(NH_4)_2SO_4$ at pH 3.0 (FERGUSON and KNYPL 1974 and KNYPL and FERGUSON 1975). These compounds were ineffective inducers at pH 6.0. Although the physiological significance of the induction at pH 3.0 is not known (FERGUSON and KNYPL 1974), these results, in conjunction with the observation that ascorbate induces nitrate reductase in *Spirodela oligorrhiza,* indicate that the control of induction of nitrate reductase does not specifically require nitrate under all conditions.

## 8.2 Hormonal

Various growth regulators influence the tissue level of nitrate reductase. In *Agrostemma githago* embryo and cucumber cotyledons the enzyme is induced by cytokinins (HIRSCHBERG et al. 1972, KENDE and SHEN 1972, KNYPL 1973). The growth retardant succinic acid 2-2 dimethyl hydrazide stimulates enzyme production in cucumber cotyledons. Inhibitor studies and density labeling studies demonstrate that enzyme production in response to cytokinins is due to de novo synthesis (HIRSCHBERG et al. 1972, KENDE and SHEN 1972).

## 8.3 Molybdenum

The production of an active nitrate reductase in response to nitrate depends upon a supply of molybdate. Molybdenum-deficient fungi (DANTZIG et al. 1978), higher plants (NOTTON et al. 1974), algae (VEGA et al. 1971) and bacteria (ENOCH and LESTER 1972) or organisms grown in conditions where tungstate or vanadium replaces molybdenum produce aberrant nitrate reductases which possess only partial enzyme activity. The enzyme from molybdenum-deficient plants, which possesses NADH-cytochrome c reductase activity, cross-reacts with antiserum produced against active nitrate reductase. The molybdenum-free apoenzyme can be activated by addition of a molybdenum-containing complex obtained by acid washing the native enzyme (NOTTON and HEWITT 1979).

## 8.4 Ammonium

Ammonium inhibits the nitrate-stimulated production of nitrate reductase in barley roots (SMITH and THOMPSON 1971), *Spirodela oligorrhiza* (FERGUSON 1969), *Lemna minor* (JOY 1969, OREBAMJO and STEWART 1974, 1975a) and in roots but not shoots of cotton plants (RADIN 1975) or in shoot tissues of other higher plants (OAKS et al. 1977). The repression of nitrate reductase in *Lemna minor* is thought to be mediated by some product of ammonium assimilation (e.g. by glutamine rather than by ammonium per se) (STEWART and RHODES 1978). Work with rose tissue culture (MOHANTY and FLETCHER 1980) and mung bean hypocotyls (MOHANTY 1979) also indicated that the ammonium effect on nitrate reductase was mediated via glutamine. When *Spirodela* was supplied with equimolar levels of nitrate and glutamine, nitrate and asparagine or nitrate, the levels of extractable nitrate reductase were suppressed by the presence of glutamine and enhanced by the presence of asparagine relative to the nitrate control (FERGUSON 1970). Utilization of nitrate was largely prevented by the presence of ammonium (FERGUSON 1969) and to a lesser extent by the amino compounds (FERGUSON 1970).

## 8.5 Light

Leaves of plants grown in shaded conditions accumulate nitrate and have low levels of extractable nitrate reductase. When plants cultured in nitrate are transferred to darkness, extractable nitrate reductase declines and during reillumination enzyme level increases. The mechanism by which light influences the nitrate-mediated production of nitrate reductase is not understood, but two proposals have been advanced. The light stimulation may be associated with increased capacity for protein synthesis in illuminated leaves (TRAVIS and KEY 1971). ASLAM et al. (1973) suggest that the main effect of light may be to supply photosynthate to support respiration, which in turn drives the induction process. Alternatively or coincidentally light may regulate the availability of nitrate at the induction site (BEEVERS et al. 1965, JONES and SHEARD 1975). Maximal light stimulation is dependent on the presence of functional chloroplasts (SLUITERS-SCHOLTEN 1975, SAWHNEY and NAIK 1972) and appears to be phytochrome-mediated (JONES and SHEARD 1975, JOHNSON 1976).

## 8.6 Genetic

Nitrate reductase mutants have been identified and characterized in several bacteria (STOUTHAMER 1976). The physiological properties of the mutants have been described and the location, on the circular bacterial "chromosome" of the genes regulating nitrate reductase have been determined. Mutants deficient in various components of the complex nitrate reductase molecule have been identified and in some instances active forms of the enzyme can be formed by mixing components extracted from appropriate mutants (STOUTHAMER 1976).

A similar complex series of mutations affecting nitrate reductase has been described in the fungi *Neurospora* and *Aspergillus* (CODDINGTON 1976, COVE 1970). Mutants deficient in the diaphorase (NADPH cytochrome c reductase) component or lacking the molybdenum cofactor have been characterized and appropriate complementation experiments with *Neurospora* have resulted in the formation of active enzymes (LEE et al. 1974).

Mutants of the alga *Chlamydomonas reinhardii* and the higher plant *Arabidopsis* deficient in their capacity to utilize nitrate have been developed by the use of the mutagen N-methyl-N′-nitro-N-nitrosoguanidine. Biochemical and physiological studies indicate that the mutations variously influence nitrate uptake (SOSA et al. 1978, OOSTINDIER-BRAAKSMA and FEENSTRA 1973), as well as activities of various components of the enzyme (DODDEMA et al. 1978, SOSA et al. 1978). Mutants deficient in partial activities of nitrate reductase have been developed in barley (TOKAREV and SHUMNYI 1977, WARNER et al. 1977) and tobacco (MENDEL and MULLER 1978). The barley mutants characterized by WARNER et al. (1977), although possessing only traces of nitrate reductase in the leaf, still possessed nitrite reductase activity and were able to grow on nitrate as sole nitrogen source. By homogenizing appropriate mutants together, active enzymes could be detected in extracts of tobacco, indicating in vitro reconstitution of the enzyme from components with only partial activities (MENDEL and MULLER 1978).

Varietal variations in the level of extractable nitrate reductase have been demonstrated in many agronomic species. In corn, the variability has been explained on a two-gene model with the genes regulating rate of synthesis and rate of decay (WARNER et al. 1969).

## 8.7 In Vivo Controls

The level of extractable nitrate reductase activity fluctuates in response to drought, and leaf water potential, temperature, light intensity and time of day. These changes may occur without any detectable alterations in endogenous nitrate level. SHANER and BOYER (1976) suggest that it is the flux of nitrate, from the apoplast medium into the cytoplasm, rather than the total nitrate which controls in vivo levels of activity. It is generally considered that variations in extractable activity are due to changes in the level of enzyme protein. The observed turnover of nitrate reductase in tobacco cells (ZIELKE and FILNER 1971) and the detection of proteases that preferentially attack nitrate reductase (WALLACE 1974, 1975) are consistent with enzyme degradation rather than inactivation, accounting for declines in enzyme activity. In some but not all instances a decline in nitrate reductase can be prevented by cycloheximide, implying a requirement for the appropriate protease in order to degrade the reductase.

The reports of reduced enzyme level in response to environmental conditions are based on measurements of activity and not on actual enzyme protein. However, reduction of in vitro activity could equally well be brought about by enzyme inactivation rather than destruction of the enzyme. Although convincing evidence for inactivation of nitrate reductase in algae by NADH and $CN^-$

has been provided, evidence for modulation of activity of higher plant nitrate reductase by changes in oxido-reduction status is not so apparent. In most higher plant systems the presence of reductants appears to stabilize the enzyme (HAGEMAN and HUCKLESBY 1971). Nitrate reductase activity may be modulated by endogenous inhibitors which have been identified in extracts of soybeans (JOLLY and TOLBERT 1978).

Although variations in oxido-reduction status and presence of proteases and inhibitors may be associated with fluctuations in measured levels of extractable nitrate reductase, it is not established to what extent these controls operate in situ. Attempts to estimate the in situ rate of nitrate reduction have been based on comparisons of the amount of reduced nitrogen made available as estimated by the anaerobic in vivo assay of nitrate reductase of the entire seedling with the actual increase in reduced N recorded in the seedling (BRUNETTI and HAGEMAN 1976). CHANTAROTWONG et al. (1976) have devised an aerobic in vivo assay for nitrate reductase based on the difference between total nitrate uptake by the intact plant and the residual nitrate found in the plant over timed intervals. No comparisons were made between the estimated input of reduced N and the actual changes in reduced N. OAKS et al. (1979) indicate that the in vivo assay based on nitrite production under anaerobiosis does not coincide with nitrate reduction based on incorporation from $^{15}NO_3^-$ and it is this latter measurement which most reliably measures nitrate metabolism. With leaves of some species at certain stages of development, products other than nitrite may be produced in the in vivo assay.

## 9 Concluding Thoughts

Current work indicates that nitrate uptake is an active process that may involve a proteinaceous carrier that is rather specific for nitrate. The understanding of the precise mechanism of uptake and the distribution and partitioning of nitrate among the various organs and within the cell between cytoplasm and vacuole is woefully inadequate.

The improvements in enzyme purification and characterization of nitrate reductase and the recent findings that siroheme and non-heme iron centers are the active components of nitrite reductase, and that glutamine rather than glutamate is the first product of ammonia assimilation, have enhanced our understanding of mechanisms and pathways of nitrate assimilation in higher plants. The immediate electron donor to nitrite reductase in roots and the primary sources of energy for regeneration of the reductants for nitrate and nitrite reductase require additional clarification.

Although the outline of the pathway and mode of action of enzymes involved in nitrate assimilation are reasonably well established, it is anticipated that future studies will result in an understanding of the mechanism of the regulation of enzyme level and methods by which manipulation of nitrate reduction may regulate crop productivity.

## References


Aparicio PJ, Knaff DB, Malkin R (1975) Role of an iron-sulfur center and siroheme in spinach nitrite reductase. Arch Biochem Biophys 169:102–107

Arnon DI (1937) Ammonium and nitrate nutrition of barley at different seasons in relation to hydrogen-ion concentration, manganese, copper and oxygen supply. Soil Sci 44:91–120

Aslam M, Oaks A (1976) Comparative studies on the induction and inactivation of nitrate reductase in corn roots and leaves. Plant Physiol 57:572–576

Aslam M, Huffaker RC, Travis LC (1973) The interaction of respiration and photosynthesis in induction of nitrate reductase activity. Plant Physiol 52:137–141

Aslam M, Huffaker RC, Rains DW, Rao KP (1979) Influence of light and ambient carbon dioxide concentration on nitrate assimilation by intact barley seedlings. Plant Physiol 63:1205–1209

Bayley JM, King J, Gamborg OL (1972) The effect of the source of inorganic nitrogen on growth and enzymes of nitrogen assimilation in soybean and wheat cells in suspension cultures. Planta 105:15–24

Beevers L, Hageman RH (1965) The role of light and nitrate in the induction of nitrate reductase in radish cotyledons and maize seedlings. Plant Physiol 40:691–698

Beevers L, Hageman RH (1980) Nitrate and nitrite reduction in the biochemistry of plants. In: Miflin BJ (ed) A comprehensive treatise, vol V. Academic Press, London New York

Beevers L, Schrader LE, Flesher D, Hageman RH (1965) The role of light and nitrate in induction of nitrate reductase in radish cotyledons and maize seedlings. Plant Physiol 40:691–698

Ben Zioni A, Vaadia Y, Lips SH (1971) Nitrate uptake by roots as regulated by nitrate reduction products of the shoot. Physiol Plant 24:288–290

Blevins DG, Hiatt AJ, Lowe RH (1974) The influence of nitrate and chloride uptake on expressed sap pH, organic acid synthesis, and potassium accumulation in higher plants. Plant Physiol 54:82–87

Bollard EG (1966) A comparative study of the ability of organic nitrogen compounds to serve as sole sources of nitrogen for the growth of plants. Plant Soil 25:153–166

Brunetti N, Hageman RH (1976) Comparison of in vivo and in vitro assays of nitrate reductase in wheat (*Triticum aestivum* L.) seedlings. Plant Physiol 58:583–587

Burke KA, Lascelles J (1979) Partial purification and some properties of the *Staphylococcus aureus* cytoplasmic nitrate reductase. J Bacteriol 139:120–125

Butt VS, Beevers H (1961) The regulation of pathways of glucose catabolism in maize roots. Biochem J 80:21–27

Butz RG, Jackson WA (1977) A mechanism for nitrate transport and reduction. Phytochemistry 16:409–417

Campbell WH (1976) Separation of soybean leaf nitrate reductases by affinity chromatography. Plant Sci Lett 7:239–244

Campbell WH (1978) Isolation of NAD(P)H: nitrate reductase from the scutellum of maize. Z Pflanzenphysiol 88:357–361

Campbell WH, Smarelli J (1978) Purification and kinetics of higher plant NADH: nitrate reductase. Plant Physiol 61:611–616

Campbell WH, Smarelli J (1979) Observations on the physical state of NADH: nitrate reductase. Plant Physiol 63 (Suppl) 249

Chantarotwong W, Huffaker RC, Miller BL, Granstedt RC (1976) In vivo nitrate reduction in relation to nitrate uptake, nitrate content and in vitro nitrate reductase activity in intact barley seedlings. Plant Physiol 57:519–522

Coddington A (1976) Biochemical studies on the Nit mutants of *Neurospora crassa*. Mol Gen Genet 145:195–206

Coleman KJ, Cornish-Bowden A, Cole JA (1978) Purification and properties of nitrite reductase from *Escherichia coli* K12. Biochem J 175:483–493

Cove DJ (1970) Control of gene action in *Aspergillus nidulans*. Proc R Soc London Ser B 176:267–275

Cove DJ, Pateman JA (1969) Autoregulation of the synthesis of nitrate reductase in *Aspergillus nidulans*. J Bacteriol 97:1374–1378
Cox GB, Payne WJ (1973) Separation of soluble denitrifying enzymes and cytochromes from *Pseudomonas perfectomarinus*. Can J Microbiol 19:861–872
Cox WJ, Reisenauer HM (1973) Growth and ion uptake by wheat supplied nitrogen as nitrate, or ammonium or both. Plant Soil 38:363–380
Dalling MJ, Tolbert NE, Hageman RH (1972a) Intracellular location of nitrate reductase and nitrite reductase I. Spinach and tobacco leaves. Biochim Biophys Acta 283:505–512
Dalling MJ, Tolbert NE, Hageman RH (1972b) Intracellular location of nitrate reductase and nitrite reductase II. Wheat roots. Biochim Biophys Acta 283:513–519
Dalling MJ, Hucklesby DP, Hageman RH (1973) A comparison of nitrite reductase enzymes from green leaves, scutella and roots of corn (*Zea mays* L.). Plant Physiol 51:481–484
Daniel RM, Gray J (1976) Nitrate reductase from anaerobically grown *Rhizobium japonicum*. J Gen Microbiol 96:247–251
Dantzig AH, Zurowsk WK, Ball TM, Nason A (1978) Induction and repression of nitrate reductase in *Neurospora crassa*. J Bacteriol 133:671–679
Davies DD (1973) Control of and by pH. In: D.D. Davies: Rate control of biological processes. SEB Symp 27:513–530
Dijkshoorn W (1962) Metabolic regulation of the alkaline effect of nitrate utilization of plants. Nature 194:165–166
Doddema H, Telkamp GP (1979) Uptake of nitrate by mutants of *Arabidopsis thaliana* disturbed in uptake or reduction of nitrate II. Kinetics. Physiol Plant 45:332–338
Doddema H, Hoffstra JJ, Feenstra WJ (1978) Uptake of nitrate by mutants of *Arabidopsis thaliana* disturbed in uptake and reduction of nitrate I. Effect of nitrogen source during growth on uptake of nitrate and chlorate. Physiol Plant 43:343–350
Driesche van den R (1978) Response of Douglas fir seedlings to nitrate and ammonium nitrogen sources at different levels of pH and iron supply. Plant Soil 49:607–623
Eaglesham AR, Hewitt EJ (1975) Inhibition of nitrate reductase from spinach (*Spinacea oleracea*) leaf by adenosine nucleotides. Plant Cell Physiol 16:1137–1149
Emes MJ, Fowler MW (1979) The intracellular location of the enzymes of nitrate assimilation in the apices of seedling pea roots. Planta 144:249–253
Emes MJ, Ashihara H, Fowler MW (1979) The influence of $NO_3^-$ on particulate 6-phosphogluconate dehydrogenase activity in pea roots. FEBS Lett 105:370–373
Enoch HG, Lester RL (1972) Effects of molybdate, tungstate, and selenium compounds on formate dehydrogenase and other enzyme systems in *Escherichia coli*. J Bacteriol 110:1032–1040
Epstein E (1953) Absorption of nitrate affected by bromide. Nature 171:83
Epstein E (1972) In: Mineral nutrition of plants, principles, and perspectives. Wiley and Sons, New York
Evans HJ, Nason A (1953) Pyridine nucleotide nitrate reductase from extracts of higher plants. Plant Physiol 28:233–254
Ferguson AR (1969) Nitrogen metabolism of *Spirodela oligorrhiza*. II. Control of the enzymes of nitrate assimilation. Planta 88:353–363
Ferguson AR (1970) Nitrogen metabolism of *Spirodela oligorrhiza*. III. Amino acids and the utilization of nitrate. Planta 90:365–369
Ferguson AR, Bollard EG (1969) Nitrogen metabolism of *Spirodela oligorrhiza* I. Utilization of ammonium nitrate and nitrite. Planta 88:344–352
Ferguson AR, Knypl JS (1974) Specificity of induction of nitrate reductase in plants. Proc 7th Int Colloq, Hannover
Frith GJT (1972) Effect of ammonium nutrition on the activity of nitrate reductase in roots of apple seedlings. Plant Cell Physiol 13:1085–1090
Guerrero MG, Vega JM, Leadbetter E, Losada M (1973) Preparation and characterization of a soluble nitrate reductase from *Azotobacter chroococcum*. Arch Mikrobiol 91:287–304
Guerrero MG, Jetschman K, Volker W (1977) The stereospecificity of nitrate reductase for hydrogen removal from reduced pyridine nucleotides. Biochim Biophys Acta 482:19–26

Guerrero MG, Vega JM, Losada M (1981) The assimilatory nitrate-reducing system and its regulation. Annu Rev Plant Physiol 32:169–204
Hageman RH (1979) Integration of nitrogen assimilation in relation to yield. In: Hewitt EJ, Cutting CV (eds) Nitrate assimilation of plants. Academic Press, London New York
Hageman RH (1980) Effect of form of nitrogen on plant growth. In: Meisinger JJ, Randall GW, Vitosh ML (eds) Nitrification inhibitors – potentials and limitations, Spec Publ Am Soc Agron Madison 38:47–62
Hageman RH, Hucklesby DP (1971) In: San Pietro A (ed) Methods of enzymology, vol 23. Academic Press, London New York
Hageman RH, Reed AJ, Femmer RA, Sherrard JE, Dalling MJ (1980) Some new aspects of the in vivo assay for nitrate reductase in wheat (*Triticum aestivum* L.) leaves. I. Reevaluation of nitrate pool size. Plant Physiol 65:27–32
Harel C, Lea PJ, Miflin BJ (1977) The localization of enzymes of nitrogen assimilation in maize leaves and their activities during greening. Planta 134:195–200
Haynes RJ, Goh KM (1978) Ammonium and nitrate nutrition of plants. Biol Rev 53:465–510
Heimer YM, Filner P (1970) Regulation of the nitrate assimilation pathway of cultured tobacco cells II. Properties of a variant cell line. Biochim Biophys Acta 215:152–165
Heimer YM, Filner P (1971) Regulation of the nitrate assimilation pathway in cultured tobacco cells III. The nitrate uptake system. Biochim Biophys Acta 230:362–372
Hewitt EJ (1974) MTP Int Rev Sci Biochem SerI 11:199–245
Hewitt EJ (1975) Assimilatory nitrate-nitrite reduction. Annu Rev Plant Physiol 26:73–100
Hewitt EJ, Cutting CV (1979) In: Nitrogen assimilation of plants. Academic Press, London New York
Hewitt EJ, Hucklesby DP, Notton BA (1976) In: Bonner J, Varner JE (eds) Plant biochemistry. Academic Press, London New York
Hirschberg K, Hubner G, Borriss H (1972) Cytokinin-induzierte de novo Synthese der Nitratreductase in Embryonen von *Agrostemma githago*. Planta 108:333–337
Honert van den TH, Hooymans JJ (1955) On the absorption of nitrate by maize in water culture. Acta Bot Neerl 4:376–384
Huber VM, Warren HC, Nelson DW, Tsai CY (1977) Nitrification inhibitors – New tools for food production. BioScience 27:523–529
Hucklesby DP, Dalling MJ, Hageman RH (1972) Some properties of two forms of nitrite reductase from corn (*Zea mays* L.) scutellum. Planta 104:220–233
Hucklesby DP, James DM, Banwell MJ, Hewitt EJ (1976) Properties of nitrite reductase from *Cucurbita pepo*. Phytochemistry 15:599–603
Hucklesby DP, Cammack R, Hewitt EJ (1979) Properties and mechanisms of nitrite reductase. In: Hewitt EJ, Cutting CV (eds) Nitrogen assimilation in plants. Academic Press, London New York
Hutchinson GL, Millington RJ, Peters DB (1972) Atmospheric ammonia: Absorption by plants. Science 175:771–772
Ida S (1977) Purification to homogeneity of spinach nitrite reductase by ferredoxin-Sepharose affinity chromatography. J Biochem (Tokyo) 82:915–918
Ingle J, Joy KW, Hageman RH (1966) The regulation of activity of the enzymes involved in the assimilation of nitrate by higher plants. Biochem J 100:577–588
Israel DW, Jackson WA (1982) Ion balance, uptake and transport processes in $N_2$-fixing and nitrate and urea-dependent soybean plants. Plant Physiol 69:171–178
Jackson WA, Williams DC (1968) Nitrate-stimulated uptake and transport of strontium and other cations. Soil Sci Soc Am Proc 32:689–704
Jackson WA, Flesher D, Hageman RH (1973) Nitrate uptake by dark-grown corn seedlings: Some characteristics of apparent induction. Plant Physiol 51:120–127
Jacob F, Monod J (1961) Genetic regulatory mechanisms in the synthesis of proteins. J Mol Biol 3:318–356
Jansson SL (1971) In: McClaren AP, Skujinis J (eds) Soil biochemistry, vol II. Dekker, New York
Johnson CB (1976) Rapid activation by phytochrome of nitrate reductase in the cotyledon of *Sinapis alba*. Planta 128:127–131

Jolly SO, Tolbert NE (1978) NADH-Nitrate reductase inhibitor from soybean leaves. Plant Physiol 62:197–203
Jolly SO, Campbell WH, Tolbert NE (1976) NADPH and NADH nitrate reductases from soybean leaves. Arch Biochem Biophys 174:431–439
Jones RW, Sheard RW (1975) Phytochrome, nitrate movement and induction of nitrate reductase in etiolated pea terminal buds. Plant Physiol 55:954–959
Joy KW (1969) Nitrogen metabolism of *Lemna minor* II. Enzymes of nitrate assimilation and some aspects of their regulation. Plant Physiol 44:849–853
Kaplan D, Mayer AJ, Lips SH (1978) Nitrite activation of nitrate reductase in higher plants. Planta 138:205–209
Kende H, Shen TC (1972) Nitrate reductase in *Agrostemma githago* comparison of the inductive effects of nitrate and cytokinin. Biochim Biophys Acta 286:118–125
Kessler E (1976) Metabolism of inorganic nitrogen compounds. Proc Bot 38:108–117
Kirkby EA, Hughes AD (1979) In: Kirkby EA (ed) Nitrogen nutrition of the plant. Waverley Press, Leeds
Kirkby EA, Knight AH (1977) Influence of the level of nitrate nutrition on ion uptake and assimilation, organic acid accumulation and cation-anion balance in whole tomato plants. Plant Physiol 60:349–353
Klepper LA, Flesher D, Hageman RH (1971) Generation of reduced nicotinamide adenine dinucleotide for nitrate reduction in green leaves. Plant Physiol 48:580–590
Knypl JS (1973) Synergistic induction of nitrate reductase activity by nitrate and benzylamino purine in detached cucumber cotyledons. Z Pflanzenphysiol 70:1–11
Knypl JS, Ferguson AR (1975) pH-Dependent induction of nitrate reductase in cucumber cotyledons by citric acid and other compounds. Z Pflanzenphysiol 74:434–439
Kuo T, Kleinhoff A, Warner RL (1980) Purification and partial characterization of nitrate reductase from barley leaves. Plant Sci Lett 17:371–381
Lancaster JR, Vega JM, Kamin H, Orme-Johnson NR, Orme-Johnson WH, Krueger RJ, Siegel LM (1979) Identification of the iron-sulfur center of spinach ferredoxin nitrite reductase as a tetra nuclear center and preliminary EPR studies of mechanisms. J Biol Chem 254:1268–1272
Lee KY, Pan SS, Erikson R, Nason A (1974) Involvement of molybdenum and iron in the in vitro assembly of assimilatory nitrate reductase utilizing *Neurospora* mutant nit-l. J Biol Chem 249:3941–3952
Lee RB (1979) The release of nitrite from barley roots in response to metabolic inhibitors uncoupling agents and anoxia. J Exp Bot 30:119–133
Losada M, Guerrero MG (1979) The photosynthetic reduction of nitrate and its regulation. In: Barker J (ed) Photosynthesis in relation to model systems. Elsevier/North-Holland Biomedical Press, Amsterdam
Losada M, Guerrero MG, Vega JM (1981) The assimilatory reduction of nitrate. In: Bothe H, Trebst A (eds) Biology of inorganic nitrogen and sulfur. Springer, Berlin Heidelberg New York
Lycklama JC (1963) The absorption of ammonium and nitrate by perennial ryegrass. Acta Bot Neerl 12:361–423
Magalhaes AC, Neyra CA, Hageman RH (1974) Nitrite assimilation and amino nitrogen synthesis in isolated spinach chloroplasts. Plant Physiol 58:411–415
Mann AF, Hucklesby DP, Hewitt EJ (1978) Sources of reducing power for nitrate reduction in spinach leaves. Planta 140:261–263
Mendel RR, Muller AJ (1978) Reconstitution of NADH-nitrate reductase in vitro from nitrate reductase deficient *Nicotiana tabacum* mutants. Mol Gen Genet 161:77–80
Miflin BJ (1967) Distribution of nitrate and nitrite reductase in barley. Nature 214:1133–1134
Miflin BJ (1974) Nitrite reduction in leaves; studies on isolated chloroplasts. Planta 116:187–192
Miflin BJ, Lea PJ (1976) The pathway of nitrogen assimilation in plants. Phytochemistry 15:873–885
Minotti PL, Williams DC, Jackson WA (1968) Nitrate uptake and reduction as affected by calcium and potassium. Soil Sci Am Proc 32:692–698

Minotti PL, Williams DC, Jackson WA (1969a) The influence of ammonium on nitrate reduction in wheat seedlings. Planta 86:267–271
Minotti PL, Williams DC, Jackson WA (1969b) Nitrate uptake by wheat as influenced by ammonium and other cations. Crop Sci 9:9–14
Mohanty B (1979) Effects of ammonium on the activities of nitrate reductase and glutamine synthetase in mung bean hypocotyls. Plant Physiol Suppl 63:253
Mohanty B, Fletcher JS (1976) Ammonium influence on the growth and nitrate reductase activity of Paul's scarlet rose suspension cultures. Plant Physiol 58:152–155
Mohanty B, Fletcher JS (1980) Ammonium influence on nitrogen-assimilating enzymes and protein accumulation in suspension cultures of Paul's scarlet rose. Physiol. Plant 48:453–459
Mulder EG, Boxma R, van Veen WL (1959) The effect of molybdenum and nitrogen deficiencies on nitrate reduction in plant tissues. Plant Soil 10:335–355
Murphy MJ, Siegel LM, Tove SR, Kamin H (1974) Siroheme: A new prosthetic group participating in six electron reduction reactions catalyzed by both sulfite and nitrite reductases. Proc Natl Acad Sci USA 71:612–616
Neyra CA, Hageman RH (1974) Dependence of nitrite reduction on electron transport in chloroplasts. Plant Physiol 54:480–483
Neyra CA, Hageman RH (1975) Nitrate uptake and induction of nitrate reductase in excised corn roots. Plant Physiol 56:692–695
Neyra CA, Hageman RH (1976) Relationship between carbon dioxide, malate and nitrate accumulation and reduction in corn (*Zea mays* L.) seedlings. Plant Physiol 58:726–730
Notton BA, Hewitt EJ (1979) Structure and properties of higher plant nitrate reductase especially *Spinacia oleracea* L. In: Hewitt EJ, Cutting CV (eds) Nitrogen assimilation of plants. Academic Press, London New York
Notton BA, Graf L, Hewitt EJ, Pover RC (1974) The role of molybdenum in the synthesis of nitrate reductase in cauliflower (*Brassica oleracea L.* var Botrytis) and spinach (*Spinacea oleracea* L.). Biochim Biophys Acta 364:45–58
Notton BA, Fido RJ, Hewitt EJ (1977) The presence of functional haem in a higher plant nitrate reductase. Plant Sci Lett 8:165–170
Notton BA, Fido RJ, Watson EF, Hewitt EJ (1979) Presence of haem in the tungsten analogue of nitrate reductase and its relationship to dehydrogenase function. Plant Sci Lett 14:85–90
Oaks A, Aslam M, Boesel I (1977) Ammonium and amino acids as regulators of nitrate reductase in corn roots. Plant Physiol 59:391–394
Oaks A, Stulen I, Boesel IL (1979) Influence of amino acids and ammonium on nitrate reduction in corn seedlings. Can J Bot 57:1824–1828
Oostindier-Braaksma FJ, Feenstra WJ (1973) Isolation and characterization of chlorate resistant mutants of *Arabidopsis thaliana.* Mutat Res 19:175–185
Orebamjo TO, Stewart GR (1974) Some characteristics of nitrate reductase induction in *Lemna minor* L. Planta 117:1–10
Orebamjo TO, Stewart GR (1975a) Ammonium repression of nitrate reductase formation in *Lemna minor* L. Planta 122:27–36
Orebamjo TO, Stewart GR (1975b) Ammonium inactivation of nitrate reductase in *Lemna minor* L. Planta 122:37–44
Pate JS (1973) Uptake, assimilation and transport of nitrogen compounds in plants. Soil Biol Biochem 5:109–119
Payne WJ (1973) Reduction of nitrogenous oxides by microorganisms. Bacteriol Rev 37:409–452
Payne WJ, Riley PS, Cox CD (1971) Separate nitrite, nitric oxide, and nitrous oxide-reducing fractions from *Pseudomonas perfectomarinus.* J Bacteriol 106:356–361
Pichinoty F (1970) Les nitrate-reductases bactériennes IV. Régulation de la biosynthèse et de l'activité de l'enzyme B. Arch Mikrobiol 71:116–122
Pichinoty F, Piechaud M (1968) Detection and characterization of bacterial nitrate reductases A and B. Ann Inst Pasteur 114:77–98
Prasad R, Rajale GB, Lakkhdive BA (1971) Nitrification retarders. Adv Agron 23:337–387

Radin JW (1975) Differential regulation of nitrate reductase induction in roots and shoots of cotton plants. Plant Physiol 55:178–182
Rao KP, Rains DW (1976a) Nitrate absorption by barley. I. Kinetics and energetics. Plant Physiol 57:55–58
Rao KP, Rains DW (1976b) Nitrate absorption by barley II. Influence of nitrate reductase activity. Plant Physiol 57:59–62
Raven JA, Smith FA (1976) Nitrogen assimilation and transport in vascular plants in relation to intracellular pH regulation. New Phytol 76:415–431
Redinbaugh MG, Campbell WH (1981) Purification and characterization of NAD(P)H: nitrate reductase and NADH: nitrate reductase from corn roots. Plant Physiol 68:115–120
Riet Vant J, Wientjes FB, van Doorn J, Planta RJ (1979) Purification and characterization of the respiratory nitrate reductase of *Bacillus licheniformis*. Biochim Biophys Acta 576:347–360
Rogers HH, Campbell JC, Volk RJ (1979) $^{15}NO_2$ uptake and incorporation by *Phaseolus vulgaris* L. Science 206:333–335
Rucklidge G, Notton BA, Hewitt EJ (1976) Reconstitution in vitro of nitrate reductase from apoprotein of molybdenum-deficient spinach. Biochem Soc Trans 4:77–80
Rufty TW, Volk RJ, McClure PR, Israel DW, Raper CD (1982) Relative content of $NO_3^-$ and reduced N in xylem exudate as an indicator of root reduction of concurrently absorbed $^{15}NO_3^{-1}$. Plant Physiol 69:166–170
Sawhney SK, Naik MS (1972) Role of light in the synthesis of nitrate reductase and nitrite reductase in rice seedlings. Biochem J 130:475–485
Sawhney SK, Naik MS, Nicholas DJD (1978) Regulation of NADH supply for nitrate reduction in green plants via photosynthesis and mitochondrial respiration. Biochem Biophys Res Commun 81:1209–1217
Schrader LE, Ritenour GL, Eilrich GL, Hageman RH (1968) Some characteristics of nitrate reductase in higher plants. Plant Physiol 43:930–940
Schrader LE, Domska D, Jung PE, Peterson LA (1972) Uptake and assimilation of ammonium-N and nitrate-N and their influence on the growth of corn (*Zea mays* L.). Agron J 64:690–695
Shaner DL, Boyer JS (1976) Nitrate reductase activity in maize (*Zea mays* L.) leaves. I. Regulation by nitrate flux. Plant Physiol 58:499–504
Shen TC (1969) The induction of nitrate reductase and the preferential assimilation of ammonium in germinating rice seedlings. Plant Physiol 44:1650–1655
Shen TC (1972) Nitrate reductase of rice seedlings and its induction by organic-nitro compounds. Plant Physiol 49:546–549
Shen TC, Funkhouser EA, Guerrero MG (1976) NADH- and NAD(P)H-nitrate reductases in rice seedlings. Plant Physiol 58:292–294
Sims AP, Folkes AP (1964) A kinetic study of the assimilation of $^{15}N$ ammonia and the synthesis of amino acids in an exponentially growing culture of *Candida utilis*. Proc R Soc London Ser B 159:479–502
Sluiters-Scholten CMT (1975) Photosynthesis and the induction of nitrate reductase and nitrite reductase in bean leaves. Planta 123:175–184
Smith FW, Thompson JF (1971) Regulation of nitrate reductase in excised barley roots. Plant Physiol 48:219–223
Solomonson LP, Lorimer GH, Hall RL, Borchers R, Bailey JL (1975) Reduced nicotinamide adenine dinucleotide-nitrate reductase of *Chlorella vulgaris*. Purification, prosthetic groups and molecular properties. J Biol Chem 250:4120–4127
Sosa FM, Ortega T, Bara JL (1978) Mutants from *Chlamydomonas reinhardii* affected in their nitrate assimilation capability. Plant Sci Lett 11:51–58
Stewart GR, Orebamjo TO (1979) Some unusual characteristics of nitrate reduction in *Erythrina senegalensis* D.C. New Phytol 83:311–320
Stewart GR, Rhodes D (1978) Control of enzyme levels in the regulation of nitrogen assimilation. In: Smith H (ed) Regulation of enzyme synthesis and activity. Academic Press, London New York
Stewart GR, Lee JA, Orebamjo TO, Havill DC (1974) Ecological aspects of nitrogen

metabolism. In: Bieleski RL, Ferguson AR, Cresswell MM (eds) Mechanisms of regulation of plant growth. R Soc N Z Bull 12:41–47

Street HE, Sheat DEG (1958) The absorption and availability of nitrate and ammonia. In: Ruhland W (ed) Encyclopedia of plant physiology, vol VIII. Springer, Berlin Göttingen Heidelberg

Stouthamer AJ (1976) Biochemistry and genetics of nitrate reductase in bacteria. Adv Microbiol Physiol 14:315–375

Stouthamer AH (1977) In: Haddock BA, Hamilton EA (eds) Microbial energetics. Cambridge Univ Press, Cambridge

Tempest DW, Meers JL, Brown CM (1973) In: Prusiner S, Stadtman ER (eds) The enzymes of glutamine metabolism. Academic Press, London New York

Thauer RK, Jungerman K, Decker K (1977) Energy conservation in chemotrophic anaerobic bacteria. Bacteriol Rev 41:100–180

Tokarev BI, Shumnyi VK (1977) Clarification of barley mutants with lowered nitrate reductase activity after treatment of the grain with ethylmethanesulfonate. Genetika (Moscow) 13:2097–2103

Travis RL, Key JL (1971) Correlation between polyribosome level and the ability to induce nitrate reductase in dark grown corn seedlings. Plant Physiol 48:617–620

Vega JM, Kamin H (1977) Spinach nitrite reductase. Purification and properties of a siroheme-containing iron-sulfur enzyme. J Biol Chem 252:896–909

Vega JM, Herrera J, Aparicio PJ, Paneque A, Losada M (1971) Role of molybdenum in nitrate reduction in *Chlorella*. Plant Physiol 48:294–299

Vennesland B, Guerrero MG (1980) Reduction of nitrate and nitrite. In: Gibbs M, Latzko E (eds) Photosynthesis II. Encyclopedia of plant physiology. new ser vol 6. Springer, Berlin Heidelberg New York

Vila R, Barcena JA, Llobell A, Paneque A (1977) Characterization of a membrane-bound nitrate reductase from *Azotobacter chroococcum*. Biochem Biophys Res Commun 75:682–688

Vincent SP, Bray RC (1978) Electron paramagnetic resonance studies on nitrate reductase from *Escherichia coli* K12. Biochem J 171:639–647

Wallace W (1974) Purification and properties of a nitrate reductase-inactivating enzyme. Biochim Biophys Acta 341:267–276

Wallace W (1975) Effects of a nitrate reductase inactivating enzyme and NAD(P)H on the nitrate reductase from higher plants and *Neurospora*. Biochim Biophys Acta 377:239–250

Wallace W, Johnson CB (1978) Nitrate reductase and soluble cytochrome c reductase(s) in higher plants. Plant Physiol 61:748–752

Warburg O, Negelein E (1920) The reduction of nitrate in green cells. Biochem Z 110:66–115

Warncke DE, Barber SA (1973) Ammonium and nitrate uptake by corn (*Zea mays* L.) as influenced by nitrogen concentration and $NH_4^+/NO_3^-$ ratio. Agron J 65:950–953

Warner RL, Hageman RH, Dudley JW, Lambert RJ (1969) Inheritance of nitrate reductase activity in *Zea mays*. Proc Natl Acad Sci USA 62:785–792

Warner RL, Lin CJ, Kleinhofs A (1977) Nitrate reductase-deficient mutants in barley. Nature 269:406–407

Weissman GS (1964) Effect of ammonium and nitrate nutrition on protein level and exudate composition. Plant Physiol 39:947–952

Woo KC, Jukinen M, Canvin DT (1980) Reduction of nitrate via a dicarboxylate shuttle in a reconstituted system of supernantant and mitochondria from spinach leaves. Plant Physiol 56:433–436

Wood PM (1978) Periplasmic location of the terminal reductase in nitrite respiration. FEBS Lett 92:214–218

Wray JL, Filner P (1970) Structural and functional relationships of enzyme activities induced by nitrate in barley. Biochem J 119:715–725

Zielke HR, Filner P (1971) Synthesis and turnover of nitrate reductase induced by nitrate in cultured tobacco cells. J Biol Chem 246:1772–1779

Zumft WG (1976) Anorganische Biochemie des Stickstoffs; die Mechanismen der Stickstoffassimilation. Naturwissenschaften 63:457–464

# II.5 Uptake and Reduction of Nitrate: Algae and Fungi

W.R. ULLRICH

## 1 Introduction

Algae have been used for nitrate and nitrite assimilation studies since the beginning of this century (WARBURG and NEGELEIN 1920). In the meantime a large body of knowledge has been accumulated and requires separate treatment in this chapter of the Volume, although many processes in nitrate assimilation are common in algae and higher plants. Some differences are due to the different evolutionary backgrounds of the groups or to different structural properties of the enzymes, but some are merely due to the relatively large surface of contact that algal cells have with the external medium. For reasons of evolution, the structural characteristics and the enzymes of blue-green algae (cyanobacteria) are very different from those of eucaryotes and require special paragraphs. However, the differences between e.g. green algae or diatoms and higher plants

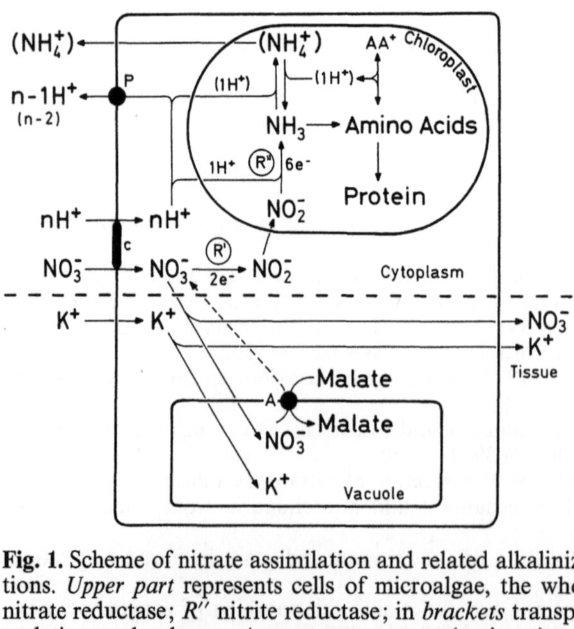


**Fig. 1.** Scheme of nitrate assimilation and related alkalinization under steady state conditions. *Upper part* represents cells of microalgae, the whole scheme vacuolated cells. *R'* nitrate reductase; *R''* nitrite reductase; in *brackets* transport related to ammonium accumulation and release; *A* counter-transport (antiport); *c* co-transport; *P* $H^+$ extrusion pump (ATPase). Protein synthesis in cytoplasm omitted. Stoichiometry of alkalinization in microalgae in steady state: $H^+$ net uptake (or net $OH^-$ release)$=NO_3^-$ uptake $-$ $NO_2^-$ release$+NH_4^+$ release

seem to be much less clearly defined with respect to nitrate assimilation, and many of them can be explained with regard to morphological organization.

As a main nutrient, nitrate belongs to the metabolizable anions, but it can also be stored to a large extent in its original form. The process of nitrate assimilation comprises several steps: uptake, storage in vacuoles, transport to other cells or organs, reduction, and assimilation of the reduced nitrogen. In higher plants this system is normally complete and involves the participation of the whole plant. In microalgae (e.g. blue-green algae or green algae with small single cells) little or no storage of nitrate is possible, because the cells do not contain storage vacuoles, nor is there any transport to other cells. As a result, nitrate uptake and nitrate reduction are tightly coupled in these cells and can easily control each other (Fig. 1). An intermediate situation is encountered in the vacuolate unicellular algae, for example the diatoms, the green algae *Hydrodictyon* or *Platymonas,* and the Characeae, in which translocation from one giant cell to another still plays only a minor role, but where storage in the large vacuoles may account for a large proportion of the nitrate uptake rates under appropriate conditions. In some cases the regulation of nitrate uptake can be separated from that of nitrate reduction, leading to a rather complicated pattern of induction/repression and activation/inactivation processes. In many cases the direct linkage between uptake and reduction and the complicated response to external factors has made investigation very difficult, and has not allowed us to apply the biophysical equations as used for uptake studies with chloride or potassium ions (WALKER 1976, RAVEN 1976).

## 2 Nitrate and Nitrite Reduction in Algae

As in higher plants and fungi, nitrate reduction in algae occurs in two steps, first the reduction of nitrate to nitrite, and then the reduction of nitrite to ammonia. A more detailed description of the processes and the enzymes involved was given by VENNESLAND and GUERRERO (1979) and recent literature was reviewed by GUERRERO et al. (1981), so the following part of this chapter is restricted to a short survey and to selected literature only.

### 2.1 Nitrate Reductase of Eucaryotic Algae

The enzyme shows many properties common with that of higher plants (see Chap. II.4). Whereas NADH is the predominant electron donor in nitrate reduction by higher plants, for several species of algae NADPH has been reported to be as good or better for algal nitrate reductases, for example those from *Ankistrodesmus braunii* (AHMED and SPILLER 1976, DIEZ et al. 1977), *Dunaliella* (LECLAIRE and GRANT 1972), and *Chlamydomonas* (SOSA and CÁRDENAS 1977). According to this specificity, algal nitrate reductase has been coded as EC 1.6.6.1 when NADH-specific, EC 1.6.6.2 when specific for both, and EC 1.6.6.3 when

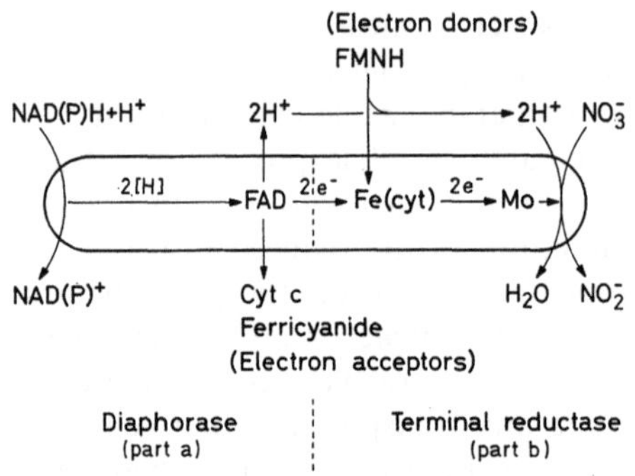


**Fig. 2.** Schematic illustration of electron transport in nitrate reduction, regardless of subunit structure of nitrate reductase. Positions of Fe (cytochrome b-557) and Mo in the chain, as assumed recently (cf. GUERRERO et al. 1981)

specific for NADPH. Generally the preference of the enzyme for one or the other electron donor is only relative. Purification of algal nitrate reductase was achieved more or less completely following classical procedures (for literature see HEWITT 1975, VENNESLAND and GUERRERO 1979, BEEVERS and HAGEMAN 1980), but in some cases serious difficulties of instability remained and did not allow high specific activities. The recent technique of affinity chromatography using blue-dextran sepharose or agarose columns has led to much shorter and easier procedures and to the preparation of homogeneous enzyme proteins in a few steps (SOLOMONSON 1975, DE LA ROSA et al. 1980).

Algal nitrate reductase is characterized as a protein complex of molecular weight between 356,000 (*Chlorella*) and 467,000 (*Ankistrodesmus*) and is composed of a varying number of subunits (between 2 and 8, DE LA ROSA et al. 1981); it contains flavin, molybdenum and heme iron. In nitrate reductase from *Chlorella* the flavin is apparently more tightly bound than in reductase from other plants and algae. The enzyme activities are normally divided into two parts: (a), diaphorase activity concerning the transfer of electrons from NAD(P)H to the flavin group (FAD) of the enzyme, and thence to artificial electron acceptors such as cytochrome c, ferricyanide or others, without reducing nitrate; (b), terminal nitrate reductase activity concerning the transfer of electrons from part (a) or from artificial electron donors such as reduced flavins or methylviologen to nitrate (Fig. 2). The diaphorase moiety of the enzyme has a different sensitivity to various inhibitors, activating or inactivating conditions. It contains the flavin. The terminal nitrate reductase moiety contains the heme iron, in most cases identified as cytochrome b-557, and the molybdenum, and exhibits the typical sensitivity to metal-binding inhibitors. A physical separation of the two enzyme moieties has not been achieved as yet.

In algae, as in higher plants, nitrate reductase is an enzyme with a short half-life of only a few hours, and therefore is subject to a rapid turnover. Hence there are many reports on its induction, mainly by light and nitrate (see reviews by HEWITT 1975, LOSADA and GUERRERO 1979, GUERRERO et al. 1981, and also DIEZ et al. 1977) as well as on its repression by ammonia or amino acid nutrition (MORRIS and SYRETT 1963, and the same reviews). In several species of algae at least part of the nitrate reductase available in a normal cell seems to be constitutive, i.e. its formation does not require special induction (VEGA et al. 1971, DIEZ et al. 1977, HIPKIN and SYRETT 1977) or the formation of an inactive precursor protein occurs even in the presence of ammonium (FUNKHOUSER and

RAMADOSS 1980). In addition to this regulation of amounts by protein synthesis and gene activity, the actual activity of existing nitrate reductase can be drastically reduced by inhibitors. Chlorate acts as a competitive inhibitor, competing with nitrate and forming chlorite, which is more toxic for the cells than nitrite (TROMBALLA and BRODA 1971, SOLOMONSON and VENNESLAND 1972a, VEGA et al. 1972). Azide, cyanate, and thiocyanate are also competitive with nitrate, while cyanide and sulfide inhibit by complexing molybdenum or iron. In the presence of the electron donor NADH, cyanide is very effective both in vivo and in vitro at inhibiting terminal nitrate reductase activity (VEGA et al. 1972, SOLOMONSON and VENNESLAND 1972b). A stoichiometric HCN-nitrate reductase complex is formed and has been studied for *Chlorella vulgaris* (LORIMER et al. 1974). This cyanide complex, however, retains its full diaphorase activity and is stable under reducing conditions. The enzyme can be reactivated easily by oxidants such as ferricyanide, and less efficiently by $O_2$, nitrate or redox dyes, but $O_2$ can also contribute to inactivation (CHAPARRO et al. 1979). Nitrate can prevent the formation of this cyanide-nitrate reductase complex, probably by maintaining the enzyme in an oxidized form. A competitive inhibition of nitrate reduction by nitrite has been reported for the enzyme from *Chlorella vulgaris* in vitro, but the affinity for nitrite was lower than for nitrate (SOLOMONSON and VENNESLAND 1972b). The diaphorase moiety is sensitive to mercuribenzoate (pCMB), an SH-group inhibitor.

Rather complex results have been shown in algae for the in vivo effect of ammonia on nitrate reduction, despite the fact that ammonia does not inhibit the isolated enzyme in vitro. Repression of the formation of nitrate reductase has been postulated for many years (MORRIS and SYRETT 1963), reversible inactivation in vivo only more recently. The inactive enzyme can be isolated from cells and then its reactivation achieved with ferricyanide. Alternatively, reactivation occurs very quickly with blue light in the presence of FAD (APARICIO et al. 1976). There is still some discussion whether the inactive form generated by ammonia inhibition may possibly result from the formation of the aforementioned cyanide complex or not (CHAPARRO et al. 1979). Since inactivation in vitro also occurs by reduction in the presence of NADH and ADP, an uncoupling effect by ammonia in vivo was thought to be responsible, but in some algal strains cyanide was always found after inactivation in vivo. Extensive studies in various species of algae, recently with the blue-green alga *Anacystis nidulans,* showed that cyanide is actually formed in considerable amounts in algal cells, mainly via amino acid oxidases (PISTORIUS et al. 1979).

An irreversibly inactive form of nitrate reductase can be produced with tungstate instead of molybdate. An exchange of molybdenum and tungsten in vitro is not possible, but tungsten is incorporated in vivo, e.g. in *Chlorella* enzyme when added to the nutrient medium instead of molybdenum (VEGA et al. 1971). Vanadium is reported to produce a similar inactive enzyme, but its action is not quite clear yet (HEWITT 1975).

The total activity of nitrate reductase found in algal cells normally considerably exceeds the activity necessary to account for the rates of nitrate assimilation in vivo. $K_m$ values for nitrate shown by nitrate reductase preparations from green algae or diatoms normally resemble those found for higher plants (100 to 200 μM).

### 2.2 Nitrate Reductase in Blue-Green Algae

It differs from that in eucaryotic photosynthetic organisms in various respects. It can be isolated very efficiently by affinity chromatography (MANZANO et al. 1978). Nitrate reductase of *Anabaena* or *Anacystis* is particle-bound, but can be solubilized by some treatments during the isolation procedure. It cannot accept electrons from reduced pyridine nucleotides, but only from single-electron donors (HATTORI and MYERS 1967, MANZANO et al. 1976). According to reconstitution experiments with chlorophyll-containing particles, ferredoxin in the photosynthetic electron transport chain is probably the direct source of electrons for nitrate reduction in these cells (MANZANO et al. 1976). In contrast to the enzyme from other algae and higher plants, nitrate reductase from blue-green algae is a single polypeptide chain with a molecular weight of about 75,000. It is inactivated by ammonia, even in vitro when bound to particles, and by reduction in the light, and it is reactivated by $O_2$ (ORTEGA et al. 1977).

### 2.3 Nitrite Reductase in Algae

Nitrite reductase has been purified to a different extent from various algae as from *Chlorella* (ZUMFT 1972), *Anacystis* (MANZANO et al. 1976), and even from the red alga *Porphyra* (HO et al. 1976). That from *Chlorella* has a molecular weight of 63,000, as is the case for several higher plants. Iron is present in the enzyme molecule in a siroheme group and most probably also (as shown for higher plant nitrite reductase) in an iron-sulfur center. The reductant for nitrite reduction is reduced ferredoxin, which is also a very effective stabilizing agent on sepharose columns during the purification procedure. Reduced ferredoxin can be replaced as the electron donor by flavodoxin, which is found also in vivo (ZUMFT 1972, BOTHE 1977), and by several artificial one-electron donors such as reduced viologens. Although hydroxylamine does not appear as a free intermediate in the reduction from nitrite to ammonia, it can be reduced to ammonia by nitrite reductase at a low rate (CZYGAN 1965, ZUMFT 1972). Algal nitrite reductase is a soluble enzyme except in blue-green algae (e.g. *Anacystis*), where a special treatment is required to dissolve it from the particles. In most algae, nitrite reductase has a higher affinity to its substrate than nitrate reductase, the reported $K_m$ values lying between 20 and 200 (up to 450) μM.

Nitrite reductase of algae is sensitive to a series of inhibitors among which cyanide, carbon monoxide, and sulfhydryl group blockers as pCMB or mersalylate are the most important. Nitrite can protect the enzyme against CO inhibition. The iron-sulfur center is completely reduced only when CO binds to siroheme and prevents its re-oxidation. In vivo in several algae, an inhibition of nitrite and nitrate assimilation by uncouplers as DNP or CCCP was observed (HOFMANN 1972, ULLRICH 1974, LARSSON and ANDERSSON 1982), but not with the isolated enzymes. For more biochemical information see the article by VENNESLAND and GUERRERO (1979) and the preceding (Chap. II.4, this Vol.) by BEEVERS and HAGEMAN.

## 2.4 Location of Nitrate and Nitrite Reduction in Algal Cells

In higher plants, according to DALLING et al. (1972), nitrate reductase seems to be generally located in the cytoplasmic compartment (cytosol), while nitrite reductase is in chloroplasts or other plastids. Similar attempts have been made to localize these enzymes in the algae, but the results were less clear, perhaps due to the much greater difficulties encountered in isolating intact chloroplasts from most algal cells. The limited evidence suggests that nitrate reductase occurs in the cytosol, but the recovery of nitrate reductase in the soluble fraction after homogenization of the cells is, of course, not a final proof that the enzyme is also soluble in vivo. A hypothesis proposed by BUTZ and JACKSON (1977) suggesting that the nitrate uptake carrier located in the plasmalemma also functions as the nitrate reductase is not in accordance with many results from algae, where independent activity and regulation of nitrate uptake and reduction were shown. There is some evidence supporting a chloroplast localization for nitrite reductase in contrast to nitrate reductase, especially the quenching of photosystem II chlorophyll fluorescence by nitrite, not by nitrate, in *Ankistrodesmus* under anaerobic conditions (KESSLER and ZUMFT 1973). This compartmentation has some implications in intracellular transport. Both processes require reducing power, but in a different form. Photosynthetic reducing power can be used directly for nitrite reduction with reduced ferredoxin. Under non-photosynthetic conditions, respiratory reducing power available as NADH must pass the chloroplast envelope by special transport shuttles (HEBER 1974) and then be transformed into reduced ferredoxin. Nitrite also has to pass through the chloroplast envelope from the cytosol, and nitrite reduction causes alkalinization in the chloroplast. In algae the rates of nitrate and especially of nitrite reduction are normally much lower in the dark than in the light, which is consistent with the more complicated transport steps needed to provide reducing power at the site of nitrite reduction and also with the much lower rates of reducing power production in the dark. According to studies with intact chloroplasts (HEBER 1974) the shuttle systems for transport of intermediates have very high affinities to their substrates, so photosynthetic electron transport will not easily become limiting, even for nitrate reduction in the cytoplasmic compartment.

## 2.5 Stoichiometry Between Nitrate Reduction and $O_2$ Exchange

The consumption of reducing power for nitrate (and nitrite) reduction leads to changes in the $O_2/CO_2$ ratios in respiration and photosynthesis. In some cases respiratory $O_2$ consumption and especially $CO_2$ evolution are increased, indicating a change in the metabolic rates, since nitrate reduction will compete with $O_2$ for NADH. Also in photosynthesis the results are variable and differ from species to species. GRANT and TURNER (1969) found a regular stoichiometry between extra-$O_2$ evolution and nitrate assimilation only in *Chlorella pyrenoidosa*, not in five other species from various groups of algae. Similarly MORRIS and AHMED (1969) reported the expected ratio of 1.5 with nitrite only for *Chlorella* and otherwise rather variable ratios, most of them exceeding the expected values of $O_2/NO_3^- = 2$, also in *Ankistrodesmus*. For this alga KESSLER (1957) reported a ratio of 1.5 with nitrite, and ULLRICH and EISELE (1977) found relatively stable ratios with nitrate which were lower (about 1.65) in the absence of $CO_2$, but higher (about 2.1) at $CO_2$ saturation. The varying ratios and the lack of extra-$O_2$ evolution in many algal species shows the involvement of various metabolic processes and limitations when nitrate is reduced.

# 3 Nitrate Uptake in Algae

## 3.1 General Remarks

Unicellular microalgae and giant algal cells differ to some degree in their properties of nitrate uptake, mainly due to the fact that the microalgae are devoid of storage vacuoles, the existing microvacuoles being specialized for the storage of polyphosphates (volutin granula) or lipids. This makes it extremely difficult to measure reliable pool sizes of nitrate in such cells. Many algal species contain vacuoles, but the storage still plays a minor role. In most microalgae, uptake and reduction of nitrate are closely coupled processes which become limiting to each other and which can be separated only under special conditions. Nevertheless, nitrate reduction can sometimes be shown to be not limiting, when the production of reducing power and the activity of nitrate reductase are considerably higher than the observed rates of nitrate assimilation.

On the basis of our present knowledge nitrate uptake has some characteristic properties that are distinctly different from those of isolated nitrate reductase: a far lower $K_m$ for nitrate (0.5 to 50 μM instead of 100 to 200 μM); different sensitivity to various inhibitors such as tungstate, ammonium or methylammonium and other uncouplers; and an immediate response to various conditions, whereas nitrate reductase inactivation or reactivation may take many minutes or even several hours.

As to the use of Michaelis-Menten kinetics in uptake studies, a few general remarks may be required. Nitrate uptake as a function of concentration follows a saturation curve in almost all organisms. This is regarded as characteristic for uptake via a "carrier" or "permease". In many instances the rate limitation for uptake could just as well be due to a limited driving force on $H^+$-anion co-transport, presumably through the electrochemical proton gradient ($\Delta\tilde{\mu}H^+$, see Sect. 3.10). At higher nitrate concentrations the saturation curve may remain unchanged or become linear with nitrate concentration (e.g. SERRA et al. 1978a, for *Skeletonema*); but in some cases high nitrate concentration leads to a serious inhibition of nitrate assimilation, probably due to control by nitrate uptake, especially in the absence of carbon sources.

Another problem in using Michaelis-Menten equations and constants arises from the frequent observation that once a rate of uptake has established at a certain initial concentration, this rate remains constant until almost all nitrate is consumed, irrespective of the $K_m$ calculated from the initial rates obtained over a wide range of concentrations (CALERO et al. 1980, see also figures in TISCHNER and LORENZEN 1979). This requires further investigation and indicates that the uptake system is complex and may involve a binding protein. The same phenomena have been studied for various ions in erythrocytes. There is sometimes interference of uptake with nitrate release from storage vacuoles as in *Platymonas* (RICKETTS and EDGE 1977, ULLRICH and GRÖGER unpublished).

### 3.2 Substrate Affinity

The nitrate affinity of the uptake system differs from species to species, but it varies also with the external conditions and with cell age or pretreatment. FALKOWSKI (1977) presents a model for a bisubstrate kinetics (nitrate and light) as an example to show how various external factors may interfere with nitrate uptake and complicate a kinetic treatment. Thus under optimal conditions $K_m$ may appear much higher than under restricted conditions in the same algal culture. As mentioned before, in microalgae and not only there, nitrate uptake may be directly limited by its consumption within the cells. These problems have led many authors to use the term $K_s$ instead of $K_m$.

There are many reports of $K_s$ values in algae, mainly measured for ecological purposes, and usually in oceanic species. In general there seems to exist a weak inverse correlation between cell size and $K_s$ (EPPLEY et al. 1969). As another general tendency, the freshwater algae (mostly from meso- or eutrophic habitats) show lower affinity (higher $K_s$) to nitrate than those algae which occur in the littoral zones of the sea or in the free ocean. We can speculate that this is related to the much higher nitrate concentrations found in most freshwaters than in the open sea, where nitrate concentrations are so low that the plankton algae mainly feed on other nitrogen sources (EPPLEY et al. 1969). There is, however, little correlation with taxonomy. According to HATTORI (1962) *Anabaena cylindrica,* a freshwater blue-green alga, shows a rather high apparent $K_s$ of 70 µM. A number of green algae species from very different ecological systems have been studied. The highly saline *Dunaliella tertiolecta* shows a medium $K_s$ value of 1.4 µM (EPPLEY et al. 1969), *Chlorella sorokiniana* when adapted to high nutrient levels a $K_s$ of 4.3 µM (TISCHNER and LORENZEN 1979). In some neritic and oceanic species $K_s$ can be as low as 0.2 µM, irrespective of whether they belong to Dinophyceae, centric or pennate diatoms, Coccolithinae or Chrysophyceae (EPPLEY et al. 1969, FALKOWSKI 1975, SERRA et al. 1978a). As mentioned before, some algae show considerable changes in nitrate uptake affinity under various external conditions; for example, the marine diatom *Skeletonema costatum* (EPPLEY et al. 1969), and the seaweeds *Codium fragile* (HANISAK and HARLIN 1978) and *Laminaria longicruris* (HARLIN and CRAIGIE 1978) show a decrease of $K_s$ towards lower temperature, or an increase of $K_s$ with increasing light intensity (GRANT and TURNER 1969, FALKOWSKI 1977). In synchronously grown *Ankistrodesmus braunii* (Chlorophyceae), the $K_s$ for nitrate seems to be extremely variable. In the absence of $CO_2$, the apparent $K_s$ (the concentration for half-saturation of uptake) is about 500 µM (ULLRICH

1974, EISELE 1976), while at $CO_2$ saturation in the light, values below 10 μM were found (EISELE 1976, ULLRICH unpublished data).

### 3.3 Light Dependence

For many years and for many species of algae, light stimulation of nitrate uptake has been reported (KESSLER 1964); however, low nitrate assimilation rates in the dark or at low light intensities do not necessarily reflect a limitation by the uptake system but may be due to the lower rates of intracellular consumption. Correspondingly, light stimulation is most pronounced in microalgae without storage vacuoles. Thus in *Ankistrodesmus* or *Chlorella* light-dark ratios of 20:1 to 50:1 can easily be measured, especially after a short nitrate-starvation period (ULLRICH-EBERIUS 1973, CALERO et al. 1980, Fig. 3). By contrast, phosphate uptake is much less stimulated by light in *Ankistrodesmus,* probably because phosphate is metabolized without reducing power (ULLRICH-EBERIUS 1973). GRANT and TURNER (1969) found a light stimulation of nitrate uptake between 2- and 15-fold in six algal species of different taxonomic groups, but they were not grown in synchronous cultures. With *Anabaena cylindrica* HATTORI (1962) showed a very strong initial light stimulation in nitrate and nitrite uptake, but the rates soon decreased. In *Chlamydomonas reinhardii* light stimulation was nine fold in non-starved cells, but only about three fold in nitrogen-starved cells which generally showed higher uptake rates (THACKER and SYRETT 1972).

Like the $K_s$ values, light saturation rates ($V_{max}$) of nitrate uptake depend on various factors, for example on adaptation to "sun" or "shade" conditions

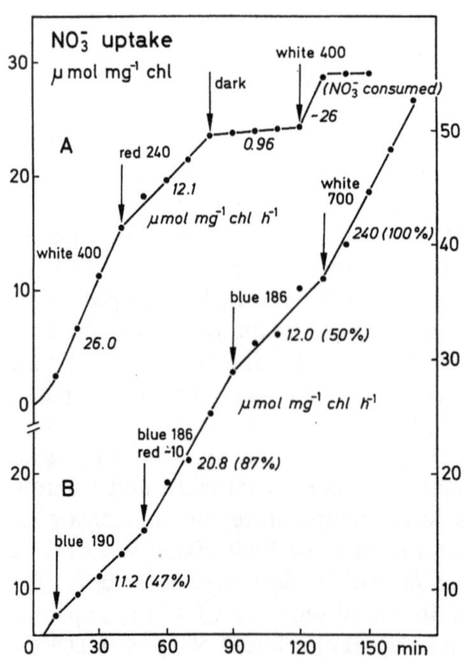


**Fig. 3.** Effect of white and monochromatic light on nitrate uptake in *Chlorella fusca* with immediate response to light changes. Light intensities in W $m^{-2}$; absolute and relative rates (%) indicated in italics. *A* comparison between saturating white light, red light and dark; *B* comparison between saturating blue light, in the absence and presence of complementary weak red light, and saturating white light. (CALERO et al. 1980)

in phytoplankton species (BATES 1976). Light saturation may be reached earlier in the absence of $CO_2$ than in its presence. Use of monochromatic light revealed a complicated light dependence in *Chlorella fusca* (CALERO et al. 1980, see Fig. 3). In the presence of $CO_2$ the rates of nitrate uptake in saturating blue or red light were only about one half of those in saturating white light, but an additional low intensity of the complementary light led to rates approaching those in white light. Nitrate uptake rates responded immediately upon changes from one light quality to the other or to dark. Nitrite or ammonia uptake and photosynthetic $O_2$ evolution were unaffected by monochromatic light at saturation. Also nitrate reductase activity remained unchanged and in excess to that necessary for the in vivo rates of nitrate assimilation. At low light intensities or in the presence of DCMU only blue light stimulated nitrate uptake, suggesting an effect on respiratory metabolism with its energy and reducing power supply (VOSKRESENSKAYA 1972). Studies with nitrate reductase in vitro (ROLDÁN et al. 1978) and the uptake experiments in vivo (CALERO et al. 1980), which suggested that the most effective wavelength was 457 nm, imply that flavin could be involved in the stimulation of the respiration-supplied rates by blue light.

Light stimulation of nitrate uptake can also be accompanied by or due to an increase in nitrate reductase activity, as has been shown for *Chlorella* (TISCHNER and LORENZEN 1979), but such changes are not immediate. They were found after the onset of the light period in synchronous cultures.

### 3.4 pH-Dependence

Most of the few data available in the literature show or suggest a pH optimum for nitrate uptake in the neutral or slightly alkaline range. It is found at pH 7.5 to 8.5 for many species in the light, and at pH 6 in the dark in *Ankistrodesmus* (ULLRICH-EBERIUS 1973), but RIGANO (1978) for good reasons grows his acidophilic unicellular red alga (*Cyanidium caldarium*) at pH 1.9 with nitrate as sole nitrogen source. In synchronously grown *Ankistrodesmus braunii*, pH dependence with the optimum between 7.8 and 9.0 was similar to that of many other species when measured in the absence of $CO_2$, but this pH dependence was completely abolished by $CO_2$ or glucose as a carbon source leading to pH independence between pH 5 and 9 (EISELE and ULLRICH 1975, 1977, Fig. 4 A). It is likely that similar effects can be found also in other species, but most experiments were carried out at the low $CO_2$ concentration of the air. The varying pH effects may reflect the intracellular pH control dependent on energy supply (SMITH and RAVEN 1979), but the internal pH seems to be very stable in some algae (LANE and BURRIS 1981, LARSSON and ANDERSSON 1982).

### 3.5 Dependence on Carbon Sources

There is some interdependence between the influence of pH and that of carbon sources in the light. DAVIS (1953) found considerable stimulation by glucose

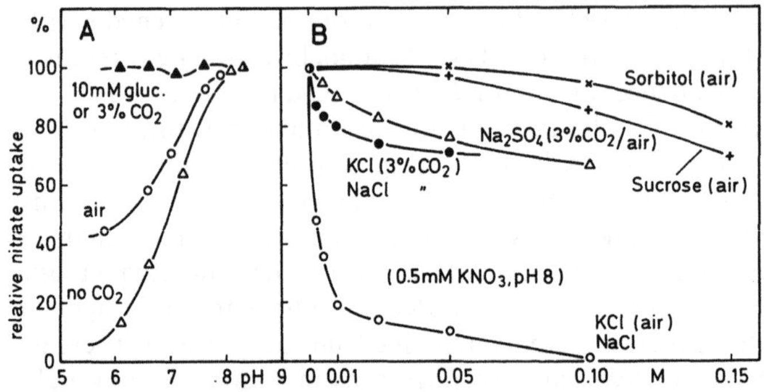


**Fig. 4A, B.** Effect of carbon sources on pH dependence (**A**) and on anion sensitivity (**B**) of nitrate uptake in *Ankistrodesmus braunii.* Relative uptake rates, in **A** related to pH 8, in **B** to 0.5 mM $KNO_3$ in dilute medium, both in the light. (Data for **A** from EISELE and ULLRICH 1977, for **B** from ULLRICH and EISELE 1977)

of nitrate uptake in *Chlorella* at pH 6.5 in the light. MORRIS and AHMED (1969) confirmed this with *Ankistrodesmus,* but not with their strain of *Chlorella.* In *Ankistrodesmus braunii* in the light at pH 5 to 6, $CO_2$ increases the rates of nitrate uptake up to tenfold or more, at pH 8 only 1.5- to 3-fold (EISELE and ULLRICH 1977). Glucose can replace $CO_2$ in *Ankistrodesmus* in that it abolishes pH dependence, but uptake is not enhanced to the same degree. In the dark, nitrate uptake rates are strongly dependent on respiration rates so that under anaerobic conditions they may approach zero after a short time (GRANT and TURNER 1969). $CO_2$ or glucose can also change the sensitivity of nitrate uptake rates to nitrate and chloride at higher concentrations (ULLRICH and EISELE 1977, Fig. 4B) and considerably lower the $K_s$ in *Ankistrodesmus* (cf. Sect. 3.2). On the other hand, carbon sources increase the regulatory inhibition of nitrate uptake by ammonium (EISELE 1976, SYRETT and LEFTLEY 1976). As shown by GRANT and TURNER (1969) the stimulatory effect on nitrate uptake of glucose is restricted to rather few species, probably to those possessing a specific glucose uptake system. For *Chlorella fusca* CALERO et al. (1980) report a five- to tenfold stimulation of nitrate uptake by glucose in the dark. In *Chlamydomonas,* acetate instead of glucose enhances nitrate uptake (THACKER and SYRETT 1972). While the stimulation by $CO_2$ in the light is obviously due to the manyfold increase of all synthetic rates by photosynthesis, organic carbon sources, in the light or in the dark, may contribute to the supply of reducing power, ATP, and carbon skeletons for amino acid formation. There is no corresponding increase in phosphate uptake by $CO_2$ in the light in *Ankistrodesmus* (ULLRICH 1971, 1972), possibly because phosphate is not reduced after it has been taken up.

### 3.6 Inhibition by Anions

Chlorate inhibits nitrate uptake in *Chlorella fusca* in a competitive way (TROMBALLA and BRODA 1971), but an indirect effect by inhibition of nitrate reduction cannot be excluded. Nitrite may, however, act as a competitive inhibitor of nitrate uptake rather than of nitrate reduction, for which higher concentrations of both nitrate and nitrite may be necessary (EPPLEY and ROGERS 1970, HANISAK and HARLIN 1978). A peculiarity in *Ankistrodesmus braunii* is the sensitivity of nitrate assimilation to nitrate at high concentrations in the simultaneous absence of a carbon source. Then nitrate at 30 to 100 mM strongly inhibits nitrate assimilation (ULLRICH 1974). Chloride has the same effect, phosphate less, sulfate is only slightly inhibitory, and mere osmotics are not at all inhibitory (ULLRICH and EISELE 1977, Fig. 4 B). Nitrate reductase in vitro shows almost no response to these salts, and there is no difference between enzymes of freshwater organisms and those of high salt tolerance (HEIMER 1973). Thus the response seems to be due to the uptake system itself.

### 3.7 Inhibition by Ammonia and Amino Compounds

Ammonia is the main alternative nitrogen source for most plants, and it is taken up and assimilated by many algae at similar or even higher rates than nitrate. In many algae, ammonia inhibits or completely suppresses nitrate assimilation. The interpretation is given on three different levels: (a), repression of nitrate reductase on the level of gene action, requiring several hours even in fast-growing algae; (b), reversible inactivation of nitrate reductase in vivo but not in vitro (except in particles from blue-green algae), requiring many minutes up to a few hours; (c), inhibition of nitrate uptake, taking place within a few minutes or even less (PISTORIUS et al. 1976, TISCHNER and LORENZEN 1979, ULLRICH unpublished results). The mechanism of this proposed uptake inhibition is unknown as yet, but a rapid membrane depolarization could be involved, as shown for *Neurospora* (SLAYMAN 1977). In *Ankistrodesmus* the inhibition occurs only in the additional presence of a carbon source. Thus in the absence of $CO_2$ or glucose, this alga can take up nitrate in the light and reduce it to nitrite and ammonia, both of which then accumulate in the medium; the process is only slightly inhibited by addition of external ammonium (EISELE 1976, SYRETT and LEFTLEY 1976). The ammonium inhibition of nitrate uptake in the presence of $CO_2$ is immediately relieved, once most of the ammonium in the medium has been consumed (EISELE 1976, TISCHNER and LORENZEN 1979). In the presence of glucose instead of $CO_2$, the ammonium inhibition is much less pronounced but still visible. Ammonium was less inhibitory in *Codium* (HANISAK and HARLIN 1978). A very clear example of inhibition at the uptake site is shown for the diatom *Phaeodactylum* which accumulates nitrate in the absence of ammonium (CRESSWELL and SYRETT 1979). The special ammonium effect on nitrate uptake must be attributed to some amino acid as the product of ammonium assimilation rather than to ammonium itself (SYRETT and LEFTLEY 1976), or to strong and continuing membrane depolarization (see Sect. 3,10).

### 3.8 Effect of Metabolic Inhibitors and Uncouplers

Most inhibitors effective in reducing nitrate uptake are those which affect the energy metabolism of the cells. They prevent either nitrate reduction by blocking the production of reducing power or else the general energy supply for all uptake processes. Another site of action could be the ATP-dependent proton pump at the plasmalemma. Cyanide, azide, sulfide, and pCMB are also known to be direct inhibitors of nitrate reductase (see Sect. 2 and VENNESLAND and GUERRERO 1979). The inhibition of nitrate and nitrite assimilation by uncouplers (HOFMANN 1972, ULLRICH 1974), which cannot be explained by sensitivity of the enzymes, could reflect a break-down of transport and of the ability of the cells to maintain pH levels (SMITH and RAVEN 1979) but in *Scenedesmus* it was independent of the stable internal pH (LARSSON and ANDERSSON 1982). Tungstate and vanadate apparently do not block nitrate uptake completely (SERRA et al. 1978b), nor do they inhibit or inactivate nitrate reductase in a direct way (see Sect. 2, and Chap. II.4, this Vol.). More recently, nitrobenzaldehyde and derivatives have been found to specifically compete with nitrate at its uptake site (TISCHNER and LORENZEN 1981).

### 3.9 Stoichiometry Between the Uptake of Nitrate and that of Other Ions

In all plants nutrition with nitrate or nitrite produces an alkalinization of the surrounding soil or water medium, whereas with ammonium the environment is acidified. In higher plants and in algae with nitrate storage capacity, stoichiometric relations of this alkalinization are not constant due to simultaneous transport of cations or other anions for storage in the vacuoles or for transport to other parts of the plant (cf. Fig. 1). According to the chemical equations of nitrate and nitrite reduction one proton is consumed per nitrate or nitrite reduced, as long as the ammonia so formed is used for polypeptide synthesis. If ammonium or free amino acids are the accumulated products and ammonium is released to the medium, another proton is consumed to form $NH_4^+$ and hence the alkalinization is increased. Thus the stoichiometry of 1 $OH^-$ for 1 $NO_3^-$, when carbon sources are present to support protein synthesis, increases to almost 2 $OH^-$ for 1 $NO_3^-$ when ammonium is accumulated in the medium. An undisturbed stoichiometry of this type was found in the medium of the microalga *Ankistrodesmus braunii* (EISELE and ULLRICH 1975, 1977). There was no significant concomitant uptake of cations as $K^+$, $Na^+$, or $Ca^{2+}$. The pH regulation in long-term and short-term analysis was calculated in a careful study for the vacuolate alga *Hydrodictyon africanum* (RAVEN and DE MICHELIS 1979).

### 3.10 Transport Mechanism

Such stoichiometric ratios are interesting not only with respect to the biochemistry of nitrate reduction and to pH regulation in the respective compartments of the cells, but also with respect to the transport mechanism. The existence of an alkaline pH optimum, and the excess production of $OH^-$ ions in nitrite reduction, together may suggest that the uptake of nitrate proceeds via a counter-transport of $OH^-$ against $NO_3^-$. So far in unicellular algae no transient pH changes or membrane potential changes have been measured with nitrate, but corresponding membrane potential changes in frond cells of the higher water plant *Lemna gibba* show a transient depolarization of the highly negative interior upon addition of nitrate which is similar to that found upon addition of glucose, amino acids, or phosphate (NOVACKY et al. 1980, ULLRICH-EBERIUS et al. 1981). The nitrate signal is small in non-induced material (NOVACKY et al. 1978), but has a saturation value of 65 mV in induced plants (ULLRICH and NOVACKY 1981). These data obtained with *Lemna* can be best explained by a nitrate-proton co-transport, in which nitrate is transported with an excess of at least one proton over neutralization by a carrier in a passive process following the electrochemical gradient across the plasmalemma, $\Delta\tilde{\mu}H^+$ (inside negative). The excess protons would then be extruded through the plasmalemma by the active, ATP-dependent $H^+$-extrusion pump (Fig. 1). In the case of nitrate, its reduction would contribute by consuming one or even two of the protons transported with the nitrate, so that the $H^+$-extrusion pump would be at least partly relieved of its work. In the marine diatom *Phaeodactylum,* sodium-dependence of nitrate uptake suggests a $Na^+/NO_3^-$ co-transport (REES et al. 1980). On the other hand, FALKOWSKI (1975) reported on the presence of a nitrate- or chloride-activated membrane ATPase in six different species of marine phytoplankton (one species gave negative results) and proposed a mechanism involving directly ATP dependent transport for nitrate.

# 4 Nitrite Uptake in Algae

In contrast to nitrate and ammonium, nitrite is of minor importance for nitrogen nutrition of algae and higher plants, due to its lower stability and its presence only at low concentrations under natural conditions. As an intermediary product of nitrate reduction, it is sometimes accumulated in small amounts within the cells, more often released to the medium and readily taken up again as soon as the conditions have improved for its reduction. In many algae studied, nitrite uptake reached higher rates in the light than nitrate uptake, especially under anaerobic conditions (KESSLER 1957, HATTORI 1962, GRANT and TURNER 1969, MORRIS and AHMED 1969, ULLRICH 1974, CALERO et al. 1980), consistent with the fact that nitrite reduction is more directly linked with photosynthetic electron transport than nitrate reduction (KESSLER and ZUMFT 1973).

In the dark, or in the absence of carbon sources in the light, nitrite is often formed in excess from nitrate and released to the medium (KESSLER 1957, ULLRICH 1974, OHMORI 1978, RIGANO 1978). In most algae studied, the $K_s$ values for the uptake of nitrite are slightly higher than those for nitrate. This implies that in most cases nitrite uptake also follows a saturation kinetics curve, indicating the involvement of a carrier. At high concentrations of nitrite, deviations from this saturation kinetics towards diffusion kinetics may occur (ULLRICH 1974).

When present in combination with nitrate, nitrite plays a variable role. In some cases it inhibits nitrate uptake in a competitive type of reaction, in which the combined uptake rates of both ions may greatly exceed that of nitrate alone (EISELE 1976). For unknown reasons nitrite assimilation in vivo is often more sensitive than nitrate uptake and reduction to uncouplers such as dinitrophenol (KESSLER 1964, HOFMANN 1972), so that dinitrophenol can be used in the dark for the in vivo assay of nitrate reductase (e.g. FISCHER and SIMONIS 1979). The competitive inhibition of nitrate uptake by nitrite and vice versa may lead to the conclusion that both anions are taken up via the same transport system, but in the algae this is unlikely: in *Chlorella fusca* the effects of monochromatic light were limited to nitrate (CALERO et al. 1980) and in many algae the effect of ammonia is less pronounced on nitrite uptake than on nitrate uptake. There are also often differences in pH dependence, but the interpretation of these data has to include the possibility of effects on nitrite reduction. Therefore, the location of nitrite reduction within the chloroplasts may play a role. In *Ankistrodesmus* (ULLRICH 1974) and also in *Chlorella* (CALERO et al. 1980) a non-linear time course of nitrite uptake is observed, with a rapid initial uptake and reduction followed by an almost complete stand-still, followed in turn by a recovery to high rates. This time course can be explained in accordance with experiments in intact chloroplasts of higher plants. In such chloroplasts nitrite causes acidification and thus a drastic but transient inhibition of photosynthetic $O_2$ evolution (PURCZELD et al. 1978), but in *Scenedesmus* no internal pH changes were observed (LARSSON and ANDERSSON 1982).

## 5 General Remarks on Regulation of Nitrate and Nitrate Uptake

If the intracellular nitrate or nitrite pool approaches saturation through the operation of the accumulation process, and at the same time nitrate or nitrite reduction is low or zero, then the transport mechanism will slow down or stop net influxes. If we assume the functioning of a proton co-transport system (see Sect. 3.10), the slow-down of transport may also be due to intracellular acidification or depolarization at rates that cannot be sufficiently counteracted by the active and probably ATP-dependent $H^+$ extrusion pump. Even if we assume the operation of a specific nitrate-transport ATPase as proposed by FALKOWSKI (1975) the energy supply to the pumping system will still be strongly regulatory in ion uptake. Nitrate reductase activity, in turn, is likely to be very dependent on nitrate accumulation by the uptake system, since all known preparations of nitrate reductase have a relatively high $K_m$. Although direct effects of light are possible, the main reason for light stimulation of nitrate or nitrite uptake may be found in the energy supply from photosynthesis or respiration. Even the lower saturation rates in monochromatic light, observed only for nitrate uptake, can be explained in this manner (VOSKRESENSKAYA 1972, CALERO et al. 1980). But in many cases nitrate uptake, especially by algae without storage vacuoles, is certainly regulated by nitrate reduction. This applies also to light stimulation.

Although it is not always easy to separate regulation of nitrate uptake from that of nitrate reduction, nitrate uptake can also be regulated at the level of protein synthesis. During nitrate starvation in *Ankistrodesmus* and *Platymonas,* nitrate uptake capacity increases several-fold during the first 3 h (EISELE 1976, ULLRICH et al. 1981) and this stimulation is sensitive to the protein synthesis inhibitor, cycloheximide (ULLRICH et al. 1981). However, in synchronous *Chlorella* cultures, nitrate uptake is stimulated soon after the start of the light period, whereas the main light-induced increase of nitrate reductase activity follows only after a lag phase (TISCHNER and LORENZEN 1979).

As described before, ammonium when added in the presence of carbon sources, immediately stops nitrate assimilation without affecting nitrate reductase activity in the first 30 to 60 min. Although it has been postulated for a long time that an intermediate of ammonium assimilation rather than ammonium itself must be responsible for this inhibition at the uptake site, no special inhibitor has yet been identified. In *Ankistrodesmus,* in contrast to many other algae, ammonium has no regulatory effect in the absence of carbon sources, but nitrate uptake and reduction still proceed at considerable rates (see Sect. 3.5). In the unicellular red alga, *Cyanidium caldarium,* RIGANO et al. (1979) could completely rule out regulation of nitrate uptake by ammonium by adding methionine sulfoximine, an inhibitor of glutamine synthetase. Like *Ankistrodesmus* in the absence of carbon sources, *Cyanidium* in the presence of this inhibitor and in the simultaneous presence of $CO_2$, shows uptake and reduction of nitrate without ammonium control. The same was found recently with *Anacystis,* and also with azaserine, an inhibitor of glutamate synthase, instead of methionine sulfoximine (FLORES et al. 1980). Some more details on regulation have been dealt with in the preceding parts of this chapter.

## 6 Uptake and Reduction of Nitrate and Nitrite in Fungi

There has been much research on fungal nitrate and nitrite assimilation, which can be only briefly summarized in this chapter. Fungi are interesting organisms, because at least some of them are completely autotrophic in nitrogen nutrition in spite of their heterotrophy in carbon nutrition. They have special uptake systems for nitrate and nitrite subject to genetic or metabolic regulation, nitrate and nitrite reductases with a well-studied regulation system and, for ammonium assimilation, glutamine synthetase, glutamate synthase and also glutamate dehydrogenase. Recent reviews were given by HEWITT (1975), by GARETT and AMY (1978), by BEEVERS and HAGEMAN (1980), and by GUERRERO et al. (1981).

Nitrate reductase has been isolated and purified from various species of fungi, but the most complete picture has been elaborated for *Neurospora crassa* and *Aspergillus nidulans* (GARRETT and NASON 1967, 1969, for more literature see GARRETT and AMY 1978, VENNESLAND and GUERRERO 1979). The procedure is similar to that described for the enzyme of algae or higher plants. More recently, affinity chromatography on blue dextran sepharose columns has been used (GUERRERO and GUTIÉRREZ 1977, DOWNEY and STEINER 1979). The properties of nitrate reductase from *Neurospora* and *Aspergillus* were similar to those from algae and higher plants. The enzyme is a metalloflavoprotein containing FAD, molybdenum, and a hemoprotein identified as cytochrome b-557. It is sensitive to SH-group inhibitors as mercuribenzoates (pCMB), and also to metal-complexing agents as cyanide and azide. The most conspicuous difference to the nitrate reductase of algae and higher plants is the substrate specificity for NADPH (code number EC 1.6.6.3). The enzyme can also catalyze several partial reactions and consists of the two moieties, diaphorase and terminal nitrate reductase. The diaphorase is sensitive to heat and SH-group inhibitors, the terminal reductase to cyanide and azide. The sensitivity to cyanide, sulfide, or azide can be prevented by preincubation of the enzyme with nitrate and can be reversed by oxidizing NADPH via the diaphorase moiety using ferricyanide or cytochrome c as electron acceptors. AMY et al. (1977) proposed a cyclic system in *Neurospora* with four forms of more or less active nitrate reductase based on oxidation or reconstitution of SH-groups and complexing with FAD or NADP. Molecular weights vary betwen 196,000 for *Aspergillus,* where four subunits of 49,000 could be separated (DOWNEY and STEINER 1979) and more than 228,000 for *Neurospora* (GARRETT and NASON 1969), where the enzyme was split into two subunits of 130,000 and 115,000. Genetical control of nitrate reductase has been widely studied in both species by various mutants whose extracts allowed recombination of enzyme activities (see GARRETT and AMY 1978).

Fungal nitrite reductase differs much more from that of photosynthetic organisms than does nitrate reductase. It is specific for NADH or NADPH and requires FAD, but does not accept electrons from reduced ferredoxin, probably because ferredoxin is not usually available in these heterotrophic organisms. Fungal nitrite reductase has a higher molecular weight (290,000), perhaps due to the additional diaphorase, similar to that of nitrate reductase, by which electrons from two-electron donors become available. Over-reduction in the presence of FAD and NAD(P)H as well as oxidation, e.g. by hydrogen peroxide, can lead to inactivation of the enzyme, which is generally labile in vitro and must be stabilized by addition of FAD and dithionite. The terminal part of the enzyme that transfers the electrons to nitrite, contains siroheme, as shown by spectral analysis and comparison with bacterial sirohemes (VEGA et al. 1975). Correspondingly, nitrite reductase is sensitive to cyanide, CO, and other iron-complexing agents. CO was shown to compete with $NO_2^-$ as a ligand in siroheme. Also the reduction of nitrite, a six-electron process, seems to occur at a siroheme-$NO_2^-$ complex.

An independent nitrate uptake system in *Neurospora crassa* was reported to cause 50-fold accumulation of nitrate within the mycelium against the medium with a $K_s$ of 0.25 mM for nitrate (SCHLOEMER and GARRETT 1974). The uptake system was apparently only formed in the presence of nitrate or nitrite; its formation was inhibited by casamino acids, cycloheximide, puromycin, and 6-methylpurine. Ammonium and nitrite are non-competitive inhibitors of nitrate uptake only, but uptake of both nitrate and nitrite is inhibited by metabolic poisons such as cyanide and dinitrophenol, indicating that both processes are somehow related to active transport, but do not proceed via the same carrier. *Neurospora* mutants devoid of nitrate reductase still show normal uptake kinetics, thus demonstrating that uptake and reduction are completely independent processes (SCHLOEMER and GARRETT 1974).

Also in fungi regulation of nitrate and nitrite reduction seems to act at several levels. A series of genes was found to be responsible for nitrate and nitrite reductase activities (COVE 1966, GARRETT and AMY 1978). Ammonium may act as a repressor in *Aspergillus* (COVE 1966), nitrate and nitrite as inducers. In some fungal species, ammonium and few amino acids promote the decay of nitrate reductase in vivo and even more that of the nitrate uptake system. This effect can be prevented by protein synthesis inhibitors (LEWIS and FINCHAM 1970, GOLDSMITH et al. 1973). This suggests the involvement of special proteins in the enzyme decay, probably proteolytic inactivators (SORGER et al. 1978). In *Neurospora*, the repression of nitrate reductase seems to be not due to ammonia itself but to glutamine (PREMAKUMAR et al. 1979). In addition to the genetic control, strong and rapid membrane depolarization upon addition of ammonium, as shown also in *Neurospora* (SLAYMAN 1977), could lead to a break-down of anion-proton cotransport and thus cause some of the ammonium effects by direct inhibition of nitrate uptake.

Also in fungi light may contribute to the regulation of nitrate reduction. Methylviologen-nitrate reductase activity in *Neurospora crassa* is stimulated by blue light, whereas the NADPH-nitrate reductase activity is decreased (KLEMM and NINNEMANN 1979), an observation that can be explained by light sensitivity of the flavin and siroheme constituents of the enzyme.

## References

Ahmed J, Spiller H (1976) Purification and some properties of the nitrate reductase from *Ankistrodesmus braunii*. Plant Cell Physiol 17:1–10

Amy NK, Garrett RH, Anderson MB (1977) Reactions of the *Neurospora crassa* nitrate reductase with NAD(P) analogs. Biochim Biophys Acta 480:83–95

Aparicio PJ, Roldán JM, Calero FS (1976) Blue light reactivation of nitrate reductase from green algae and higher plants. Biochem Biophys Res Commun 70:1071–1077

Bates SS (1976) Effects of light and ammonium on nitrate uptake by two species of estuarine phytoplankton. Limnol Oceanogr 21:212–218

Beevers L, Hageman RH (1980) Nitrate and nitrite reduction. In: Stumpf PK, Conn EE (eds) The biochemistry of plants, vol V. Academic Press, London New York

Bothe H (1977) Flavodoxin. In: Trebst A, Avron M (eds) Encyclopedia of plant physiology, New Ser, vol V. Springer, Berlin Heidelberg New York

Butz RG, Jackson WA (1977) A mechanism for nitrate transport and reduction (review). Phytochemistry 16:409–417

Calero F, Ullrich WR, Aparicio PJ (1980) Regulation by monochromatic light of nitrate uptake in *Chlorella fusca*. In Senger H (ed) The blue light syndrome. Springer, Berlin Heidelberg New York, pp 411–421

Chaparro A, Rosa de la MA, Vega JM (1979) Involvement of oxygen in *Chlorella fusca* nitrate reductase inactivation by reduced nicotinamide adenine dinucleotide. Z Pflanzenphysiol 95:77–85

Cove DJ (1966) The induction and repression of nitrate reductase in the fungus *Aspergillus nidulans*. Biochim Biophys Acta 113:51–56
Cresswell RC, Syrett PJ (1979) Ammonium inhibition of nitrate uptake by the diatom, *Phaeodactylum tricornutum*. Plant Sci Lett 14:321–325
Czygan FC (1965) Zur Frage der Zwischenprodukte der Nitratreduktion bei Grünalgen. Planta 64:301–311
Dalling MJ, Tolbert NE, Hageman RH (1972) Intracellular location of nitrate reductase and nitrite reductase. Biochim Biophys Acta 283:505–512
Davis EA (1953) Nitrate reduction by *Chlorella*. Plant Physiol 28:539–544
Diez J, Chaparro A, Vega JM, Relimpio AM (1977) Studies on the regulation of assimilatory nitrate reductase in *Ankistrodesmus braunii*. Planta 137:231–234
Downey RJ, Steiner FX (1979) Further characterization of the reduced nicotinamide adenine dinucleotide phosphate: nitrate oxidoreductase in *Aspergillus nidulans*. J Bacteriol 137:105–114
Eisele R (1976) Photosynthetische Nitrataufnahme und Nitratreduktion bei *Ankistrodesmus braunii*. Ph D thesis, Tech Hochsch Darmstadt
Eisele R, Ullrich WR (1975) Stoichiometry between photosynthetic nitrate reduction and alkalinisation by *Ankistrodesmus braunii* in vivo. Planta 123:117–123
Eisele R, Ullrich WR (1977) Effect of glucose and $CO_2$ on nitrate uptake and coupled $OH^-$ flux in *Ankistrodesmus braunii*. Plant Physiol 59:18–21
Eppley RW, Rogers JN (1970) Inorganic nitrogen assimilation of *Ditylum brightwellii*, a marine plankton diatom. J Phycol 6:344–351
Eppley RW, Rogers JN, McCarthy JJ (1969) Half-saturation constants for uptake of nitrate and ammonium by marine phytoplankton. Limnol Oceanogr 14:912–920
Falkowski PG (1975) Nitrate uptake in marine phytoplankton: Comparison of half-saturation constants from seven species. Limnol Oceanogr 20:412–417
Falkowski PG (1977) A theoretical description of nitrate uptake kinetics in marine phytoplankton based on bisubstrate kinetics. J Theor Biol 64:375–379
Fischer S, Simonis W (1979) Tagesperiodische Schwankungen und lichtinduzierte Zunahme der Nitratreduktase-Aktivität bei Synchronkulturen von *Ankistrodesmus braunii*. Z Pflanzenphysiol 92:143–152
Flores E, Guerrero MG, Losada M (1980) Short-term ammonium inhibition of nitrate utilization by *Anacystis nidulans* and other cyanobacteria. Arch Microbiol 128:137–144
Funkhouser EA, Ramadoss CS (1980) Synthesis of nitrate reductase in *Chlorella*. II. Evidence for synthesis in ammonia-grown cells. Plant Physiol 65:944–948
Garrett RH, Amy NK (1978) Nitrate assimilation in fungi. Adv Microb Physiol 18:1–65
Garrett RH, Nason A (1967) Involvement of a b-type cytochrome in the assimilatory nitrate reductase of *Neurospora crassa*. Proc Natl Acad Sci USA 58:1603–1610
Garrett RH, Nason A (1969) Further purification and properties of *Neurospora* nitrate reductase. J Biol Chem 244:2870–2882
Goldsmith J, Livoni JP, Norberg CL, Segel IH (1973) Regulation of nitrate uptake in *Penicillium chrysogenum* by ammonium ion. Plant Physiol 52:362–367
Grant BR, Turner IM (1969) Light-stimulated nitrate assimilation in several species of algae. Comp Biochem Physiol 29:995–1004
Guerrero MG, Gutiérrez M (1977) Purification and properties of the NAD(P)H: nitrate reductase of the yeast *Rhodotorula glutinis*. Biochim Biophys Acta 482:272–285
Guerrero MG, Vega JM, Losada M (1981) The assimilatory nitrate-reducing system and its regulation. Annu Rev Plant Physiol 32:169–204
Hanisak MD, Harlin MM (1978) Uptake of inorganic nitrogen by *Codium fragile* subspec. *tomentosoides* (Chlorophyta). J Phycol 14:450–453
Harlin MM, Craigie JS (1978) Nitrate uptake by *Laminaria longicruris* (Phaeophyceae). J Phycol 14:464–467
Hattori A (1962) Light-induced reduction of nitrate, nitrite and hydroxylamine in a blue-green alga, *Anabaena cylindrica*. Plant Cell Physiol 3:355–369
Hattori A, Myers J (1967) Reduction of nitrate and nitrite by subcellular preparations of *Anabaena cylindrica*. II. Reduction of nitrate to nitrite. Plant Cell Physiol 8:327–337

Heber U (1974) Metabolite exchange between chloroplasts and cytoplasm. Annu Rev Plant Physiol 25:393–421

Heimer YM (1973) The effects of sodium chloride, potassium chloride and glycerol on the activity of nitrate reductase of a salt-tolerant and two non-tolerant plants. Planta 113:279–281

Hewitt EJ (1975) Assimilatory nitrate-nitrite reduction. Annu Rev Plant Physiol 26:73–100

Hipkin CR, Syrett PJ (1977) Some effects of nitrogen-starvation on nitrogen and carbohydrate metabolism in *Ankistrodesmus braunii*. Planta 133:209–214

Ho CH, Ikawa T, Nisizawa K (1976) Purification and properties of a nitrate reductase from *Porphyra yezoensis* Ueda. Plant Cell Physiol 17:417–430

Hofmann A (1972) Über die energetische Kopplung der Nitratassimilation von Grünalgen an Atmung und Photosynthese. Planta 102:72–84

Kessler E (1957) Untersuchungen zum Problem der photochemischen Nitratreduktion in Grünalgen. Planta 49:505–523

Kessler E (1964) Nitrate assimilation by plants. Annu Rev Plant Physiol 15:57–72

Kessler E, Zumft WG (1973) Effect of nitrite and nitrate on chlorophyll fluorescence in green algae. Planta 111:41–46

Klemm E, Ninnemann H (1979) Nitrate reductase – a key enzyme in blue light-promoted conidiation and absorbance change of *Neurospora*. Photochem Photobiol 29:629–632

Lane AE, Burris JE (1981) Effects of environmental pH on the internal pH of *Chlorella pyrenoidosa, Scenedesmus quadricauda,* and *Euglena mutabilis*. Plant Physiol 68:439–442

Larsson C-M, Andersson M (1982) Uptake and photoreduction of $NO_3^-$ and $NO_2^-$ in *Scenedesmus*: Interactions with $CO_2$ fixation. In: Akoyunoglou G (ed) Photosynthesis Vol IV, Proc 5th Int Congr Photosynthesis, Halkidiki, Greece

LeClaire JA, Grant BR (1972) Nitrate reductase from *Dunaliella tertiolecta*, purification and properties. Plant Cell Physiol 13:899–907

Lewis CM, Fincham JRS (1970) Regulation of nitrate reductase in the basidiomycete *Ustilago maydis*. J Bacteriol 103:55–61

Lorimer GH, Gewitz HS, Völker W, Solomonson LP, Vennesland B (1974) The presence of bound cyanide in the naturally inactivated form of nitrate reductase of *Chlorella vulgaris*. J Biol Chem 249:6074–6079

Losada M, Guerrero MG (1979) The photosynthetic reduction of nitrate and its regulation. In: Barber J (ed) Photosynthesis in relation to model systems. Amsterdam, Elsevier, pp 365–408

Manzano C, Candau P, Gómez-Moreno C, Relimpio AM, Losada M (1976) Ferredoxin-dependent photosynthetic reduction of nitrate and nitrite by particles of *Anacystis nidulans*. Mol Cell Biochem 10:161–169

Manzano C, Candau P, Guerrero MG (1978) Affinity chromatography of *Anacystis nidulans* ferredoxin-nitrate reductase and NADP reductase on reduced ferrodoxin-sepharose. Anal Biochem 90:408–412

Morris I, Ahmed J (1969) The effect of light on nitrate and nitrite assimilation by *Chlorella* and *Ankistrodesmus*. Physiol Plant 22:1166–1174

Morris I, Syrett PJ (1963) The development of nitrate reductase in *Chlorella* and its repression by ammonium. Arch Mikrobiol 47:32–41

Novacky A, Fischer E, Ullrich-Eberius CI, Lüttge U, Ullrich WR (1978) Membrane potential changes during transport of glycine as a neutral amino acid and nitrate in *Lemna gibba* G1. FEBS Lett 88:264–267

Ohmori M (1978) Nitrite excretion by a blue-green alga, *Oscillatoria rubescens* DC. Arch Hydrobiol 83:485–493

Ortega T, Rivas J, Cárdenas J, Losada M (1977) Metabolic interconversion of ferrodoxin-nitrate reductase and NADP reductase of *Nostoc muscorum*. Biochem Biophys Res Commun 78:185–193

Pistorius EK, Gewitz HS, Voss H, Vennesland B (1976) Reversible inactivation of nitrate reductase in *Chlorella vulgaris* in vivo. Planta 128:73–80

Pistorius EK, Jetschmann K, Voss H, Vennesland B (1979) The dark respiration of

*Anacystis nidulans*. Production of HCN from histidine and oxidation of basic amino acids. Biochim Biophys Acta 585:630–642
Premakumar R, Sorger GJ, Gooden D (1979) Nitrogen metabolite repression of nitrate reductase in *Neurospora crassa*. J Bacteriol 137:1119–1126
Purczeld P, Chon CJ, Portis AR, Heldt HW, Heber U (1978) The mechanism of the control of carbon fixation by the pH in the chloroplast stroma. Studies with nitrite-mediated proton transfer across the envelope. Biochim Biophys Acta 501:488–498
Raven JA (1976) Transport in algal cells. In: Lüttge U, Pitman MG (eds) Transport in plants II. Encyclopedia of plant physiology new ser, vol 2. Springer, Berlin Heidelberg New York
Raven JA, DeMichelis MI (1979) Acid-base regulation during nitrate assimilation in *Hydrodictyon africanum*. Plant Cell Environ 2:245–257
Rees TAV, Cresswell RC, Syrett PJ (1980) Sodium-dependent uptake of nitrate and urea by a marine diatom. Biochim Biophys Acta 596:141–144
Ricketts TR, Edge PA (1977) The effect of nitrogen refeeding of starved cells of *Platymonas striata* Butcher. Planta 134:169–176
Rigano C (1978) Studies in vivo on the control by ammonia of nitrate reduction to nitrite in the unicellular alga *Cyanidium caldarium*. Plant Sci Lett 13:301–307
Rigano C, DiMartino-Rigano V, Vona V, Fuggi A (1979) Glutamine synthetase activity, ammonia assimilation and control of nitrate reduction in the unicellular red alga *Cyanidium caldarium*. Arch Microbiol 121:117–120
Roldán JM, Calero F, Aparicio PJ (1978) Photoreactivation of spinach nitrate reductase: role of flavins. Z Pflanzenphysiol 90:467–474
Rosa de la MA, Diez J, Vega JM, Losada M (1980) Purification and properties of assimilatory NAD(P)H-nitrate reductase from *Ankistrodesmus braunii*. Eur J Biochem 106:249–256
Rosa de la MA, Vega JM, Zumft WG (1981) Composition and structure of assimilatory nitrate reductase from *Ankistrodesmus braunii*. J Biol Chem 256:5814–5819
Schloemer RH, Garrett RH (1974) Nitrate transport system in *Neurospora crassa*. J Bacteriol 118:259–269
Serra JL, Llama MJ, Cadenas E (1978a) Nitrate utilization by the diatom *Skeletonema costatum*. I. Kinetics of nitrate uptake. Plant Physiol 62:987–990
Serra JL, Llama MJ, Cadenas E (1978b) Nitrate utilization by the diatom *Skeletonema costatum*. II. Regulation of nitrate uptake. Plant Physiol 62:991–994
Slayman CL (1977) Energetics and control of transport in *Neurospora*. In: Jungreis AM, Hodges TK, Kleinzeller A, Schultz SG (eds) Water relations in membrane transport in plants and animals. Academic Press, London New York
Smith FA, Raven JA (1979) Intracellular pH and its regulation. Annu Rev Plant Physiol 30:289–312
Solomonson LP (1975) Purification of NADH-nitrate reductase by affinity chromatography. Plant Physiol 56:853–855
Solomonson LP, Vennesland B (1972a) Nitrate reductase and chlorate toxicity in *Chlorella vulgaris* Beijerinck. Plant Physiol 50:421–424
Solomonson LP, Vennesland B (1972b) Properties of a nitrate reductase of *Chlorella*. Biochim Biophys Acta 267:544–557
Sorger GJ, Premakumar R, Gooden D (1978) Demonstration in vitro of two intracellular inactivators of nitrate reductase from *Neurospora*. Biochim Biophys Acta 540:33–47
Sosa FM, Cárdenas J (1977) NADPH as electron donor for nitrate reduction in *Chlamydomonas reinhardi*. Z Pflanzenphysiol 85:171–175
Syrett PJ, Leftley JW (1976) Nitrate and urea assimilation by algae. In: Sunderland N (ed) Perspectives in experimental biology, vol II. Botany. Pergamon, Oxford New York
Thacker A, Syrett PJ (1972) The assimilation of nitrate and ammonium by *Chlamydomonas reinhardi*. New Phytol 71:423–433
Tischner R, Lorenzen H (1979) Nitrate uptake and nitrate reduction in synchronous *Chlorella*. Planta 146:287–292

Tischner R, Lorenzen H (1981) Nitrate uptake and reduction in *Chlorella* – Characterization of nitrate uptake in nitrate-grown and nitrogen-starved *Chlorella sorokiniana*. In: Bothe H, Trebst A (eds) Biology of inorganic nitrogen and sulfur. Springer, Berlin Heidelberg New York

Tromballa HW, Broda E (1971) Das Verhalten von *Chlorella fusca* gegenüber Perchlorat und Chlorat. Arch Mikrobiol 78:214–223

Ullrich WR (1971) Nitratabhängige nichtcyclische Photophosphorylierung bei *Ankistrodesmus braunii* in Abwesenheit von $CO_2$ und $O_2$. Planta 100:18–30

Ullrich WR (1972) Der Einfluß von $CO_2$ und pH auf die $^{32}P$-Markierung von Polyphosphaten und organischen Phosphaten bei *Ankistrodesmus braunii* im Licht. Planta 102:37–54

Ullrich WR (1974) Die nitrat- und nitritabhängige photosynthetische $O_2$-Entwicklung in $N_2$ bei *Ankistrodesmus braunii*. Planta 116:143–152

Ullrich WR, Eisele R (1977) Relations between nitrate uptake and nitrate reduction in *Ankistrodesmus braunii*. In: Thellier M, Monnier A, Demarty M, Dainty J (eds) Echanges ioniques transmembranaires chez les végétaux. Colloq CNRS

Ullrich WR, Novacky A (1981) Nitrate-dependent membrane potential changes and their induction in *Lemna gibba* G1. Plant Sci Lett 22:211–217

Ullrich WR, Schmitt H-D, Arntz E (1981) Regulation of nitrate uptake in green algae and duckweeds. Effect of starvation and induction. In: Bothe H, Trebst A (eds) Biology of inorganic nitrogen and sulfur. Springer, Berlin Heidelberg New York

Ullrich-Eberius CI (1973) Beziehungen der Aufnahme von Nitrat, Nitrit und Phosphat zur photosynthetischen Reduktion von Nitrat und Nitrit und zum ATP-Spiegel bei *Ankistrodesmus braunii*. Planta 115:25–36

Ullrich-Eberius CI, Novacky A, Fischer E, Lüttge U (1981) Relationship between energy-dependent phosphate uptake and the electrical membrane potential in *Lemna gibba* G1. Plant Physiol 67:797–801

Vega JM, Herrera J, Aparicio PJ, Paneque A, Losada M (1971) Role of molybdenum in nitrate reduction by *Chlorella*. Plant Physiol 48:294–299

Vega JM, Herrera J, Relimpio A, Aparicio PJ (1972) NADH-nitrate réductase de *Chlorella*: nouvelle contribution à l'étude de ses propriétés. Physiol Vég 10:637–652

Vega JM, Garrett RH, Siegel LM (1975) Siroheme: a prosthetic group of the *Neurospora crassa* assimilatory nitrite reductase. J Biol Chem 250:7980–7989

Vennesland B, Guerrero MG (1979) Reduction of nitrate and nitrite. In: Gibbs M, Latzko E (eds) Photosynthesis II Encyclopedia of plant physiology new ser, vol 6. Springer, Berlin Heidelberg New York

Voskresenskaya NA (1972) Blue light and carbon metabolism. Annu Rev Plant Physiol 23:219–234

Walker NA (1976) Membrane transport: theoretical background. In: Lüttge U, Pitman MG (eds) Transport in plants II. Encyclopedia of plant physiology new ser, vol 2, part A. Springer, Berlin Heidelberg New York

Warburg O, Negelein E (1920) Über die Reduktion der Salpetersäure in grünen Zellen. Biochem Z 110:66–115

Zumft WG (1972) Ferredoxin: nitrite oxidoreductase from *Chlorella*. Purification and properties. Biochim Biophys Acta 276:363–375

# III. Metabolism of Sulfur and Phosphorus

# III.1 Reduction and Other Metabolic Reactions of Sulfate

J.A. SCHIFF

## 1 Introduction

Sulfur, like nitrogen, is required to form the ubiquitous proteins and coenzymes of plants. It is not surprising, then, that this element is fairly uniformly distributed throughout the tissues and organs of higher plants and in the cells of plant-like microorganisms. Since sulfur is indispensable for plant growth and development, sulfur deficiency results in discoloration and abnormal growth of plant tissues. Sulfur deficiency influences nitrogen metabolism as well (FRIEDRICH and SCHRADER 1978). The importance of sulfur in the nutrition of plants and, therefore, in agriculture is discussed in several reviews (MCLACHLAN 1975, THOMPSON et al. 1970, ÅSLANDER 1958, BAUMEISTER 1958) and in the quarterly periodical *Sulphur Research and Development*, published by The Sulphur Institute, Washington, D.C.

This chapter is intended to summarize our present understanding of sulfate metabolism in the biological world, particularly the biochemical and enzymatic reactions in cells which are involved in sulfate utilization and reduction. The better we understand these cellular reactions the better will be our understanding of the role of sulfur in inorganic plant nutrition.

A concise chapter of this sort cannot deal with all the details of the subject. Useful general reviews of sulfate metabolism are those of ROY and TRUDINGER (1970), SCHIFF and HODSON (1973), SIEGEL (1975) and ANDERSON (1980); nutritional and ecological aspects are discussed by ANDERSON (1978). The sulfated polysaccharides are described by PERCIVAL and MCDOWELL (1967) in HARVEY and MCLACHLAN (1973) and by CRAIGIE and MCCANDLESS (1979); further references will be found in SCHIFF and HODSON (1973).

The accumulation of sulfuric acid in the vacuoles of the marine brown alga *Desmerestia* has been summarized (SCHIFF 1963) and the production of volatile sulfur compounds by algae was discussed by CHALLENGER (1959). The metabolism of reduced sulfur compounds formed from sulfate including the sulfur-containing coenzymes, peptides such as glutathione, and the sulfur-containing amino acids are reviewed by various authors in GREENBERG (1975); further information on glutathione and the $\gamma$-glutamyl cycle is available (GRIFFITH et al. 1979).

Regulation of the metabolic pathways leading to the sulfur-containing amino acids is comprehensively discussed by UMBARGER (1978) and by GIOVANELLI et al. (1980).

The remainder of this chapter will be concerned with our current understanding of the reactions of sulfate metabolism with particular emphasis on the enzymology, biochemistry and evolution of the sulfate-reducing pathways.

## 2 The Place of Sulfate Reduction in the Sulfur Cycle

Unlike most of the other elements required by living systems, carbon, nitrogen and sulfur undergo extensive metabolic transformations. Sulfur, unlike carbon and nitrogen, can be utilized in its most highly oxidized naturally occurring form, sulfate; sulfate reduction is necessary for the formation of sulfur-containing amino acids and proteins (SCHIFF and HODSON 1973, ROY and TRUDINGER 1970, SIEGEL 1975).

Figure 1 shows the relation of sulfate to the many reactions which sulfur undergoes in the biosphere. Sulfate is used by living systems to form sulfate esters of polysaccharides, phenols, steroids and other organic compounds through a series of activation and transfer reactions. Sulfate is reduced by certain anaerobic organisms such as the bacterium *Desulfovibrio* in their energy metabolism where, during anaerobic respiration, sulfate is used in place of oxygen to oxidize substrates releasing energy and hydrogen sulfide (PECK 1959, PECK 1962a, b, SIEGEL 1975, ROY and TRUDINGER 1970). Many organisms reduce sulfate to the thiol level found in several coenzymes, and the amino acids cysteine and methionine which serve as building blocks for peptides and proteins (GREENBERG 1975); this process is called assimilatory sulfate reduction. Most organisms (including higher animals and higher plants) oxidize reduced sulfur to sulfate (SCHIFF and HODSON 1973, SINGER 1975) although the aerobic chemosynthetic bacteria are the only organisms which have been observed to couple the energy released (some 180 kcal per mol) to the reduction of carbon dioxide (SIEGEL 1975). The anoxygenic photosynthetic bacteria use reduced sulfur compounds as photosynthetic electron donors thereby oxidizing them to sulfate (SIEGEL 1975). Recent reviews will be found in BOTHE and TREBST (1981).

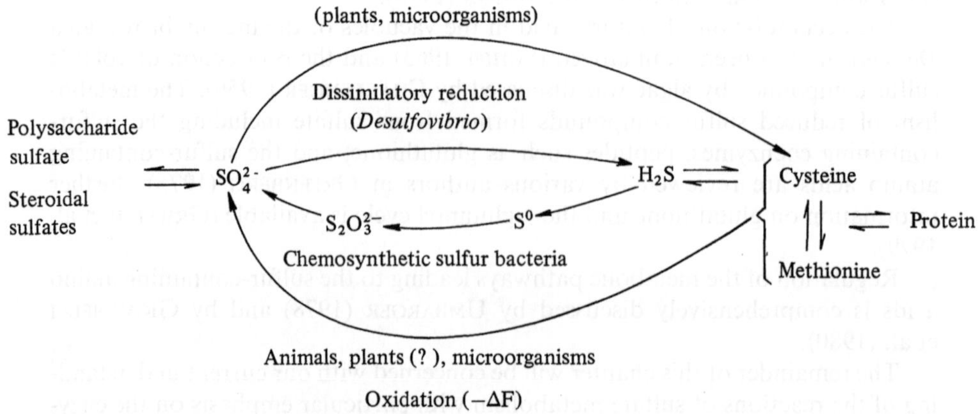


**Fig. 1.** Reactions of sulfur in the biosphere

## 3 Phylogenetic Distribution of Reactions Involving Sulfate Transfer and Reduction

Reactions resulting in the esterification of organic compounds by sulfate are widely distributed (SCHIFF and HODSON 1973, DEMEIO 1975). Sulfate esters of organic compounds are formed by higher animals in the form of steroid sulfates, phenol sulfates, and polysaccharide sulfates such as chondroitin sulfate. Luciferyl sulfate formed by sulfate transfer from PAPS serves as a reservoir of luciferin for bioluminescence (CORMIER et al. 1970, 1974). The algae produce sulfated polysaccharide as wall constituents, frequently in copious amounts (SCHIFF and HODSON 1973, CALLOW et al. 1978, BINGHAM and SCHIFF 1979, RAMUS 1974, RAMUS and GROVES 1974, QUATRANO 1978, CRAIGIE and MCCANDLESS 1979). The better-known sulfated polysaccharides of commercial importance such as carrageenan and agar are produced by marine members of the red algae or Rhodophyta. There is at least one report of sulfated polysaccharides

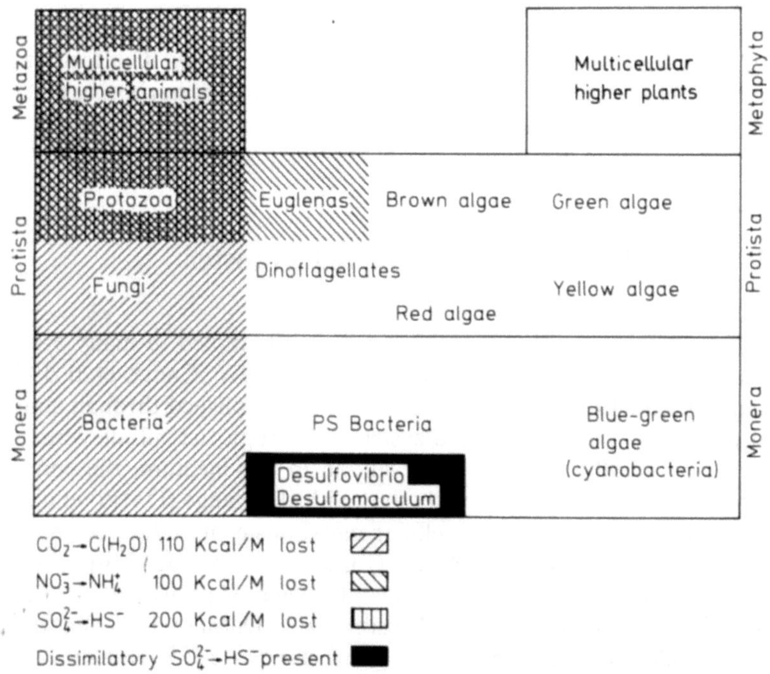


**Fig. 2.** Loss of energetically expensive reduction reactions in various contemporary groups of organisms. Dissimilatory sulfate reduction is restricted to two small genera of anaerobic bacteria (*Desulfovibrio* and *Desulfomaculum*). *Clear areas* indicate organisms that can carry out the reduction of sulfate, nitrate and carbon dioxide [a small group of aerobic chemosynthetic bacteria (not shown) are capable of carrying out carbon dioxide reduction given oxidizable inorganic substrates as energy sources]. The various *patterns* indicate the loss of these reduction reactions in various groups. The loss of all three reduction pathways has occurred in the evolution of the protozoa and higher animals

among the prokaryotes (bacteria and cyanobacteria or blue-green algae) in *Halococcus* (STEBER and SCHLEIFER 1975), but none in higher plants. These gelatinous materials may have evolved as adaptations to periodic wetting and drying as water conservation mechanisms, a particular problem for benthic and terrestrial algae. In the higher animals, the sulfated polysaccharides are used to provide a matrix for cellular structures such as cartilage and skin. From our present knowledge the occurrence of sulfated polysaccharides seems to represent the adaptation of particular groups of organisms to special needs.

The distribution of reduction reactions, however, seems to follow a more regular evolutionary pattern (Fig. 2) (SCHIFF and HODSON 1973). Reduction reactions are expensive, energetically, and require the investment of hundreds of kcal per mol to allow the reaction to take place (Fig. 2). The reduction of nitrate, carbon dioxide and sulfate seems to have been lost in the evolution of the primitive animals, the protozoa, and this deficiency persists in the evolution of the multicellular animals. These reactions may have been lost when photosynthetic reactions trapping light energy were lost. Animals require energy for movement, and having lost photosynthetic energy found themselves with a tight energy budget. Being predators, they could fulfill their architectural and energy requirements for reduced sulfur, nitrogen and carbon by eating organisms which could perform these reductions, such as microorganisms or plants, originating the food chains we know today. Contemporary animals obtain their reduced sulfur compounds by eating or harboring sulfate-reducing plants or microorganisms, or by eating other animals which do so.

## 4 Sulfate Uptake, Activation and Transfer

Sulfate uptake and its control has been well studied in many systems (SCHIFF and HODSON 1973, SIEGEL 1975, SMITH 1975, 1976, CACCO et al. 1977, RENASTO and FERRARI 1975). Uptake seems to be accomplished through active transport mediated by a carrier enzyme system. Although the sulfate content of *Lemna* increased greatly with increased sulfate in the medium, the concentration of other cellular sulfur compounds showed little change (DATKO et al. 1978).

As far as is known, the cellular metabolism of sulfate begins with a series of activation reactions with ATP to form the nucleoside phosphosulfates adenosine 5′-phosphosulfate (APS) and adenosine 3′-phosphate 5′-phosphosulfate (PAPS) (Fig. 3) (SCHIFF and HODSON 1973, DEMEIO 1975, PASTERNACK et al. 1965, DEMEIO et al. 1955, HILZ and LIPMANN 1955, ROBBINS and LIPMANN 1957, 1958, REUVENY and FILNER 1977, REUVENY 1977, FARLEY et al. 1978). (In crabgrass, *Digitaria sanguinalis*, a $C_4$ plant, 90% of the enzyme activity forming APS was found in extracts from bundle-sheath cells (GERWICK and BLACK 1979)). APS is formed against an unfavorable equilibrium since the free energy of hydrolysis of the phosphosulfate bond is higher than that of the pyrophosphate linkage (ROY and TRUDINGER 1970). To offset this unfavorable equilibrium, pyrophosphate removal through inorganic pyrophosphatase activity and/or reaction of APS with another molecule of ATP to form PAPS

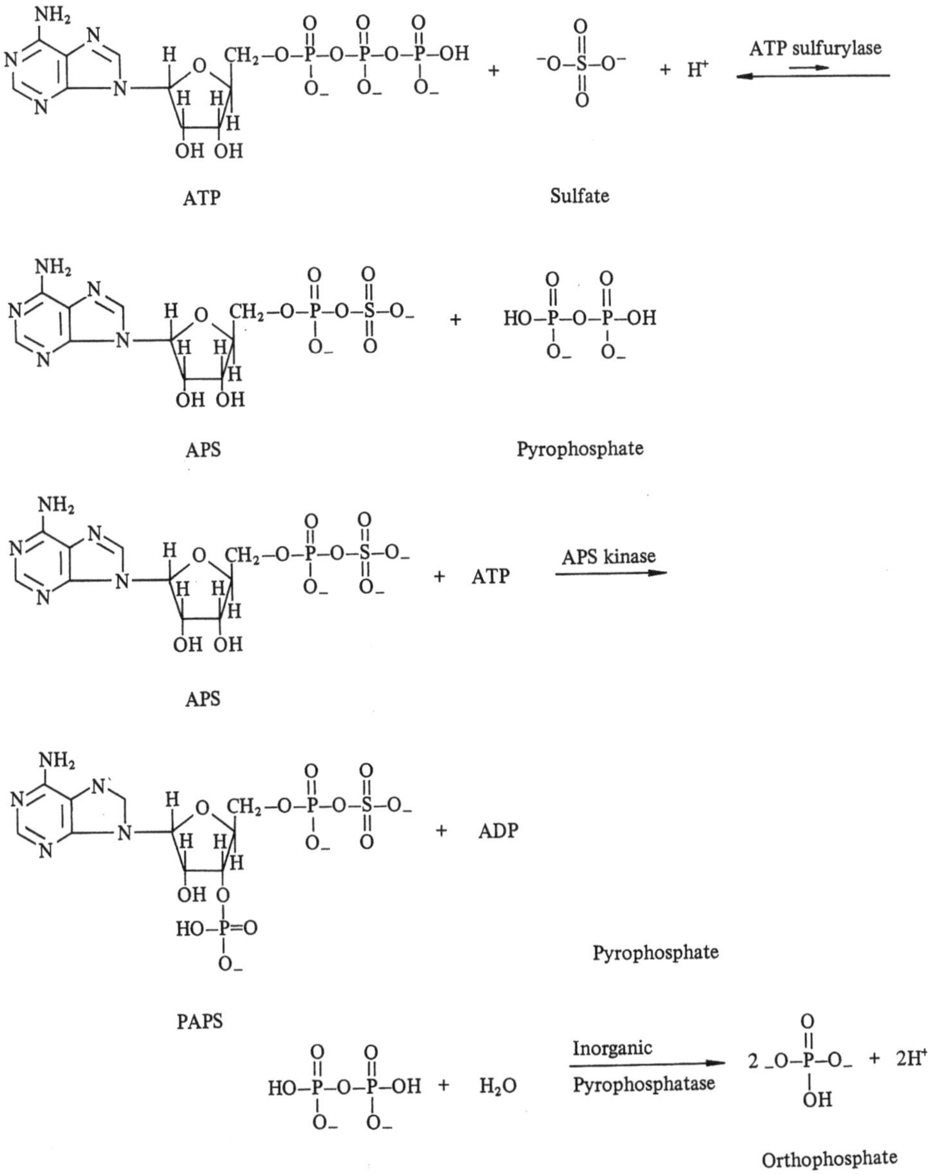


**Fig. 3.** Enzymatic reactions involved in sulfate activation. The enzymes shown are ATP sulfurylase (ATP:sulfate adenylyl transferase 2.7.7.4), APS kinase (ATP:adenylyl sulfate 3′ phosphotransferase 2.7.1.25), and inorganic pyrophosphatase (pyrophosphate phosphohydrolase 3.6.1.1)

are used to allow an accumulation of nucleoside phosphosulfates. A widely distributed 3′(2′), 5′-diphosphonucleoside 3′(2′)-phosphohydrolase (DPNPase) converts PAPS to APS and, together with inorganic pyrophosphatase, may facilitate APS formation despite the unfavorable equilibrium of the first reaction forming APS (Fig. 4) (TSANG and SCHIFF 1976c). This enzyme may also provide control of APS and PAPS levels in organisms where both are required for different purposes.

All sulfate transfer reactions resulting in sulfate esterification appear to use PAPS as the sulfate donor as far as is known (DEMEIO 1975, SCHIFF and HODSON 1973). APS is a substrate for hydrolases that release sulfate (TSANG and SCHIFF 1976c) and for a novel enzyme reaction (adenylyl sulfate: $NH_3$ adenylyl transferase (APSAT) which converts APS and ammonia to adenosine 5′-phosphoramidate (FANKHAUSER et al. 1981a) (this activity was previously thought to be a cyclase-forming cAMP; cAMP shares many properties with adenosine 5′-phosphoramidate; TSANG and SCHIFF 1976c). A molecule with the properties of adenosine 5′-phosphoramidate has been isolated from *Chlorella* (FANKHAUSER et al. 1981b). Sulfonic acids which contain sulfur at the redox level of sulfite, occur in many living systems, the plant sulfolipid of chloroplast thylakoids being an important example (SCHIFF and HODSON 1973, HARWOOD and NICHOLLS 1979, HARWOOD 1980, BENSON 1963, HAINES 1973, HOPPE and SCHWENN 1981). Although the formation of the sulfonic acid group of compounds such as taurine or isethionate (HOSKIN and KORDIK 1977) can arise through oxidation of reduced sulfur in animals (SIEGEL 1975) or, perhaps, from sulfate via PAPS and serine (MARTIN et al. 1974), the formation of the sulfonic acid group of the plant sulfolipid appears to come more readily from sulfate via as yet unknown reactions. The formation of sulfonic acids is an area requiring more study. Sulfur dioxide and sulfite are readily metabolized by plants and their chloroplasts (ZIEGLER 1977, ZIEGLER and HAMPP 1977, PLESNIČAR 1977, SILVIUS et al. 1976) but $SO_2$ can also cause changes in metabolism (PAUL and BASSHAM 1978) and produce injury (BRESSAN et al. 1978).

Natural products containing both oxidized and reduced sulfur are formed by plants including glucosinolates and polyacetylenes (see HODSON and SCHIFF 1973).

## 5 Sulfate Reduction

Dissimilatory sulfate reduction occurs in certain anaerobic bacteria such as *Desulfovibrio* which use sulfate as an oxidant in place of molecular oxygen. In this process APS is the nucleotide sulfate donor for reduction, resulting in the accumulation of hydrogen sulfide and ATP synthesis through oxidative phosphorylation (ROY and TRUDINGER 1970, SIEGEL 1975, PECK 1970, ISHIMOTO and FUJIMOTO 1959, PECK et al. 1965). A light-dependent emission of hydrogen sulfide from higher plants has been described (WILSON et al. 1978).

Assimilatory sulfate reduction, where sulfate is reduced to form the thiol groups of sulfur-containing amino acids, coenzymes and other organic compounds, now appears to exhibit two distinct patterns (TSANG and SCHIFF 1975, SCHIFF and HODSON 1973):

a) In most oxygen-evolving photosynthesizers including all eukaryotic algae and some prokaryotic blue-green algae (GOLDSCHMIDT et al. 1975, TSANG and SCHIFF 1975, SCHMIDT 1977), all higher plants tested (SCHMIDT 1975a) and

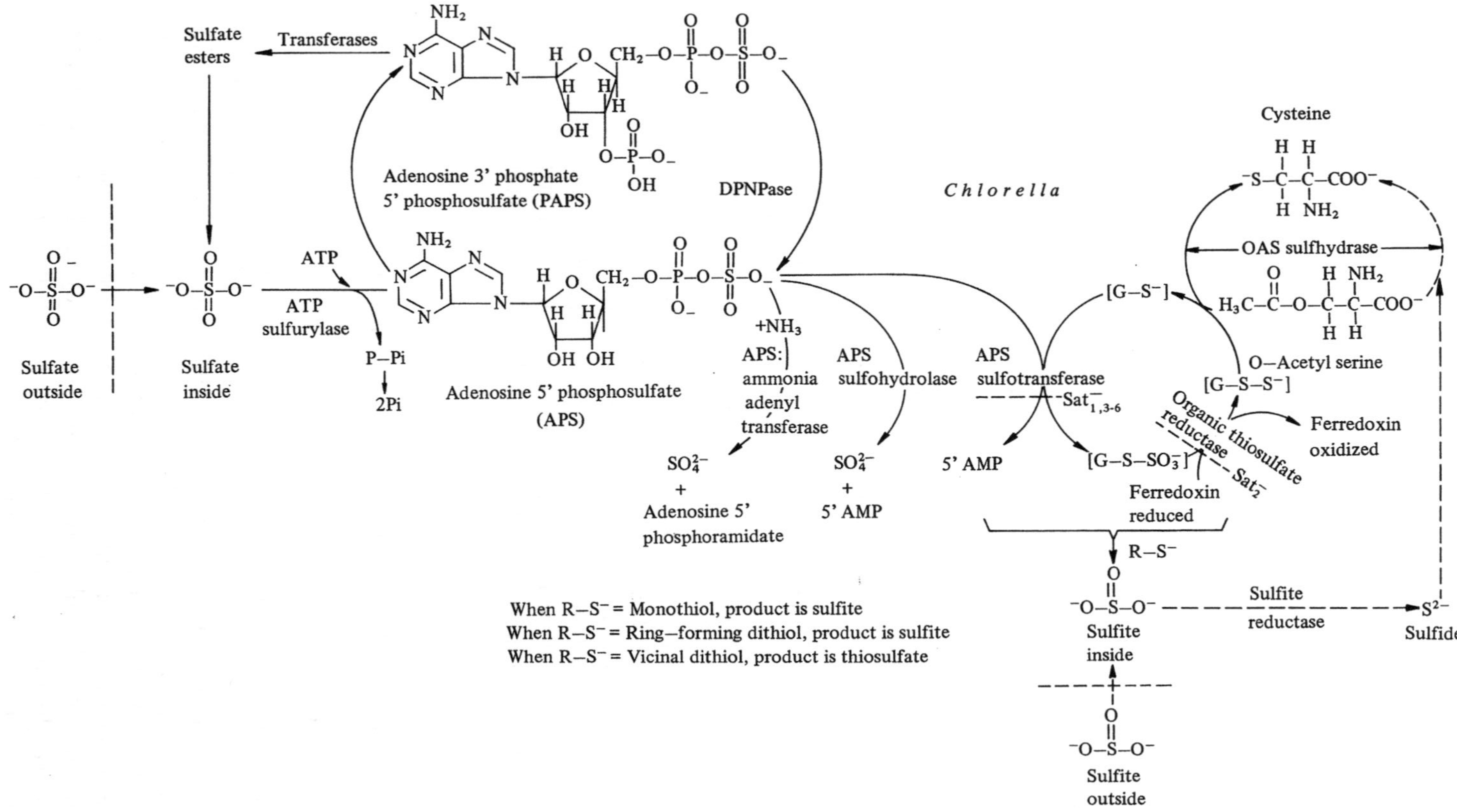


**Fig. 4.** APS pathway of sulfate reduction in *Chlorella*. Reactions thought to be on the main path of sulfate reduction in vivo are shown as *solid lines*. Reactions blocked in various mutants that cannot grow on sulfate (*Sat*$^-$ mutants) are shown by *dashes* through the reactions deleted. Side reactions are shown by *alternate dashes and dots*. Sulfite (or thiosulfate) is only produced through chemical reactions requiring thiols and probably only occurs in vivo when sulfite (or thiosulfate) is available to the cells from outside. G-S$^-$ designates reduced glutathione

spinach chloroplasts (SCHMIDT and SCHWENN 1971, SCHMIDT 1975b, FANKHAUSER and BRUNOLD 1978), the nucleotide sulfate donor for reduction is APS via a highly specific APS sulfotransferase (adenylyl sulfate: thiol sulfotransferase) which will not use PAPS (Fig. 4) (TSANG and SCHIFF 1976a). Where this form of reduction has been studied in detail, e.g. *Chlorella, Euglena* and spinach chloroplasts (SCHMIDT and SCHWENN 1971, BRUNOLD and SCHIFF 1976, SCHMIDT et al. 1974, ABRAMS and SCHIFF 1973) a low molecular weight thiol carrier (which may be glutathione in *Chlorella*; TSANG and SCHIFF 1976a) is also involved, but the APS sulfotransferase will transfer promiscuously to other added thiols in the presence or absence of the carrier (TSANG and SCHIFF 1976a). Ferredoxin (in the *Chlorella* and higher plant system) or NADPH (in *Euglena* extracts) must be added to achieve reduction of the sulfo group of APS to the thiol level of amino acids.

b) In organisms which lack oxygen-evolving photosynthesis [besides animals, which do not reduce sulfate to the thiol level (SCHIFF and HODSON 1973)], such as yeast and *E. coli* and other bacteria (TSANG and SCHIFF 1975, TORRI and BANDURSKI 1967, WILSON et al. 1961, ASAHI et al 1961, LEINWEBER and MONTY 1963, DREYFUSS and MONTY 1963a, b, FUJIMOTO and ISHIMOTO 1961, TSANG and SCHIFF 1976b, 1978b, WILSON and BIERER 1976) and a few oxygen-evolving blue-green algae (SCHMIDT 1977), PAPS is the nucleoside phosphosulfate donor for reduction via a specific PAPS sulfotransferase (3′ phosphoadenylyl sulfate: thiol sulfotransferase) which will not use APS (TSANG and SCHIFF 1976b, PECK 1961) (Fig. 5). The transfer of the sulfo group of PAPS and further reduction requires thioredoxin in yeast and *E. coli* (PORQUÉ et al. 1970, TSANG and SCHIFF 1976b, 1978b). Only reduced pyridine nucleotide needs to be added to obtain reduction, but the PAPS sulfotransferase will only function in the presence of thioredoxin (TSANG and SCHIFF 1976b) or glutaredoxin (TSANG and SCHIFF 1976b, TSANG 1981); other added thiols in the absence of thioredoxin or glutaredoxin are inactive. Interestingly, APS sulfotransferases requiring thioredoxin have been found recently in certain cyanobacteria (SCHMIDT 1982). Thioredoxin has also been implicated in the regulation of chloroplast enzymes by light (BUCHANAN 1980).

## 5.1 Detailed Reactions of the Two Assimilatory Pathways

### 5.1.1 The APS Pathway

Our most detailed information comes from work on the enzymes from *Chlorella, Euglena,* spinach chloroplasts (SCHMIDT and SCHWENN 1971, SCHMIDT 1975b, BRUNOLD and SCHIFF 1976, SCHMIDT et al. 1974, ABRAMS and SCHIFF 1973, GOLDSCHMIDT et al. 1975, TSANG and SCHIFF 1976a, 1978a, FANKHAUSER and BRUNOLD 1978) and duckweed (*Lemna*) (BRUNOLD and SCHMIDT 1978) (Fig. 4). It has taken some time to sort out the reactions of APS sulfotransferase because the enzyme transfers the sulfo group of APS not only to the normal carrier [which may be glutathione (TSANG and SCHIFF 1978a) in *Chlorella* or a somewhat larger molecule in spinach chloroplasts (SCHMIDT and SCHWENN 1971)] but to practically any thiol added (TSANG and SCHIFF 1976a). Thiols forming rings on oxidation (such as dithiothreitol) form sulfite in this reaction, monothiols (such as glutathione and thioethanolamine) form the Bunte salt ($R-S-SO_3^-$) of the thiol and, in the presence of excess thiol, sulfite. Vicinal dithiols such as 2,3-dithiopropanol (BAL) form thiosulfate ($^-S-SO_3^-$) where the oxidized sulfur comes from APS and the reduced sulfur from the thiol. Of all the thiols, glutathione stands out because it has a much higher activity with the enzyme than any other monothiol and shows kinetics suggestive of a regulatory interaction. Glutathione seems to be the only small molecule in *Chlorella* extracts which is active in the APS sulfotransferase reaction (TSANG and SCHIFF 1978a). If thiols are omitted from the reaction, it is possible to demonstrate that the enzyme transfers the sulfo group of APS to an enzyme-bound acceptor in the crude extracts (ABRAMS and SCHIFF 1973). This labeled material shows the properties of a Bunte salt or organic thiosulfate ($R-S-SO_3^-$) and is thought to be the sulfo form of the thiol carrier. Reconstruction experiments using glutathione$-S-SO_3^-$ show that sulfite and thiosulfate can be formed chemically by a non-enzymatic reaction with the appropriate thiol (TSANG and SCHIFF 1976a). While the transfer of the sulfo group from APS to the enzyme-bound carrier is heat-labile (and, therefore, enzymatic), the subsequent reaction of the enzyme-bound Bunte salt with added thiols is non-enzymatic and the same products are formed with each thiol as in the model experiments (TSANG and SCHIFF 1976a, ABRAMS and SCHIFF 1973).

Thus we view the reaction catalyzed by APS sulfotransferase as the transfer of the sulfo group of APS to the thiol of a carrier such as glutathione ($G-S^-$) to form the Bunte salt ($G-S-SO_3^-$). As will be discussed shortly, further physiological reduction appears to involve this bound sulfo group. The reductive release of sulfite or thiosulfate by addition of thiols results from chemical side reactions of this carrier-bound sulfo group. Regulation of APS sulfotransferase by $H_2S$ and cysteine has been described in *Lemna* (BRUNOLD and SCHMIDT 1976, 1978).

There are two candidates for enzymes catalyzing reduction to the thiol level. One is sulfite reductase (hydrogen-sulfide:$NADP^+$ oxidoreductase 1.8.1.2) which is present in these systems and will bring about the reduction of free sulfite to free sulfide with reduced pyridine nucleotides (SIEGEL 1975, SIEGEL et al. 1973, 1974, SIEGEL and DAVIS 1974, SCHMIDT et al. 1974, TSANG and SCHIFF 1976b, ASADA et al. 1969). The other enzyme is an organic thiosulfate reductase (ferredoxin: sulfoglutathione oxidoreductase) (formerly called thiosulfonate reductase) which reduces glutathione-$S-SO_3^-$ to bound sulfide (probably glutathione-$S-S^-$) or, dithionite to sulfide, with reduced ferredoxin (SCHMIDT 1973, SCHMIDT et al. 1974). Studies of a mutant *Chlorella* ($Sat_2^-$) which cannot grow on or reduce sulfate have shown that organic thiosulfate reductase activity is extremely low or absent while normal levels of sulfite reductase activity are present (SCHMIDT et al. 1974). Thus the presence of sulfite reductase alone does not allow sulfate reduction to proceed in vivo and shows that organic thiosulfate reductase is the preferred enzyme for sulfate reduction in this system. We view the physiological reaction as the reduction of $G-S-SO_3^-$ to $G-S-S^-$ which seems to be the primary reduction reaction in vivo in the APS system. Sulfite reductase would probably act on free sulfite only if it arose through side reactions in vivo or if it was taken up from outside by the cell.

These systems contain a very active O-acetyl serine sulfhydrase [O-acetyl-L-serine acetate-lyase (adding $H_2S$) 4.2.99.8] which forms cysteine from free sulfide or from carrier-bound sulfide (BRUNOLD and SCHIFF 1976, SCHMIDT et al. 1974). Two electrons would be required to convert $G-S-S^-$ to the thiol group of cysteine in this reaction.

An important advance in studies of this system was finding conditions for the reduction of APS to form cysteine in cell-free extracts without the addition of thiols (SCHMIDT et al. 1974) which form free sulfite and sulfide through chemical side reactions (TSANG and SCHIFF 1976a). Studies of this cell-free system from wild-type *Chlorella* and mutants blocked for organic thiosulfate reductase and APS sulfotransferase have proven to be consistent with the interpretations offered above and have led to a scheme in which what are thought to be the normal reactions in vivo are shown in solid lines and the side reactions in broken lines (Fig. 4).

Studies of the cellular location and synthesis of the various enzymes of this pathway have been undertaken. The activating enzymes for sulfate, APS sulfotransferase, organic thiosulfate reductase (reductant, ferredoxin) and O-acetyl serine sulfhydrase have been found in spinach chloroplasts (SCHMIDT and SCHWENN 1971, SCHWENN and TREBST 1976, FANKHAUSER and BRUNOLD 1979). It is likely that this is also the case in *Chlorella,* judging from the close similarity of green algal and higher plant systems. In *Euglena,* however, which is an animal-like cell resembling the protozoa but containing chloroplasts, APS sulfotransferase is found in the mitochondria and microbodies and the organic thiosulfate reductase (reductant, NADPH) and OAS sulfhydrase are found in the mitochondria (BRUNOLD and SCHIFF 1976); thus the mitochondria of *Euglena* contain all of the presently known enzymes of the bound pathway of APS reduction. It is interesting that green algal and higher plant cells cannot be induced to lose their chloroplasts, perhaps because certain enzyme systems (such as sulfate reduction) which are essential for viability are plastid-localized. In *Euglena,* however, chloroplasts and plastid DNA can be lost (SCHIFF 1972). *Euglena* does not seem to localize essential reactions (other than those essential for photosynthesis) in the plastids.

In the course of studies on the APS sulfotransferase another activity was found which acts on APS (TSANG and SCHIFF 1976c), an APS sulfohydrolase (adenylyl sulfate sulfohydrolase 3.6.2.1) which forms sulfate and AMP from APS. Other sulfohydrolases are known (DODGSON and ROSE 1975). PAPS is much more stable than APS in extracts; perhaps this is why the DPNPase reaction forming APS from PAPS is advantageous as an APS source.

### 5.1.2 The PAPS Pathway

This system has been best studied in yeast and *Escherichia coli* (TSANG and SCHIFF 1975, TORII and BANDURSKI 1967, WILSON et al. 1961, FUJIMOTO and ISHIMOTO 1961, LEINWEBER and MONTY 1963, DREYFUSS and MONTY 1963a, b, TSANG and SCHIFF 1976b, WILSON and BIERER 1976) (Fig. 5). PAPS sulfo-

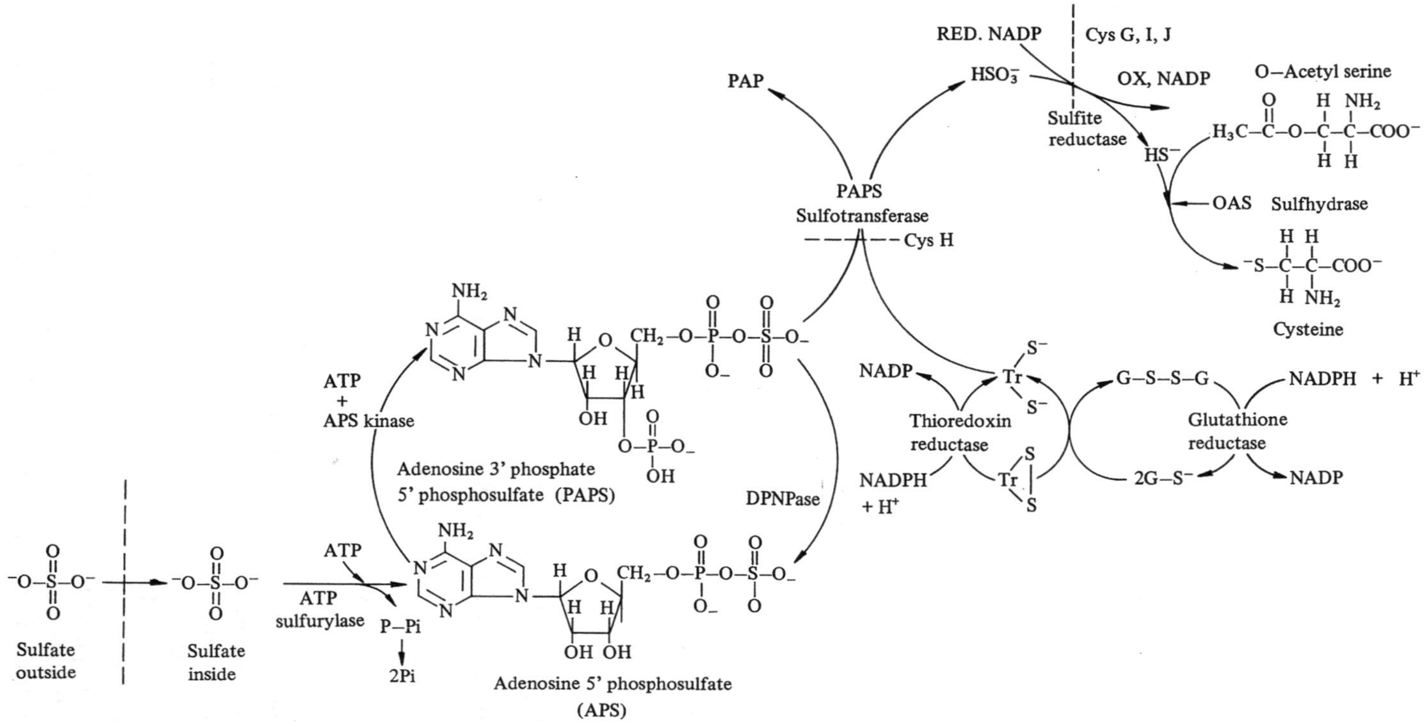


**Fig. 5.** PAPS pathway of sulfate reduction in *E. coli*. Thioredoxin is required in the PAPS sulfotransferase reaction and formation of sulfite, but the reactions it undergoes have not been elucidated fully. Mutants (*Cys*) blocked in various enzyme reactions are shown by dashes through the reactions deleted. G-S$^-$ designates reduced glutathione. Tr<$^{S^-}_{S^-}$ designates reduced thioredoxin

transferase (also called PAPS reductase before thioredoxin was identified as a component) is specific for PAPS as the sulfate donor and is also completely specific for thioredoxin; other thiols will not replace thioredoxin (TSANG and SCHIFF 1976b) or glutaredoxin (TSANG and SCHIFF 1978b, TSANG 1981). As in *Chlorella,* an enzyme-bound intermediate with the properties of a Bunte salt ($R-S-SO_3^-$) can be detected as a product of the PAPS sulfotransferase reaction (TSANG and SCHIFF 1976b, TORII and BANDURSKI 1967). The formation of free sulfite requires thioredoxin reductase (NADPH:oxidized-thioredoxin oxidoreductase 1.6.4.5) and reduced pyridine nucleotide (TORII and BANDURSKI 1967, WILSON et al. 1961, ASAHI et al. 1961, PORQUÉ et al. 1970) but these can be replaced by appropriate thiols (TSANG and SCHIFF 1976b) to yield sulfite or thiosulfate as in the APS system. A glutaredoxin can accept electrons from the glutathione-glutathione reductase system [NAD(P)H:oxidized-glutathione oxidoreductase 1.6.4.2)] (TSANG and SCHIFF 1978b, HOLMGREN et al. 1978, TSANG 1981). The product of this PAPS sulfotransferase-thioredoxin system is sulfite when NADPH is used as the reductant. Free sulfite is reduced to sulfide with reduced pyridine nucleotides via sulfite reductase (SIEGEL 1975).

*E. coli* and *Salmonella* mutants blocked for sulfate reduction lack sulfite reductase activity, suggesting that this enzyme system is the physiological reaction in these organisms (DREYFUS and MONTY 1963a, b, JONES-MORTIMER 1968a, b, LEINWEBER and MONTY 1963, HODSON and SCHIFF 1971, TSANG and SCHIFF 1978b). O-acetyl serine sulfhydrase is also present, (KREDICH and TOMPKINS 1966) which can form cysteine from O-acetyl serine and sulfide. Taking together the information presented here, and suggestions made previously by various workers, it is possible to write a scheme consistent with current knowledge (Fig. 5).

The role of thioredoxin in this system is still under study. Reduced thioredoxin ($Tr\langle^{S^-}_{S^-}$) could act as the thiol carrier acceptor for the sulfo group of PAPS in the PAPS sulfotransferase reaction forming $Tr\langle^{S-SO_3^-}_{S^-}$ (sulfothioredoxin). Elimination of sulfite would yield oxidized thioredoxin ($Tr\langle^{S}_{S}$) which could then be reduced as shown in Fig. 5. In view of the ready elimination of sulfite from Bunte salts of non-vicinal dithiols (TSANG and SCHIFF 1976a), some mechanism would be necessary to prevent the immediate release of sulfite from sulfothioredoxin since the bound intermediate formed from PAPS in *E. coli* is rather stable (TSANG and SCHIFF 1976b). Perhaps linkage of one of the thiol groups of thioredoxin to a thiol group of the transferase via an S–S bond accomplishes this. Another possibility is suggested by recent work which shows that a large proportion of the thioredoxin has a thiol group substituted with phosphate (PIGIET and CONLEY 1978). Perhaps this substitution of one thiol group by phosphate prevents spontaneous elimination of sulfite from sulfothioredoxin until timely enzymatic removal of the phosphate. It is possible, of course, that thioredoxin is not the carrier and that the bound intermediate contains the sulfo group bound to a different acceptor. In this case, thioredoxin would act as the immediate reductant.

It is worth noting that although the bound intermediates in both the APS and PAPS systems show many reactions of Bunte salts, frequently not all of the activity is exchangeable with sulfite (TSANG and SCHIFF 1976b, TORII and BANDURSKI 1967) and further work is needed to prove that these are definitely organic thiosulfates and that they account for all of the activity in both types of systems. Also unexplained is why thiosulfate can be formed from APS in *Chlorella* extracts in the absence of added thiols (SCHIFF and LEVINTHAL 1968, LEVINTHAL and SCHIFF 1968) when NADPH is used as the reductant; a donor of reduced sulfur must be present in the extracts themselves.

### 5.2 Location of Sulfate Reduction in Tissues and Organs of Multicellular Plants

We have already mentioned that sulfate reduction is localized in the chloroplasts of higher plants. However, cell, tissue and organ cultures from a wide variety of plant parts including roots can be grown on media where sulfate is the sole source of sulfur (WHITE 1963, WHITE and GROVE 1965, THOMAS and DAVEY 1975, BUTENKO 1968). Non-green tissues of plants such as beet discs (ELLIS 1963) and roots (PATE 1965) reduce sulfate. Enzymes of sulfate reduction such as APS sulfotransferase (SCHMIDT 1976, FANKHÄUSER and BRUNOLD 1978) and O-acetylserine sulfhydrylase (FANKHAUSER and BRUNOLD 1979, TAMURA et al. 1976) are found in roots and hypocotyls. Enzymes which may be involved in sulfate reduction but which may participate in other functions, such as ATP sulfurylase (ELLIS 1969, ONAJOBI et al. 1973, CACCO et al. 1977) and sulfite reductase (MAYER 1967) are also found in roots and hypocotyls. It appears that sulfate reduction may be widely distributed throughout the plant. There may be large quantitative differences, however, in the rates, since the activity of APS sulfotransferase in spinach roots, for example, is far lower than in fully green leaves (FANKHAUSER and BRUNOLD 1978). In green leaves and chloroplasts, sulfate reduction is a light-dependent process, but in non-green tissues energy and reducing power for sulfate reduction must come from non-photosynthetic sources. It remains for future work to determine whether the sulfate reduction system in non-green tissues is localized in plastids or in other organelles, but O-acetyl serine sulfhydrylase is found in plastids in spinach roots (FANKHAUSER and BRUNOLD 1979). In this connection it is worth remembering that sulfate reduction in higher plant chloroplasts requires reduced ferredoxin but when sulfate reduction is localized in mitochondria as in *Euglena*, NADPH is the reductant.

## 6 Speculations on the Origin and Evolution of Pathways of Sulfate Reduction

It is likely that the most primitive form of sulfate reduction is the dissimilatory pathway of the anaerobes (ROY and TRUDINGER 1970, SIEGEL 1975, PECK 1970) (Fig. 6) which may have survived with little modification from the anaerobic

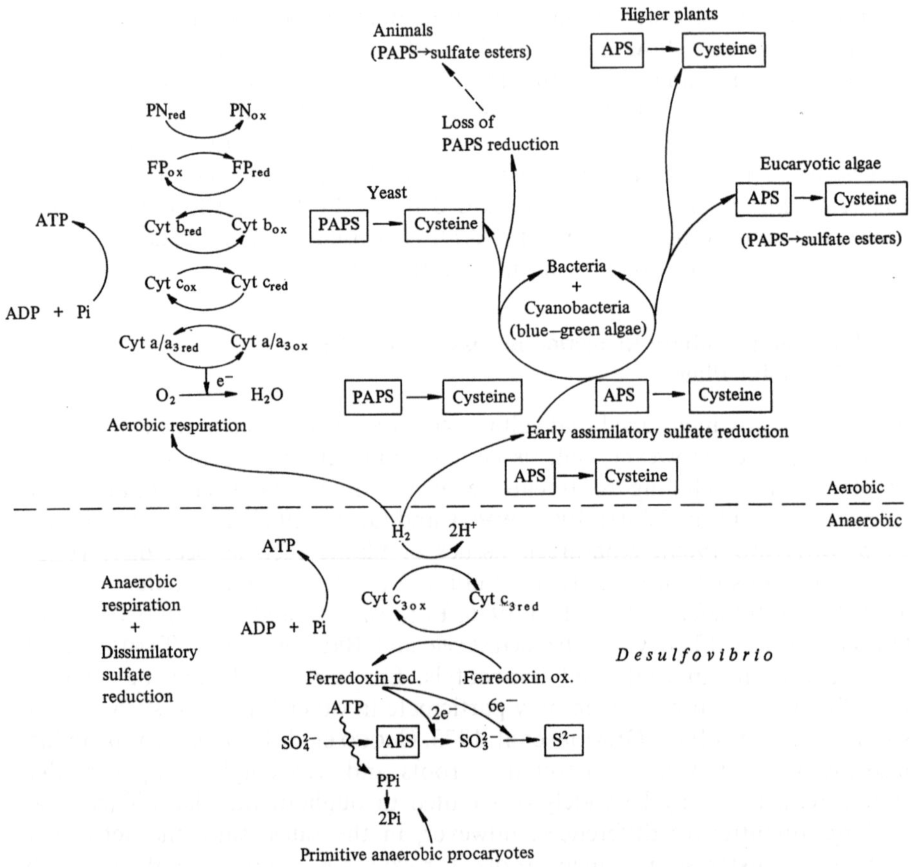


**Fig. 6.** Suggested evolutionary relationships of sulfate-reducing pathways in various groups. Dissimilatory sulfate reduction, as part of anaerobic respiration, probably arose in the anaerobic phase of the origin of life and has persisted in the modern procaryotic anaerobes such as *Desulfovibrio* which uses sulfate as an oxidant in respiration forming ATP and sulfide. This pathway begins with APS. With the release of oxygen into the atmosphere, oxygen became the electron acceptor in respiration and a separation of respiration and sulfate reduction occurred leading to the establishment of assimilatory sulfate reduction, still using APS, as the activated sulfate for reduction. For as yet unknown reasons, the PAPS pathway for reduction evolved from the APS pathway and modern prokaryotes (bacteria and blue-green algae or cyanobacteria) have one or the other. The APS pathway has persisted in the evolution of the eukaryotic photosynthetic oxygen evolvers including the eukaryotic algae and higher plants. The PAPS pathway leads to yeast, among the fungi. The animals have lost the ability to reduce PAPS to the thiol level of the amino acids but retain PAPS as a donor of the sulfo group in esterification reactions via specific sulfotransferases. The eukaryotic algae also use PAPS as the donor of the sulfo group in esterification reactions although they use APS as the substrate for reduction

phase of the origin of life and evolution. In the absence of molecular oxygen, other oxidants were pressed into service to oxidize substrates. The result was anaerobic respiration employing sulfate or nitrate as oxidants with the consequent formation of their reduced forms, sulfide and nitrogen gas. In organisms such as *Desulfovibrio* the formation of ATP from the oxidation of substrates is coupled to dissimilatory sulfate reduction which serves as a sink for the electrons removed in the oxidation of the substrate (hydrogen in the present example). This dissimilatory pathway uses APS as the activated form of sulfate for reduction and the pathway proceeds through the formation of free sulfite and sulfide. In an anaerobic environment, there is no danger from chemical oxidation of these intermediates by molecular oxygen. Anaerobic conditions also allow the undisturbed participation of autoxidizable low potential reductants such as ferredoxin in the sulfate reduction pathway. Multielectron reductions utilizing low potential reductants such as those involved here, in sulfite reduction, in nitrite reduction and in nitrogen fixation very likely evolved under the early anaerobic conditions in which life originated.

When oxygen came into the atmosphere through the activities of the first oxygenic photosynthesizers (very likely early prokaryotes related to the cyanobacteria or blue-green algae) (see SCHIFF 1972, SCHIFF and LYMAN 1982) aerobic respiration using oxygen as the electron acceptor evolved and the organisms carrying out anaerobic respiration were forced into anaerobic niches (such as deep mud) to survive. With the availability of oxygen, I suppose, came the separation of respiration and sulfate reduction in the aerobic organisms. Respiration now employed the reduction of oxygen to water (Fig. 6) in the formation of ATP. Sulfate reduction, although no longer required for respiration, was still necessary for the formation of reduced sulfur compounds including cysteine and methionine of protein. This separation of sulfate reduction from respiration led to the establishment of assimilatory sulfate reduction as a separate process to provide these reduced sulfur compounds.

This assimilatory sulfate reduction process was, at first I suppose, similar to the process of dissimilatory sulfate reduction in the anaerobes in that it employed APS as the activated intermediate and reduced ferredoxin as the reductant. Free sulfite and sulfide react with oxygen, however, and this may have been the point at which carrier-bound intermediates such as organic thiosulfates ($G-S-SO_3^-$) and persulfides ($G-S-S^-$) which are not autoxidizable, were introduced into the APS pathway. This process would have been present in the early aerobic prokaryotes and is found among contemporary prokaryotes such as the cyanobacteria or blue-green algae (TSANG and SCHIFF 1976b, SCHMIDT 1977). Chloroplasts and mitochondria of modern eukaryotes may have originated from the invasion of non-photosynthetic non-respiratory eukaryotes by appropriate procaryotic endosymbionts such as primitive blue-green algae and bacteria (SCHIFF 1972, SCHIFF and LYMAN 1982). In view of this, it is not surprising that the APS pathway of assimilatory sulfate reduction with ferredoxin as reductant is found in eukaryotic algae such as *Chlorella* (TSANG and SCHIFF 1975, SCHMIDT et al. 1974) and higher plants (SCHMIDT 1975a, BRUNOLD and SCHMIDT 1978) and in spinach has been shown to be localized in the chloroplasts (SCHMIDT and SCHWENN 1971, SCHWENN and TREBST 1976). This is probably one of the reasons why chloroplast function can never be eliminated completely from green algae and higher plants even when reduced carbon is available; essential functions such as sulfate reduction (and nitrite reduction) are localized in the plastids. Organisms more closely related to the animals such as *Euglena* do not put essential functions other than carbon dioxide fixation in their plastids and when provided with a reduced carbon source plastid function can be eliminated completely. In *Euglena,* sulfate reduction is localized in the mitochondria and microbodies (BRUNOLD and SCHIFF 1976). In this organism, the APS pathway characteristic of oxygenic eukaryotes is present but the reductant is NADPH rather than reduced ferredoxin which is available in eukaryotes only in the chloroplasts as far as we know.

Although the APS pathway is pervasive among eukaryotes having oxygenic photosynthesis, another pathway appears to have evolved at the same time. While some prokaryotes use the APS pathway of assimilatory sulfate reduction, others use a PAPS pathway (TSANG and SCHIFF 1975, SCHMIDT 1977), suggesting that the PAPS pathway evolved from the APS pathway at or after the establishment of assimilatory sulfate reduction (Fig. 6). This pathway is found in bacteria such as *E. coli* (TSANG and SCHIFF 1975, 1976b, 1978b, DREYFUSS and MONTY 1963a, b, FUJIMOTO and ISHIMOTO 1961) and in fungi such as yeast (TORII and BANDURSKI 1967, WILSON and BIERER 1976) where the reductant is NADPH. Since PAPS is formed from APS it presumably arose later in evolution than APS but why PAPS was selected for assimilatory sulfate reduction in these organisms is not readily explained. As we have already noted (Fig. 2) sulfate reduction was lost in the evolution of the protozoa and higher animals; the pathway from sulfate to PAPS formation has been conserved, however, and PAPS serves as the source of sulfate for esterification reactions via PAPS sulfotransferases of various specificities (DEMEIO 1975). PAPS also serves as the donor for esterification reactions in organisms that use APS for sulfate reduction, suggesting that PAPS formation was fairly common among primitive prokaryotes, but was selected for use in assimilatory sulfate reduction only in certain bacteria and in non-photosynthetic eukaryotes such as the fungi.

*Acknowledgments.* The support of a series of grants from the National Science Foundation (PCM 76-21486) is gratefully acknowledged. The technical assistance of Mrs. LINDA CORRADO is appreciated as is the secretarial help of Mrs. MARGARET KING. NANCY O'DONOGHUE prepared the illustrations. HEINZ FANKHAUSER provided several references and helpful suggestions.

## References

Abrams WR, Schiff JA (1973) Studies of sulfate utilization by algae. 11. An enzyme-bound intermediate in the reduction of adenosine 5′ phosphosulfate (APS) by cell-free extracts of wild-type *Chlorella* and mutants blocked for sulfate reduction. Arch Mikrobiol 94:1–10

Anderson JW (1978) Sulphur in biology. Arnold, London

Anderson JW (1980) Assimilation of inorganic sulfate into cysteine. In: Stumpf PK, Conn EE (eds) Plant biochemistry, vol V. Academic Press, London New York

Asahi T, Bandurski RS, Wilson LG (1961) Yeast sulfate-reducing system. II. Enzymatic reduction of protein disulfide. J Biol Chem 236:1830–1835

Asada K, Tamura G, Bandurski RS (1969) Methyl viologen-linked sulfite reductase from spinach leaves. J Biol Chem 244:4904–4915

Åslander A (1958) Nutritional requirements of crop plants. Encycl Plant Physiol IV:977–1019

Baumeister W (1958) Hauptnährstoffe. In: Ruhland W (ed) Encyclopedia of plant physiology, vol IV. Springer, Berlin Göttingen Heidelberg

Benson AA (1963) The plant sulfolipid. Adv Lipid Res 1:387–394
Bingham S, Schiff JA (1979) Conditions for attachment and development of single cells released from mechanically disrupted thalli of *Prasiola stipitata* Suhr. Biol Bull 156:257–271
Bothe H, Trebst A (eds) (1981) Biology of inorganic nitrogen and sulfur. Springer, Berlin Heidelberg New York
Bressan RA, Wilson LC, Filner P (1978) Mechanisms of resistance to sulfur dioxide in the Cucurbitaceae. Plant Physiol 61:761–767
Brunold C, Schiff JA (1976) Studies of sulfate utilization by algae. 15. Enzymes of assimilatory sulfate reduction in *Euglena* and their cellular localization. Plant Physiol 57:430–436
Brunold C, Schmidt A (1976) Regulation of adenosine 5′ phosphosulfate sulfotransferase activity by $H_2S$ in *Lemna minor*. Planta 133:85–88
Brunold C, Schmidt A (1978) Regulation of sulfate assimilation in plants. 17. Cysteine inactivation of adenosine 5′ phosphosulfate sulfotransferase in *Lemna minor* L. Plant Physiol 61:342–347
Buchanan BB (1980) Role of light in the regulation of chloroplast enzymes. Annu Rev Plant Physiol 31:341–374
Butenko RG (1968) Plant tissue culture and plant morphogenesis. Isr Program Sci Transl US Dep Commerce, Springfield, VA
Cacco G, Saccomani M, Ferrari G (1977) Development of sulfate uptake capacity and ATP-sulfurylase activity during root elongation in maize. Plant Physiol 60:582–584
Callow ME, Coughlan SJ, Evans LV (1978) The role of Golgi bodies in polysaccharide sulphation in *Fucus* zygotes. J Cell Sci 32:337–356
Challenger F (1959) Aspects of the organic chemistry of sulfur. Academic Press, London New York
Cormier MJ, Hori K, Karkhanis YD (1970) Studies on the bioluminescence of *Renilla reniformis*. VII. Conversion of luciferin into luciferyl sulfate by luciferin sulfokinase. Biochemistry 9:1184–1190
Cormier MJ, Hori K, Anderson JM (1974) Bioluminescence in coelenterates. Biochim Biophys Acta 346:137–164
Craigie JS, McCandless E (1979) Sulfated polysaccharides in red and brown algae. Annu Rev Plant Physiol 30:41–53
Datko AH, Mudd SH, Giovanelli J, Macnicol PK (1978) Sulfur-containing compounds in *Lemna perpusilla* 6746 grown at a range of sulfate concentrations. Plant Physiol 62:629–635
DeMeio RH (1975) Sulfate activation and transfer. In: Greenberg DM (ed) Metabolism of sulfur compounds. Academic Press, London New York
DeMeio RH, Wizerkaniuk M, Schreibman I (1955) Enzymatic system-synthesizing sulfuric acid esters of phenols. J Biol Chem 213:439–443
Dodgson KS, Rose FA (1975) Sulfohydrolases. In: Greenberg DM (ed) Metabolism of sulfur compounds. Academic Press, London New York
Dreyfuss J, Monty KJ (1963a) The biochemical characterization of cysteine-requiring mutants of *Salmonella typhimurium*. J Biol Chem 238:1019–1024
Dreyfuss J, Monty KJ (1963b) Coincident repression of the reduction of 3′ phosphoadenosine 5′ phosphosulfate, sulfite, and thiosulfate in the cysteine pathway of *Salmonella typhimurium*. J Biol Chem 238:3781–3783
Ellis RJ (1963) Cysteine biosynthesis in beet discs. Phytochemistry 2:129–136
Ellis RJ (1969) Sulphate activation in higher plants. Planta 88:34–42
Fankhauser H, Brunold C (1978) Localization of adenosine 5′ phosphosulfate sulfotransferase in spinach leaves. Planta 143:285–289
Fankhauser H, Brunold C (1979) Localization of O-acetyl-L-serine sulfhydrylase in *Spinacia oleracea* L. Plant Sci Lett 14:185–192
Fankhauser H, Schiff JA, Garber L (1981a) Purification and properties of adenylyl sulfate: ammonia adenylyl transferase from *Chlorella* catalyzing the formation of adenosine 5′ phosphoramidate from adenosine 5′ phosphosulphate and ammonia. Biochem J 195:545–560

Fankhauser H, Berkowitz GA, Schiff JA (1981b) A nucleotide with the properties of adenosine 5′ phosphoramidate from *Chlorella* cells. Biochem Biophys Res Commun 101:524–532
Farley JR, Nakayama G, Cryns D, Segel IH (1978) ATP sulfurylase from *Penicillium chrysogenum*. Equilibrium binding, substrate hydrolysis, and isotope exchange studies. Arch Biochem Biophys 185:376–390
Friedrich JW, Schrader LE (1978) Sulfur deprivation and nitrogen metabolism in maize seedlings. Plant Physiol 61:900–903
Fujimoto D, Ishimoto M (1961) Sulfate reduction in *E. coli*. J Biochem 50:533–537
Gerwick BC, Black CC (1979) Sulfur assimilation in $C_4$ plants. Intercellular compartmentation of ATP sulfurylase in crabgrass leaves. Plant Physiol 64:590–593
Giovanelli J, Mudd SH, Datko AH (1980) Sulfur amino acids in plants. In: Stumpf PK, Conn EE (eds) Plant biochemistry, vol V. Academic Press, London New York
Goldschmidt EE, Tsang ML-S, Schiff JA (1975) Studies of sulfate utilization by algae. 13. Adenosine 5′ phosphosulfate (APS) as an intermediate in the conversion of adenosine 3′ phosphate 5′ phosphosulfate (PAPS) to acid volatile radioactivity. Plant Sci Lett 4:293–299
Greenberg DM (ed) (1975) Metabolic pathways, vol VII. Metabolism of sulfur compounds. Academic Press, London New York
Griffith OW, Anderson ME, Meister A (1979) Inhibition of glutathione biosynthesis by prothionine sulfoximine (S-n-propyl homocysteine sulfoximine) a selective inhibitor of γ-glutamylcysteine synthetase. J Biol Chem 264:1205–1210
Haines TW (1973) In: Erwin JA (ed) Lipids and biomembranes of eucaryotic microorganisms. Academic Press, London New York
Harvey MJ, McLachlan J (eds) *Chondrus crispus*. Nova Scotian Inst Sci Halifax, Nova Scotia
Harwood JL (1980) Sulfolipids. In: Stumpf PK, Conn EE (eds) The biochemistry of plants, vol IV. Academic Press, London New York
Harwood JL, Nicholls RG (1979) The plant sulpholipid – a major component of the sulphur cycle. Biochem Soc Trans 7:440–447
Hilz H, Lipmann F (1955) The enzymic activation of sulfate. Proc Natl Acad Sci USA 41:880–890
Hodson R, Schiff JA (1971) The ubiquity of sulfate reduction to thiosulfate. Plant Physiol 47:296–299
Holmgren A, Ohlsson I, Grankvist M-L (1978) Thioredoxin from *Escherichia coli*. Radioimmunological and enzymatic determinations in wild-type cells and mutants defective in Phage $T_7$ DNA replication. J Biol Chem 253:430–436
Hoppe W, Schwenn JD (1981) In vitro biosynthesis of the plant sulpholipid: on the origin of the sulphonate group. Z Naturforsch 36C:820–826
Hoskin FCG, Kordik ER (1977) Hydrogen sulfide as a precursor for the synthesis of isethionate in the squid giant axon. Arch Bioch Biophys 180:583–586
Ishimoto M, Fujimoto D (1959) Adenosine 5′ phosphosulfate as an intermediate in the reduction of sulfate by a sulfate-reducing bacterium. Proc Jpn Acad 35:243–245
Jones-Mortimer MC (1968a) Positive control of sulphate reduction in *Escherichia coli*. Isolation, characterization and mapping of cysteine-less mutants of *E. coli* K12. Biochem J 110:589–595
Jones-Mortimer MC (1968b) Positive control of sulphate reduction in *Escherichia coli*. Biochem J 110:597–602
Kredich N, Tomkins G (1966) The enzymic synthesis of L-cysteine in *Escherichia coli* and *Salmonella typhimurium*. J Biol Chem 241:4955–4965
Leinweber FJ, Monty K (1963) The metabolism of thiosulfate in *Salmonella typhimurium*. J Biol Chem 238:3775–3780
Levinthal M, Schiff JA (1968) Studies of sulfate utilization by algae. 5. Identification of thiosulfate as a major acid-volatile product formed by a cell-free sulfate-reducing system from *Chlorella* extracts. Plant Physiol 43:555–562
Martin WG, Truex CR, Tarka SM, Hill LJ, Groby WG (1974) The synthesis of taurine

from sulfate. VIII. A constitutive enzyme in mammals. Proc Soc Exp Biol Med 147:563–565

Mayer AM (1967) Subcellular location of sulphite reductase in plant tissues. Plant Physiol 42:324–326

McLachlan KD (ed) (1975) Sulphur in Australasian agriculture. Sydney Univ Press, Sydney

Moller ME, Evans LV (1976) Sulphate activation in the unicellular red alga *Rhodella*. Phytochemistry 15:1623–1626

Onajobi FD (1975) Effects of adenine nucleotides on rice-root adenosine triphosphate sulfurylase activity in vitro. Biochem J 149:301–304

Onajobi FD, Cole CV, Ross C (1973) Adenosine 5′-triphosphate-sulfurylase in corn roots and its partial purification. Plant Physiol 52:580–584

Pasternak CA, Ellis RJ, Jones-Mortimer MC, Crichton CE (1965) The control of sulphate reduction in bacteria. Biochem J 96:270–275

Pate JS (1965) Roots as organs of assimilation of sulfate. Science 149:547–548

Paul JS, Bassham JA (1978) Effects of sulfite on metabolism in isolated mesophyll cells from *Papaver somniferum*. Plant Physiol 62:210–214

Peck HD Jr (1959) The ATP-dependent reduction of sulfate with hydrogen in extracts of *Desulfovibrio desulfuricans*. Proc Natl Acad Sci USA 45:701–708

Peck HD Jr (1961) Enzymatic basis for assimilatory and dissimilatory sulfate reduction. J Bacteriol 82:933–939

Peck HD Jr (1962a) The role of adenosine 5′ phosphosulfate in the reduction of sulfate to sulfite by *Desulfovibrio desulfuricans*. J Biol Chem 237:198–203

Peck HD Jr (1962b) Symposium on metabolism of inorganic compounds. V. Comparative metabolism of inorganic sulfur compounds in microorganisms. Bacteriol Rev 26:67–94

Peck HD Jr (1970) Sulfur requirements and metabolism of microorganisms. In: Muth OH, Oldfield JE (eds) Symposium: Sulfur in nutrition. Avi, Westport

Peck HD Jr, Doacon TE, Davidson JT (1965) Studies on adenosine 5′-phosphosulfate reductase from *Desulfovibrio desulfuricans* and *Thiobacillus thioparus*. 1. The assay and purification. Biochim Biophys Acta 96:429–446

Percival E, McDowell RH (1967) Chemistry and enzymology of marine algal polysaccharides. Academic Press, London New York

Pigiet V, Conley RR (1978) Isolation and characterization of phosphothioredoxin from *Escherichia coli*. J Biol Chem 253:1910–1920

Pitman MG (1976) Ion uptake by plant roots. Encycl Plant Physiol IIB:95–128

Plesničar M (1977) The uptake and distribution of $^{35}SO_2$ in bean (*Phaseolus vulgaris* L. cv. Pasuljica). Plant Sci Let 10:205–211

Porqué PG, Baldesten A, Reichard P (1970) The involvement of the thioredoxin system in the reduction of methionine sulfoxide and sulfate. J Biol Chem 245:2371–2374

Quatrano R (1978) Development of cell polarity. Annu Rev Plant Physiol 29:487–510

Ramus J (1974) In vivo molybdate inhibition of sulfate transfer to *Porphyridium* capsular polysaccharide. Plant Physiol 54:945–949

Ramus J, Groves ST (1974) Precursor-product relationships during sulfate incorporation into *Porphyridium* capsular polysaccharide. Plant Physiol 54:434–439

Renosto F, Ferrari G (1975) Mechanism of sulfate transport inhibition by cycloheximide in plant tissues. Plant Physiol 56:478–480

Reuveny Z (1977) Derepression of ATP sulfurylase by the sulfate analogs molybdate and selenate in cultured tobacco cells. Proc Natl Acad Sci USA 74:616–622

Reuveny Z, Filner P (1977) Regulation of adenosine triphosphate sulfurylase in cultured tobacco cells. J Biol Chem 252:1858–1864

Robbins PW, Lipmann F (1957) Isolation and identification of active sulfate. J Biol Chem 229:837–851

Robbins PW, Lipmann F (1958) Separation of the two enzymatic phases in active sulfate synthesis. J Biol Chem 233:681–685

Roy AB, Trudinger PA (1970) The biochemistry of inorganic compounds of sulfur. Cambridge Univ Press

Schiff JA (1963) The metabolism of sulfur in algae. In: Lewin R (ed) Biochemistry and physiology of the algae. Academic Press, London New York

Schiff JA (1972) The development, inheritance and origin of the plastid in *Euglena*. In: Advances in morphogenesis, vol X. Academic Press, London New York

Schiff JA, Hodson RC (1973) The metabolism of sulfate. Annu Rev Plant Physiol 24:381–414

Schiff JA, Levinthal M (1968) Studies of sulfate utilization by algae. 5. Identification of thiosulfate as a major acid-volatile product formed by a cell-free sulfate-reducing system from *Chlorella* extracts. Plant Physiol 43:555–562

Schiff JA, Lymann H (eds) (1982) On the origins of chloroplasts. Elsevier, Amsterdam

Schmidt A (1973) Sulfate reduction in a cell-free system of *Chlorella*. The ferredoxin-dependent reduction of a protein-bound intermediate by a thiosulfonate reductase. Arch Mikrobiol 93:29–52

Schmidt A (1975a) Distribution of APS sulfotransferase activity among higher plants. Plant Sci Lett 5:407–415

Schmidt A (1975b) A sulfotransferase from spinach leaves using adenosine 5′ phosphosulfate. Planta 124:267–275

Schmidt A (1976) Development of the adenosine 5′ phosphosulfate sulfotransferase in sunflower (*Helianthus annuus* L.). Z Pflanzenphysiol 78:164–168

Schmidt A (1977) Assimilatory sulfate reduction via 3′ phosphoadenosine 5′ phosphosulfate (PAPS) and adenosine 5′ phosphosulfate (APS) in blue-green algae. FEMS Microbiol Lett 1:137–140

Schmidt A (1982) Assimilation of sulfur. In: Schiff JA, Lyman H (eds) On the origins of chloroplasts. Elsevier, Amsterdam

Schmidt A, Schwenn JD (1971) On the mechanism of photosynthetic sulfate reduction. Proc 2nd Int Congr Photosynthesis. Junk, The Hague

Schmidt A, Abrams WR, Schiff JA (1974) Studies of sulfate utilization by algae. 12. Reduction of adenosine 5′ phosphosulfate to cysteine in extracts from *Chlorella* and mutants blocked for sulfate reduction. Eur J Biochem 47:423–434

Schwenn JD, Trebst A (1976) Photosynthetic sulfate reduction by chloroplasts. In: Barber J (ed) The intact chloroplast. Elsevier, Amsterdam

Siegel LM (1975) Biochemistry of the sulfur cycle. In: Greenberg DM (ed) Metabolism of sulfur compounds. Academic Press, London New York

Siegel LM, Davis PS (1974) Reduced nicotinamide adenine dinucleotide phosphate-sulfite reductase of enterobacteria. IV. The *Escherichia coli* hemoflavoprotein: subunit structure and dissociation into hemoprotein and flavoprotein components. J Biol Chem 249:1587–1598

Siegel LM, Murphy MJ, Kamin H (1973) Reduced nicotinamide adenine dinucleotide phosphate-sulfite reductase of enterobacteria. J Biol Chem 248:251–264

Siegel LM, Davis PS, Kamin H (1974) Reduced nicotinamide adenine dinucleotide phosphate-sulfite reductase of enterobacteria. III. The *Escherichia coli* hemoflavoprotein: catalytic parameters and the sequence of electron flow. J Biol Chem 249:1572–1586

Silvius JE, Baer CH, Dodrill S, Patrick H (1976) Photoreduction of sulfur dioxide by spinach leaves and isolated spinach chloroplasts. Plant Physiol 57:799–801

Singer TP (1975) Oxidative metabolism of cysteine and cystine in animal tissues. In: Greenberg DM (ed) Metabolism of sulfur compounds. Academic Press, London New York

Smith IK (1975) Sulfate transport in cultured tobacco cells. Plant Physiol 55:303–307

Smith IK (1976) Characterization of sulfate transport in cultured tobacco cells. Plant Physiol 58:358–362

Steber J, Schleifer KH (1975) *Halococcus morrhuae:* A sulfated heteropolysaccharide as the structural component of the bacterial cell wall. Arch Microbiol 105:173–177

Tamura G, Iwasawa T, Masada M, Fukushima K (1976) Some properties of cysteine synthase from radish roots. Agric Biol Chem 40:637–638

Thomas E, Davey MR (1975) From single cells to plants. Springer, Berlin Heidelberg New York

Thompson JF, Smith IK, Moore DP (1970) Sulfur requirement and metabolism in plants. In: Muth OH, Oldfield JE (eds) Symposium: Sulfur in nutrition. Avi, Westport

Torii K, Bandurski RS (1967) Yeast sulfate-reducing system. III. An intermediate in the reduction of 3′ phosphoryl 5′ adenosine phosphosulfate to sulfite. Biochim Biophys Acta 136:286–295

Tsang ML-S (1981) Assimilatory sulfate reduction in *Escherichia coli:* Identification of the alternate cofactor for adenosine 3′ phosphate 5′ phosphosulfate reductase as glutaredoxin. J Bacteriol 146:1059–1066

Tsang ML-S, Schiff JA (1975) Studies of sulfate utilization by algae. 14. Distribution of adenosine 5′ phosphosulfate (APS) and adenosine 3′ phosphate 5′ phosphosulfate (PAPS) sulfotransferases in assimilatory sulfate reducers. Plant Sci Lett 4:301–307

Tsang ML-S, Schiff JA (1976a) Studies of sulfate utilization by algae. 17. Reactions of the adenosine 5′ phosphosulfate (APS) sulfotransferase from *Chlorella* and studies of model reactions which explain the diversity of side products with thiols. Plant Cell Physiol 17:1209–1220

Tsang ML-S, Schiff JA (1976b) Studies of a sulfate-reducing pathway in *Escherichia coli* involving bound intermediates. J Bacteriol 125:923–933

Tsang ML-S, Schiff JA (1976c) Properties of enzyme fraction A from *Chlorella* and copurification of 3′(2′), 5′ bisphosphonucleoside 3′(2′) phosphohydrolase, adenosine 5′ phosphosulfate sulfohydrolase, adenosine 5′ phosphosulfate cyclase activities. Eur J Biochem 65:113–121

Tsang ML-S, Schiff JA (1978a) Studies of sulfate utilization by algae. 18. Identification of glutathione as a physiological carrier in assimilatory sulfate reduction by *Chlorella* extracts. Plant Sci Lett 11:177–183

Tsang ML-S, Schiff JA (1978b) Assimilatory sulfate reduction in a mutant of *Escherichia coli* lacking thioredoxin activity. J Bacteriol 134:131–138

Umbarger E (1978) Amino acid biosynthesis and its regulation. Annu Rev Biochem 47:533–606

White PR (1963) The cultivation of animal and plant cells. Ronald, New York

White PR, Grove AR (eds) (1965) In: Proc Int Conf Plant Tissue Cult. McCutchan, Berkeley

Wilson LG, Bierer D (1976) The formation of exchangeable sulphite from adenosine 3′ phosphate 5′ sulphatophosphate in yeast. Biochem J 158:255–270

Wilson LG, Asahi T, Bandurski RS (1961) Yeast sulfate-reducing system. 1. Reduction of sulfate to sulfite. J Biol Chem 236:1822–1829

Wilson LG, Bressan RA, Filner P (1978) Light-dependent emission of hydrogen sulfide from plants. Plant Physiol 61:184–189

Ziegler I (1977) Subcellular distribution of $^{35}$S-sulfur in spinach leaves after application of $^{35}SO_3^{2-}$, and $^{35}SO_2$. Planta 135:25–32

Ziegler I, Hampp R (1977) Control of $^{35}SO_4^{2-}$ and $^{35}SO_3^{2-}$ incorporation into spinach chloroplasts during photosynthetic $CO_2$ fixation. Planta 137:303–307

# III.2 Physiology and Metabolism of Phosphate and Its Compounds

R.L. BIELESKI and I.B. FERGUSON

## 1 Introduction

The distribution of phosphorus[1] in the world, unlike that of all other elements but carbon, is dominated by the present and past activities of living organisms. Thus it was first isolated as an element from that preeminently biological fluid, urine, by the Arabian alchemists in the 12th century and then by H. BRAND in 1669; while the next source to be discovered was bone, in 1770 (CORBRIDGE 1978). It is widely distributed in the Earth's crust, where it comprises 0.1% by weight of the elements present. Igneous deposits are known, but most of the phosphate used by man has been formed either as guano and its end-product, phosphatized coral, or as sedimentary deposits laid down under marine conditions in a combination of biological and physicochemical processes. In each case, the key event in formation of the deposit has been the ability of living organisms to scavenge phosphate from their surroundings, so that the concentration within the organism is increased one thousand fold or more (see Sect. 2). With guano-based products, phosphate has passed through a long food chain (marine microorganism → crustacean → fish → sea bird) and has finally been drawn into one place as excreta and as fish and bird remains. With sedimentary rocks, the process has been more direct, in that the microorganisms themselves settle to the bottom of the sea where, as they die, they create a zone in which the phosphate ion concentration is many hundreds of times higher than in the surrounding waters. If the combination of pH, temperature, and concentration of $Ca^{2+}$ and $F^-$ is right, the solubility product of various calcium phosphates is exceeded, and precipitates form. Alternatively, pre-existing calcium carbonate deposits in the sea bed, arising from shells of marine organisms, are progressively phosphatized, a process wherein the phosphate ion being brought in by the microorganisms replaces the carbonate one of the shell. Typically, these events have taken place in the shallow waters of the eastern seaboards of continental masses, and the subsequent uplifting of the seabeds has given us our phosphate deposits. Now that the more accessible sources of phosphate are disappearing, geologists are considering whether it might be economically feasible to mine phosphate deposits directly from the sea bed in their formative state as phosphate nodules (CORBRIDGE 1978).

Because it is part of the DNA and RNA molecule, P is part of every living thing. In typical plant tissues it is present at about 0.04% of the fresh weight

1 The following abbreviations will be used throughout. P, phosphorus; Pi, inorganic phosphate; P-lipid, phospholipid; P-esters, acid-soluble and water-soluble simple phosphate esters; RNA, ribonucleic acid; DNA, deoxyribonucleic acid; G6P, glucose-6-phosphate; 3PGA, 3-phosphoglyceric acid. Standard nucleotide abbreviations are used

or 0.3% of the dry weight. If we exclude carbohydrate, however, P forms about 4% of the nutrient content, with N, K, Ca, Mg, P and S being present in the molar ratio 20:5:2.5:2:1:0.5 (EPSTEIN 1972). At values below 1/3 and above three times those quoted above, P deficiency and P toxicity respectively may be operating. Usually, somewhere in excess of 50% of the P present is in inorganic form, so that the Pi concentration of living plant tissues is typically around 5–20 mM (see Sect. 4).

With only very minor exceptions, all P in both geological and biological systems is in the fully oxidized form; geologically, mainly as the various insoluble calcium salts, and biologically as the free Pi ion and the various phosphate esters (Sect. 4). Soils represent something of a half-way house in that, while P coming from the breakdown of rock particles is inorganic (and held at very low concentration by the presence of Fe, Al and Ca ions) much of the phosphate deposited in plant remains can stay in its original organic form, or can be rapidly taken up by soil microorganisms and converted to other organic forms. Thus, soils high in humus can often contain over half their P in organic form and as a consequence, those processes which lead to the recycling of organic P in the soil can be of major importance to plants (COLE et al. 1978, GOULD et al. 1979). Whether in inorganic or organic form, P is very easily locked up and made unavailable to the plant, and so P deficiency is a common phenomenon. Availability of P can be the single most important factor limiting plant growth: for example, life in the world's seas is dominated by the upwelling of phosphate-rich waters from the ocean deeps, particularly in Antarctic regions, leading to massive phytoplankton blooms and the forging of the first links in the food chain (HARVEY 1969).

Man causes similar processes to occur when he discards phosphate-rich materials such as detergents, sewage and processing wastes into inland waters. Again, by supplying P (and/or nitrogen) we remove the existing limit, and the resulting burst of plant growth is what we see as eutrophication and pollution. In general, very little P is lost from the soil by leaching because phosphate is strongly bound. Thus in a moderate rainstorm of 34 mm spread over 3 days, only 0.03 g P $ha^{-1}$ was removed from a cropping catchment, corresponding to an annual loss of around 1 g $ha^{-1}$ $year^{-1}$. In a more intense storm of 63 mm over 12 h, the loss was much greater, 0.8 g P $ha^{-1}$ (corresponding to 12 g P $ha^{-1}$ $year^{-1}$), about half being sorbed to fine particles of sediment (KUNISHI et al. 1972). Thus the rate of loss is greatly increased when the soil particles are lost, particularly from the upper soil layers where P tends to accumulate (HEALY and MCCOLL 1974, MCCOLL et al. 1975). Severe erosion by wind and water, as with loess soils in China and dust-bowl soils in the USA, has caused P losses estimated at more than 20 kg P $ha^{-1}$ $year^{-1}$. Another factor which can greatly increase rate of P loss is the application of P fertilizer: under particularly severe conditions of a 76 mm rainfall immediately following fertilizer application, about a quarter of the applied P was lost (GILCHRIST and GILLINGHAM 1970). A high level of animal activity, with much of the P excreted in organic or bound form, can also lead to a higher rate of loss, estimated at 0.5–2 kg P $ha^{-1}$ $year^{-1}$ for stable high-yielding pasture/herd systems (COOKE 1981). Thus, P loss is extremely sensitive to land management.

All these factors combine to make phosphate one of the key nutrients in horticultural and agricultural management, and fertilizer use accounts for more than 90% of total P use in the world. Countries with a highly-developed agriculture use an average of 20–50 kg arable $ha^{-1}$ $year^{-1}$, developing countries as little as a tenth of this level or in population terms, less than 1 kg $head^{-1}$ $year^{-1}$. Apart from the requirements of crops and stock, man himself excretes some 0.4 kg P $head^{-1}$ $year^{-1}$, and this must be replaced by P in food.

## 2 Uptake and Transport of Phosphate

More than for any other mineral nutrient, the uptake of phosphate has to be a metabolically driven process. This is apparent if we compare the nutrient compositions of a typical plant tissue, its xylem sap, and the soil solution bathing the roots (Table 1). If there were no accumulation, the xylem sap would have the same concentration and composition as the soil water, whilst the tissue would have the elements in the same ratio but at about 50–150 times higher concentration (because of the concentration of nutrients due to water loss through transpiration). In practice, most of the elements show the observed pattern but phosphate does not: its concentration in the xylem sap is much higher than would be expected (Table 1). In the step from the soil solution to the xylem sap, most elements do not increase their concentration markedly, so that their uptake could be passive, but phosphate increases its concentration 400-fold both with respect to that in the soil solution and in relation to the other elements. The phenomenon is also shown in model systems: the phosphate concentration of xylem sap exuding from excised roots is at least two orders of magnitude higher than in the surrounding nutrient solution (COLLINS and REILLY 1968). More detailed experiments show that the root cortical cells are

**Table 1.** Comparison of nutrient concentrations in plant tissue, xylem sap and soil solution. Tissue nutrient concentrations (T, μmol $g^{-1}$ fresh wt.) are given for the kiwifruit vine (fruit plus leaf) (FERGUSON and EISEMAN 1983), and resemble those of many plant tissues (EPSTEIN 1972). Xylem sap concentrations (X, μmol $ml^{-1}$) are those obtained for sap of kiwifruit vines, averaged over the main growing period (Dec-Apr, see FERGUSON et al. 1983). Soil solution concentrations (S, μmol $ml^{-1}$) are those found for a range of about 250 soils (BARBER et al. 1963, REISENAUER 1966)

| Element | Tissue conc. (T) | Sap conc. (X) | Ratio (T/X) | Soil conc. (S) | Ratio (X/S) |
|---|---|---|---|---|---|
| N[a] | 170 | 2.7 | 63 | 1–3 | 1.3 |
| K | 80 | 2.8 | 29 | 1–2 | 1.8 |
| Ca | 28 | 1.1 | 25 | 0.5–1.5 | 1.1 |
| Mg | 14 | 0.6 | 23 | 2–4 | 0.2 |
| S | – | 0.2 | – | 0.3–0.7 | 0.4 |
| P | 9 | 0.4 | 23 | 0.0005–0.002 | 400 |

[a] N in xylem sap is the sum of nitrate plus arginine plus amino acid

the major site of this phosphate accumulation step, and that phosphate is then transported across the root in the symplast (FERGUSON and CLARKSON 1975). Within most plant cells, the Pi concentration is 1–10 mM (BIELESKI 1973) so that the final gradient is as much as 10,000-fold. If we finally take into account the negative membrane potential of the cell (in the region of $-120$ mV in clover roots; BOWLING and DUNLOP 1978), then the effective gradient to be overcome is about $10^6$-fold (BIELESKI 1976).

It is clear from the work of ASHER and EDWARDS (Chap. I.3, this Vol.) that the soil water measurements used in Table 1 represent a realistic situation, and that optimum plant growth is obtained at very low external phosphate concentrations (3–13 μM) despite the high concentrations in the cell. Epiphytic plants apparently encounter a similar situation, as the P concentration of rainwater is in the range 0.7–3 μM (GROVES 1981). Thus it is not surprising that phosphate uptake shows to a marked degree the features to be expected of a metabolically driven accumulation process. It is very sensitive to temperature, to anoxia, to metabolic inhibitors (SMITH 1966, RAVEN 1974b, FERGUSON and CLARKSON 1975) and to competitive inhibition by the very similar anion, arsenate (JUNG and ROTHSTEIN 1965). A close relationship between the cell PD and phosphate uptake suggests the operation of some type of electrogenic pump (BOWLING and DUNLOP 1978); and recent work with fungi (BEEVER and BURNS 1980) and *Lemna* (ULLRICH-EBERIUS et al. 1981) points to a proton extrusion pump of the type envisaged by Mitchell.

There has been some consideration given to the ionic form entering the plant. At normal soil pH, the $H_2PO_4^-$ ion is predominant (Chap. I.6, this Vol.), and there is a general agreement that this can be transported. However, at higher pH's, still within the biological range, there is almost no $H_2PO_4^-$, and $HPO_4^{2-}$ is the major ion, yet Pi is still taken up. One view is that there is a separate uptake mechanism for the divalent anion (HAGEN and HOPKINS 1955). A factor which tends to be overlooked in such discussions is that the pH in the soil solution need not be anything like that existing at the uptake site itself. The uptake system, by its very operation, must have a powerful effect on the effective pH nearby. If, for example, Pi or any other anion is taken up without the corresponding transport of an exactly balancing amount of cation, then the pH at the uptake site must change. Indeed, if current theories on uptake of Pi prove correct, an outward proton pump must lower the pH at the outer surface of the membrane at the same time as it creates a potential difference; Pi will actually enter as a positively charged complex (ULLRICH-EBERIUS et al. 1981, BEEVER and BURNS 1980). The question has also been raised as to whether it is energetically possible for Pi to overcome the major concentration and electrical gradient if in the divalent form (ULLRICH-EBERIUS et al. 1981). For these reasons, the $H_2PO_4^-/HPO_4^{2-}$ ratio in the bulk solution is likely to have limited significance. This is reflected in practice, as Pi uptake is often not markedly influenced by pH over a range that drastically alters the $H_2PO_4^-/HPO_4^{2-}$ ratio, and the pH relationship that occurs can never be simply interpreted by having a $HPO_4^{2-}$ uptake mechanism operating additively.

A feature of Pi uptake systems is that the rate of uptake is seldom a simple function of concentration over the full concentration range ($10^{-7}$–$10^{-2}$ M).

Surprisingly often, the complex uptake kinetics that are observed can be satisfactorily interpreted in terms of the simultaneous presence and activity of *two* systems showing Michaelis-Menten kinetics (uptake a rectangular hyperbolic function of concentration), one having a high affinity ($K_m$ around 5 µM) and one of low affinity ($K_m$ around 500 µM) (LEGGETT et al. 1965, CARTER and LATHWELL 1967). This pattern, often called the "dual isotherm" or "double hyperbola" or "two-phase kinetics", has been variously interpreted and is beyond the scope of this chapter; but the most detailed study to date merits separate attention as it brings out the importance to the plant of maintaining a constant Pi intake in the face of a varying external concentration (BURNS and BEEVER 1977, BEEVER and BURNS 1977). *Neurospora* spore germlings supplied with Pi in the concentration range 1–3,200 µM have an uptake rate/concentration behaviour that fits a clearly defined double hyperbola pattern (Fig. 1). Depending on the prior history of the germlings – that is, the Pi concentration in which they were grown prior to the actual uptake experiments – the kinetic constants differed, but the overall double-hyperbola pattern was still maintained. As the Pi concentration used to grow the germlings was increased from 50 µM to 10 mM, the affinity of the high affinity system remained fixed ($K_m$ about 2.5 µM) but the affinity of the low affinity system fell by a factor of 3 ($K_m$ rising from 350 to 1,000 µM); the capacity of the high affinity system decreased 8× ($V_{max}$ from 2.4 to 0.3 µmol $g^{-1}$ dry wt. $min^{-1}$) and the capacity of the low affinity system decreased 2× ($V_{max}$ from 6.8 to 3.4) (Fig. 1B). The effect of these changes was to adjust the total uptake capacity of the cell in such a way that regardless of the growth conditions (Pi concentration range 50 µM–10 mM), the Pi uptake rate *at that concentration* was fixed at about 3.3 µmol $g^{-1}$ dry wt. $min^{-1}$ (Fig. 1B). That is, by relatively small modulations to two uptake systems, the plant achieved a constant Pi supply regardless of the concentration in the environment. Very much larger modulations would have been needed if only a single Michaelis-Menten uptake system had been available to the cell. It remains to be seen whether this explanation for two-phase kinetics will also apply to higher plant tissues.

In a macroscopic land plant, uptake is only the first stage in utilizing P from the environment. Accumulative uptake of Pi by the root cortical cells must then be followed by transfer across the root to the xylem. This occurs through the symplast; thus a barrier to apoplastic transport such as endodermal suberization has little effect on Pi transport, whereas a barrier to cortical cell uptake such as hypodermal suberization reduces it (FERGUSON and CLARKSON 1975, 1976). The vacuoles of the root cells are seen as being to one side of the main pathway (GREENWAY and KLEPPER 1968).

Although a case has been made for a diffusive release of ions into the xylem vessels (DUNLOP and BOWLING 1971) most current evidence supports a release which is energy-dependent (PITMAN 1977), allowing a greater control by xylem parenchyma cells of ion levels in the xylem fluid. There is certainly a need for some secreting or accumulating mechanism in the xylem. Under normal transpiration conditions with a steady xylem flow, the xylem Pi concentration is about 2%–5% of that in the root cells. However, under conditions of xylem sap immobility such as in a deciduous plant just before bud break, the sap Pi concentration can rise to levels as high or higher than in the root

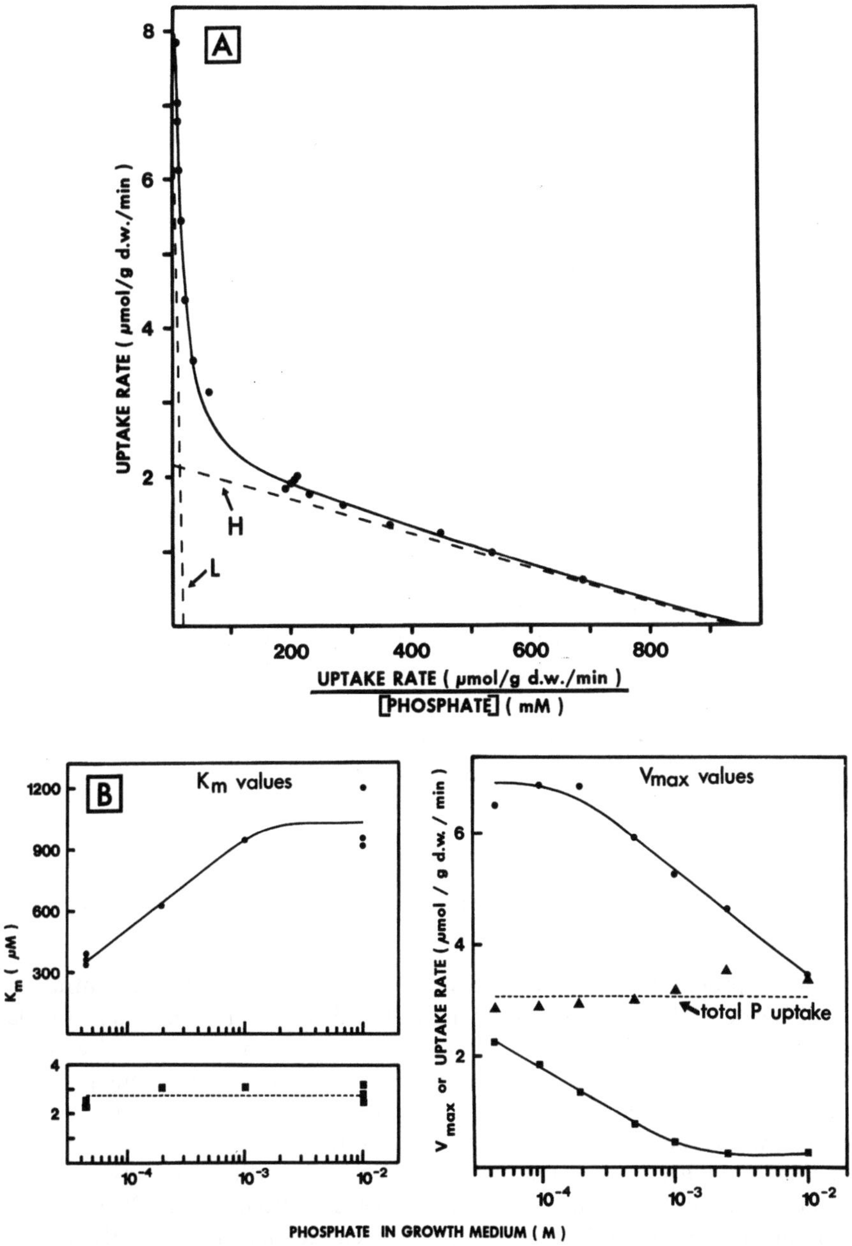


**Fig. 1 A, B.** Dual phosphate uptake mechanisms in *Neurospora*. (BEEVER and BURNS 1980). **A** The Hofstee plot of the relationship between Pi concentration and uptake rate for germlings previously grown in low Pi concentration, 50 μM. The data were analyzed assuming simultaneous operation of two uptake systems, one of high affinity (*H*) and one of low affinity (*L*). The *three lines* (which are the calculated lines based on the kinetic constants) give the uptake due to the two systems operating individually (*dotted lines*) and operating together (*solid line*). Note the goodness of fit of the model curve to the experimental data. **B** Adaptive changes in the kinetic constants brought about by growing the germlings in a range of Pi concentrations (50–10,000 μM). *Left* $K_m$ values; *right* $V_{max}$ values; *top* and –●– low affinity system; *bottom* and –■– high affinity system. The total calculated Pi uptake that would result at each Pi concentration from the combined operation of both uptake mechanisms (each having the kinetic constants characteristic of that concentration) is shown by –▲–. Note that it is almost constant

cells themselves (15 mM – FERGUSON et al. 1983). This property is shared by other xylem tissues: excised shoot segments of a dormant plant, induced to leaf in water in a warm room, can increase the Pi concentration in the xylem fluid from 0.4 to 3.4 mM in the absence of roots (FERGUSON et al. 1983). Such observations strongly suggest an energy-dependent secretion, although direct evidence for this process is still lacking.

P transferred to the xylem stream is mostly or completely present as Pi, and the occasional presence of phosphatyl choline still requires explanation (BIELESKI 1973). Once there, Pi is swept to all parts of the plant. Nor does it stop moving when it reaches a leaf or growing point because P is, with N, the most readily retranslocated nutrient (Chap. I.6, this Vol.). Thus if Pi is supplied to the leaf surface, or when the leaf ages, up to 60% of the P can be retranslocated through the phloem to other plant parts, particularly the growing points and developing fruits (BARRIER and LOOMIS 1957, BIDDULPH et al. 1958, GREENWAY and GUNN 1966). When P enters the sieve tubes it is rapidly metabolized to a wide range of P-esters, particularly ATP; but Pi is also present, and it has been suggested (BIELESKI 1969, KLUGE et al. 1970) that this is the form being transported in the mobile phase, with the P-esters being associated with the metabolically active stationary cytoplasmic phase.

Thus uptake into the cell, secretion and transport in the xylem, and retranslocation within the phloem, all seem to occur with P in the form of Pi, and this seems to be the normal transport form. The only place where P-ester transport is likely to be important is between organelles and the cytoplasm.

## 3 Efflux of Phosphate, and Aspects of Phosphate Deficiency

The uptake processes discussed in Section 2 create a steep Pi concentration gradient between the inside of the living cell and the external medium. In the cells near the root surface, where the apoplast is in direct diffusional contact with the soil solution, the plasmalemma is the only barrier to the strong physical forces tending to drive Pi out of the cell. Given this situation, it is to be expected that there would be some actual outward movement or efflux occurring even while there is a *net* uptake taking place. Whilst Pi efflux from plant roots has been recognized for some time (HEVESY 1946, EMMERT 1959, WEIGL 1968, MAZEL and FOKIN 1977) detailed study has been lacking until recently. In the water plants *Spirodela, Lemna* and *Azolla* (MCPHARLIN 1981, BIELESKI unpublished data), P efflux occurs at a significant rate and shows the following features. The rate is increased by raising the external Pi concentration (Table 2A), by the presence of metabolic inhibitors such as CCCP, azide and cyanide, and by lowering the temperature (Table 2B). When the plant starts to become P-deficient, there is a marked decrease in efflux rate, accompanying the expected increase in uptake rate (Table 2A). The P-containing material lost to the medium as a consequence of efflux is entirely in the form of Pi.

**Table 2.** Effect of various conditions on Pi efflux and uptake in *Spirodela* (rates in nmol $g^{-1}$ fresh wt. $h^{-1}$)

| | Control plants | | | – P plants | | |
|---|---|---|---|---|---|---|
| | Uptake | Efflux | Net change | Uptake | Efflux | Net change |
| A. External Pi conc. (μM) | | | | | | |
| 1000 | 500 | 40 | 460 | 1200 | 16 | 1184 |
| 100 | 220 | 29 | 191 | 950 | 12 | 938 |
| 10 | 100 | 24 | 76 | 280 | 10 | 270 |
| 1 | 20 | 14 | 6 | 40 | 4 | 36 |
| B. Temperature | | | | | | |
| 25° | 460 | 38 | 422 | | | |
| 15° | 180 | 58 | 122 | | | |
| 5° | 80 | 84 | –4 | | | |

The effect of external Pi concentration (A) was measured at 25 °C; the effect of temperature (B) was measured at 1,000 μM external Pi concentration. "– P plants" are in incipient deficiency, having been 4 days without external Pi, leading to a 30% reduced growth rate but no marked visual symptoms of deficiency. (Data from McPharlin 1981)

Most of the responses are opposite to those obtained with the uptake process, but not all: for example, increased external Pi concentration increases both uptake and efflux. The nature of the responses is consistent with at least part of the efflux being under metabolic control, as would occur, for example, if metabolism was needed to maintain the physical impermeability of the cell membrane. The actual rates of loss measured have been in the range 4–50 nmol $g^{-1}$ fresh wt.$h^{-1}$, which can be compared with uptake rates in the range 20–1,000 nmol $g^{-1}$ fresh wt.$h^{-1}$.

We can compare the measured uptake and efflux rates with the change in total P content of exponentially-growing *Spirodela* plants. Under exponential growth, the instantaneous net P uptake rate U (nmol $g^{-1}$ fresh wt.$h^{-1}$) can be calculated from $U = C \ln 2/t$ where C is the P content of the tissue (in nmol $g^{-1}$ fresh wt.), and t is the doubling time (h). For plants such as those described in Table 2, growing under optimal conditions in medium containing 1 mM Pi, $t = 53$ h and $C = 37{,}000$ nmol $g^{-1}$ fresh wt., so $U = 484$ nmol $g^{-1}$ fresh wt.$h^{-1}$. The close agreement with the measured net uptake rate, 460 (Table 2) indicates that the experimentally measured net phosphate uptake rates do have meaning in terms of the growing plant.

In such optimally growing plants, the efflux rate is 8% of the uptake rate. But if such plants are suddenly put into less favourable conditions (lowered external Pi concentration, lowered temperature), efflux can equal or even exceed uptake, leading to net loss of P from the plant. In contrast, a plant that has been grown for a short time under P-deficient conditions, upon being returned to optimal growing conditions, has a decreased efflux but is able to take up Pi about 2.5 times faster than is required to maintain optimal growth, and so is able to restore its P status within about 12 h. Because P efflux remains much lower than in the normal plant, the efflux rate is now only 1.3% of the uptake rate. Thus when a plant responds to P deficiency, it not only increases P uptake, it also decreases P loss, and the second effect is not insignificant.

The importance to a plant of maintaining a constant P intake over a wide range of external P supply conditions has already been discussed (Sect. 2). Equally as important is the ability of a plant to adjust the P intake to the rate of utilization (CHAPIN and BIELESKI 1982). At a doubling time of 2.2 days, *Spirodela* needs a net intake of P at 480 nmol $g^{-1}$ fresh wt.$h^{-1}$. If a change in conditions (e.g. low temperature, poor light, reduced N supply) slows growth so that the doubling time is now 6.6 days, the net P intake must fall to 160 nmol $g^{-1}$ fresh wt.$h^{-1}$ if the P content of the tissue is to remain the same. In the example quoted, an unchanged P intake would lead to a tissue P content of 100 nmol $g^{-1}$ fresh wt. – about 2% of dry weight, and well into the region of severe toxicity. If the plant is able to actively modulate the P loss through the efflux mechanism as well as the P uptake itself, a much more effective total control of net P intake would be achieved.

We believe that future research will confirm that Pi efflux is part of the mechanism for maintaining the P balance of the plant. As yet, the signal the plant uses in responding to deficiency and excess of P has not been identified, but on present evidence it is most likely to be the Pi concentration in the cytoplasm. For both *Neurospora* (BURNS and BEEVER 1979) and *Spirodela* and *Lemna* (MCPHARLIN 1981), a depletion of cellular Pi is the fastest and largest response to P deficiency conditions that has been detected in the various P compounds. Much of the cell Pi in green plants is in the vacuole, and passage of Pi from vacuole to cytoplasm may be slow and restricted (BIELESKI 1968b, 1973). The depletion of Pi arising out of P deficiency would be expected to take effect in the cytoplasm, where Pi is being withdrawn for growth: thus an 8% depletion of total P will represent a 12% depletion of cellular Pi, but an 80% depletion of cytoplasmic Pi. These are the sorts of conditions that obtain in *Spirodela* when, after 12 h of P deficiency, the capacity of the Pi uptake mechanism has already increased by about 50% (MCPHARLIN 1981).

One final point emerges from these efflux experiments. It is apparent from the data in Table 2 that lowering the Pi concentration decreases Pi uptake much more markedly than Pi efflux. This means that eventually an *equilibrium concentration* must be reached where efflux and uptake are equal: below this concentration, there must be a net loss of P from the plant.

The equilibrium point will not be immutable but will partly depend on other factors (for example, in *Spirodela* it is below 1 μM P at 25 °C but approximately 1,000 μM Pi at 5 °C; see Table 2B). We can find the value of this equilibrium concentration either by separately measuring uptake and efflux rates at different Pi concentrations as in Table 2A, and finding where the uptake and efflux lines cross; or by putting plants into a Pi-containing solution and allowing uptake and efflux to proceed until no further concentration change occurs in the medium, then measuring the concentration. Or, finally, we can supply Pi at different concentrations to plants, and find the minimum concentration that will support growth (ASHER and LONERAGAN 1967). The limited information to date from all these procedures suggests that for higher plants, the equilibrium point is around 0.3 μM Pi for control plants; but that in P deficient plants, the combination of increased uptake with decreased efflux lowers the equilibrium point to about 0.07 μM Pi.

The ecological implications of this are of interest, particularly in light of the possibility that temperature, with its opposite effect on uptake and efflux, will have a powerful influence on the equilibrium concentration. As they stand, the results suggest that P deficiency will occur much more readily at suboptimal temperatures.

## 4 Phosphorus Compartments and Pools

We can use the terms "pool" and "compartment" to describe the situation where different parts of the cell are physically isolated from each other and behaving in somewhat different ways, and also to describe the situation where different portions of a single compound are metabolically isolated from one another, whether or not this is due to their actual physical separation in the organelles. The situation is of particular importance with the P-esters. Metabolic compartmentation shows up in the following ways. (a) When tissues supplied with $^{32}$Pi are returned to non-radioactive medium, the rate of $^{32}$P reappearance in the medium can show two or more distinct phases, due to differences in the rates of movement from different compartments. The most obvious is the distinction between the cell wall (apoplast) and the rest of the cell inside the plasmalemma (symplast) (MACROBBIE 1971). (b) When subcellular fractions are isolated and analyzed we detect differences in P-ester composition and concentration. Thus phosphatidyl glycerol is largely restricted to the chloroplast (BIELESKI 1973, see Sect. 5) and RNA is absent from the amyloplast (LIU and SHANNON 1981). (c) By comparing the rates of movement of various solutes into isolated organelles, we can uncover selective permeability behaviour in the membrane. Thus the chloroplast membrane is readily permeable to 3-PGA, but not to ribulose 1,5-diphosphate (JOHNSON and BRUFF 1967, BASSHAM et al. 1968). (d) There can be radioactive labelling anomalies with unequal labelling in different parts of the cell, or unexpected patterns of specific activity in different compounds. Thus when $^{32}$Pi is supplied for a short time to a root, the specific activity of Pi recovered in the xylem sap is higher than that in the root tissues, showing that the $^{32}$Pi entering the root did not mix with all of the Pi already present in the root tissue (GREENWAY and KLEPPER 1968). Within the tissue itself, the P-esters have a specific activity about 10 times higher than the Pi extracted from the tissue (BIELESKI 1968b, WEIGL 1963, LOUGHMAN 1960). Again, the anomalous labelling suggests that much of the Pi within the tissue does not equilibrate readily with $^{32}$Pi that is entering, and does not contribute to P-ester synthesis or to xylem transport. This inactive portion has been called the "non-metabolic pool" and is estimated to contain 85%–95% of the total Pi present (BIELESKI 1973). The small Pi pool, 5%–15%, which receives the entering $^{32}$Pi and which contributes to P-ester synthesis is termed the "metabolic

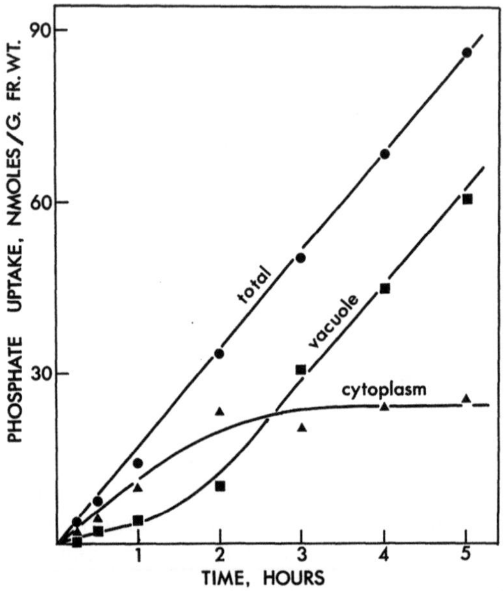


**Fig. 2.** Changes in cytoplasmic and vacuolar Pi concentration in the cell of *Hydrodictyon*. (From data of RAVEN 1974a)

pool". Although the two pools could conceivably be in different types of cell (a "storage" cell and an "ester-producing" cell), the accepted view is that the non-metabolic pool represents Pi in the vacuole, and the metabolic pool is the Pi in the cytoplasm plus its organelles. Thus when ULLRICH et al. (1965) pulse-labelled leaves with $^{32}$Pi then isolated various subcellular fractions, 75% of the Pi was found to be in the vacuole, and very slow to label; 15% was in the cytoplasm, and rapidly labelled; and 10% was in the chloroplast, rapidly labelled. Recently, high-resolution NMR studies have shown that 90% of the Pi in living maize root tips is at a pH characteristic of the vacuole, while the remaining 10% of Pi, along with all the ATP and glucose-6-phosphate, is at a much higher pH characteristic of the cytoplasm (LOUGHMAN personal communication). Finally, in an elegant experiment with the giant alga *Hydrodictyon*, RAVEN (1974a) has separately analysed samples of the vacuole and cytoplasm, and shown that the P-esters are confined to the cytoplasm; that about 10% of the Pi is in the cytoplasm and 90% in the vacuole; and that when $^{32}$Pi is supplied to the alga, it labels the cytoplasmic Pi and P-esters long before the vacuolar Pi becomes significantly labelled (Fig. 2).

Such compartmentation is to be expected. The various P-esters are centrally involved in many biochemical pathways, and often control the activities of those paths. The same P-ester can take part in an anabolic pathway in one part of the cell and a catabolic pathway in another (ROWAN 1966). The need for separate regulation of the paths creates the need for separate pools. It is of some interest that P-deficient *Spirodela* plants behave experimentally as though they have lost their vacuolar Pi pool but have retained their metabolic Pi (BIELESKI 1968b). In view of the central involvement of Pi as an end-product of many enzyme reactions, a substrate of many others, and as a controller of others again, it is logical to expect that the cell would use compartmentation either to maintain cytoplasmic Pi concentration constant at the expense of that in the vacuole, or by changing Pi fluxes between compartments to create Pi concentration changes as triggers for some metabolic events. An example of the latter might be the triggering of phosphofructokinase activity in the cytoplasm of fruit cells as a prelude to the respiratory burst of the climacteric. This activity can respond to increased Pi levels in the cytoplasm which might result from changes in Pi flux across both the tonoplast and the plasmalemma (CHALMERS and ROWAN 1971, WOODROW and ROWAN 1979).

There needs to be separate mention about the compartmentation of the phosphatases (phosphomonoesterase, RNAase, DNAase, P-lipase). Most of the activity in normal cells appears to be localized in the vacuole (MATILE 1978), but under the stress of P deficiency, there can be a considerable increase in phosphatase activity which may now be located in the apoplast or outer cell membrane (RIDGE and ROVIRA 1971, BIELESKI and JOHNSON 1972). Not much is known about the significance of these different pools of phosphatase: by analogy with the animal lysosome, the normal plant vacuole may supply a salvage mechanism whereby structural P-containing compounds such as P-lipids and RNA are hydrolyzed to their basic units (MATILE and WIEMKEN 1976). Phosphatase located in the outer apoplast is demonstrably able to release Pi from externally supplied P-esters, making Pi available for growth (KNYPL 1978). Thus the apoplastic phosphatase could potentially release Pi from humus materials in the soil. However, the roles of the phosphatases and the significance of their location need a lot more study.

## 5 The Form of Phosphorus in the Cell

Though aminoethyl phosphonate is known from protozoa (BEEVER and BURNS 1976), all P in higher plants is as phosphate, either free as Pi, or esterified through a hydroxyl group to a carbon chain (C–O–P) as in the simple phosphate esters, or attached to another phosphate by a pyrophosphate bond, as in the nucleoside di- and triphosphates and in polyphosphate (Fig. 3). Individual compounds where phosphate is differently linked (e.g. by N-P bonding in carbamyl phosphate) are biologically important but present in very small amounts. It is a characteristic of the ester phosphate bond (particularly when associated with a carboxyl or enol group), and even more so the pyrophosphate bond, that it has a high negative free energy of hydrolysis and yet is reasonably stable under biological temperatures and pH. Under the control of various enzyme systems, the breaking of such a bond, which is highly exergonic, can be coupled to various endergonic steps, driving on otherwise unfavourable synthesis in what is now an overall exergonic and favourable reaction. An example here is the reaction, catalyzed by PEP carboxylase, where phosphoenolpyruvate plus $CO_2$ gives oxaloacetate plus Pi. Similarly, where there are existing ester bonds to cleave, phosphorolysis by Pi can replace hydrolysis by $H_2O$, conserving some of the bond energy in the newly created phosphate ester bond, at the

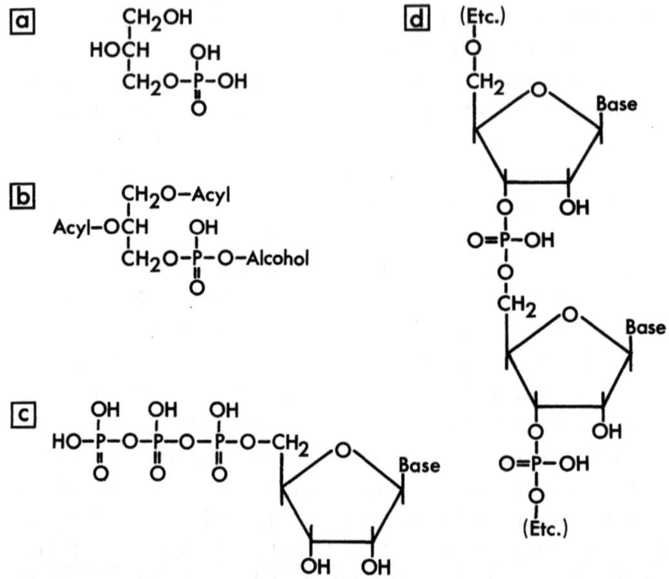


**Fig. 3a–d.** General structures of phosphorus-containing compounds. The unit structure of RNA is shown in **d**, where "base" can be adenine, guanine, cytosine or uracil. In DNA the free OH on the pentose is replaced by H, and uracil by thymine. The structure of a phospholipid is shown in **b**, where *Acyl* can be one of several fatty acids (e.g. linolenic acid), and *Alcohol* can be one of the hydroxyl-bearing compounds choline, ethanolamine, inositol, serine or glycerol. The monoester structure of α-glycerophosphate (1-phosphoglycerol) is shown in **a**, as an example of a simple phosphate ester. The pyrophosphate structure of ATP is shown in **c**

expense of making the reaction *less* exergonic. An example here is the phosphorylase-catalyzed formation of glucose-1-phosphate from starch. Thus energy which is carried in the phosphate ester and pyrophosphate bond (the so-called "energy-rich bond") is used to drive many of the anabolic processes in the cell, and to conserve the energy released by many of the catabolic ones. In evolution, the phosphate ester and pyrophosphate bonds have become the energy currency of the cell, to the extent that almost all the metabolic pathways involve one or another of the nucleotides and phosphate esters (ROWAN 1966).

There is a second, contrasting role that phosphate can play, based on the relative stability of its diester state, where there are *two* C chains linked to *two* of the hydroxyls of a single phosphate. In this state, phosphate forms a bridging group, connecting units together and thus helping to build up macromolecular structures. Note that the third hydroxyl, which remains free, is still readily ionizable, with a pKa around 5, so that even as a bridge, the phosphate remains a hydrophilic, moderately acidic centre.

In analytical fractionation procedures, the P-containing material in the tissue tends to fall into five groups; and conveniently, these groups match the rather distinct roles that are served by the different compounds. Two, DNA and RNA, remain in the tissue residue when the tissue is extracted with simple solvents;

P-lipid is soluble in and extracted by lipid solvents such as chloroform; while the simple P-esters plus Pi are extracted by ordinary aqueous solvents. Pi can then be selectively precipitated as its barium salt or trimethyl ammonium complex (SUGINO and MIYOSHI 1964), or selectively dissolved in butanol and assayed as its phosphomolybdate complex (MARSH 1959). The absolute and relative amounts of each fraction will depend on the nature of the tissue. Over the years there have been many studies of varying degrees of thoroughness on plant tissue P fractions, and we have noted an overall pattern in these data. Thus tissues which have a high proportion of cytoplasm and very small vacuoles (e.g. root tips) have high levels of RNA, P-lipid and particularly the P-esters (TSUJI 1964); tissues that are metabolically active but vacuolate (young to mature leaves, young fruits) have moderately high amounts of all five groups (BARKER et al. 1962, WESTE et al. 1974, CHAPIN and BIELESKI 1982); storage tissues (seeds, tubers) may have high amounts of Pi, or phytic acid (MAKOWER 1969, MUKHERJI et al. 1971), or in the case of fungi, polyphosphate (HAROLD 1966); while woody tissues of low metabolic activity will have a generally low P level with a relatively high proportion of Pi. Taking the young leaf as a representative tissue, the proportions of the different groups are (values in μg atoms P $g^{-1}$ fresh wt.): Pi, 10; RNA, 2; DNA, 0.15; P-lipid, 1.5; P-ester, 1.0 (BIELESKI 1973). Values for actively growing fungi are similar in proportion but about ten times higher on a fresh weight basis: Pi (plus polyphosphate), 100; RNA, 45; DNA, 2.5; P-lipid, 10; P-ester, 10 (BEEVER and BURNS 1980).

It is possible to define the roles of the different P fractions reasonably confidently. DNA is of course macromolecular, with a molecular weight exceeding $10^6$, and is responsible for carrying the genetic information of the cell. In theory, the amount of DNA in a cell is fixed, being that amount required to equip one nucleus with one set of chromosomes. Although the amount of DNA per cell has been measured and used to indicate whether the cells are polyploid, the situation is rather a special one. There are at least two reasons why plant tissues do not have a fixed amount of DNA per cell; firstly, polyploid cells can develop quite easily during cell differentiation and are common in differentiated organs; secondly, DNA is not confined to the nucleus. There is now ample evidence that the chloroplasts and mitochondria separately contain their own DNA, responsible for carrying part of the genetic information required by the organelles themselves (KUNG 1977, BORST 1972). In a photosynthetic cell, this can be a substantial proportion of the total DNA in the cell, e.g. about 3% in *Chlamydomonas* (SAGER and ISHIDA 1963). The main function of the P in DNA is to form a structural bridge between the deoxyribonucleosides that are the letters in the genetic code.

RNA is similar to DNA in structure, macromolecular ($23{,}000–1.3 \times 10^6$ m.w.), and again the role of P is to form a bridge between the ribonucleoside units. (An important exception is the terminal phosphate of tRNA, which has an energy-transfer function as a carrier and activator of amino acids.) The RNA is involved in the translation of the genetic information (via mRNA) and in the synthesis of protein, both as a structural part of the ribosomes (rRNA, 25S and 18S) and as the amino acid carrier which recognizes the nucleotide triplet code (tRNA, 4S). Thus the amount of RNA in the cell tends to be much higher when the tissue is actively synthesizing protein. RNA is distributed throughout the cytoplasmic space of the cell, but each of the different organelles isolated by cell fractionation techniques is characterized by its own particular pattern and proportion of the various RNA's: thus the nucleus, the mitochondrion, the "general cytoplasm", the amyloplast and the chloroplast are all distinctive. In particular, the chloroplast apparently possesses its own protein-synthesizing machinery, can contain 5%–50% of the cellular RNA, and has a ribosomal RNA fraction (23S RNA and 16S RNA) which can be distinguished from the RNA of the main cell (25S and 18S) (LOENING and INGLE 1967).

The P-lipid fraction is recovered in the various analytical procedures by its ability to dissolve in a lipid solvent such as chloroform or ether. The phosphate, though in the diester or "bridging" form, is readily ionized and highly hydrophilic, and so for the whole molecule to be lipid-soluble in this way, there needs to be a lipid-like part also present. This is in fact what occurs: a phospholipid can be thought of as a lipid which has one of its three fatty acids replaced by a phosphate-containing group, the esterification being through the phosphate itself (Fig. 3). The glycerol group can be thought of as the centre of the molecule. Each of the two fatty acids attached to it can come from eight or so long-chain fatty acids, $C_{12}$-$C_{18}$, saturated or mono-, di- or tri-unsaturated, the most important being palmitic, linoleic and linolenic, and giving in excess of 50 possibilities. The phosphate-containing group esterified to the third hydroxyl is one out of only five main possibilities. In all classical separation techniques, the nature of this phosphate-containing group dominates the separation, the nature of the fatty acids playing very little part, so that we recognize five main "phospholipids" – phosphatidyl choline, phosphatidyl ethanolamine, phosphatidyl glycerol, phosphatidyl inositol and phosphatidyl serine. There is also a sixth of more complex structure, diphosphatidyl glycerol, which has *two* diacylated glycerols, or fatty parts, connected by *two* phosphates to a glycerol which now forms a bridge. Strictly speaking, each of these phospholipids so defined is a whole series of compounds depending on the fatty acids present, and though the nature of these has little effect on the broad physical properties, there is not much doubt that it has considerable biological significance, since different membranes yield phospholipids with characteristically different fatty acid patterns (DONALDSON and BEEVERS 1977).

There are other compounds encountered in plant extracts which are related to the phospholipids, and which can be obtained from them as a result of hydrolytic cleavage by one of the phospholipases found in plant tissues. Taking phosphatidyl choline as an example, if one fatty acid is removed from the C1 position (by phospholipase B), we obtain lyso-phosphatidyl choline; if the second fatty acid is also removed from the C2 position (by phospholipase A) we have glycerophosphatyl choline; if the whole glycerol-plus-two-fatty-acid unit is split off (by phospholipase C) we obtain phosphatyl choline; if on the other hand, the choline is split from the rest of the phospholipid (by phospholipase D) we obtain phosphatidic acid. (Phosphatidic acid is also obtained when the other four phospholipids have their characteristic end-group removed in the same way.) Phosphatidic acid is one of the main precursors in phospholipid synthesis (MAZLIAK 1973).

The P-lipid compositions of the various plant tissues are relatively uniform, the biggest single factor being whether or not the tissue is photosynthetic. In both green and non-green tissues, phosphatidyl choline is predominant (40%–50% of total P-lipid), and phosphatidyl ethanolamine is also important (20%–30%), but the proportion of phosphatidyl glycerol is very much higher in photosynthetic tissues (15%–25%) than in non-green ones (below 5%) (ABDELKADER 1968, BIELESKI 1972, ASHWORTH et al. 1981). The remaining P-lipids are present in generally small amounts; phosphatidyl inositol (5%–10%), phosphatidyl serine (1%–4%), diphosphatidyl glycerol (1%–3%) and phosphatidic acid (1%–3%). Because phospholipases can so easily create artefacts during tissue extraction, it is uncertain whether some P-lipid related compounds such as the lysophospholipids are normally present in tissues, but labelling patterns and other data make it clear that others such as phosphatidic acid (0.5%–3% of total P-lipid), phosphatyl choline (0.3%–1.5%) and α-glycerophosphate (0.1%–1%) are genuine constituents. The various cell organelles have more clearly distinguished P-lipid compositions than the bulk tissues (FALK and STOCKING 1976). The most characteristic pattern is that of the chloroplast fraction, which has very little phosphatidyl ethanolamine (4%–8% of total P-lipid), and where the amount of phosphatidyl choline (30%–40%) is exceeded by that of phosphatidyl glycerol (35%–45%) (ONGUN et al. 1968, BIELESKI 1972). The phosphatidyl glycerol is itself unusual in having a very high proportion of its fatty acid present as trans hexadec-3-enoic acid, rare in or absent from the other P-lipids. The microsome fraction appears to be high in phosphatidyl inositol (DONALDSON et al. 1972, FALK and STOCKING 1976) and to have P-lipid with a relatively high proportion of linolenic acid (ABDELKADER and MAZLIAK 1970). However, there is a great deal yet to be done in relating specific phospholipids to specific organelles and to specific functions.

While the detailed function of specific phospholipids is uncertain, the broad function is clear. The structure of the phospholipid is very like that of a detergent: there is a hydrophilic and a lipophilic region combined in one molecule, so that where there is a lipid–water interface, the phospholipid molecules can orient themselves across the boundary and stabilize it. This can lead to the formation of a stable emulsion. More important to living systems is the situation where an oriented layer of phospholipid molecules folds back on itself to create a laminate or bilayer with two hydrophilic outer surfaces sandwiching an inner hydrophobic region; this is the basic structure of the various cell membranes (FALK and STOCKING 1976). It is no surprise that the chloroplast, with its highly developed membrane (thylakoid) system, can contain over 40% of the total P-lipid in photosynthetic cells (BIELESKI 1972). The proteins which perform the various functions of metabolism and transport are thought to be held in place (whether on one face or crossing the membrane from one side to the other) by the interaction of their own lipophilic regions with that of the membrane. The interaction should be made more precise and specific by the nature of the phospholipid (and the nature of its fatty acids) located in a given region.

This is one way in which the specific structure of each individual phospholipid could be of significance in the living cell, despite its great similarity to other phospholipids.

The simple P-esters form the fourth of our five phosphate fractions. Together they represent the metabolic machinery of the cell. Upwards of 50 individual esters have been identified at various times in various tissues and by various methods: amongst the compounds recognized are the 4 nucleoside mono-, di- and triphosphates; about 8 nucleotide diphosphosugars and related compounds, zeatin derivatives, about 12 pentose, hexose and heptose phosphates and diphosphates, about 6 polyol phosphates, 4 triose phosphates, 3PGA, phosphoenolpyruvate, phosphatyl choline, and 6 phosphogluconate amongst them, but there is no point in making a full list. The following general conclusions emerge. P-esters that have been found in animal tissues are, in general, also found in plant tissues (creatine phosphate may be an exception), but at much lower concentrations. Even though the individual enzymes may differ somewhat, the central metabolic pathways involving energy transfer and carbohydrate transformation are the same in plants and animals, and involve the same P-esters (ROWAN 1966), presumably reflecting a very early evolutionary development of the various roles for P-esters. Within different plants there is again a very conservative pattern in the P-esters, to the extent that it is possible to write down a generalized P-ester pattern for most plant tissues (see Table 1 in BIELESKI 1973). About 70% of the P in the P-ester fraction is usually contained in nine compounds: G6P, fructose-6-P and mannose-6-P (20%, 6% and 4%), ATP and ADP (10% and 3%), UTP, UDP and UDPG (4%, 5% and 9%) and 3PGA (8%). [A special situation can arise in seeds and tubers where phytic acid – i.e. inositol hexaphosphate – can be the major P-ester present: functionally, it needs to be considered as a storage compound like polyphosphate rather than as a metabolically active P-ester (SAMOTUS and SCHWIMMER 1962, MATHESON and STROTHER 1969)]. In normal intact tissues, the ATP/ADP/AMP ratio is around 10/3/1 and the "energy charge" remains in the region 0.8–0.9 (PRADET and RAYMOND 1981). The significance of the energy charge as a prime controller of plant cell processes is discussed in a forthcoming review (PRADET 1983: Annu Rev. Plant Physiol.).

There have been very few attempts to compare total P-ester patterns for different subcellular fractions, partly because of the difficulty in measuring the very small amounts of the many P-esters present, and partly because the rapid turnover of P-esters (see Sect. 6) must inevitably allow for drastic changes in P-ester pattern during isolation of the organelles. Special rapid-freezing and non-aqueous isolation techniques have been used to try and avoid this problem (WALKER 1976). There is some hope that new $^{31}$P-NMR techniques will allow us to estimate the concentrations of some P-esters in different intracellular compartments within an intact, living tissue (ROBERTS et al. 1980, LOUGHMAN personal communication). Useful information has also come from studying the behaviour of individual P-esters in isolated fractions: not surprisingly, mitochondria show a major capability for ATP synthesis and are readily permeable to Pi, ADP and ATP (HELDT 1976); amyloplasts lack RNA but contain relatively high amounts of the triose and hexose phosphates plus Pi (LIU and SHANNON

1981); chloroplasts contain Pi (10–15 mM), 3PGA and the two triose phosphates, and though not permeable to Pi (MOURIOUX and DOUCE 1981), readily exchange it across the chloroplast membranes for triose-P, suggesting that this is the mechanism by which carbon fixed in photosynthesis passes into the cytoplasm to be further elaborated to carbohydrate (FLIEGE et al. 1978). The vacuole is another recognizable cell compartment: here, Pi is the main P-compound and P-esters are absent (Sect. 4).

Inorganic phosphate, Pi, is the last of the five fractions to be considered. Though the most important single compound, it is by far the most variable in proportion and amount of all the P fractions, ranging from about 1 $\mu mol\ g^{-1}$ fresh wt. and 15% of total P in a P-deficient tissue to more than 40 $\mu mol\ g^{-1}$ fresh wt. and 70% of total P in a tissue approaching phosphate toxicity conditions (BIELESKI 1968a, b, WESTE et al. 1974). There are two points of interest about this: firstly the presence of a significant amount of Pi even under conditions of extreme P stress implies that some Pi is required for the continued metabolic activity of the cell; secondly, the variability in amount of Pi and the very high tissue contents reached suggest that much of the Pi in a normal cell is surplus to requirements, and serves a storage function (see Sect. 7 for a fuller discussion). In a normal tissue, much of the Pi appears to be in the vacuole, and as already noted, it is apparently the only form in which P is transported in the higher plant (Sect. 2). There is evidence that individual cell organelles (chloroplast, mitochondrion) can transport Pi, and require Pi at concentrations around 10 mM for their normal functioning (SANTARIUS and HEBER 1965, WALKER 1976, WISKICH 1977).

Two inorganic P compounds that are related to Pi require separate comment, pyrophosphate and polyphosphate (in which many Pi units are linked in a chain by pyrophosphate bonds). Pyrophosphate has only occasionally been detected in plant cells, but is a known control agent for some plant enzyme systems and a product of others (BEEVER and BURNS 1980). It must therefore be present, but in functional terms it should be considered as one of the P-esters, along with the nucleotides. The situation with polyphosphate is different. Claims have been made for its presence in higher plant tissues (COLEMAN and SPECHT 1981), but we feel that, because of possible contamination by mycorrhizal or other fungi, the situation is still a "not proven" one. There is no doubt, however, about its importance in the fungi (BEEVER and BURNS 1980) and some algae (RAVEN 1974a). A survey of the earlier literature on polyphosphates can be found in by KUHL (1960) and KULAEV (1979). Polyphosphate can be virtually absent from fungal cells that are starved of P or that are actively growing, but can comprise over 60% of the total P and as much as 3% of the dry weight (250 $\mu mol\ g^{-1}$ fresh wt.) in mature hyphae (BEEVER and BURNS 1980). Polyphosphate has been found in both the vacuole and the cytoplasm, possibly in different forms: it can sometimes be soluble in the cell, but it readily becomes associated with positively charged macromolecules as microscopically visible granules under various conditions of study (BEEVER and BURNS 1980). It may be the main form in which phosphate is transported in the fungal hypha (COX et al. 1980), though in mycorrhizas, the transfer from hypha to host root is probably as Pi (MARX et al. 1982). In its sensitivity to Pi supply, its transport role and much of its general behaviour, polyphosphate can be thought of as a special form of Pi. In most plant cells, Pi can be segregated from the cytoplasm by being stored in the vacuole; polyphosphate, however, represents a form in which Pi can be segregated chemically within the cytoplasm itself as well as physically in the vacuole. Phytic acid, though a P-ester of inositol, also serves a storage role, particularly in seeds where it can form 50% or more of the total P. As germination of the seed begins, phytase is formed, phytic acid is hydrolyzed and Pi is released (GUARDIOLA and SUTCLIFFE 1971, FERGUSON and BOLLARD 1976). Most of the P used during early shoot and root growth is derived from this source; only later does Pi uptake from the medium take over.

Thus, a reasonable amount is known about the function and distribution of the various P-containing compounds. We are helped by the relative insolubility of four of them – DNA, RNA, P-lipid and polyphosphate – and the specific staining procedures that can be used to recognize them under both the optical and electron microscopes. Further effort is now needed to obtain more precise information on the distribution of the remaining fractions – Pi and the various P-esters. Enzyme-coupled stains, selective precipitation methods, electron probe and NMR techniques will all play their part in the near future.

# 6 Synthesis and Turnover of Phosphorus Compounds

A frequent procedure for studying the metabolic pathways followed by a particular compound is to supply it in radioactive form to the tissue or organelle preparation for a very brief time, to follow this up by the same compound in non-radioactive form, and to identify the radioactive products – that is, to carry out a "pulse-chase" experiment. There are some basic difficulties in using this tactic to discover what happens to Pi in intact plants. Firstly, Pi is metabolized very rapidly and the pulse time needs to be very short. Secondly, there is a very large amount of Pi already present in the tissue, and it is divided between two pools which behave very differently (see Sect. 4). Thirdly, in standard pulse-chase experiments, the procedure is to study how the labelling of a particular intermediate falls to zero: this cannot be done with the P-esters because an inevitable end-product of the reactions is the original precursor, Pi itself, and so the P-ester radioactivity never falls close to zero, but comes to equilibrium. This "time to equilibrium" can be used as an alternative although less satisfactory estimate of the rate and course of reactions, and despite its poorer pedigree has given some useful information. The actual parameter involved is the "half-time of labelling" (the time taken for a given ester to reach half the equilibrium radioactivity), and related to it, the "turnover time" (the time taken for the amount that is present of that ester to be broken down and resynthesized). In a simple case, turnover time = half-time/ln 2 (see BIELESKI and LATIES 1963, BIELESKI 1973 for further discussion).

We can look at what happens when $^{32}$Pi enters intact plant tissues (LOUGHMAN 1960, WEIGL 1963, BIELESKI 1968b, 1973). The most rapidly labelled compounds detected are ATP and UTP, supporting the view that in plants, as in animals, the main entry for P into organic compounds is through esterification of Pi as the $\gamma$-phosphate group of ATP. It further suggests that the exchange reaction between ATP and UTP, catalyzed by nucleoside diphosphate kinases, is very rapid, so that the effective pool size of ATP is in fact (ATP+UTP). The estimated turnover time for these two nucleotide triphosphates is always very short, 10–40 s: that is to say, the entire supply of ATP+UTP in the cell is broken down (to ADP+UDP) and resynthesized in less than a minute. From this turnover time and the pool size, we can then calculate the rate at which the $\gamma$-phosphate link is synthesized. Data obtained for *Spirodela* are given in Table 3, showing the relative rates of synthesis of the different P-esters. It has not been possible to directly measure a turnover time for the cytoplasmic Pi pool, but a maximum value (i.e. minimum rate of Pi utilization) can be estimated on the reasonable basis that one Pi is used for each $\gamma$-phosphate link synthesized (see below). The P-lipid turnover time is an average value for all five major P-lipids (BIELESKI 1972). The data available on RNA turnover take no recognition of different RNA fractions and are based on "non-extractable P" rather than P that has been shown to be in RNA (BIELESKI unpublished data); but a turnover time of 15 h is indicated. Again, a maximum time of 47 h for RNA and DNA can be deduced on the basis that in the absence of true turnover (i.e. cycling) there will still be an increase in the *amount* of each during growth, and this will give an *apparent* turnover time (Table 3).

Such data are direct estimates of P-ester production. It is also possible to indirectly estimate what the rate of ATP production should be from the respiration rate. If the tissue produces all its $\gamma$-phosphate as ATP by normal aerobic respiration, then the utilization of 1 µmole of oxygen yields 6.2 µmol ATP.

The data presented in Table 3 are subject to a number of assumptions and approximations, and should be viewed in this light. However, they do allow us to make some useful general comments on the probable behaviour of P in the various metabolic pathways of a plant cell. Firstly, $\gamma$-phosphate production as ATP or UTP measured in *Spirodela* is very similar to that expected if it all came from respiration through the normal pathway. The reasonable match between the two sets of figures allows us to hope that the turnover rates which have been measured do have some significance. Secondly, the relatively high rate of $\beta$-phosphate turnover suggests that about one in four ATP or UTP molecules is utilized in reactions that take it through to the mononucleotide. One important pathway of this type is involved in carbohydrate synthesis and metabolism: UTP+G6P→UDPG→UMP+carbohydrate synthesis. Thirdly, only 1 in 6 G6P molecules that are produced is used in respiration: the rest are presumably drawn off in processes that do not yield ATP, such as starch synthesis. A significant part of cellular G6P has been shown to be localized in the amyloplast, along with other esters of starch synthesis, as a separate pool (LIU and SHANNON 1981). Fourthly, 3PGA has a high rate of turnover, which we can take as an index of the relative importance of the photosynthetic pathway in terms of total metabolism in the cell. (Note that *Spirodela* used in this study was grown on medium containing glucose and so carbohydrate metabolism could exceed photosynthesis.) Fifthly, the turnover of phosphatyl choline is consistent with it being largely involved in P-lipid

**Table 3.** Turnover times and rates of synthesis of various P fractions in *Spirodela*. (Data from BIELESKI 1968a, b, 1972 and unpublished data)

| Phosphorus fraction | Amount (nmol $g^{-1}$ fresh wt.) | Turnover time (min) | Synthesis rate (nmol P $g^{-1}$ $min^{-1}$) |
|---|---|---|---|
| From tissue analysis: | | | |
| ATP+UTP (γP)[a] | 170 | 0.50 | 340 |
| NTP+NDP (βP)[b] | 300 | 3.5 | 85 |
| G6P | 670 | 7 | 95 |
| 3PGA | 330 | 6.5 | 50 |
| Adenosine αP[c] | 200 | ca. 160 | 1[e] |
| Uridine αP | 320 | ca. 160 | 2[e] |
| Phosphoryl choline | 70 | 6 | 12 |
| P-lipid | 2700 | 130 | 20 |
| RNA (est) | 4900 | ca. 900 | 5[e] |
| RNA (min. rate)[d] | 4900 | 2800 | 2 |
| DNA (min. rate)[d] | 560 | 2800 | 0.2 |
| From respiration rate:[f] | | | |
| "ATP" | 170 | 0.25 | 650 |
| G6P | 670 | 40 | 17 |
| Pi[g] | (3400) | 5 | 650 |

[a] "γ" is the γP of the nucleotide pool. The turnover time is given by the labelling pattern of ATP and UTP, and the amount is $^1/_3$ (total P in ATP+UTP)

[b] "βP" is the βP of the nucleotide pool. The turnover time is given by the labelling pattern of ADP and UDP, and the amount is $^1/_3$ (P in ATP+UTP)+$\frac{1}{2}$ (P in ADP+UDP)

[c] "Adenosine αP" is the P directly attached to adenosine. The turnover time is given by the labelling pattern of AMP, and the amount is $^1/_3$ (P in ATP)+$\frac{1}{2}$ (P in ADP)+(P in AMP). "Uridine αP" is calculated in a similar way

[d] Minimum synthesis rates for RNA and DNA are calculated on the basis that, as the amount of each must increase with growth, the maximum turnover time (minimum synthesis rate) is the same as the doubling time for the tissue, 1.9–2.0 days. The estimated RNA time is from unpublished data

[e] If all the "adenosine αP" and "uridine αP" synthesis was used to replace nucleotide lost in RNA synthesis, the synthesis rate of each should be about 25% of the RNA value (as 1 in 4 P of RNA, approximately, is contributed by adenylic and uridylic acids)

[f] Values can be calculated from the known respiration rate of *Spirodela*, 2.3 µl $O_2$ $g^{-1}$ fresh wt. $min^{-1}$ (105 nmol $O_2$ $g^{-1}$ fresh wt. $min^{-1}$) assuming normal aerobic respiration rate and 6.2 mol ATP produced (or 6.2 mol Pi utilized) per 1 mol oxygen consumed and $^1/_6$ mol glucose utilized. The phosphate uptake rate is 7.5 nmol $g^{-1}$ fresh wt. $min^{-1}$

[g] The size of the Pi fraction assumes that 12% of the tissue Pi is in the metabolic pool and turning over, and that the remaining 88% is non-metabolic

synthesis; and the turnover rate of adenosine- and uridine-αP is similarly consistent with the turnover rate found for RNA. Taken all together, the data indicate the following rough balance sheet: for every 1,000 P atoms involved in ATP (and UTP) production, about 250 are involved in carbohydrate synthesis, 150 in photosynthesis, 60 in P-lipid turnover and synthesis, 15 in RNA turnover and synthesis, and 1 in DNA turnover and synthesis. About 15–25 new atoms of P are imported into the cell to replace P withdrawn for growth and storage. But for every P atom taken up, 40–80 are cycled through ATP – that is, each Pi molecule which enters the cell makes on average 60 passages through ATP and the metabolic cycle before it is withdrawn in growth, storage or transport. We can conclude that the entire Pi pool in the cytoplasm must turn over every 5 min or so. This is a remarkably short time, and indicates the very important role played by Pi in cellular metabolism.

## 7 Dynamics of Phosphate Use in the Plant

At this point it is instructive to look at the dynamics of cellular and plant utilization of P. We can consider the plant in terms of four compartments. The apoplast, which is the non-living tissue lying outside the plasmalemma, can be divided into the outer apoplast, in ready communication with the soil, and the inner apoplast (e.g. the xylem) in communication with the main body of the plant but separated from the outer apoplast by a diffusion barrier (endodermal suberin or the casparian band). The region inside the plasmalemma can be divided into the cytoplasm or metabolic space (the symplast) and the vacuole or non-metabolic space, inside a second membrane, the tonoplast. Pi moves readily between these four compartments, and Fig. 4 summarizes the nature of the interplay.

Pi in the soil is released into solution by the dissolving of soil minerals or by the death and decomposition of other organisms (BEEVER and BURNS 1976). It diffuses into the outer apoplast. The uptake mechanism at the plasmalemma pumps Pi into the cytoplasm where it is rapidly esterified into ATP. This is the metabolic function of P. A small portion is drawn off in the synthesis of P-lipid, DNA and RNA for growth – this is the structural function of P – but most passes through the metabolic paths and reappears as Pi in the metabolic pool. If the amount withdrawn for growth is not as great as the uptake, the Pi concentration in the cytoplasm will rise. An active homeostatic mechanism at the tonoplast comes into play, restoring the Pi concentration in the cytoplasm by transferring Pi to the vacuole. Though this process has not been directly demonstrated in higher plants, it is well shown in the data for *Hydrodictyon* (RAVEN 1974a) (Fig. 2). This movement into the vacuole represents the storage function of P in the cell (incorporation into polyphosphate may play a similar role in fungi). A second option is that Pi may instead be moved to other cells far away from the source of supply – this is the transport function of P. The major movement upwards in the plant occurs by Pi being unloaded at the plasmalemma into the inner apoplast, thence to be taken upwards in the xylem stream. An opposite situation arises if the Pi demands of the cell exceed the supply from the environment. Pi must now be returned to the cytoplasm from the vacuole, or brought in by retranslocation from distant tissues: both are known responses to P deficiency. Such retranslocation typically

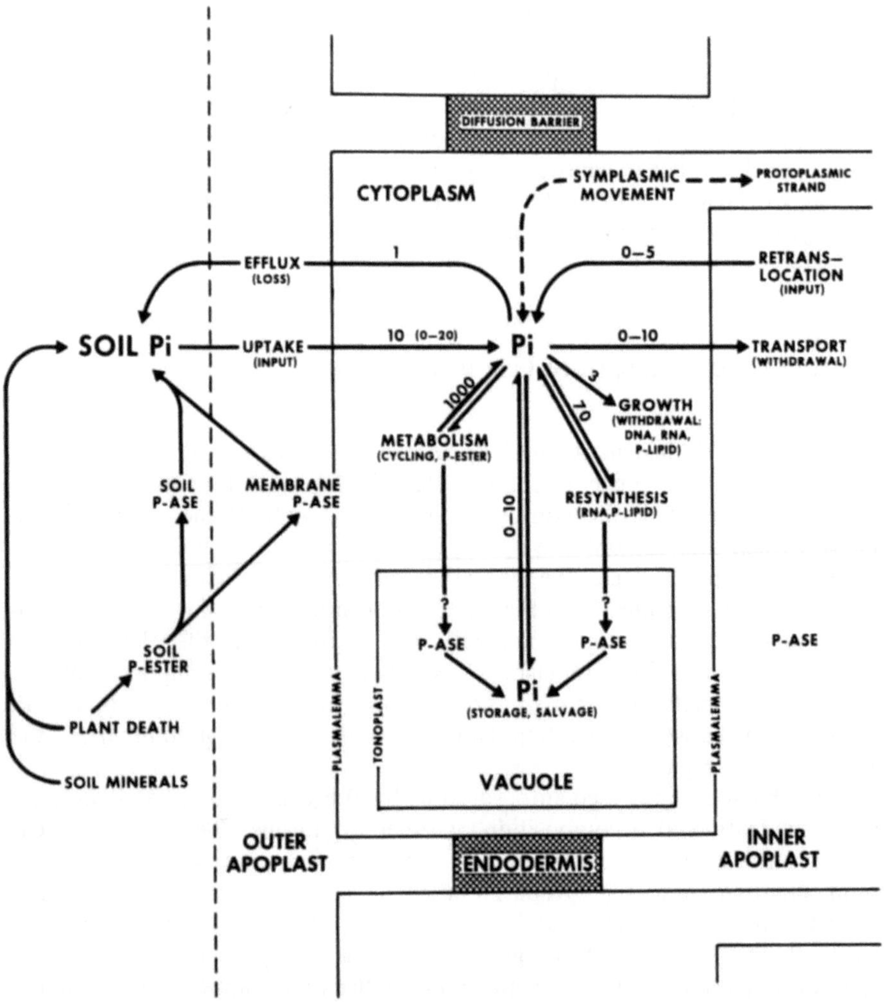


**Fig. 4.** Dynamics of phosphorus use in the cell. *Numbers relating to arrows* on the diagram give the relative significance of the different pathways of P movement and in a plant like *Spirodela,* are approximately equivalent to (nmol $g^{-1}$ fresh wt. $min^{-1}$). *P-ase is* "phosphatase", and in the context of this diagram, includes non-specific acid and alkaline phosphomonoesterases, plus ribonucleases, plus phospholipases

occurs in the phloem, and apparently involves both apoplastic and symplastic transport steps.

In summary: the cytoplasmic pool of Pi appears to be the hub of all cellular P metabolism and of the whole P economy of the plant. It is a major controller of cell processes, but what we now need to know is how this control is achieved and regulated. That will be a major research area for the future.

## 8 Conclusions

With a subject as broad as the P nutrition of plants, it is too much to hope for a compact body of information without any large gaps of knowledge. In this chapter, we used scattered data on P metabolism for relatively few plants, in order to draw a picture for "plants" in general. We do not apologize for doing so: we had a choice of either presenting the subject in a very broken fashion, reviewing only those parts where a hundred researchers have gone before, or else extending the little knowledge we do have to fill in those gaps, and present a coherent if tentative and sketchy picture. For this volume, which aims to synthesize current knowledge rather than review the literature, we have chosen the second course. There is some hope that our picture will not be too far wrong, partly because a lot of the data, though scattered, do correspond quite well and can be accounted for by our interpretations. A second factor is that the behaviour of plants towards P seems to be very conservative, going back to the fundamental importance of P in all three of the major life processes – reproducing oneself (DNA), segregating oneself from the hotch-potch world outside (P-lipid), and acquiring and using energy to beat the second law of thermodynamics in creating the ordered state of living things (P-esters). Thus the same compounds are present in all tissues, plant and animal, they serve the same role, and they usually seem to be present in similar proportions, so there is a lot of common ground in the processes that occur. But despite these reassuring thoughts, the chapter we have presented must be taken as a starting point for future research, not a museum full of past knowledge.

## References


Abdelkader AB (1968) La lipogenèse dans le tubercule de pomme de terre (*Solanum tuberosum* L.). 1. Analyse et biosynthèse des lipides dans le parenchyme central. Influence de la "survie" (ageing) de rondelles de parenchyme sur cette biosynthèse. Physiol Veg 6:417–442

Abdelkader AB, Mazliak P (1970) Echanges de lipides entre mitochondries, microsomes et surnageant cytoplasmique de cellules de pomme de terre ou de chou-fleur. Eur J Biochem 15:250–262

Asher CJ, Loneragan JF (1967) Responses of plants to phosphate concentration in solution culture: I. Growth and phosphorus content. Soil Sci 103:225–233

Ashworth EN, St John JB, Christiansen MN, Patterson GW (1981) Characterization of the phospholipid composition of wheat roots using high-performance liquid chromatography. J Agric Food Chem 29:879–881

Barber SA, Walker JM, Vasey EH (1963) Mechanisms for the movement of plant nutrients from the soil and fertilizer to the plant root. J Agric Food Chem 11:204–207

Barker J, Isherwood FA, Jakes R, Solomos T, Younis ME (1962) Determination of certain phosphate compounds in plant extracts. Nature 196:1115

Barrier GE, Loomis WE (1957) Absorption and translocation of 2,4-dichlorophenoxyacetic acid and $P^{32}$ by leaves. Plant Physiol 32:225–231

Bassham JA, Kirk M, Jensen RG (1968) Photosynthesis by isolated chloroplasts. I. Diffusion of labelled photosynthetic intermediates between isolated chloroplasts and suspending medium. Biochim Biophys Acta 153:211–218
Beever RE, Burns DJW (1976) Microorganisms and the phosphorus cycle: some physiological considerations. In: Blair GJ (ed) Prospects for improving efficiency of phosphorus utilization. Reviews in rural science, vol III. Univ New England, Armidale
Beever RE, Burns DJW (1977) Adaptive changes in phosphate uptake by the fungus *Neurospora crassa* in response to phosphate supply. J Bacteriol 132:520–525
Beever RE, Burns DJW (1980) Phosphate uptake, storage and utilization by fungi. Adv Bot Res 8:127–219
Biddulph O, Biddulph S, Cory R, Koontz H (1958) Circulation patterns for phosphorus, sulfur and calcium in the bean plant. Plant Physiol 33:293–300
Bieleski RL (1968a) Levels of phosphate esters in *Spirodela*. Plant Physiol 43:1297–1308
Bieleski RL (1968b) Effect of phosphorus deficiency on levels of phosphorus compounds in *Spirodela*. Plant Physiol 43:1309–1316
Bieleski RL (1969) Phosphorus compounds in translocating phloem. Plant Physiol 44:497–502
Bieleski RL (1972) Turnover of phospholipids in normal and phosphorus-deficient *Spirodela*. Plant Physiol 49:740–745
Bieleski RL (1973) Phosphate pools, phosphate transport, and phosphate availability. Annu Rev Plant Physiol 24:225–252
Bieleski RL (1976) Passage of phosphate from soil to plant. In: Blair GJ (ed) Prospects for improving efficiency of phosphorus utilization. Reviews in rural science, vol III. Univ New England, Armidale
Bieleski RL, Johnson PN (1972) The external location of phosphatase activity in phosphorus-deficient *Spirodela oligorrhiza*. Aust J Biol Sci 25:707–720
Bieleski RL, Laties GG (1963) Turnover rates of phosphate esters in fresh and aged slices of potato tuber tissue. Plant Physiol 38:586–594
Borst P (1972) Mitochondrial nucleic acids. Annu Rev Biochem 41:333–376
Bowling DJF, Dunlop J (1978) Uptake of phosphate by white clover. I. Evidence for an electrogenic phosphate pump. J Exp Bot 29:1139–1146
Burns DJW, Beever RE (1977) Kinetic characterization of the two phosphate uptake systems in the fungus *Neurospora crassa*. J Bacteriol 132:511–519
Burns DJW, Beever RE (1979) Mechanisms controlling the two phosphate uptake systems in *Neurospora crassa*. J Bacteriol 139:195–204
Carter OG, Lathwell DJ (1967) Effects of temperature on orthophosphate absorption by excised corn roots. Plant Physiol 42:1407–1412
Chalmers DJ, Rowan KS (1971) The climacteric in ripening tomato fruit. Plant Physiol 48:235–240
Chapin FS, Bieleski RL (1982) Mild phosphorus stress in barley and a related low-phosphorus-adapted barleygrass: phosphorus fractions and phosphate absorption in relation to growth. Physiol Plant 54:309–317
Cole CV, Elliott ET, Hunt HW, Coleman DC (1978) Trophic interactions in soils as they affect energy and nutrient dynamics. V. Phosphorus transformations. Microb Ecol 4:381–387
Coleman RG, Specht RL (1981) Mineral nutrition of heathlands: The possible role of polyphosphate in the phosphorus economy of heathland species. In: Specht RL (ed) Ecosystems of the world, vol 9B. Heathlands and related shrublands. Analytical studies. Elsevier, Amsterdam
Collins JC, Reilly EJ (1968) Chemical composition of the exudate from excised maize roots. Planta 83:218–222
Cooke JG (1981) Pollution from our pastures. Soil Water 17:13–15
Corbridge DEC (1978) Phosphorus. Elsevier, Amsterdam
Cox G, Moran KJ, Sanders F, Nockolds C, Tinker PB (1980) Translocation and transfer of nutrients in vesicular-arbuscular mycorrhizas. III. Polyphosphate granules and phosphorus translocation. New Phytol 84:649–659

Donaldson RP, Beevers H (1977) Lipid composition of organelles from germinating castor bean endosperm. Plant Physiol 59:259–263

Donaldson RP, Tolbert NE, Schnarrenberger C (1972) A comparison of microbody membranes with microsomes and mitochondria from plant and animal tissue. Arch Biochem Biophys 152:199–215

Dunlop J, Bowling DJF (1971) The movement of ions to the xylem exudate of maize roots. I. Profiles of membrane potential and vacuolar potassium activity across the root. J Exp Bot 22:434–444

Emmert FH (1959) Loss of phosphorus-32 by plant roots after foliar application. Plant Physiol 34:449–454

Epstein E (1972) Mineral nutrition of plants. Principles and perspectives. Wiley and Sons, New York

Falk RH, Stocking CR (1976) Plant membranes. In: Stocking CR, Heber U (eds) Transport in plants III Encyclopaedia of plant physiology new ser, vol 3. Springer, Berlin Heidelberg New York

Ferguson AR, Eiseman JA (1983) Estimated annual removal of macronutrients in fruit and prunings from a kiwifruit orchard. NZJ Agric Res 26:115–117

Ferguson AR, Eiseman JA, Leonard JA (1983) Xylem sap from *Actinidia chinensis:* seasonal changes in composition. Ann Bot (in press)

Ferguson IB, Bollard EG (1976) The movement of calcium in germinating pea seeds. Ann Bot 40:1047–1055

Ferguson IB, Clarkson DT (1975) Ion transport and endodermal suberization in the roots of *Zea mays*. New Phytol 75:69–79

Ferguson IB, Clarkson DT (1976) Ion uptake in relation to the development of a root hypodermis. New Phytol 77:11–14

Fliege R, Flügge U-I, Werdan K, Heldt HW (1978) Specific transport of inorganic phosphate, 3-phosphoglycerate, and triosephosphates across the inner membrane of the envelope in spinach chloroplasts. Biochim Biophys Acta 502:232–247

Gilchrist AN, Gillingham AG (1970) Phosphate movement in surface run-off water. NZ J Agric Res 13:225–231

Gould WD, Coleman DC, Rubenk AJ (1979) Effect of bacteria and amoebae on rhizosphere phosphatase activity. Appl Environ Microbiol 37:943–946

Greenway H, Gunn A (1966) Phosphorus retranslocation in *Hordeum vulgare* during early tillering. Planta 71:43–67

Greenway H, Klepper B (1968) Phosphorus transport to the xylem and its regulation by water flow. Planta 83:119–136

Groves RH (1981) Nutrient cycling in heathlands. In: Specht RL (ed) Ecosystems of the world, vol 9B. Heathlands and related shrublands. Analytical studies. Elsevier, Amsterdam

Guardiola JL, Sutcliffe JF (1971) Mobilisation of phosphorus in the cotyledons of young seedlings of the garden pea (*Pisum sativum* L.). Ann Bot 35:809–823

Hagen CE, Hopkins HT (1955) Ionic species in orthophosphate absorption by barley roots. Plant Physiol 30:193–199

Harold FM (1966) Inorganic polyphosphates in biology: structure, metabolism and function. Bacteriol Rev 30:772–794

Harvey HW (1969) The chemistry and fertility of sea waters. Cambridge Univ Press, Cambridge

Healy WB, McColl RHS (1974) Clay particles as sources of phosphorus for *Lemna* and their role in eutrophication. NZJ Sci 17:409–420

Heldt HW (1976) Transport of metabolites between cytoplasm and the mitochondrial matrix. In: Stocking CR, Heber U (eds) Transport in plants III. Encyclopaedia of plant physiology, new ser, vol 3. Springer Berlin Heidelberg New York

Hevesy G (1946) Interaction between phosphorus atoms of the wheat seedling and the nutrient solution. Ark Bot 33:1–16

Johnson EJ, Bruff BS (1967) Chloroplast integrity and ATP-dependent $CO_2$ fixation in *Spinacia oleracea*. Plant Physiol 42:1321–1328

Jung C, Rothstein A (1965) Arsenate uptake and release in relation to the inhibition of transport and glucolysis in yeast. Biochem Pharmacol 14:1093–1112
Kluge M, Becker D, Ziegler H (1970) Untersuchungen über ATP und andere organische Phosphorverbindungen im Siebröhrensaft von *Yucca flaccida* und *Salix triandra*. Planta 91:68–79
Knypl JS (1978) Reversal of the symptoms of phosphate deficiency in *Spirodela* by RNA and adenosine monophosphates. Z Pflanzenphysiol 90:265–277
Kuhl A (1960) Die Biologie der kondensierten organischen Phosphate. Ergeb Biol 23:144–186
Kulaev IS (1979) The biochemistry of inorganic polyphosphates. Wiley, Chichester New York Brisbane Toronto
Kung S-D (1977) Expression of chloroplast genomes in higher plants. Annu Rev Plant Physiol 28:401–437
Kunishi HM, Taylor AW, Heald WR, Gburek WJ, Weaver RM (1972) Phosphate movement from an agricultural watershed during two rainfall periods. J Agric Food Chem 20:900–905
Leggett JE, Galloway RA, Gauch HG (1965) Calcium activation of orthophosphate absorption by barley roots. Plant Physiol 40:897–902
Liu T-TY, Shannon JC (1981) Measurement of metabolites associated with nonaqueously isolated starch granules from immature *Zea mays* L. endosperm. Plant Physiol 67:525–529
Loening UE, Ingle J (1967) Diversity of RNA components in green plant tissues. Nature 215:363–367
Loughman BC (1960) Uptake and utilization of phosphate associated with respiratory changes in potato tuber slices. Plant Physiol 35:418–424
MacRobbie EAC (1971) Fluxes and compartmentation in plant cells. Annu Rev Plant Physiol 22:75–96
Makower RU (1969) Changes in phytic acid and acid-soluble phosphorus in maturing pinto beans. J Sci Food Agric 20:82–84
Marsh BB (1959) The estimation of inorganic phosphate in the presence of adenosine triphosphate. Biochim Biophys Acta 32:357–361
Marx C, Dexheimer J, Gianinazzi-Pearson V, Gianinazzi S (1982) Enzymatic studies on the metabolism of vesicular-arbuscular mycorrhizas. IV. Ultracytoenzymological evidence (ATPase) for active transfer processes in the host-arbuscle interface. New Phytol 90:37–43
Matheson NK, Strother S (1969) The utilization of phytate by germinating wheat. Phytochemistry 8:1349–1356
Matile Ph (1978) Biochemistry and function of vacuoles. Annu Rev Plant Physiol 29:193–213
Matile Ph, Wiemken A (1976) Interactions between cytoplasm and vacuole. In: Stocking CR, Heber U (eds) Transport in plants III. Encyclopaedia of plant physiology new ser, vol 3. Springer, Berlin Heidelberg New York
Mazel YT, Fokin AD (1977) Excretion of ions by roots of plants. Sov Plant Physiol 24:805–810
Mazliak P (1973) Lipid metabolism in plants. Annu Rev Plant Physiol 24:287–310
McColl RHS, White E, Waugh JR (1975) Chemical run-off in catchments converted to agricultural use. NZJ Sci 18:67–84
McPharlin IR (1981) Phosphorus transport and phosphorus nutrition in *Lemna* (*Lemna major* L.) and *Spirodela* (*Spirodela oligorrhiza* (Kurz.) Hegelm.). Ph D thesis, Univ Auckland, NZ
Mourioux G, Douce R (1981) Slow passive diffusion of orthophosphate between intact isolated chloroplasts and suspending medium. Plant Physiol 67:470–473
Mukherji S, Dey B, Paul AK, Sircar SM (1971) Changes in phosphorus fractions and phytase activity of rice seeds during germination. Physiol Plant 25:94–97
Ongun A, Thomson WW, Mudd JB (1968) Lipid composition of chloroplasts isolated by aqueous and nonaqueous techniques. J Lipid Res 9:409–415
Pitman MG (1977) Ion transport into the xylem. Annu Rev Plant Physiol 28:71–88

Pradet A, Raymond P (1982) Adenylate energy charge, an indicator of energy metabolism. In: Physiology and biochemistry of plant respiration. Palmer JM (ed) Soc Exp Biol. Cambridge Univ Press, Cambridge

Raven JA (1974a) Phosphate transport in *Hydrodictyon africanum*. New Phytol 73:421–432

Raven JA (1974b) Energetics of active phosphate influx in *Hydrodictyon africanum*. J Exp Bot 25:221–229

Reisenauer HM (1966) Mineral nutrients in soil solution. In: Altman PL, Dittmer DS (eds) Environmental biology. Fed Am Soc Exp Biol, Bethesda

Ridge EH, Rovira AD (1971) Phosphatase activity of intact young wheat roots under sterile and non-sterile conditions. New Phytol 70:1017–1026

Roberts JKM, Ray PM, Wade-Jardetzky N, Jardetzky O (1980) Estimation of cytoplasmic and vacuolar pH in higher plant cells by $^{31}P$ NMR. Nature 283:870–872

Rowan KS (1966) Phosphorus metabolism in plants. Int Rev Cytol 19:301–390

Sager R, Ishida MR (1963) Chloroplast DNA in *Chlamydomonas*. Proc Natl Acad Sci USA 50:725–730

Samotus B, Schwimmer S (1962) Phytic acid as a phosphorus reservoir in the developing potato tuber. Nature 194:578–579

Santarius KA, Heber U (1965) Changes in intracellular levels of ATP, ADP, AMP, and Pi and regulatory function of the adenylate system in leaf cells during photosynthesis. Biochim Biophys Acta 102:39–54

Smith FA (1966) Active phosphate uptake by *Nitella translucens*. Biochim Biophys Acta 126:94–99

Sugino Y, Miyoshi Y (1964) The specific precipitation of orthophosphate and some biological applications. J Biol Chem 239:2360–2364

Tsuji H (1964) Acid-soluble phosphate ester contents of developing rice and oat shoots. Bot Mag 77:247–252

Ullrich W, Urbach W, Santarius KA, Heber U (1965) Die Verteilung des Orthophosphates auf Plastiden, Cytoplasma und Vacuole in der Blattzelle und ihre Veränderung im Licht-Dunkel-Wechsel. Z Naturforsch 20B:905–910

Ullrich-Eberius CI, Novacky A, Fischer E, Lüttge U (1981) Relationship between energy-dependent phosphate uptake and the electrical membrane potential in *Lemna gibba* G 1. Plant Physiol 67:797–801

Walker DA (1976) Plastids and intracellular transport. In: Stocking CR, Heber U (eds) Transport in plants III. Encyclopaedia of plant physiology new ser, vol 3. Springer, Berlin Heidelberg New York

Weigl J (1963) Die Bedeutung der energiereichen Phosphate bei der Ionenaufnahme durch Wurzeln. Planta 60:307–321

Weigl J (1968) Austauschmechanismus des Ionentransports in Pflanzen am Beispiel des Phosphat- und Chloridtransports bei Maiswurzeln. Planta 79:197–207

Weste JE, Rowan KS, Chalmers DJ (1974) The distribution of phosphorus-containing compounds in tomato plants during the development of phosphorus deficiency. In: Bieleski RL, Ferguson AR, Cresswell MM (eds) Mechanisms of regulation of plant growth. Bull 12. R Soc NZ, Wellington

Wiskich JT (1977) Mitochondrial metabolite transport. Annu Rev Plant Physiol 28:45–69

Woodrow IE, Rowan KS (1979) Change of flux of orthophosphate between cellular compartments in ripening tomato fruits in relation to the climacteric rise in respiration. Aust J Plant Physiol 6:39–46

MIX
Papier aus verantwortungsvollen Quellen
Paper from responsible sources
FSC® C105338

If you have any concerns about our products,
you can contact us on
**ProductSafety@springernature.com**

In case Publisher is established outside the EU,
the EU authorized representative is:
**Springer Nature Customer Service Center GmbH**
**Europaplatz 3, 69115 Heidelberg, Germany**

Printed by Libri Plureos GmbH
in Hamburg, Germany